Magellanic Sub-Antarctic Ornithology

First Decade of Long-Term Bird Studies at the Omora Ethnobotanical Park, Cape Horn Biosphere Reserve, Chile

RICARDO ROZZI AND JAIME E. JIMÉNEZ

Editors

SUB-ANTARCTIC BIOCULTURAL CONSERVATION PROGRAM

University of North Texas - Universidad de Magallanes

Library of Congress Cataloging-in-Publication Data

Magellanic Sub-Antarctic Ornithology: First Decade of Long-Term Bird Studies at the Omora Ethnobotanical Park, Cape Horn Biosphere Reserve, Chile / Ricardo Rozzi & Jaime E. Jiménez, editors. -- First edition.

pages cm

"Sub-Antarctic Biocultural Conservation Program. University of North Texas -- Universidad de Magallanes."

Includes bibliographical references and index.

ISBN 978-1-57441-531-5 (pbk. : alk. paper) -- ISBN 978-1-57441-543-8 (ebook)

1. Birds--Chile--Cabo de Hornos Biosphere Reserve. 2. Forest birds--Habitat--Chile--Parque Etnobotánico Omora. 3. Habitat surveys--Chile--Parque Etnobotánico Omora. 4. Biodiversity--Chile--Cabo de Hornos Biosphere Reserve. 5. Endangered species--Chile--Cabo de Hornos Biosphere Reserve. 6. Birds--Conservation--Chile--Cabo de Hornos Biosphere Reserve. 7. Animal ecology--Chile--Cabo de Hornos Biosphere Reserve. 8. Parque Etnobotánico Omora (Chile) 9. Cabo de Hornos Biosphere Reserve (Chile) 10. Parque Etnobotánico Omora (Agency : Chile) I. Rozzi, Ricardo, 1960- editor of compilation. II. Jiménez, Jaime E., 1960- editor of compilation. III. Sub-Antarctic Biocultural Conservation Program. IV. Parque Etnobotánico Omora (Agency : Chile)

QL695.2.S83 2013

333.95'8160983--dc23

2013038603

UNT PRESS

Universidad de Magallanes

"The maps published in this document that refer to or are related to the limits and boundaries of Chile do not at all commit the Chilean state, in agreement with Article 2, letter g of DFL. 83 from 1979, of the Ministry of Foreign Relations."

FOREWORDS

MAGELLANIC SUB-ANTARCTIC ORNITHOLOGY AT OMORA PARK

The desolate and extensive sub-Antarctic landscapes are located at southwestern of South America, close to the dry zone of the Southern Cone of the Neotropics, from the primarily Patagonian pampas vegetation and peatlands, to the rainy and cold Krummholz forests of the southwestern islands and archipelagos. Since the 19th Century, this area has been of interest for its particular isolation and biodiversity. The seaway of Cape Horn, as well as its costly and risky expeditions – with numerous shipwrecks – have attracted adventurers to its coasts with calm fijords, islands, and extensive grasslands for its fauna and flora that are typical of this unique biogeographic region. Three indigenous groups that inhabited these cold environments, the Onas, Yahgans, and Alacalufes, knew and made sustainable use of its marine and terrestrial fauna. Thanks to the Bridges family and its descendants, and to the Institute of Patagonia, we know much about these brave people that are today extinct, and the other beings with whom they then co-inhabited, and the ethnic names given to them. Literature about this geographically young and windswept, farthest corner of the world is still being produced, but the interactions among fauna and flora have recently been unveiled in a more systematic way.

Hence, this thematically organized ornithological work published by the scientific team at the Omora Ethnobotanical Park, leads us to an orderly way of knowing the particularities of some of the birds of these distant environments. It is important to contribute new knowledge about the morphology and ecology of the species that have been able to adapt to these harsh sub-Antarctic environments. Also, for its annual dynamic with marked seasonality, the fauna is subjected to adaptations, of which no more than a handful of species can survive here throughout the year, while others reproduce during the spring-summer to migrate later. Through the errors that humanity will never stop committing, we should also know the impact of the introduction of invasive, exotic species that have been altering the communities or reshaping the watersheds at the end of the world. For these reasons, socio-culturally, the development should be oriented with ethics, education, ecotourism, and conservation of ecosystems increasingly accessible today to the development of policies. Among the values that should not be lost are the ethnic, traditional ecological knowledge of the original people, and these, to this end are addressed in an original, integrated chapter.

Finally, the team of foreign and national specialists that have been building the activities during the decade in which they have been collecting data, applying novel methodologies, banding species, forwarding hypothesis, and studying the behavior of the bird communities at the extreme tip of the Neotropics, deserve our appreciation for the important advancements made. I am sure that Omora Park will continue to unravel the interactions and dynamics of our southern birds, and thus safeguard the ethnic legacy to educate about and conserve these fragile and young ecosystems. To have all this synthesized in one book provides us with an integrative view of what has been the recent history of the studies on biodiversity in the most remote and isolated corner of our country.

Roberto P. Schlatter
Facultad de Ciencias
Universidad Austral de Chile, Chile

OMORA PARK AND THE EMERGENCE OF LONG-TERM ORNITHOLOGY IN THE CAPE HORN REGION

The name Cape Horn cannot be read without immediately thinking of a land that is distant, difficult, harsh, and perhaps even violent. Our minds go to images of sailing ships and whalers "rounding the horn" during intense gales with destructive waves, in a time when we were still exploring the world, where risk was everywhere and life was not easy. Yet rather than repel us, these difficult images captivate our imagination and we want to see this southern land, to understand it, to live it and to realize that for all its harshness it is a land of immense beauty. Amazonia may be the lungs of South America, the Andes its spine, the Pampas its heart, but the Magellanic Region is where South America gets its character! This is an incredibly diverse continent, and it can be argued that as tropical biological exploration began early, and has been so focused, today the greater mysteries of this continent lie in the temperate and sub Antarctic region where detailed ecological analysis is still in its early stage, although it is exploding in volume and prominence as we see here. It is this new systematic study of the far south that we enjoy in this volume, and again the far south captivates us with images of explorers, although this time they are biologists, not seamen.

This work is a story, one that is still being written; in which science and discovery are part of the plot, and the scientific team at the Omora Ethnobotanical Park, the sub-Antarctic forests, and birds the protagonists. What is particularly exciting is discovering that in merely ten years we have found out so much about the southernmost Neotropical avifauna. At the start of the millennium there was very little known about the natural history and interrelationships of the birds of the far south, and now we have abundant details of morphology, temporal distributions, migration, food, ecological roles, and interrelationships of the birds of South America's sub-Antarctic. As expected, as answers are published, more questions emerge. A case in point being that many of the sub-Antarctic birds are migratory as one would predict in such a harsh and temperate situation, but contrary to expectations, some of the migrants visit in winter more so than summer? The interrelationship between the ocean, mountains and moisture do not make for a simple pattern in these southernmost forests in the world! This is a unique biogeographical region, and fortunately it is less of a mystery than it once was. Yet as this story unfolds, it does not do so without some drama. One of the major findings presented is that as discoveries are being made, the future of this avifauna is in peril from introduced Mink. The impact of this predator on ground nesting waterbirds is significant, and what is considered one of the most pristine habitats on earth is being eroded in its quality. But this is step one to finding a solution to the problem, so perhaps the next chapter will be a positive one.

As this story ends we find that in its conclusion, science and discovery are indeed deep in the plot, but it is something else that is rarer and unusual that leaves us considering the importance of the research done at Omora. This feature is the cultural aspect of the work, the "ethno" in ethnobotanical, as in this study in ecology, people are not ignored. The great legacy of this research is the integration of ornithology with our Chilean cultural history, allowing birds to be a window to our past and as Chileans and citizens of this world to cement the responsibility we have as custodians, protectors and beneficiaries of this land. This clarifies that the study of ornithology is not separate from who we are, but through the understanding of our avifauna we not only become richer from the science, but we connect to the people who have been learning from these birds thousands of years before us, and that the questions we answer are not about birds alone, but are a way to understand who we are and what future we will create as our ecological heritage.

Álvaro Jaramillo
San Francisco Bay Bird Observatory
California

PREFACE

ORNITHOLOGICAL RESEARCH AT OMORA PARK: THE LONGEST BIRD-BANDING PROGRAM IN SUB-ANTARCTIC FORESTS OF THE SOUTHERN HEMISPHERE

The Cape Horn Biosphere Reserve is a landmark for global biodiversity conservation. Today, southwestern South America still preserves an extensive area of evergreen forests that covers most of the Magellanic sub-Antarctic ecoregion (Rozzi *et al.* 2012). In this ecoregion that spans a myriad of islands, Cape Horn emerges as the southernmost forested point in the world. Metaphorically we have called it "the Southern summit of the Americas."[1]

Just as Mount Everest is the world's highest summit, Cape Horn is the highest latitude forested point in the Southern Hemisphere. Because of their extreme altitude or latitude, respectively, both peaks are subject to extreme climatic conditions, which acquire special relevance as we confront global climate change. Furthermore, in contrast to the broad longitudinal extension of temperate and boreal forests in the Northern Hemisphere, the temperate and sub-Antarctic forests of South America extend along a strip of land that is latitudinally very long (33-56ºS), but longitudinally very narrow (71-73ºW). This inter-hemispheric contrast creates unique possibilities for comparative studies on the life histories, ecology, and impacts of global environmental change on the birds, and biodiversity in general for at least five unique attributes of Cape Horn.

1. *Prevalence of birds.* In contrast to subpolar terrestrial ecosystems of the Northern Hemisphere that are inhabited by birds, large mammals, and some representatives of the herpetofauna, the subpolar Cape Horn region completely lacks the herpetofauna, and terrestrial mammals are almost completely absent. Therefore, birds are the dominant terrestrial vertebrates in the Cape Horn ecoregion, and they are the most diverse vertebrate group in both the sub-Arctic and sub-Antarctic regions. Due to their migratory behavior, the birds of the Northern Hemisphere subpolar forests are studied today as a sensitive biological indicator of global climate change. Birds of subpolar ecosystems in South America could now provide a valuable indicator group to monitor global climate change under environmental conditions that are biotically, geograpically, and climatically different from the subpolar ecosystems in the Northern Hemisphere.

2. *Cape Horn: "a geographic funnel" for migratory forest bird species.* In contrast to the Northern Hemisphere boreal forests, which offer a vast expanse of habitat to returning breeding migratory birds, Cape Horn is characterized by a narrow tip where migratory forest birds converge. This geographical attribute has methodological advantages for the study of migration, and provides a unique natural laboratory for inter-hemispheric comparative studies of the avifaunas of subpolar forests.

3. *Oceanic climate.* In contrast to the continental climate (freezing winters and hot summers) that characterizes boreal forests, the climate of the sub-Antarctic forests of Cape Horn is moderated by a strong oceanic influence that determines its mild winters and cool summers (for details see Chapter 1). The isothermal climatic conditions recorded in the sub-Antarctic forests could be a determinant factor for longevity and migration patterns of birds.

[1] The term "Magellanic sub-Antarctic ecoregion" was coined by Ricardo Rozzi as part of the proposal for the UNESCO Cape Horn Biosphere Reserve, in order to distinguish it from Patagonia. He also coined the metaphor of the "Southern Summit of the Americas" to highlight the relevance of proposing Cape Horn as a biosphere reserve to the UNESCO-MAB Committee. See page 13 in Rozzi, R. (2004) "The Cape Horn Archipelago territory: A biocultural treasure at the south of the world," pp. 11-25, in Rozzi, R., F. Massardo & C. Anderson (Eds.), *The Cape Horn Biosphere Reserve: A Proposal for Conservation and Tourism to Achieve Sustainable Development at the Southern End of the Americas*, bilingual English-Spanish Edition, Ediciones Universidad de Magallanes, Punta Arenas, Chile. For a summary of the main attributes of the Magellanic sub-Antarctic ecoregion see R. Rozzi, J.J. Armesto, J. Gutiérrez, F. Massardo, G. Likens, C.B. Anderson, A. Poole , K. Moses, G. Hargrove, A. Mansilla, J.H. Kennedy, M. Willson, K. Jax, C. Jones, J.B. Callicott & M.T. Kalin (2012) "Integrating ecology and environmental ethics: Earth stewardship in the southern end of the Americas." *BioScience* 62: 226-236.

PART I

THE LONG-TERM ORNITHOLOGICAL RESEARCH PROGRAM AT OMORA PARK: BIRD CONSERVATION, NEW MORPHOLOGICAL INFORMATION, AND ECOLOGICAL INSIGHTS

Figure 1. Map of the Magellanic sub-Antarctic ecoregion showing the full extent of the broadleaf evergreen rainforests (green) and the Magellanic moorland (gray), from Cape Horn (56ºS) to Penas Gulf (47ºS) in Chile. The satellite image below illustrates the location of the Omora Ethnobotanical Park, and the Cape Horn Biosphere Reserve area (enclosed by the light blue dashed line). Omora Park is the biocultural research, education, and conservation center of this biosphere reserve, and it is situated on Navarino Island (55ºS), south of Tierra del Fuego. Modified from Rozzi et al. 2008a. Photo Adam Wilson.

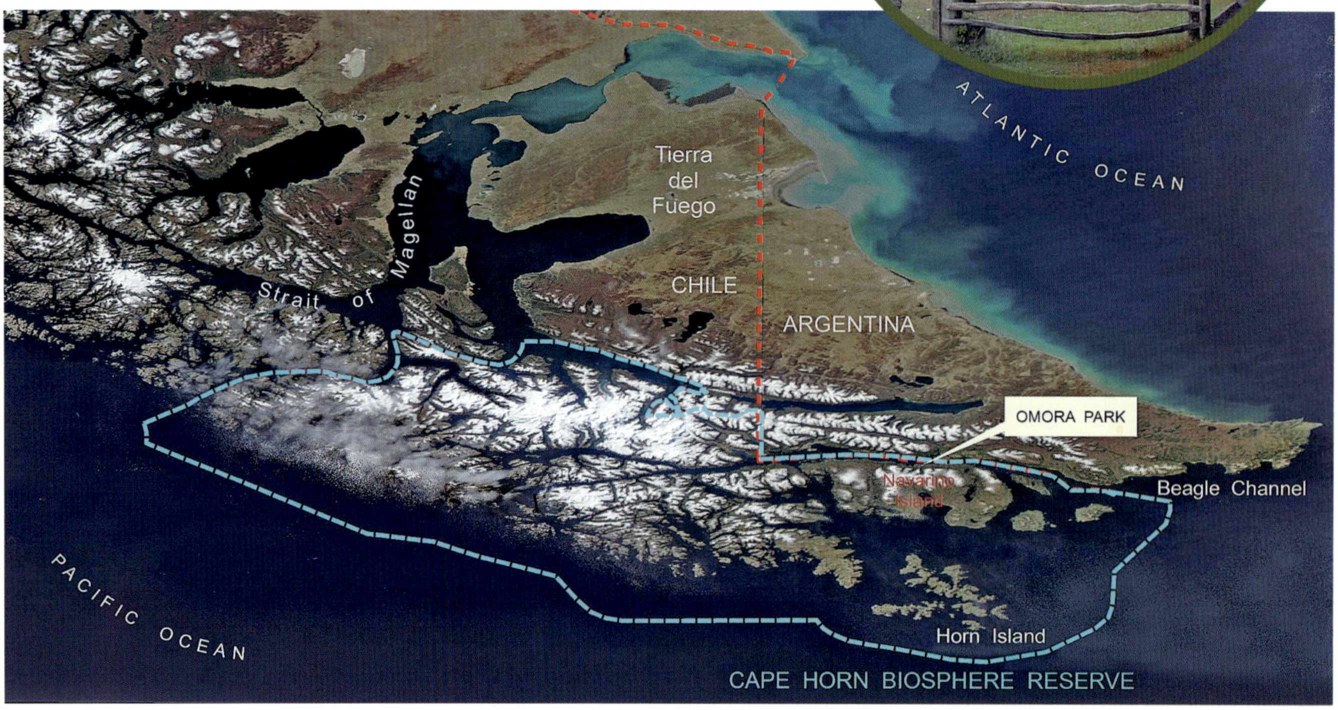

Figure 2. *Front page.* Glacier Italia, one of the numerous in the Cape Horn Biosphere Reserve that flows down from the Darwin Cordillera to the sea. Photo Jordi Plana.

1 THE OMORA PARK LONG-TERM ORNITHOLOGICAL RESEARCH PROGRAM: STUDY SITES AND METHODS

Ricardo Rozzi, Jaime E. Jiménez, Francisca Massardo, Juan Carlos Torres-Mura, and Rajan Rijal

In January 2000, we initiated a Long-term Ornithological Research Program at Omora Ethnobotanical Park in the world's southernmost forests: the sub-Antarctic forests of the Cape Horn Biosphere Reserve. In this chapter, we first present some key climatic, geographical, and ecological attributes of the Magellanic sub-Antarctic ecoregion compared to subpolar regions of the Northern Hemisphere. We then describe the study sites at Omora Park and other locations on Navarino Island and in the Cape Horn Biosphere Reserve. Finally, we describe the methods, including censuses, and present data for each of the bird species caught in mist nets during the first eleven years (January 2000 to December 2010) of the Omora Park Long-Term Ornithological Research Program.

THE MAGELLANIC SUB-ANTARCTIC ECOREGION

The contrast between the southwestern end of South America and the subpolar zone of the Northern Hemisphere allows us to more clearly distinguish and appreciate the peculiarities of an ecoregion that until recently remained invisible to the world of science and also for the political administration of Chile. So much so, that this austral region lacked a proper name, and it was generally subsumed under the generic name of Patagonia. For this reason, to distinguish it from Patagonia and from sub-Arctic regions, in the early 2000s we coined the name "Magellanic sub-Antarctic ecoregion" (Rozzi 2002).

The Magellanic sub-Antarctic ecoregion extends along the southwestern margin of South America between the Gulf of Penas (47°S) and Horn Island (56°S) (Figure 1). Its mountainous terrain includes steep slopes, and along the Andes, the Darwin Cordillera in Tierra del Fuego and numerous high summits on other islands there are ice fields with glaciers that flow down to the sea (Figure 2). Its forests are characterized by the dominance of an evergreen broadleaf tree species, the Magellanic Coigüe or Evergreen Beech (*Nothofagus betuloides*), and for being embedded in a matrix of moorlands and wetlands, known as the Magellanic tundra complex (Pisano 1977, 1980,

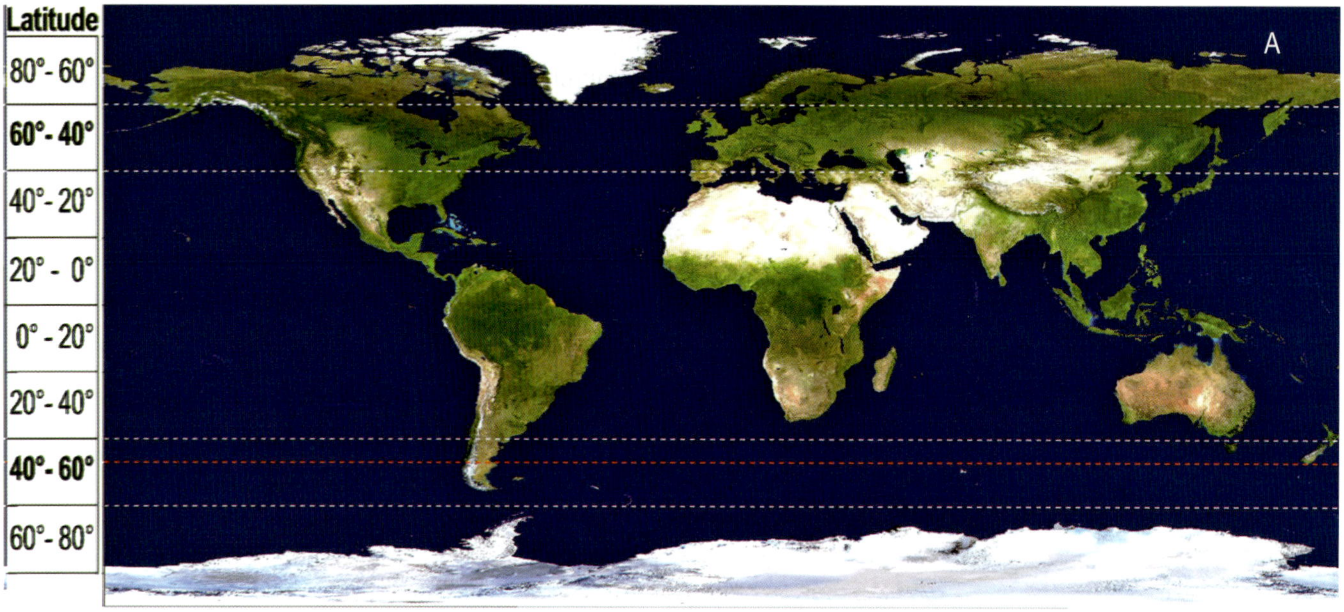

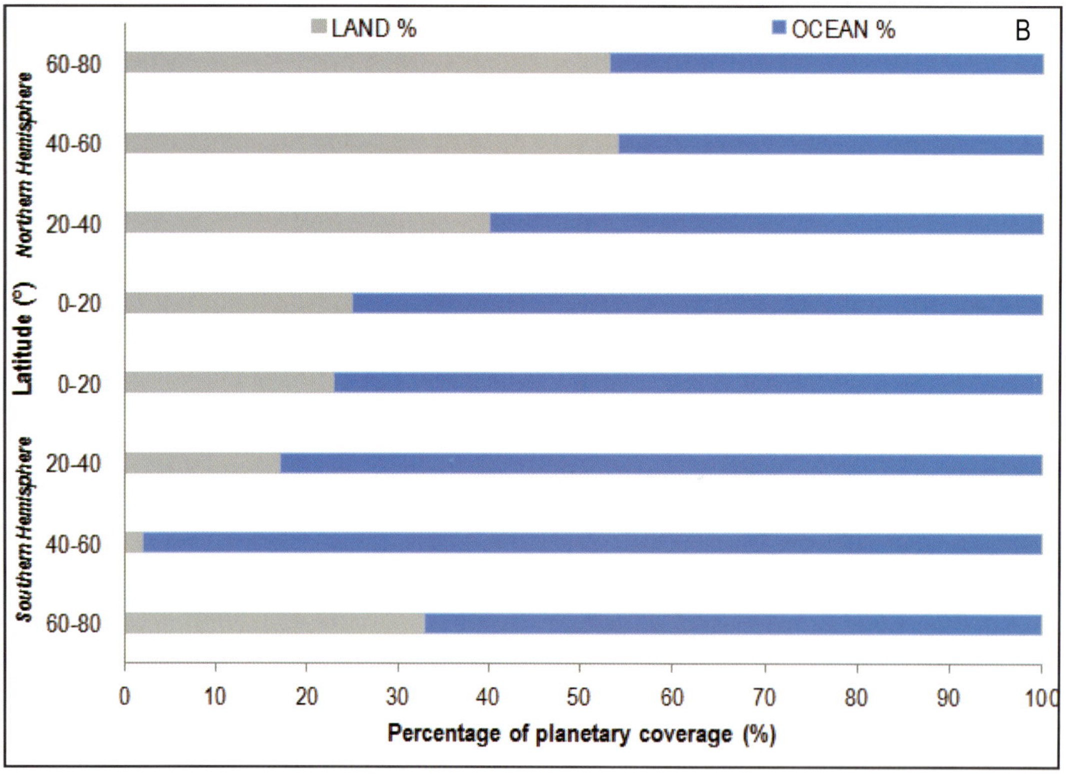

Figure 3. A. World image showing that the Magellanic sub-Antarctic forests extend nearly 10 degrees further south than the temperate forests of Stewart Island, New Zealand (47ºS, indicated by the dashed red line). Therefore, the world's southernmost forests in the sub-Antarctic Magellanic ecoregion have no eco-geographical equivalent in the Southern Hemisphere. The sub-Antarctic forests also contrast with the boreal forests, because while the former span a narrow strip of land in southwestern South America the latter cover a vast cross-continental area in the Northern Hemisphere. The white dashed lines enclose the latitudinal bands at 40–60º in both hemispheres to illustrate the conspicuous difference in land:ocean ratios at this latitudinal range (Figure prepared by Omora Park & GIS Laboratory, CERE-Universidad de Magallanes, based on "The MODIS Rapid Response System"). B. The Southern Hemisphere has a clear prevalence of oceanic surface over land surface. The prevalence of oceanic surface reaches a maximum at the 40-60°S latitudinal band where the proportion of land:ocean areas is 2%:98%. In contrast, in the Northern Hemisphere, there is a prevalence of land surface at temperate and subpolar latitudes, which reaches a maximum at the 40-60°N latitudinal band where the proportion of land:ocean areas is 54%:46%. Data from Rozzi *et al.* 2012.

Arroyo *et al.* 2005). Its climate is hyperhumid with cool temperatures and an annual rainfall that is greater than 1000 mm in most areas, and greater than 5000 mm in the western areas (Tuhkanen *et al.* 1990).

As compared to names used in earlier literature, such as "antiboreal" region (Koehler 1912, Ekman 1935, Wace 1965), "Westpatagonia" (Skottsberg 1924), "Subantarctic subregion" (Cabrera 1973, Morrone 2000), "Magellanic rainforests" (Veblen *et al.* 1983), and "Magellanic subpolar rainforests" (Mittermeier *et al.* 2002, 2003), the term "Magellanic sub-Antarctic ecoregion" (Rozzi 2002, Rozzi *et al.* 2012) has the advantage of clearly distinguishing this ecoregion from:

(i) Patagonia in southeastern South America, which is dominated by low shrub steppe, has a flat relief, and a semi-arid climate (Cabrera & Willink 1973).

(ii) Sub-Antarctic oceanic islands in the Southern Hemisphere (*e.g.*, Diego Ramirez Islands, South Georgia and the South Sandwich Islands), which completely lack forests (Pisano 1972, Pisano & Schlatter 1981a,b).

(iii) The sub-Arctic region in the Northern Hemisphere, which is characterized by its boreal forests dominated by evergreen conifers and deciduous broadleaf trees, and have a continental climate (Chapin *et al.* 2006).

Omora Park is located near the southern end of the Magellanic sub-Antarctic ecoregion, south of Tierra del Fuego, along the Beagle Channel on the northern coast of Navarino Island (Figure 1). Due to the inter-hemispheric contrast between sub-polar areas (sub-Arctic and sub-Antarctic), the Long-Term Ornithological Research (LTOR Program) developed at Omora Park and the Cape Horn Biosphere Reserve is not only of regional significance, but of global significance for the following reasons.

WORLD RELEVANCE OF THE OMORA PARK LONG-TERM ORNITHOLOGICAL RESEARCH (LTOR) PROGRAM IN THE MAGELLANIC SUB-ANTARCTIC FORESTS

1. *The LTOR-Omora Park has no eco-geographical equivalent worldwide*

The forests of Cape Horn are not only the most southern of the planet, but neither do any latitudinally equivalent forests exist in the Southern Hemisphere. Omora Park is located at 55°S, near the southern end of the Magellanic sub-Antarctic ecoregion, approximately 10 degrees of latitude further south than New Zealand southernmost forests on Steward Island at 47°S or Tasmania's forests at 43°37'S in South Cape Bay (Figure 3A). The forests of New Zealand and Tasmania are temperate forests. Therefore, there is no eco-geographically equivalent ecoregion to the Magellanic sub-Antarctic ecoregion in the southern Hemisphere. In addition, the Magellanic sub-Antarctic ecoregion has conspicuous geographic, climatic, and biological contrasts with the subpolar forests of the Northern Hemisphere. This confirms that the LTOR-Omora Park, including its mist-netting and bird banding program, is conducted in an ecoregion that has no equivalent elsewhere.

2. *Omora Park has an oceanic climate that contrast with the continental climate of sub-polar study sites in the Northern Hemisphere*

The sub-Antarctic ecosystems of southwestern South America strongly contrast with their latitudinal counterparts in the Northern Hemisphere for geographical reasons. In the Southern Hemisphere, in the latitudinal band 40-60°S, the proportion of land:ocean areas is 2%: 98%, whereas in the latitudinal band 40-60°N the proportion of land:ocean areas is 54%: 46% (Figure 3B). At a global scale, this marked difference in the proportion of terrestrial *versus* oceanic areas creates a strong macro-climatic contrast between the temperate and sub-polar regions of the north and south hemispheres.

In the Southern Hemisphere, the climate in temperate and sub-Antarctic ecosystems is moderated by the vast expanse of ocean that creates an oceanic climate characterized by very low thermal fluctuations between seasons, with winters that do not have extremely cold temperatures and mild summers (Figure 4A). This feature was well described by Rockwell Kent (the celebrated illustrator of the American frontier of Alaska) who concluded his book *Voyaging Southwards from the Strait of Magellan* (Kent 1924) by stating that "the forests are luxuriantly green, with stately trees, and violets starring their dark floors; and in them live the gentlest of wild creatures. It is a peaceful and friendly wilderness, neither intemperately hot or cold" (p. 184).

In contrast, in the Northern Hemisphere the high proportion of land at high latitudes generates a continental type of climate, which is characterized by great annual temperature ranges, with very cold winters and warm summers (Arroyo *et al.* 1996, Lawford *et al.* 1996). For example, during the years 2009-2012 the annual temperature range recorded at the northernmost forested site of the US Network of Long-Term Ecological Research (LTER), Bonanza Creek in Alaska (64°42'52"N, 148°08'43"W) was 47.9°C. The mean temperatures of the coldest (January) and the warmest month (July) were 19.2°C and -28.7°C, respectively (http://www.lter.uaf.edu/bnz_bcef.cfm) (Figure 4B). For the same period of time, the annual temperature range recorded by the microclimate station of the Chilean Long-Term Socio-Ecological Research network (LTSER-Chile) at Omora Park, 10 m above sea level, was 8.9°C. The annual temperature range is calculated

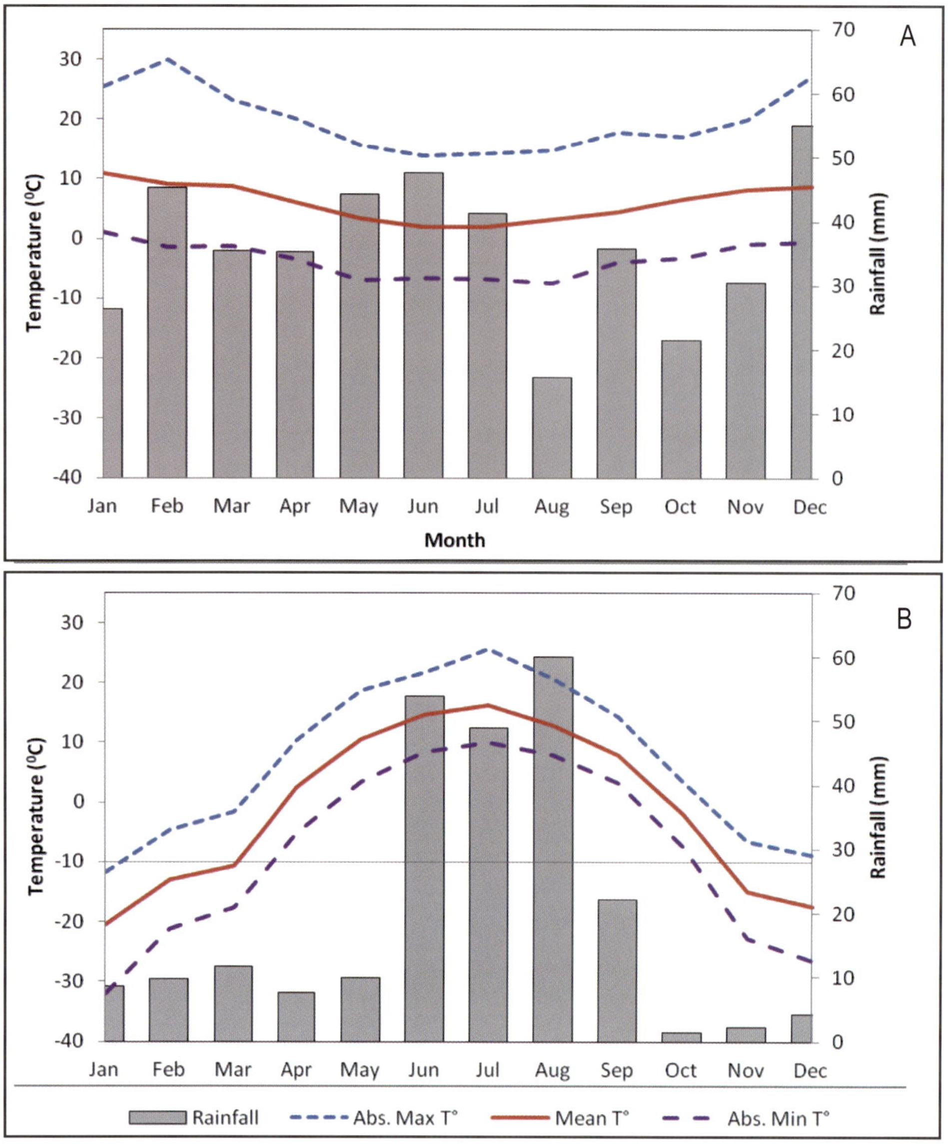

Figure 4. Average monthly rainfall (bars), mean monthly temperature (red central line), absolute maximum temperature (blue dashed line) and absolute minimum temperature (purple dashed line) at (top) the Omora Park Station, LTSER Chile, and (bottom) at Bonanza Creek in Alaska (LTER, (64º42'52"N, 148º08'43"W) for the period January 2009–December 2012. Data from LTSER-Chile (http://www.ceazamet.cl/index.php?pag=mod_estacion&e_cod=CNPW&p_cod=ceazamet), and LTER-Bonanza Creek (http://www.lter.uaf.edu/bnz_bcef.cfm).

Figure 5A. The Magellanic Woodpecker (*Campephilus magellanicus*) is the largest woodpecker in South America. About ten family groups are found in Omora Park, which offers protection to this species that is classified as Endangered in south-central Chile, and as Vulnerable in southern Chile and Argentina. The Magellanic Woodpecker is a congeneric species of the Ivory-billed (*Campehilus principalis*) and Imperial (*C. imperialis*) woodpeckers of North America. In contrast to the Magellanic Woodpecker, both North American species are probably extinct today due to habitat degradation and hunting pressure. Photo Jordi Plana.

as the difference between the mean temperature of the warmest month (10.8ºC, January) and the mean temperature of the coldest month (1.9ºC, June and July) (http://www.ceazamet.cl/index.php?pag=mod_estacion&e_cod=CNPW&p_cod=ceazamet) (Figure 4A). Therefore, the annual temperature range recorded at Bonanza Creek is 6.4 times greater than in Omora Park. In addition, during the years 2009-2012 the lowest temperature recorded at Bonanza Creek (-32ºC in January 2012, Hollingsworth 2005) was 25ºC colder than the lowest temperature recorded at Omora Park (-7.5ºC in August 2011).

As compared to Omora Park, seasonality, as characterized by thermal regimes and patterns of rainfall, is much more marked in Bonanza Creek. Whereas at Omora Park rainfall is evenly distributed throughout the year (Figure 4A), at Bonanza Creek there is a concentration of rainfall in the summer (Figure 4B). We have included the climatic comparison between Omora Park and the Bonanza Creek LTER site for two reasons:

(a) Bonanza Creek contains forests that are representative of the northernmost subpolar or boreal forests in the Americas. In this sense Bonanza Creek represents a kind of "inter-American mirror image" of the southernmost forests at Omora Park.

(b) In 2010, a formal agreement of long-term collaboration was signed between the University of Alaska Fairbanks that administrates Bonanza Creek and the University of Magallanes and the University of North Texas that co-administrate Omora Park. In the next years we plan to develop comparisons between the forest avifauna of both sites.

It is important to note, however, that in sites located on the west coast of Alaska and other coastal sites in sub-polar zones of the Northern Hemisphere, the maritime influence results in lower seasonal fluctuations in temperature and rainfall than those recorded at Bonanza Creek. Nevertheless, the oceanic climate characteristics of these boreal sites are less than those recorded in southwestern South America (Alaback 1996). Additionally, in the Magellanic sub-Antarctic ecoregion all forests are subjected to strong oceanic climatic conditions and no forests are subjected to continental climatic conditions. In contrast, in the Northern Hemisphere most boreal forests are subjected to strong continental climatic conditions (Arroyo *et al.* 1996). Given these marked inter-hemispheric climatic differences, in future ornithological research at Omora Park we plan to conduct comparative sub-Antarctic—sub-Artic studies that may generate new insights about the impacts of global climate change on the biology and population dynamics of the subpolar forest birds.

3. *The Omora Park long-term ornithological studies are conducted under conditions of minimal direct human impact*

Southwestern South America includes a vast and remote area of forest that was recently identified as one of 24 world's most pristine ecoregions (Mittermeier *et al.* 2002, 2003). The archipelago of Cape Horn is one of the few island groups in South America that has remained free of direct modern human impact (Silander 2000). Under these conditions, the Omora Park and the Cape Horn Biosphere Reserve represent a natural refuge for species that are threatened in other areas of their range, such as Magellanic Woodpecker (*Campephilus magellanicus*) (Figure 5A).

The Magellanic Woodpecker is the largest woodpecker in South America. It is also a congeneric species of the two world's largest woodpecker species: the Ivory-billed (*Campephilus principalis*) and the Imperial (*C. imperialis*) woodpeckers. These two species that inhabited the temperate forests of North America are presumed to be extinct today or are Critically Endangered (IUCN 2013). The causes of their extinction include hunting pressure and drastic habitat loss of the temperate forests in North America. In contrast, at the southern end of the Americas the Cape Horn Biosphere Reserve preserves extensive contiguous old-growth forest. For this reason, Navarino Island hosts relatively large populations of *C. magellanicus*, a species that is threatened in the rest of Chile.

The logging of native forests and the replacement of native forests by plantations of monocultures of exotic fast-growing trees such as *Eucalyptus* and the Monterrey Pine (*Pinus radiata*) in south-central Chile and Argentina has drastically affected the populations of the Magellanic Woodpecker over most of its range. Today, the Magellanic Woodpecker is considered a threatened species. It is classified as an Endangered species in the administrative regions O'Higgins and Maule of Chile (33-37ºS), and as a Vulnerable species in the administrative regions Bío-Bío to Magallanes (38ºS-56ºS) in Chile (Glade 1993, Hunting Law, SAG 2012), as well as in Argentina (Ojeda 2004). Therefore, the populations of Magellanic Woodpecker in Omora Park as well as in other sites of Navarino Island and the Cape Horn Biosphere Reserve are particularly important for the preservation of this species that has global relevance for the conservation of birds and biodiversity.

Other birds of special interest for conservation include the Rufous-legged Owl (*Strix rufipes*) and the Chilean Hawk (*Accipiter chilensis*). Like the Magellanic Woodpecker, the Rufous-legged Owl is an indicator species of old-growth forests (Figure 5B). It is also a congeneric species of the Spotted Owl (*Strix occidentalis*), which has played a central role in the conservation debate of old-growth forests in the Pacific Northwest in the USA and Canada (Martínez & Jaksic 1996). The Rufous-legged Owl is frequently sighted at Omora Park, but it is considered to have sparse populations in Chile. At the national level, it is classified as an Inadequately Known species (SAG 2009; the Chilean Hunting Law uses an earlier version of IUCN categories). The Chilean Hawk is another raptor species that nests in the forests of Omora Park, which is of conservation concern at the national level. The Chilean Hawk is classified as a Rare Species in Chile (SAG 2009, Figure 5C). For these endangered birds, including the largest survivor of the genus *Campephilus*, the forests of Omora Park and the Cape Horn Biosphere Reserve provide a critical protected area. In addition to a conservation area, Omora Park provides a *natural laboratory* for the study of the biology and ecology of the bird assemblages of the sub-Antarctic forests under conditions with a minimal degree of direct modern human disturbance.

Figure 5B. The Rufous-legged Owl (*Strix rufipes*) is a resident species in the old-growth forests of Omora Park. It is the species of owl with the highest conservation priority in Chile (SAG 2009). Photo José Tomás Ibarra.

Figure 5C. The Chilean Hawk (*Accipiter chilensis*) is frequently observed at Omora Park. In Chile it is classified as a Rare Species (SAG 2009). Photo José Tomás Ibarra.

4. *Omora Park fills a global gap in long-term ornithological and ecological studies*

The ornithology of Magellanic sub-Antarctic forests has been little studied, as compared to the ornithology of temperate and boreal forests of the Northern Hemisphere (Jarvinen & Vaisänen 1977, Jaksic & Feinsinger 1991, Meserve & Jaksic 1991, Willson 1991, Rozzi 2002, Brommer 2004, Brommer *et al.* 2012). Within the South American temperate forest biome, Magellanic sub-Antarctic forests have been also overlooked. Ornithological studies in the area of Cape Horn have been limited to short periods, mainly during the breeding season of the austral summer (see Jiménez and Rozzi, Epilogue in this book). Ecological studies and a few long-term mist-netting program have been conducted in the Mediterranean and temperate forests of south-central Chile, as well as in the relict fog forests of Fray Jorge National Park (*e.g.*, García 1982, Erazo 1984, Egli 1987, 1996, Armesto & Rozzi 1989, Armesto *et al.* 1987, 1995, 2001, Correa *et al.* 1990, López 1990, Sabag 1993, Rozzi *et al.* 1995, 1996, Smith-Ramírez & Armesto 1998, Gantz & Rau 1999, Cornelius *et al.* 2000, de Santo *et al.* 2001, Kelt 2001, Reid *et al.* 2004, Díaz *et al.* 2005, Tomacevic & Estades 2006, Ibarra *et al.* 2010). However, not a single year-round ornithological study was conducted in the Magellanic sub-Antarctic forests before the Omora Park Long-Term Ornithological research program was started. Moreover, in previous comparisons of bird assemblages of temperate and subpolar forests of North and South America, no mention was made of the Magellanic sub-Antarctic forests (cfr. Jaksic & Feinsinger 1991, Willson 1991, Willson *et al.* 1994).

Not only the ornithology, but the ecology of the austral Magellanic Region has also been understudied (Figure 6). Long-term ecological research networks and environmental observatories have conspicuously overlooked this region. For example, in 2010 the International Network of Long-Term Ecological Research (ILTER, www.ilternet.edu), the largest global network of ecological monitoring, included 543 sites in 44 countries. However, 509 of these sites (93.7%) are located in the Northern Hemisphere. Most sites of the ILTER network are located in boreal and temperate latitudes of the Northern Hemisphere with 69% of ILTER sites located at latitudes > 40°N (Figure 6). Less than 10% (n = 34) of ILTER network sites have been established within the tropical latitudes between 20°N and 20°S, where most of the diversity of birds and other groups of organisms is found (Myers *et al.* 2000). In the Southern Hemisphere as a whole, there are only 34 sites (6.3%), and until 2010 there was a complete absence of ILTER sites in temperate and sub-Antarctic latitudes in the Southern Hemisphere (Figure 6). This blind spot in the ILTER network involved the omission of the full temperate biome in South America, including its Magellanic sub-Antarctic ecoregion (Rozzi *et al.* 2012).

Other international networks of long-term global monitoring and ecological research have also omitted the 40-60ºS latitudinal range in the Southern Hemisphere. For example, FLUXNET has over 500 sites with meteorological towers continuously operating on five continents to record the flow of carbon dioxide in terrestrial ecosystems, but these sites are currently restricted to the latitudinal range between 70ºN and 30ºS (Sundareshwar *et al.* 2007). The Global Lakes Ecological Observatory Network (GLEON) includes 27 observatories in the five continents, but these sites are concentrated between 69ºN and 38ºS. GLEON omits the 40-60ºS latitudinal area, and includes a lake observatory in Antarctica at 77ºS (www.gleon.org). Terrestrial transects established by the International Geosphere-Biosphere Programme (IGBP) do not include any point in temperate and sub-Antarctic forests (Koch *et al.* 1995, Steffen *et al.* 1999). Therefore, the geographical area of temperate and sub-Antarctic forests of South America represents a blind spot in all the major systems of long-term ecological monitoring, including long-term ornithological studies.

The LTOR- Omora Park aims to contribute to address the geographical gap in long-term studies in the Southern Hemisphere (Rozzi *et al.* 2008b). To this end, Omora Park was one of the founding sites of the recently created Chilean Network of Socio-Ecological Long-Term Research Sites (LTSER-Chile) (Figure 7A). The three founding sites of the LTSER-Chile network span the whole latitudinal range of the South American temperate forests biome. These sites include a coastal fog forest in a matrix of semiarid scrublands at Fray Jorge National Park (30°S) (Gutiérrez *et al.* 2010), Valdivian and North Patagonian forests on Chiloé Island at the Senda Darwin Biological Station (42°S)

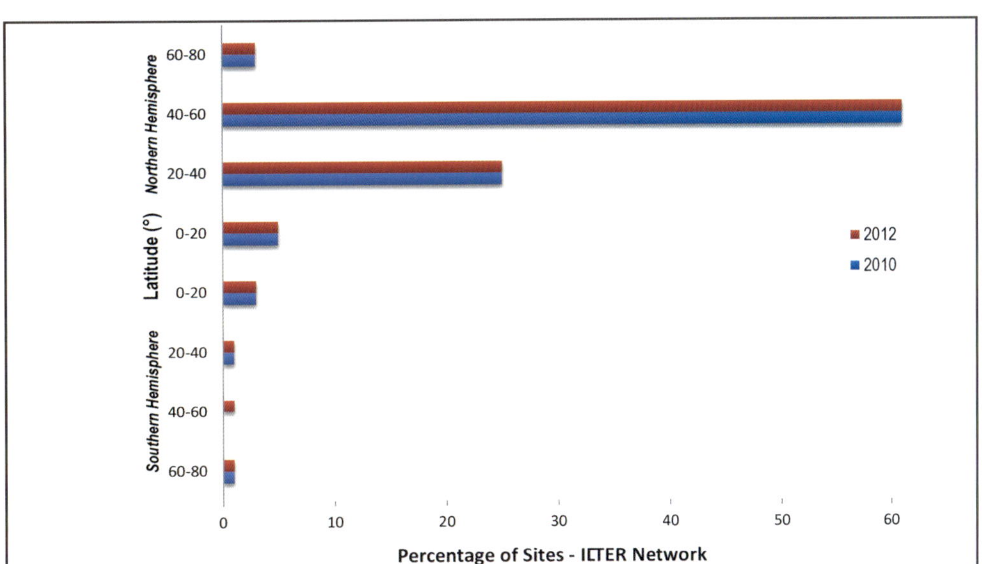

Figure 6. Relative percentage of study sites in the International Long-Term Ecological Research (ILTER, www.ilternet.edu) network at different latitudinal intervals. There is a very low proportion of sites in the Southern Hemisphere, especially in the latitudinal band of 40° -60° S that include the Magellanic sub-Antarctic ecoregion. Notice that the latter has ILTER sites only since 2011. Modified from Rozzi *et al.* 2012.

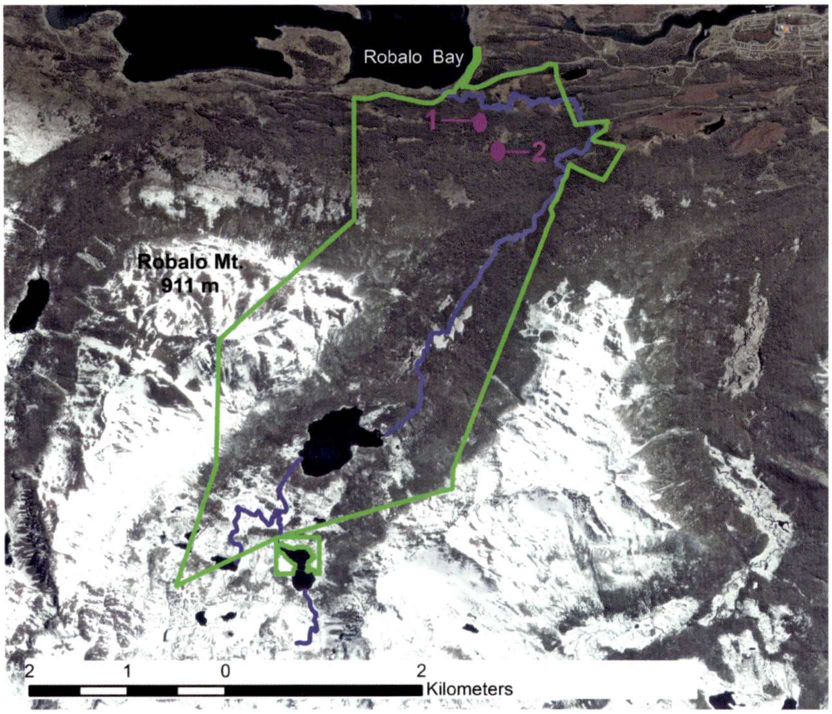

Figure 10. The Omora Park covers 1069 hectares of the Róbalo river watershed, which provides drinkable water to the city of Puerto Williams (upper right in the satellite image). The pink dots show the location of the permanent long-term mist-netting sites of forest birds (1, Canelo site) and forest edge birds (2, Notro site). Figure prepared by Omora Park & GIS Laboratory, CERE-Universidad de Magallanes, based on a Quick-Bird image taken in 2003.

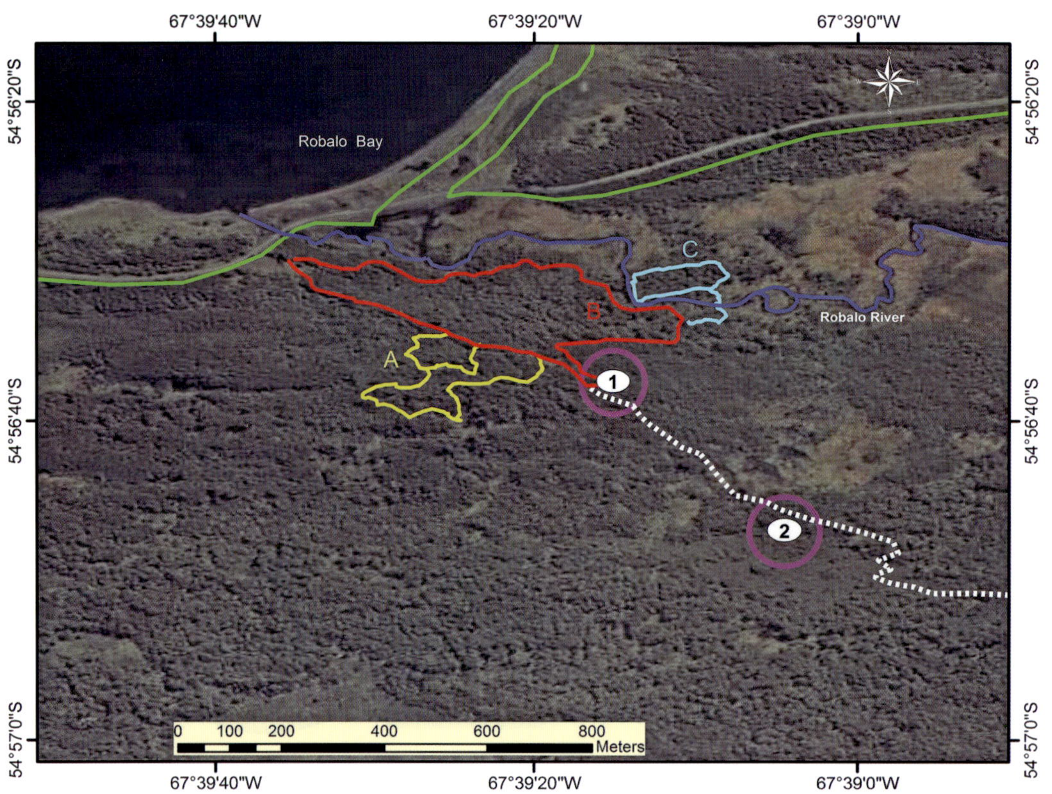

Figure 11. Close-up of the two long-term bird mist-netting sites at Omora Park: (1) "Canelo site" is a broadleaf evergreen forest site located within a large patch of old-growth forests dominated by Magellanic Coigue (*Nothofagus betuloides*), and Winter's Bark or "Canelo" (*Drimys winteri*); (2) "Notro site" is a forest edge site located at the ecotone formed by a patch of mixed evergreen and deciduous forests (dominated by Lenga (*Nothofagus pumilio*) and by a shrubland dominated by Firebush or "Notro" (*Embothrium coccineum*), Winter's Bark (*Drimys winteri*), Fashine (*Chiliotrichum diffusum*), Box-leafed Barberry (*Berberis buxifolia*), and Prickly heath (*Gaultheria mucronata*). The figure also illustrate the three main interpretive trails used for educational and ecotourism activities at Omora Park, which include bird watching. These trails are: A = Ecotourism with a Hand-Lens in the Miniature Forests of Cape Horn; B = the Southernmost Forests of the World; C = Underwater with a Hand-Lens in the cleanest freshwaters of the planet. Figure prepared by Omora Park & GIS Laboratory, CERE-Universidad de Magallanes, based on a Quick-Bird image taken in 2003.

The high Andean (or alpine) zone above the tree line on the summits of Róbalo Mountain and Dientes de Navarino Range, are inhabited by two bird species with a restricted geographic distribution, and specialized habitat requirements: the Yellow-bridled Finch (*Melanodera xanthogramma*) and the White-bellied Seedsnipe (*Attagis malouinus,* a Rare species according SAG 2012). The bodies of freshwater, such as Róbalo Lake, provide breeding habitat for the Flying Steamer-Duck (*Tachyeres patachonicus*). As mentioned above, the forests are dominated by *Nothofagus* species and provide old-growth forest habitats for three bird species of conservation interest: Magellanic Woodpecker, Rufous-legged Owl, and Chilean Hawk.

The permanent long-term mist-netting sites for forest interior birds are located at lower altitudes (Figure 11). They include two sites: 1) Canelo site located in forest interior habitats, within a large patch of old-growth forests dominated by Evergreen Beech or *Magellanic Coigüe* (*N. betuloides*), and Winter's Bark or *Canelo* (*Drimys winteri*); 2) Notro site located at the forest edge of mixed evergreen and deciduous forests, adjacent to a scrubland dominated by Firebush or *Notro* (*E. coccineum*) (Figure 11). At these two sites, bird-banding has been conducted monthly since 2000 (see Rozzi 2002). Other study sites at Omora Park, including high Andean habitats, have been used to conduct occasional supplementary mist-netting.

Outside Omora Park, on the northern coast of Navarino Island the main complementary study site is Guerrico Hill (54°55'S, 67°54'W), a gentle, north-facing slope (Figure 12), where bird-banding has been conducted at least once a year since 2000. Guerrico is a critical study site to assess site fidelity of migratory birds because it is separeted from Omora Park by only 25 km. Guerrico is dominated by a patch of *E. coccineum* adjacent to a fire regenerated forest of Evergreen Beech and High-deciduous Beech or *Lenga* (*N. pumilio*), and to the main dirt road along the northern coast of Navarino Island. The mist-netting sites have experienced a range of anthropogenic influences, mostly from timber harvest, and clearing and burning for pastureland. As in Omora Park, the landscape near Guerrico and other areas of Navarino Island includes natural and anthropogenic disturbance on north-facing slopes. The land's history of use generated a mosaic of habitats, which include:

(a) Anthropogenic park formations characterized by large trees of *N. betuloides* and *N. pumilio*, growing among burned trunks, logs, snags, regenerating saplings, and abundant shrubs of *Berberis buxifolia*, *B. ilicifolia*, and *Ribes magellanicum* that produce berries during spring and summer (November-February);

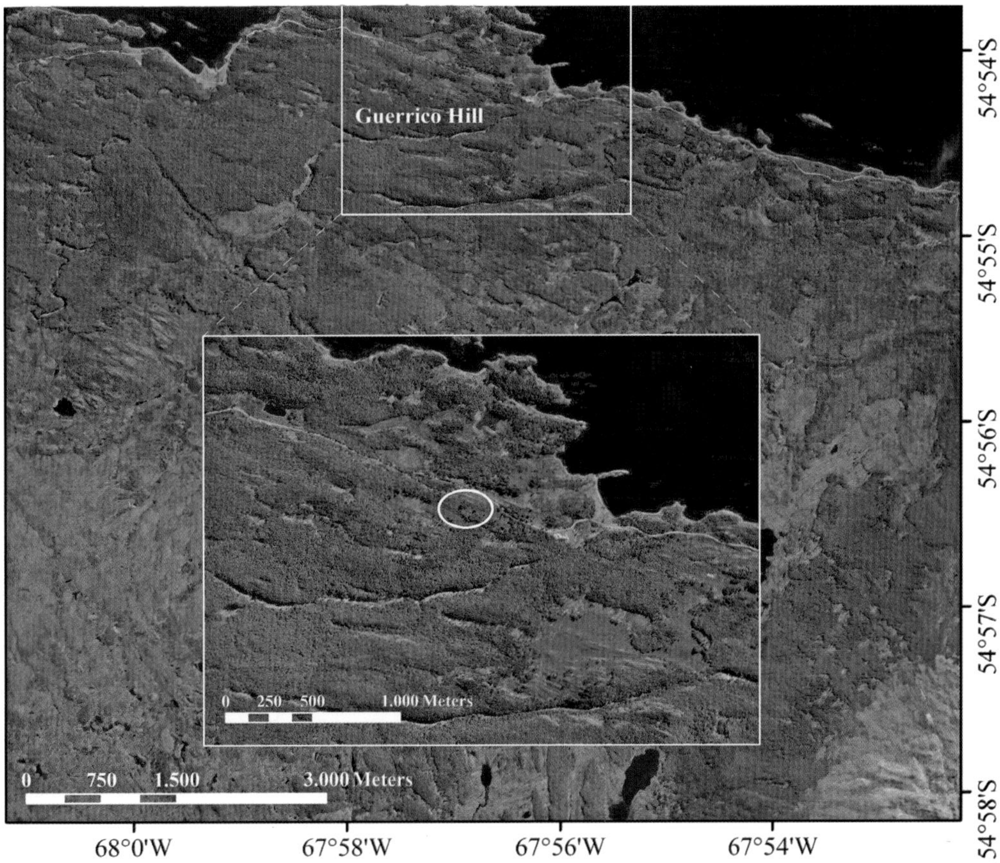

Figure 12. Aerial photograph of the Guerrico study site on the northern coast of Navarino Island. The 10 x 10 km view shows the landscape mosaic with patches of forests on the north facing slopes of the Navarino mountain range. Inset: Detail of the *Embothrium coccineum* study site, which is surrounded by patches of *Nothofagus betuloides* and *N. pumilio* forests, deforested areas, and a beaver pond on the left top corner. Image prepared by Gerardo Soto.

Nº	A) Navarino Island sites	Nº	B) Beagle Channel West sites	Nº	C) Almirantazgo Sound sites	Nº	E) Cape Horn Archipelago
1	Omora Park	19	Yendegaia Bay: Ferrari Cove - Serka Glacier	35	Bahia Sound Cien Cascadas	53	Hoste Island: Hardy Peninsula - Orange Bay
2	Guerrico	20	Holanda Glacier: Olla Cove shore & Eastern hills	36	Cien Cascadas Bay	54	Hoste Island: Hardy Peninsula - Lort Bay & Cannelliert Point
3	Mejillones Bay	21	Frances Mountain: Forested slopes west of Olla Cove	37	Ainsworth Bay: West area	55	Bayly Island: Washington Channel
4	Puerto Inutil	22	Alemania Glacier: Glacio-fluvial riparian forest	38	Ainsworth Bay: Marinelli Glacier	56	Wollaston Island: Kendall Cove
5	Wulaia Bay	23	Pia Glacier: East Arm	39	Ainsworth Bay: East area	57	Wollaston Island- Eliana Cove: Bravo Channel
6	Douglas Bay	24	Pia Glacier: West Arm	40	Parry Fjord: Morro Point	58	Herschel Island: Martial Cove
7	Grandi Sound	25	España Glacier: lowlands forests	41	Parry Fjord: Southwest Arm	59	Hermite Island: Maxwell Port
8	Windhond Lake	26	Garibaldi Sound: lowland forests	42	Parry Fjord: Southeast Arm	60	Hermite Island: Saint Martin Cove
9	Windhond Bay	27	Ventisqueros Sound: 2da Angostura West- lowland forests	43	Parry Fjord: Barria Point	61	Hornos Island: Southwest Peninsula: Espolón Point
10	Puerto Toro	28	Ventisqueros Sound: Verde Point - lowland forests	44	Parry Fjord: Blanca Bay		
11	Holger Islets	29	Chair Island: tundra - lowland forests	45	Almirantazgo Sound: Maria Cove (Caleta)		
12	Eugenia Cove	30	Gordon Island: Morning Cove - Romanche Bay	46	Paralelo River		
13	Picton Island: Piedras Cove	31	Gordon Island: Fleuriais Bay	**Nº**	**D) West Areas Biosphere Reserve**		
14	Nueva Island: Pescado Cove	32	Hoste Island: Mil Cascadas Fjord	47	de Agostini Sound: East Coast		
15	Button Island	33	Hoste Island: Coloane Bay	48	de Agostini Sound: West Coast		
16	Mascart Island	34	Hoste Island: Fouque Fjord	49	Martinez Sound – Escandallo Fjord		
17	Ponsonby Sound: Narrow Passage to Carfort Channel			50	Courtney Sound: Edwards Peninsula		
18	Ponsonby Sound: Narrow Passage to Beagle Channel			51	London Island: Horacio Mountain		
				52	Basket Island: Basket Cove & Murray Bay		

Table 1. Key to the numbers on the map of the Cape Horn Biosphere Reserve (CHBR, Figure 13), that identify each of the sites where bird censuses (transects and/or species lists) have been conducted during the breeding season (spring-summer) and non-breeding season (autumn-winter). 61 sampling sites were identified within 5 major areas of the CHBR: Navarino Island (18 sites), North-west Arm of the Beagle Channel and neighboring areas (16 sites), Almirantazgo Sound (12 sites), West Areas of the CHBR (6 sites), and the Cape Horn Archipelago area (9 sites).

(b) Open areas covered by low shrubby Ericaceae, *Gaultheria mucronata* and *Empetrum rubrum*, that produce berries between September and March, with some fruits available during the whole year;

(c) Recently burned sites dominated by *Chiliotrichium diffusum*;

(d) Naturally or anthropogenically-disturbed areas dominated by thickets of the Protecaeae *E. coccineum* which produces red tubular flowers rich in nectar twice a year, in late spring in November-December and in early fall in March; and

(e) Areas of different sizes covered by continuous old-growth forests dominated by *N. betuloides* and *N. pumilio*, with an understory of several species that produce fleshy fruits, such as *D. winteri*, and *Maytenus magellanica*.

In addition to forest burning, wood-cutting, and livestock ranching, the forests of Navarino Island have been subjected to the disturbance effects of the introduced beaver (*Castor canadensis*) since the 1950s (Sielfeld & Venegas 1980). Many of the deforested areas and ponds near the *E. coccineum* sites seen in the aerial photograph, reflect the effects of beavers, which today occupy almost every available watershed in the island, modifying drastically the hydrology and adjacent forest structure.

To complement and expand the range of study sites on Navarino Island, in the years 2000-2003 mist-netting was also conducted at Mejillones Bay (54°55´S, 67°54´W), and three sites on the western coast of Navarino Island: Puerto Inútil (54°59'S, 68°13'W), Wulaia Bay (55°03'S, 68°09'W), and Douglas Bay (55°05'S, 68°09'W). These are the first six study sites listed in the column "Navarino Island sites" in Table 1; their locations are shown in Figure 13. Beyond Navarino Island, we have conducted bird censuses on other islands to better capture the rich diversity of habitats of the Cape Horn Biosphere Reserve.

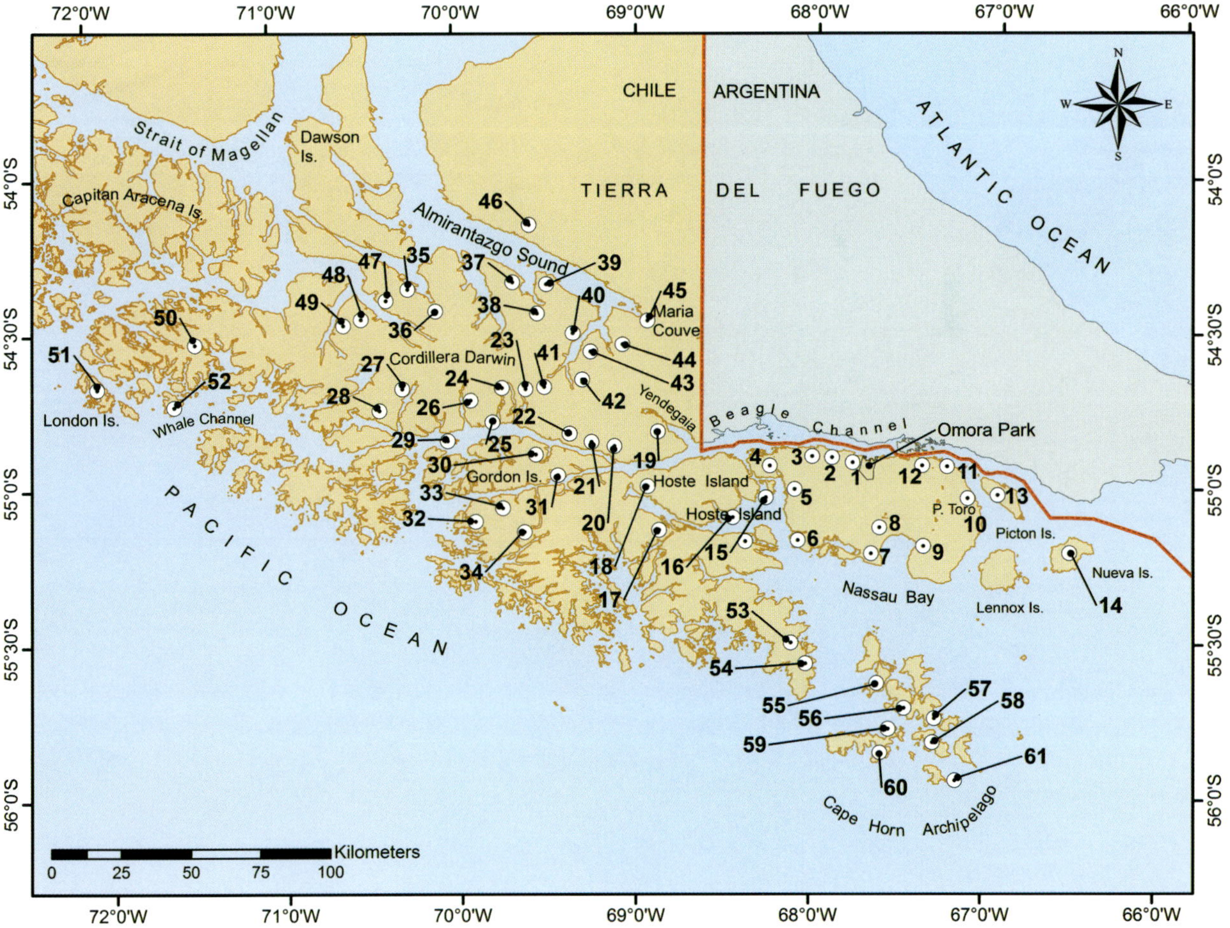

Figure 13. Map of the Biosphere Reserve Cape Horn showing the sites in which at least two bird surveys have been conducted, one during the breeding season (spring - summer) and one in the non-breeding season (autumn-winter), during the decade January 2000 - December 2010. The names of the sites are listed in Table 1. Figure prepared by Omora Park & GIS Laboratory, CERE-Universidad de Magallanes.

Figure 14. The presence of Neotropical trees and birds, such as the Winter's bark (*Drimys winteri*) and the White-Crested Elaenia (*Elaenia albiceps*), is one of the most surprising attributes of the world's southernmost forests. The White-Crested Elaenia eats the fruits of Winter's bark before its migration from Cape Horn to the Amazonian forests each autumn. Photos José Tomás Ibarra (top) and Ricardo Rozzi (bottom).

MOSAIC OF HABITATS AND STUDY SITES IN THE CAPE HORN BIOSPHERE RESERVE

At the landscape scale, Cape Horn offers a remarkable diversity of habitats with a rich biodiversity that derives its uniqueness to a great extent from the variety of its biogeographical relationships (Moore 1983). The biota of Cape Horn has affinities with five contrasting biogeographic regions.

1. ***Gondwana.*** Today, the genus *Nothofagus* is only found in the Southern Hemisphere, in Australasia (eastern and southeastern Australia, New Zealand, New Guinea, and New Caledonia) and southern South America (Chile, Argentina). Fossil trees are also preserved in Antarctica (Veblen *et al.* 1996). In most islands of the Cape Horn Biosphere Reserve, dense old-growth forests dominated by species of *Nothofagus* grow on the slopes protected from the wind. The presence of this genus provides a living example of the Gondwanic connections that southern South America had with Antarctica, New Zealand, and Australia, until the late Cretaceous 75 million years B.P.

2. ***Neotropical.*** *Drimys* is a Neotropical genus whose species are distributed throughout South and Central America reaching southern Mexico (Arroyo *et al.* 1996). The southernmost representative of this genus, the Winter's Bark or *Canelo* (*D. winteri*), grow in the broadleaf evergreen forests of Cape Horn near the coast. In coastal forests, the Winter's Bark is a dominant tree that is characterized by its large leaves. It also produces fleshy fruits, which are consumed by the Austral Thrush (*Turdus falcklandii*) and the White-crested Elaenia (*Elaenia albiceps*), which feeds on them before its migration each autumn from Cape Horn to the Amazonian forests (Figure 14, Jiménez *et al.*, in preparation). In the coastal broadleaf evergreen forests it is also possible to observe parrots and Hummingbirds that reflect the biogeographical connections that the forests of Cape Horn have with the Neotropical forests of South and Central America.

3. ***Andean***. Above the tree line, in Omora Park it is possible to observe birds of the genera *Attagis* (seedsnipes) and *Meladonera* (finches) that are distributed northward along the high Andes. The cushion plants of the genus *Azorella* that grow above the treeline as well as near the coast at Omora Park and other places in the Cape Horn Biosphere Reserve. These plants grow about 4000 m altitude in the high Andes of Bolivia, Peru, and northern Chile and Argentina (Armesto *et al.* 1980). Both the bird and cushion plants are examples of the connections that the biota of Cape Horn has with the biota of the high-Andean Puna.

4. ***Antarctica.*** On the rocks of the coast in the Cape Horn Biosphere Reserve, the bright orange lichens *Caloplaca hookeri* and *C. sublobata* (Goffinet *et al.* 2012), and the penguins *Eudyptes chrysolophus* (Macaroni Penguin) and *Pygoscelis antarctica* (Chinstrap Penguin), are an example of the close connections between the terrestrial and marine biota of the Magellanic sub-Antarctic ecoregion and the Antarctica.

5. ***Bipolar.*** The abundance of low shrubs of Red Crowberry or Diddle-dee (*Empetrum rubrum*) is an example of long dispersal between the sub-Antarctic and subarctic regions, which generates a "bipolar distribution." The genus *Empetrum* has only two species: *E. nigrum* restricted to Alaska and other subarctic zones, and *E. rubrum* restricted to Cape Horn and other sub-Antarctic zones (Donoghue 2011, Popp *et al.* 2011).

Two other factors that contribute to the great heterogeneity of habitats in the Cape Horn Biosphere Reserve (CHBR) are the complex physical geography and the strong west-east gradient of rainfall (Rozzi *et al.* 2006a,b,c). In the western and southern areas of the CHBR toward the Pacific Ocean are innumerable islands with peninsulas, bays, steep mountains, narrow valleys, and fjords. In the northern area of the CHBR, the most dominant characteristic of the physical geography is the Andean Cordillera system that bends eastwards along the southwestern extension of Tierra del Fuego giving origin to the Darwin Cordillera, also known as the Fuegian Andes. This cordillera creates an abrupt alteration in the topography that generates abrupt changes in climate over short distances. The highest peaks of the Darwin Cordillera are found to the west on Mt. Sarmiento (2404 m) and in the Darwin Cordillera (Mt. Luis de Saboya 2469 m, Mt. Darwin 2438 m) (Figure 15). The altitude of the mountains diminishes eastwards and southwards. On Navarino Island the highest peaks are the "Dientes de Navarino" (1195 m) where the Róbalo River is born. At Omora Park, the highest altitude is the summit of Róbalo Mountain (920 m). In the Cape Horn Archipelago, only Wollaston Island has peaks above 500 m.

The high altitude of the Darwin Cordillera combined with the prevailing winds carrying high moisture from the Pacific Ocean, creates a remarkable rainfall gradient from the west to the east. When the westerlies arrive on the western coast and the Darwin Cordillera their moisture condenses and precipitates as rain on the lowlands and as snow on the mountains. It is not possible to give a detailed and reliable map of the distribution of the precipitation, but the main features have been well synthesized by Tuhkanen *et al.* (1990). In less than 300 km the annual rainfall decreases approximately an order of magnitude, from over 3000 mm in the west to less than 500 mm in Puerto Williams and Omora Park (Figure 16). The combination of biological, orographic, climatic, and other factors gives origin to a terrestrial ecosystem mosaic that includes a complex of (I) forests, (II) scrublands, (III) moorlands and wetlands, and (IV) high-Andean habitats. Among the forests, four main types are present at Omora Park and in other areas of the Cape Horn Biosphere Reserve.

Figure 15. Mount Darwin peak (top left), one of the three highest peaks of the Darwin Cordillera. Photograph of Alemania (Germany) Glacier taken from the North-West Arm of the Beagle Channel, illustrating the diversity of high-Andean, moorland, and forest habitats in the Cape Horn Biosphere Reserve (top). A closer view of the evergreen Magellanic sub-Antarctic forest (bottom). Photos Ricardo Rozzi.

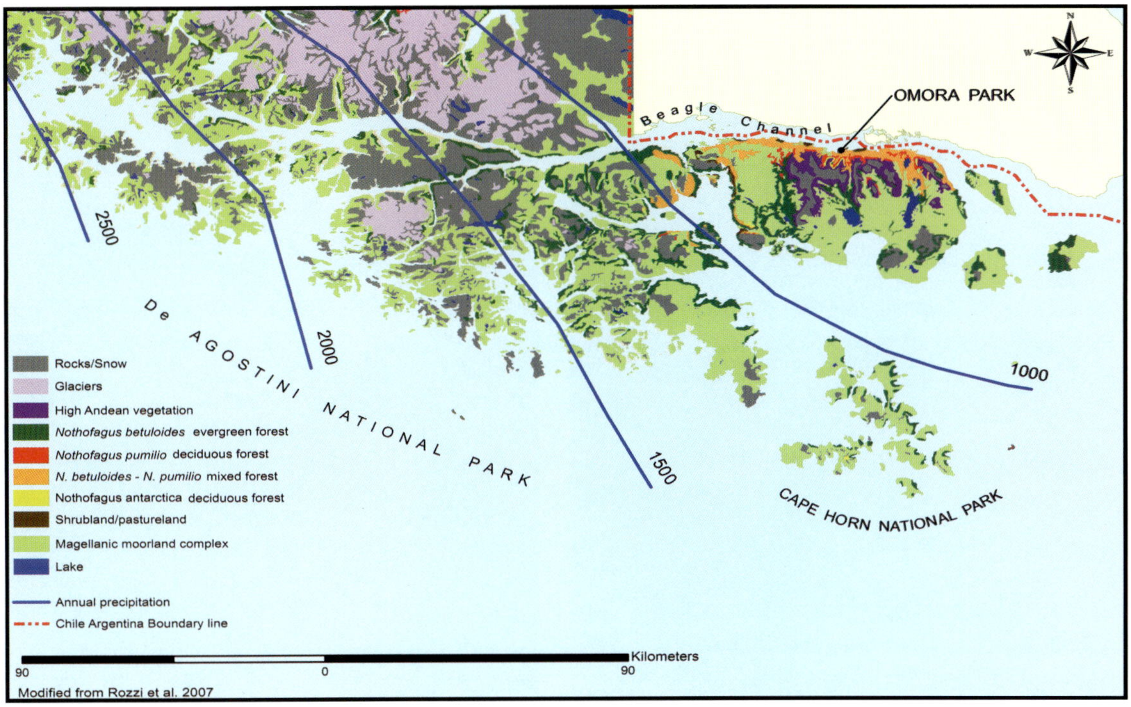

Figure 16. West-east rainfall gradient within the Cape Horn Biosphere Reserve (CHBR). The overlap of the annual precipitation curves and the land-cover depicting the major habitat types distinguishes two main ecologicalclimatic zones within the CHBR: 1) A hyper-humid zone dominated by tundra and Magellanic broadleaf evergreen rainforests (*Nothofagus betuloides* and *Drimys winteri*), which is located in the southwest area of the CHBR with over 1000 mm annual rainfall. 2) Less humid zone, located northeast of the 1000 mm isohyet, which encompasses a mosaic of habitats, including mixed evergreen and deciduous forests (*N. betuloides* and *N. pumilio*), as well as pure stands of deciduous forests dominated by *N. pumilio*. Omora Park is located within the second zone, and therefore it includes a mosaic of evergreen and deciduous forests, as well as high Andean vegetation. (Modified from Engel 1969 and Tuhkanen *et al.* 1990). (Figure prepared by Omora Park & GIS Laboratory, CERE-Universidad de Magallanes).

I. Main Sub-Antarctic Forest Types

I.1) Sub-Antarctic Magellanic broadleaf evergreen rainforest. Along the coastlines of the Magellanic islands the high oceanic influence buffers temperature fluctuations, enabling the growth of broadleaf evergreen forests at lower altitudes. These forests are dominated by the endemic tree species that defines the sub-Antarctic Magellanic ecoregion: the Evergreen Beech (*N. betuloides*). It grows together with other evergreen species such as Winter's Bark (*Drimys winteri*), and Pickwood (*Maytenus magellanica*). The understory of these forests varies in association with the rainfall gradient (Pisano 1977, Rozzi *et al.* 2006b). In the most humid areas (>2000 mm annual rainfall) the understory is dominated by liverworts (*e.g.*, *Gackstroemia magellanica*, *Schistochilla lamelata*) and lichens (*e.g.*, *Pseudocyphelaria* spp.). In intermediate humid areas (1000-2000 mm annual rainfall) the understory is dominated by mosses, woody ferns (including *Blechnum magellanica* and *Gleichenia pinnatifolia*) (Figure 17A), and shrubs that include the "Coicopihue" (*Philesia magellanica*), an endemic plant with large red tubular flowers that produces abundant nectar. *Philesia magellanica* flowers almost all year round, and it is a key species to sustain the populations of the Green-backed Firecrown hummingbird (*Sephanoides sephaniodes*) that are present in the area until the end of fall (May) (Rozzi 2002, Massardo *et al.* in preparation). On the less humid areas (<1000 mm annual rainfall), such as at Omora Park, the woody ferns (*B. magellanica* and *G. pinnatifolia*) and *P. magellanica* are absent; instead, the understory is dominated by shrubs that produce abundant berries: Holly-leafed Barberry (*Berberis ilicifolia*), Box-leafed Barberry (*B. buxifolia*), and Wild Currant (*Ribes magellanicum*). The physiognomy of these forests is complex and multi-stratified; trunks and large branches are profusely covered by mosses, liverworts, and lichens (Pisano 1977, 1980). The Thorn-tailed Rayadito (*Aphrastura spinicauda*, Figure 17B) and the White-crested Elaenia (*Elaenia albiceps*) occupy a wide variety of habitats, but they are particularly abundant in the broadleaf evergreen coastal forests.

I.2) Deciduous Lenga forest. In areas of good drainage that receive less than 1000 mm annual rainfall, such as Navarino Island, including Omora Park, High-deciduous Beech (*N. pumilio*) forests dominate the majority of the hill slopes (Figure 17C). This type of forest has the tallest trees in the Magellanic sub-Antarctic ecoregion. *Nothofagus pumilio* has straight trunks, and the understory is dominated by low shrubs, such as the Dwarf Pickwood (*Maytenus disticha*), herbs, including orchids (*Gavilea lutea*), and mosses. The Magellanic

A
B
C
D
E
F
G
H

Woodpecker (*Campephilus magellanicus*) inhabits evergreen and deciduous forests, where they frequently feed and nest. Cavities made by woodpeckers on Lenga trees are often used for nesting by the Austral parakeets (*Enicognathus ferrugineus*; Rozzi *et al.* 2006a,b).

I.3) Mixed evergreen-deciduous forest. This forest combines evergreen and deciduous species where *N. pumilio* and *N. betuloides* co-dominate. It extends over the areas that are more sheltered and have relatively good drainage, and where annual rainfall is between 550 mm and 1000 mm (Pisano 1977, Moore 1983, Rozzi *et al.* 2006a,b).

I.4) Deciduous Ñirre forests. Small forest stands dominated of *Ñirre* or Low Deciduous Beech (*N. antarctica*) are found in wet areas with poor drainage. The trees reach heights of only up to 15 m. However, they are abundantly covered with mosses, liverworts, and lichens such as "old man's beard" (*Usnea* sp.) (Rozzi *et al.* 2008a, Goffinet *et al.* 2012).The trunks often host abundant ascomycetous fungi of the genus *Cyttaria*, particularly *C. darwinii* or Darwin's Fungus. The floor of wet areas is usually covered by the Dwarf Gunnera or Devil's Strawberry (*Gunnera magellanica*) (Figure 17D). Occasionally, *N. antarctica* grows also in krumholtz formations together with *N. pumilio* at the treeline near the mountain summits.

II) Scrublands

Among scrublands, the woodland dominated by the Firebush (*E. coccineum*) is a particularly important habitat type for our bird studies (Figure 17E). The Firebush is a small evergreen tree in the family Proteaceae that produces tubular red flowers with abundant nectar. It flowers twice a year, in spring and fall, and the flowers are visited by diverse species of birds, including the Green-backed Firecrown and nectar robbers such as the Patagonian Sierra-Finch (*Phrygilus patagonicus*), and the Austral Parakeet (*E. ferrugineus*, Figure 17F). The Firebush grows among a dense cover of shrubs of *Chiliotrichum diffusum*, *Baccharis* spp., and *Berberis buxifolia*. In these thickets we have captured large number of Patagonian Sierra-Finches (*P. patagonicus*, Figures 17G, 17H). In wetter and windier areas, such as on Horn Island, scrublands are dominated by dense thickets of low shrubs of Box-leafed Barberry (*B. buxifolia*), Holly-leafed Barberry (*B. ilicifolia*), and Prickly Heath (*Gautlheria mucronata*). This habitat type is characteristically used by the Magellanic Tapaculo (*Scytalopus magellanicus*) on the islands that are more exposed to the Pacific Ocean.

III) Moorlands: The Magellanic Tundra Complex

Vast areas of the Cape Horn Biosphere Reserve are dominated by mires. Edmundo Pisano (1977) calls the set of different types of mires the "Magellanic tundra complex." Pisano distinguishes the following four main types: (i) *Sphagnum* tundra or peatlands, formed by two species of mosses, *Sphagnum magellanicum* and *S. fimbriatum*. In the peatlands, dwarf trees of *N. antarctica* and *N. betuloides*, and dwarf woody plants with berries (*e.g., Myrteola numularia*) grow. (ii) Cushion plant bogs, saturated zones with poor drainage dominated by cushion plants of the genera *Astelia*, *Azorella*, *Laretia*, *Donatia*, and *Bolax*. Small Juncaceae such as *Rostkovia magellanica* and *Juncus stipulates*, which typically grow around ponds found embedded in the layer of cushion plants. (iii) Graminoid bogs, commonly associated with cushion bogs areas there are zones dominated by grass-like plants, "graminoid," such as *Schoenus antarcticus*, *Tetroncium magellanicum* and *Uncinia kingie*. (iv) Rush wetlands, extensive wetland zones dominated by the rush *Marsippospermum grandiflorum*, which provide a critical habitat for birds such as the South American Snipe or Magellan Snipe (*Gallinago paraguaiae*, Figure 18A).

IV) High-Andean Habitats

Above the tree line, high Andean habitats are characterized by a high diversity of mosses and lichen species that grow on the rocks or associated with cushion plants or low shrubs. The high Andean habitats have three distinct altitudinal levels: i) a lower zone dominated

Figure 17. *Front page.* A. Understory of a coastal broadleaf evergreen rainforest site at the North-West Arm of the Beagle Channel, which receives 1500 mm of annual rainfall. Its understory is characterized by the presence of the woody ferns *Blechnum magellanica* and *Gleichenia pinnatifolia*. B. The Thorn-tailed Rayadito (*Aphrastura spinicauda*) is particularly abundant in the broadleaf evergreen coastal forests of the Cape Horn Biosphere Reserve. C. High-deciduous Beech or "Lenga" (*Nothofagus pumilio*) forests dominate the slopes with good drainage that receive less than 1000 mm of annual rainfall, such as the slopes of Róbalo Hill at Omora Park. The photograph taken in May 2002 on the equatorial facing slopes of Omora Park shows the dominance of the red color of the Lenga leaves during autumn. D. Low Deciduous Beech (*Nothofagus antarctica*) or "Ñirre" forests are found in wet areas. The trunks are characteristically covered with mosses, while the floor is usually covered by the Dwarf Gunnera or Devil's Strawberry (*Gunnera magellanica*), which plays a key ecological role in nitrogen fixation. E. The red tubular flowers of the Firebush (*Embothrium coccineum*) produce abundant nectar. F. The Firebush is an evergreen tree (Proteaceae) with flowers that are visited by a diversity of bird species, including nectar robbers such as the Austral Parakeet (*Enicognathus ferrugineus*). G. The Firebush blooms twice a year and its flowers are visited by another abundant nectar robber: the Patagonia Sierra-Finch (*Phrygilus patagonicus*), which is the bird most frequently captured in the mist-nets. A banded female is shown in the picture. H. Ornithologist Steve McGehee working at the "Notro" or Firebush site in Omora Park. Photos A, C, and D Ricardo Rozzi, B Paola Vezzani, E John Schwenk, F-H Steff van Dongen.

A
B
C
D
E

by low shrubs (*e.g.*, *Empetrum rubrum*); ii) an intermediate zone dominated by cushion plants (*e.g.*, *Bolax gummifera*; iii) a high zone dominated by mosses and lichens on rocky substrates (Rozzi *et al.* 2006a, Méndez 2013, Méndez *et al.* 2013). On the Darwin Cordillera, and Hoste Island the high Andean habitats include glaciers, several of which descend into the ocean. The lateral moraines of the glaciers are rapidly colonized by *Nothofagus antarctica* and *N. betuloides* trees (Figure 18B). The Rufous-collared Sparrow (*Zonotrichia capensis*) is a habitat in the CHBR that inhabits all types of habitats, including the margins of glaciers (Figure 18C). However, glaciers are no longer present in the high-Andean habitats on Navarino Island (Figure 18D) and the Cape Horn Archipelago (Figure 18E).

As mentioned above, to capture the heterogeneity of habitats we included 61 study sites widely distributed throughout the CHBR (see Figure 13 above). To capture the degree of human impact on species composition and population dynamics of birds in zones under different protection regimes in the CHBR, the study sites were located in core, buffer, and transition zones (Figure 19). Forest ecosystems of the island territory of Cape Horn are embedded in a mosaic of habitats that has remained largely free of the direct impact of modern society into the 21st century. Thus, the Cape Horn habitats provide a baseline to study sub-Antarctic forest bird assemblages under conditions similar to those that were present prior to the European colonization.

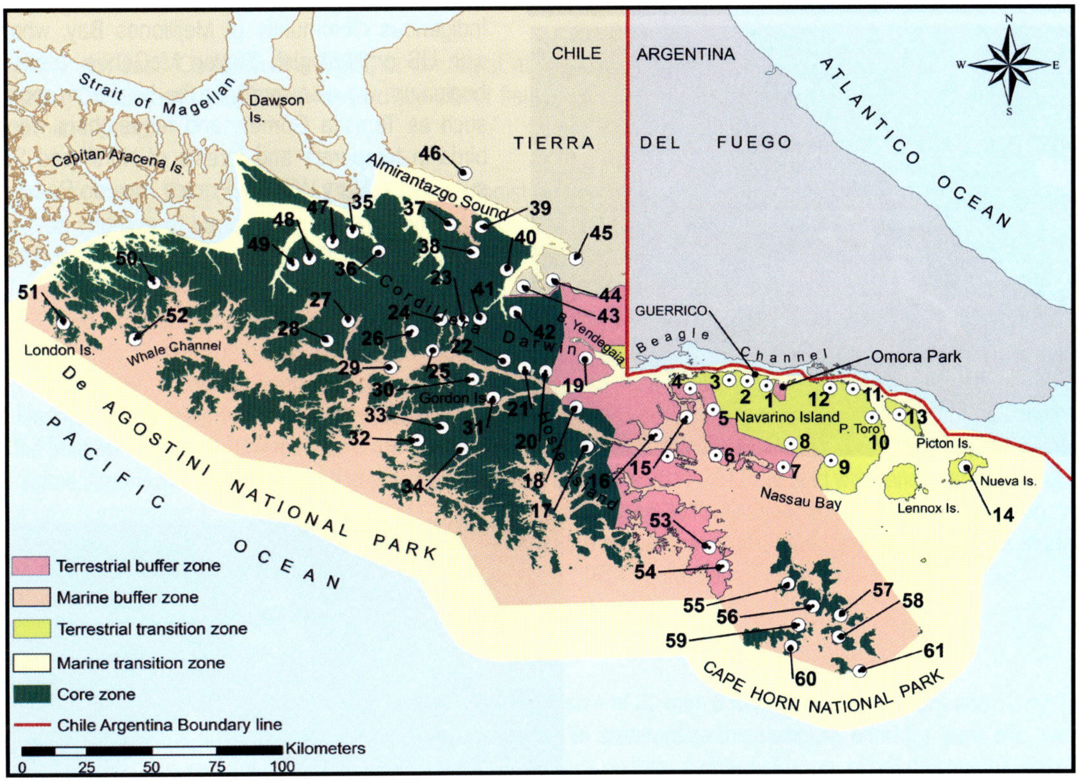

Figure 19. Zoning map of the Cape Horn Biosphere Reserve (CHBR) that illustrates its core zones (green), buffer zones (pink), and transition zones (yellow). The core zones include two national parks: Alberto de Agostini and Cape Horn. The buffer zones include the Omora Ethnobotanical Park. The transition zones include the main complementary long-term ornithological study site, Guerrico Hill, which is located on the northern coast of Navarino Island, 25 km west of Omora Park. Numbers on the map identify each of the 61 sites in which at least two bird surveys have been conducted, one during the breeding season (spring - summer) and one in the non-breeding season (autumn-winter), during the decade January 2000 - December 2010. Sampling sites are located within 5 major geographic areas of the CHBR: (i) Navarino Island (18 sites), (ii) West Arms Beagle Channel (16 sites), (iii) Almirantazgo Sound (12 sites), (iv) West Areas of the CHBR (6 sites), and (v) the Cape Horn Archipelago area (9 sites). The names of the locations are listed in Table 1 (page 16). Figure prepared by Omora Park & GIS Laboratory, CERE-Universidad de Magallanes.

Figure 18. *Front page.* A. The South American Snipe or Magellan Snipe (*Gallinago paraguaiae*), one of the species that nests in the wetland habitats dominated by the rush *Marsippospermum grandiflorum*. B. Glacier and later moraine at Parry Glacier in the northern zone of the Cape Horn Biosphere Reserve (CHBR), showing the recolonization of broadleaf evergreen forests dominated by *Nothofagus betuloides*. C. A Rufous-collared Sparrow (*Zonotrichia capensis*) in the high-Andean habitats of Marinelli Glacier; this bird is the most generalist species in terms of habitat use in the CHBR. D. The summit of Robalo Mountain at Omora Park is free of permanent ice; however, the Robalo River watershed clearly shows a glacier geomorphology. E. Pyramid Hill is the highest peak on Horn Island, and its east-facing slope has the world's southernmost forested watershed. Photos A Jordi Plana, B-D Paola Vezzani, E Ricardo Rozzi.

Figure 22. (left) Mist-nets deployed at a forest-edge site on Guerrico in the northern coast of Navarino Island. (top right) A Thorn-tailed Rayadito (*Aphrastura spinicauda*) captured in a mist-net. (bottom right) Omora Park ornithologist, Dr. Ronnie Reyes, carefully handling the captured Rayadito. Photos Adam Wilson.

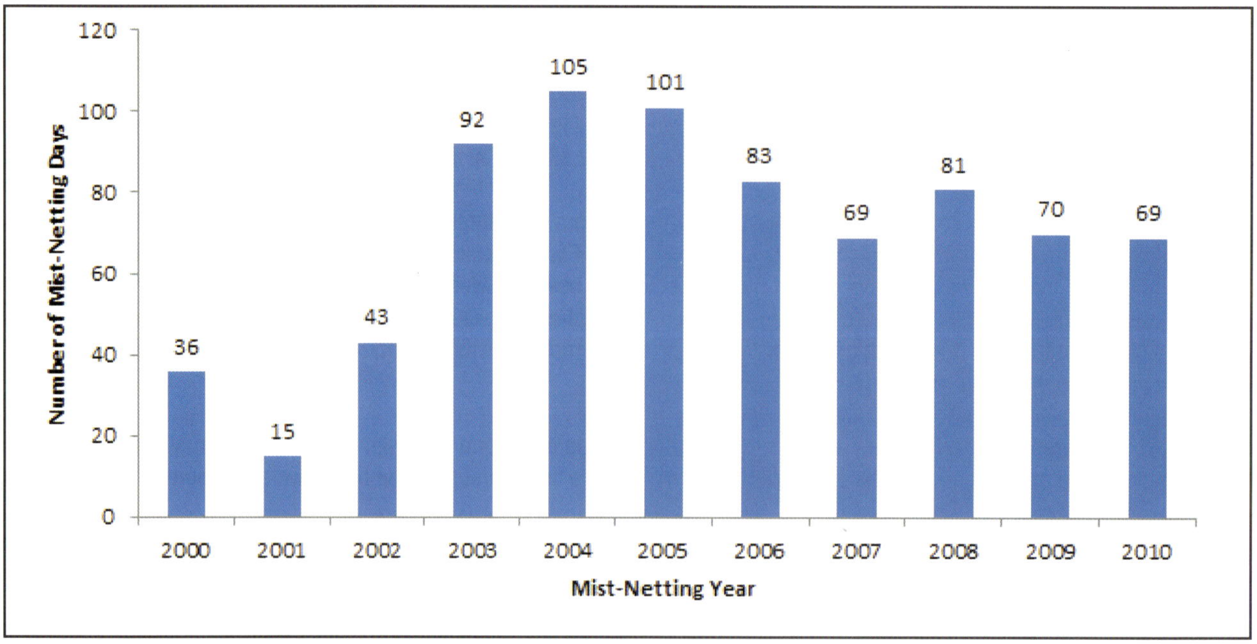

Figure 23. Bird monitoring effort quantified as the number of mist-netting days per year at Omora Park and complementary study sites on Navarino Island, conducted from January 2000 to December 2010.

provided by SAG. Occasionally, we also have used additional color bands to facilitate the identification of individuals (Figure 24). The Green-backed Firecrown or *Omora* Hummingbird (*Sephanoides sephaniodes*) required a species-specific methodology. This species requires special care, and captured individuals were initially banded with small, blue-plastic rings that were sequentially numbered, and later with a small band cut from a numbered aluminum sheet that was sanded to remove sharp edges. These bands were placed on the legs of hummingbirds using special pliers.

Analyses of captures, recaptures, and longevity of species. The information gathered from mist-nets recorded all captured individuals, including occasional cases of individuals that escaped and were not banded. It also includes all known recaptured individuals. Total Captures (TC) is the value that sums up all capture events of all captured birds. For each species we calculated its:

A. *Relative Percentage of the Total Captures (RTC).* We used the ratio: TCSn/TC, where TCSn is the sum of all capture events for species n, and TC of all captured birds. In the descriptive section of each captured species in Chapter 2, species are ordered from the highest to the lowest RTC value. The RTC value for each species is presented as "% of total captures" right to the photograph of the species. The total number of captures for each species is also recorded right to the photograph in Chapter 2.

B. *Total Number of Different Individuals Captured and Banded (TCB).* We calculated this value based on the number of captured birds that (i) were not previously banded and that (ii) we succeeded in banding for the first time. Therefore, TCB does not include multiple recaptures nor individuals that escaped before they were banded. This criterion enables TCB to be used as a basis for the estimation of the percentage of banded individuals that were recaptured. In Chapter 2, TCB for each species is presented as "Different individuals captured," to the right of the photograph of the respective species.

C. *Total Number of Different Individuals Recaptured (TRDI).* We calculated this value as the total number of individuals that were recaptured at least once. We did not distinguish between one or multiple recaptures of the same individual.

D. *Percentage of Recaptured Individuals (%RI).* We calculated this value using the ratio: % RI = (TRDI/TCB) * 100. In Chapter 2, the %RI value for each species is presented as "% individuals recaptured," right to the photograph of the respective species.

E. *Maximum Recorded Longevity (years) (MRL).* We estimated this value based on the individuals that had the longest period between their first capture and banding event, and the last recapture event. This value represents a preliminary estimate of maximum longevity. This preliminary estimate provides the first available information on longevity for bird species in the sub-Antarctic forests of Navarino Island.

Morphometric measurements. Each captured bird was identified, banded, and the following morphological measurements were taken (Figure 25; see Pyle *et al.* 1987 for more details):

a) *Total length*, defined as the length between the tip of the tail and the tip of the bill with the bird extended and supported dorsally over an ornithological ruler with 1 mm accuracy.

b) *Tail length*, defined as the length of the extended wing from the base to the tip of the longest primary feather, measured with an ornithological ruler with 1 mm accuracy.

c) *Tarsus length*, measured with a caliper with 0.1 mm accuracy.

d) *Wing length*, defined as the length of the extended wing from the wrist to the tip of the longest primary feather, with the wing extended flat over (flat, *i.e.*, arch) an ornithological ruler with 1 mm accuracy.

f) *Bill length*, defined as the length of the exposed culmen, measured with a caliper with 0.1 mm accuracy.

g) *Bill depth*, defined as the height of the upper and lower mandibles measured perpendicular to the bill´s length and at the anterior end of the nostrils, using a caliper with 0.1 mm accuracy,

h) *Bill width*, defined as the width of the culmen, measured perpendicular to the bill´s length and at the anterior end of the nostrils, using a caliper with 0.1 mm accuracy,

i) *Weight*, measured with an electronic scale with 0.01g accuracy or a Pesola scale of 30 g and 0.5 g accuracy for the majority of the species. For larger species we used a 100 g Pesola scale with 0.5 g accuracy and for hummingbirds a 10 g scale of 10 g with 0.1 g accuracy.

Figure 24. (top) Numbered aluminum bands provided by SAG and color bands on the legs of a captured White-crested Elaenia (*Elaenia albiceps*). (bottom) The captured Elaenia, banded and ready to be released by Omora Park ornithologist Dr. Sebastián Dardanelli, who carefully holds the bird from its legs. Photos Jaime Jiménez and Sebastián Dardanelli.

Figure 25. (top) "Notro site" bird-banding station at Omora Park, where each captured bird is identified, banded, and subjected to morphological measurements by ornithologists, such as the Omora Park field technician Omar Barroso and the student Aquiles Gutierrez. (bottom) One of the measurements, wing length, taken on a Patagonian Sierra-Finch (*Phrygilus patagonicus*). Photos Jaime Jiménez and Steff van Dongen.

When possible, the gender (color dimorphism and presence/absence of incubation patch and cloacal protuberance) and age (adult or juvenile; through feather development and color) of each captured individual was estimated.

Diet information for each species. We have collected the feces of birds that defecated during the mist-netting and banding handling for subsequent diet analysis. Feces samples were preserved and dried in paper envelopes. To identify the seeds and invertebrates found in bird feces, we prepared a reference collection of seeds of all plant species with fleshy fruits, as well as the main Gramineae. To identify the invertebrates found in bird feces, we mounted reference collections of the most common insect taxa, as well as other invertebrates, present in the canopy and soil of forest, scrublands, and riparian habitats. We identified the contents of feces by using dissecting microscopes,

To study the diet of potential pollinators, for captured individuals of *Elaenia albiceps* and *Sephanoides sephaniodes* that visit the flowers of *Embothrium coccineum* and of other plant species with red-tubular flowers, in the period 2000- 2002 we sampled pollen grains removed with Scotch tape from above and below the beak of the bird. Tapes were later stored on microscope slides. In order to determine the taxa of pollen, we prepared a reference collection of pollen grains of plants with red tubular flowers that are present at Omora Park and/or the Cape Horn Biosphere Reserve. We identified the pollen on slides by examining the samples under a compound microscope.

The study of diet is under way. Hence, only preliminary information is provided in the description of each species; our original preliminary information is also complemented with information from the literature.

Bird capture rates per month. In order to assess the seasonal variations in the number of captured birds in the mist-nets, we summed the number of individuals of each species captured each month, and standardized this number with the mist-netting effort: Total number captured birds/[net-area (m^2) x number of hours of net display (h)] x 10^4. Then, we calculated the monthly mean (± SE) based on the standardized monthly data for each month between 2003 and 2010. These data are represented in a histogram for each species.

CENSUSES OF FOREST BIRDS IN THE CAPE HORN BIOSPHERE RESERVE

In addition to mist-netting sampling methods, at Omora Park and other study sites on Navarino Island, we have conducted censuses along transects with observation-listening stations at different sites across the Cape Horn Biosphere Reserve (CHBR). At each site we conducted a transect of at least 10 observation-listening stations separated by at least 100 m, and at each station we:

(i) Recorded the bird species heard or seen during 5 minutes within an approximate of 50 m radius;

(ii) Counted the number of individuals of each species at each station;

(iii) Recorded the types of activity, habitat, and microhabitat used by each of the observed individuals;

(iv) Documented the habitat type at the counting point, the weather conditions, and the time of day.

During the course of the transects, and our entire stay at each of the sites, we also prepared lists of bird species recorded. Preliminary distribution maps were prepared for each species based on information gathered during forest bird censuses in spring/summer and late fall/winter at 61 sites broadly distributed throughout the CHBR (Table 1; Figure 13). For the preparation of the distribution maps of species within the CHBR, expeditions were considered developed in five main sectors:

1. Navarino Island (18 sites), including Picton Island, as well as Button Island, Mascart Island in the area of Murray Channel and sites in Ponsonby fjord in Hoste Island.

2. North-west Arm of the Beagle Channel and neighboring areas (16 sites), inciuding sites at both the North-West and the South-West arms of the Beagle Channel, as well as neighboring fjords (Yendegaia Sound, Pia Sound, Ventisqueros Sound) and islands (Chair Island, Devil Island).

3. Almirantazgo Sound (12 sites), including neighboring fjords (Parry Sound, Ainsworth Bay, Brooke Sound), valleys (Paralelo River), and Calela Maria (Maria Cove) at the eastern end of Almirantazgo Sound.

4. West Areas Biosphere Reserve (6 sites), including neighboring islands (London, Baskets).

5. Cape Horn Archipelago area (9 sites), including the islands of the Wollaston or Cape Horn Archipelago, as well as two sites on the eastern coast of Hardy Peninsula (Orange Bay, Lort Bay).

2

A SUMMARY OF THE BIRDS CAPTURED WITH MIST-NETS AND/OR CENSUSED IN THE SOUTHERNMOST FORESTS OF THE WORLD

Ricardo Rozzi and Jaime E. Jiménez

In this section we present a summary of morphological and ecological data for 26 species that were captured with mist-nets during the first eleven years (January 2000-December 2010) of the Long-Term Ornithological Research Program at the Omora Ethnobotanical Park and its associated study sites in the Cape Horn Biosphere Reserve. The birds are presented in a sequence starting with the most frequently captured to the less frequently captured species. For each species we provide a two-pages description.

The first page begins with a photograph of the species, including the following information to the right:

- Total number of captures of species *n* (TC)
- Relative percentage of the total captures of species *n* (% of TC species *n*/ ∑TC all species)
- Total number of different captured and banded individuals of species *n*
- Percentage of recaptured individuals of species *n*
- Maximum recorded longevity for species *n*

The definition of each of these terms is provided in detail in the previous chapter.

Below the picture, we include a summary table with the morphometric measurements of the birds captured with mist-nets, distinguishing between males and females in the case of species with a clear sexual dimorphism.

Below the morphometric table, we include a brief paragraph with information about the diet of each species, based on preliminary analyses of the content of the feces collected from birds captured with mist-nets, supplemented with field observations and review of the literature.

The second page begins with a distribution map of the species in the Cape Horn Biosphere Reserve (CHBR) based on 61 sites in which bird censuses along transects with observation-listening stations were conducted (see detail in previous chapter). For each site, the map indicates the presence (color circles) or the absence (white circles) of the species. When present the species is present, the map illustrates if it is found year round (green circles) or only during the breeding season (yellow circles).

Below the CHBR map, we include another map with the entire range of distribution of the species. We use the following nomenclature of colors to indicate the species' spatio-temporal patterns:

- green = year-round resident
- yellow = present only during the breeding season
- blue = present only during the non-breeding season
- violet = occasional visitor

To the right of the distribution map of the species, a graph illustrates the seasonal variations in the monthly mean (± SE) of birds captured with mist-nets (Total number captured birds / [net area (m^2) x number of hours of net display (h)] x 10^4) at Omora Park and associated study sites in the Cape Horn Biosphere Reserve.

Finally, for each species we include three short paragraphs with concise information about (i) the distribution and habitat use, (ii) the life-habits and ecology, and (iii) the conservation status. This information is based on the mist-net capture data, supplemented by field observations and a review of the literature. The conservation status considered also the Chilean Hunting Law (*La Ley de Caza y su Reglamento*, Edición 2009, Servicio Agrícola y Ganadero, Chile; http://www.avesdechile.cl/0docs/leydecaza.pdf) and the IUCN recommendations.

The 26 species of birds captured with mist-nets are organized into four groups. Because the banding program focuses on passerines and other small birds that inhabit the world's southernmost forests, the first group of birds (Section A) includes the 18 species of small birds that have been captured in the interior or at the margins of forests "Passerines and other small birds captured with mist-nets in the sub-Antarctic forests at Omora Park and complementary sites on Navarino Island."*

Because we accidentally captured with the mist-nets some larger birds in the forest interior or margins (non-passerines), the data for five species were included in Section B "Larger-sized birds incidentally captured with mist-nets in the sub-Antarctic forests at Omora Park and other sites on Navarino Island."

At Omora Park, we also mist-netted passerine birds inhabiting the high-Andean habitats. For this reason, we included data from two species captured just above the tree-line in Section C "Passerines inhabiting the high-Andean habitats above the tree-line captured with mist-nets at Robalo Mountain, Omora Park, Navarino Island."

Finally, at Omora Park and some preliminary mist-netting efforts in the capital city of the Chilean Antarctic Province, Puerto Williams, we captured one exotic bird species, which we included in the last Section D "Exotic passerines captured with mist-nets in the sub-Antarctic forests at Omora Park and Puerto Williams on Navarino Island."

At the end of this chapter we indicate the main bibliographic sources used in the preparation of the maps with the whole range of distribution of the species, and the concise information about the general description of morphology, diet, distribution and habitat use, life-habits and ecology, and conservation status.

Our quantitative data on morphology, spatial and temporal distribution of the species complement previous ornithological publications based on studies conducted mostly during the summer in Tierra del Fuego and occasionally on Navarino Island and the Cape Horn Archipelago. Most of these early publications have been limited to short-term studies or correspond to field guides that include qualitative data or quantitative data recorded in other regions of South America. A list with the main bibliographic sources of these valuable previous studies that were particularly valuable for preparing the description of the morphology, feeding, distribution and use of habitat, lifestyle and ecology, conservation status and preparing maps of the total area of each species is provided.

*Two of the species included in this first group, the Magellanic Tapaculo (*Scytalopus magellanicus*) and the Blue-and-white Swallow (*Pygochelidon cyanoleuca*), were captured with mist-nets in Omora Park only at the beginning of the long-term banding program in the year 2000. No morphological data were recorded for the few individuals captured. For this reason, for these two species we provide morphological information recorded in other study sites nearby, and we do not include a graph for monthly capture rates with mist-nets.

A

SECTION

PASSERINES AND OTHER SMALL BIRDS CAPTURED WITH MIST-NETS IN THE SUB-ANTARCTIC FORESTS AT OMORA PARK AND COMPLEMENTARY SITES ON NAVARINO ISLAND

THRAUPIDAE

Patagonian Sierra-Finch
Cometocino Patagónico
Phrygilus patagonicus

MIST-NET DATA
Number of captures: 3435
% of total captures: 38.65
Different individuals captured: 2973
% individuals recaptured: 9.4
Maximum recorded longevity (years): 7.59

	MALE			FEMALE			TOTAL		
Morphometrical measurements	Mean ± SD	Min-Max	N	Mean ± SD	Min-Max	N	Mean ± SD	Min-Max	N
Total length (mm)	141.2 ± 5.6	126 - 158	253	136.0 ± 5.5	127 - 156	287	141.0 ± 5.5	126.0 - 156.0	540
Tail length (mm)	62.4 ± 2.5	54 - 69	573	59.5 ± 2.3	54 - 68	306	60.9 ± 2.4	54.0 - 69.0	879
Tarsus length (mm)	24.6 ± 1.0	21.4 - 27.0	589	24.2 ± 1.1	21.4 - 26.6	306	24.4± 1.0	21.4 - 27.0	895
Wing length (mm)	79.9 ± 2.3	72.5 - 84.9	736	76.2 ± 2.2	71.5 - 83.0	383	78.0 ± 2.2	71.5 - 83.0	1119
Head and Bill length (mm)	32.4 ± 0.6	30.5 - 34.2	288	32.0 ± 0.7	30.4 - 33.4	155	32.2 ± 0.6	30.4 - 34.2	443
Bill length (mm)	12.6 ± 1.1	10.0 - 14.9	479	12.5 ± 1.1	9.8 - 14.9	239	12.5 ± 1.1	9.8 - 14.9	718
Bill depth (mm)	7.2 ± 0.3	6.3 - 8.0	353	7.0 ± 0.4	6.3 - 8.0	172	7.1 ± 0.3	6.3 - 8.0	525
Bill width (mm)	6.1 ± 0.5	5.0 - 7.3	359	6.0 ± 0.5	5.0 - 7.3	172	6.1 ± 0.5	5.2 - 7.3	531
Weight (g)	23.5 ± 1.4	19.7 - 27.0	810	22.6 ± 1.4	19.4 - 27.0	414	23.0 ± 1.4	19.4 - 27.0	1224

MORPHOLOGY
The Patagonian Sierra-Finch is a passerine with a robust conical bill, which in Omora Park averages 12.4 mm in length, 7.2 mm in height, and 6.1 mm in width. Display sexually dimorphic coloration: the females are paler and have duller colors; the male has a bright yellow-brown back and deep yellow abdomen; additionally, the head, neck, wings, and tail are greyish-blue.

DIET
On Navarino Island they feed primarily on seeds, and have also been recorded to consume fruits of Winter's Bark (*Drimys winteri*) and Box-leafed Barberry (*Berberis buxifolia*). They pierce the base of the Chilean Firebush (*Embothrium coccineum*) and are considered a "nectar thief" because they rob the nectar from the base of the flowers, but do not contribute to pollination.

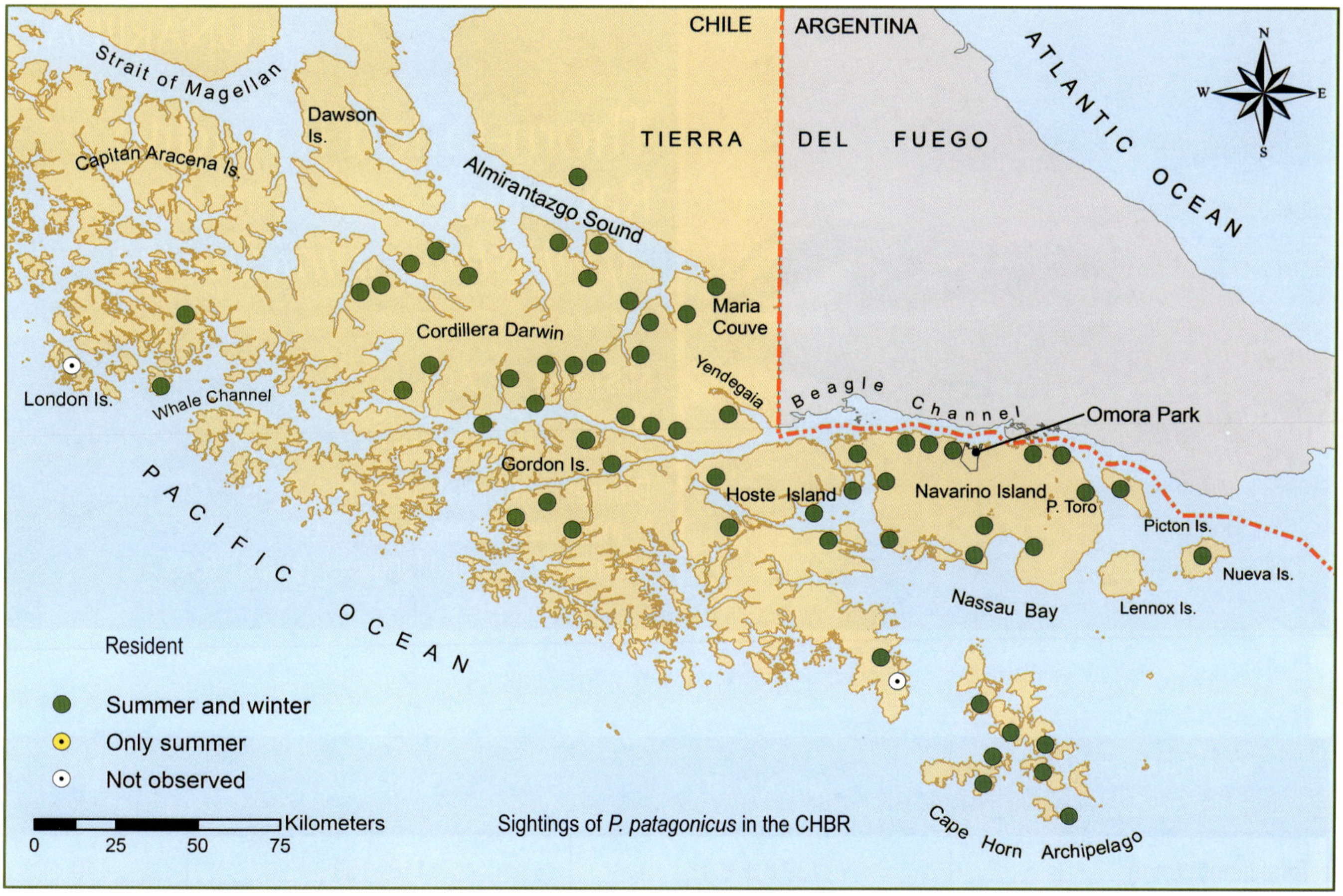

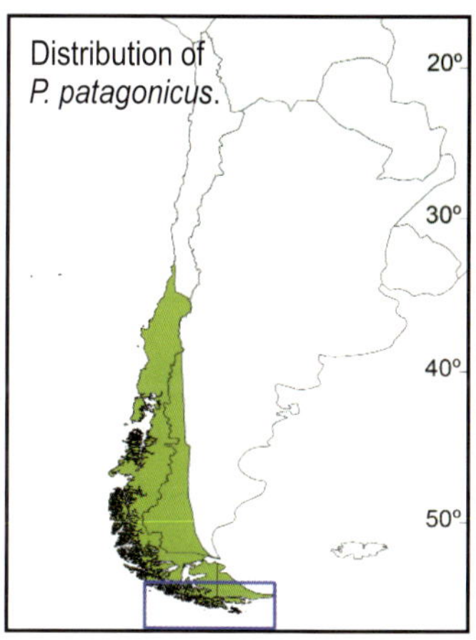

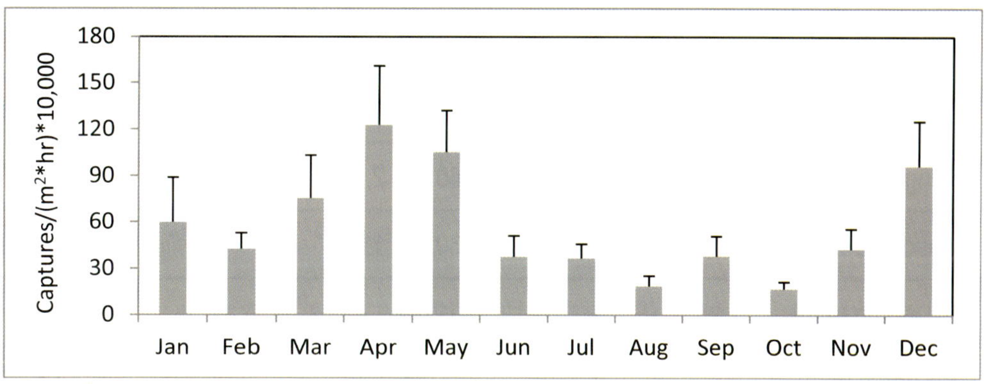

Mean number of captures of the Patagonian Sierra-Finch standardized by monthly net surface area and hours of capture effort of mist-nets between 2003 and 2010 in the Omora Park.

DISTRIBUTION AND HABITAT

Endemic to the temperate forests biome of South America. They are found from La Serena (30ºS) to the Cape Horn Archipelago. In the CHBR they display a wide distribution from coastal habitats to the treeline, inhabiting the interior and margins of the forest, as well as scrublands.

HABITS AND ECOLOGY

One of the most abundant resident birds in the sub-Antarctic forests of the archipelagos of the CHBR, where they add color and life in the snowy months. In the reproductive season, they are active singers. When they settle on a tall tree, they emit a musical and repetitious territorial song, *fwiip-fwiiio, fwiip-fwiiio, fwiip-fwiiio*. Other times, the members of the flocks communicate with a sharp cracking alarm, chic-chic-chic, as they move thorugh the foliage. They nest in dense mats of Box-leafed Barberry In Omora Park, individuals live to be more than 6 years in age.

CONSERVATION

Species of least concern (IUCN).

FRINGILLIDAE

Black-chinned Siskin
Jilguero
Carduelis barbatus

MIST-NET DATA
Number of captures: 569
% of total captures: 6.40
Different individuals captured: 543
% individuals recaptured: 4.1
Maximum recorded longevity (years): 3.54

MORPHOMETRICAL MEASUREMENTS	MEAN ± SD	MIN-MAX	N
Total length (mm)	122.6 ± 7.5	110 - 144	198
Tail length (mm)	50.2 ± 2.3	41 - 61	304
Tarsus length (mm)	18.1 ± 1.1	15.6 - 20.5	314
Wing length (mm)	72.6 ± 2.1	68 .0 - 77.0	376
Head and bill length (mm)	26.3 ± 1.7	17.0 - 29.3	82
Bill length (mm)	10.2 ± 1.0	8.0 - 12.6	289
Bill depth (mm)	7.3 ± 0.4	6.4 - 8.2	221
Bill width (mm)	5.9 ± 0.5	4.8 - 7.1	210
Weight (g)	16.5 ± 1.2	13.5 - 19.8	401

MORPHOLOGY
It is the smallest of the granivores in the Omora Park, reaching an average weight of 16.5 g. Its plumage is yellow-green. The male has brighter colors than the female and displays a distinct black chin, from which its English and scientific name, *barbatus* (bearded), originated. The female is paler and its crown is grey.

DIET
It feeds on the ground and in the canopy, primarily on seeds, but in the Omora Park we have documented that it also eats berries and insects (such as butterflies).

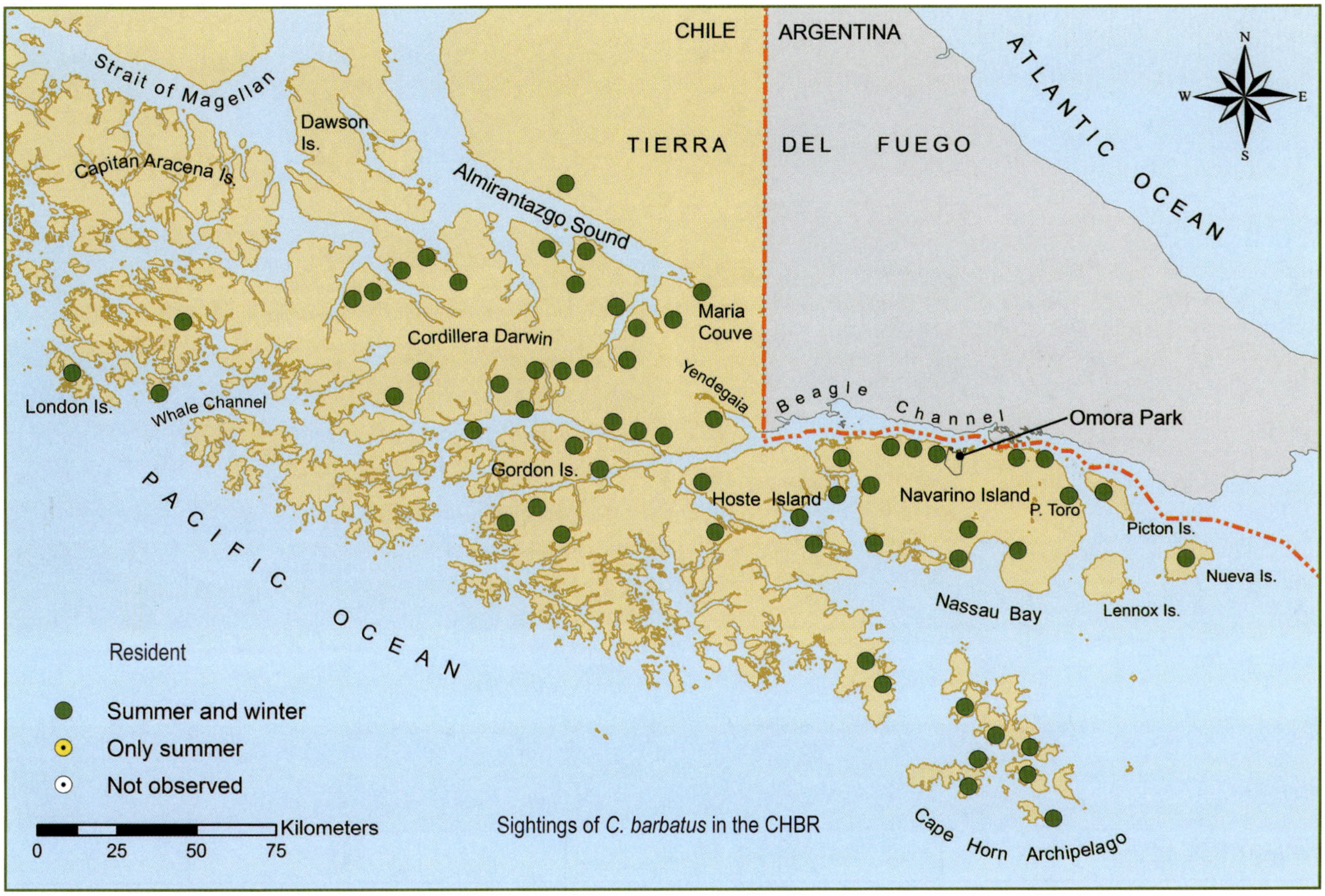

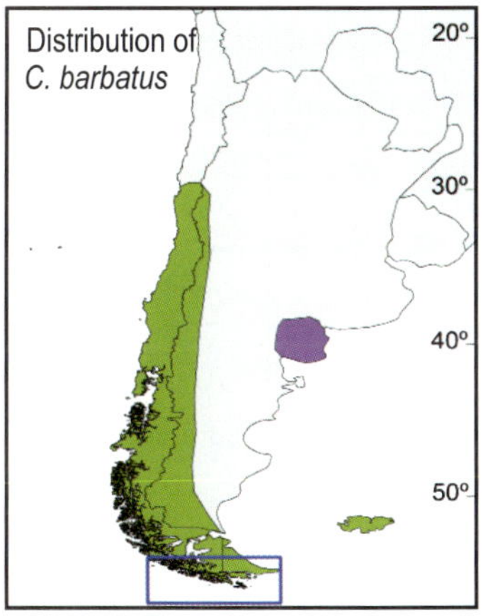

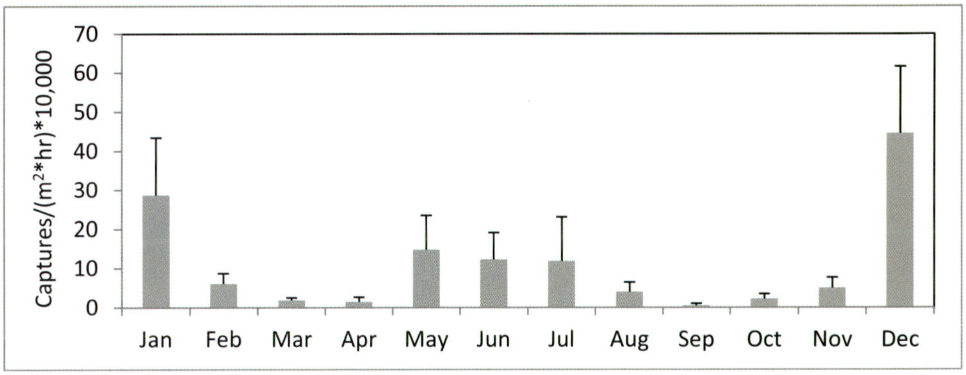

Mean number of captures of the Black-chinned Siskin standardized by monthly net surface area and hours of capture effort of mist-nets between 2003 and 2010 in the Omora Park.

DISTRIBUTION AND HABITAT

This species is endemic to southern South America, inhabiting from Huasco Valley (28ºS) to the Cape Horn Archipelago. In the CHBR, it has a broad distribution and inhabits the interior and margins of forests, scrublands, and is frequently observed in the town of Puerto Williams.

HABITS AND ECOLOGY

Resident species that is abundant throughout the year in all areas of the CHBR. However, we have recorded the highest number of captures in summer. During the breeding season, they are observed in pairs or in small groups, and form flocks during winter. We have determined that this species lives at least three and a half years.

CONSERVATION

Considered a species of Least Concern (IUCN).

EMBERIZIDAE

Rufous-collared Sparrow
Chincol
Zonotrichia capensis

MIST-NET DATA
Number of captures: 408
% of total captures: 4.59
Different individuals captured: 357
% individuals recaptured: 8.4
Maximum recorded longevity (years): 3.68

MORPHOMETRICAL MEASUREMENTS	MEAN ± SD	MIN-MAX	N
Total length (mm)	144.8 ± 5.8	132 - 157	67
Tail length (mm)	65.1 ± 3.0	59 - 72	135
Tarsus length (mm)	24.0 ± 1.1	21.3 - 26.4	209
Wing length (mm)	79.7 ± 2.9	73.7 - 86.0	242
Head and bill length (mm)	30.6 ± 0.5	29.5 - 31.8	48
Bill length (mm)	10.7 ± 1.3	8.0 - 13.4	161
Bill depth (mm)	6.1 ± 0.3	5.4 - 6.8	115
Bill width (mm)	5.5 ± 0.5	4.4 - 6.6	115
Weight (g)	23.3 ± 1.5	19.5 - 26.9	246

MORPHOLOGY
Zonotrichia capensis is characterized by its rufous necklace around its neck (from which its English name is derived), a grey head with black markings, white throat, and a noticeable crest. Juveniles are les colorful and lack the red necklace. The subspecies *Z. capensis australis* lacks the black on the crown, is of larger size than those in the north, and reaches an average weight of 23.3 g in the Omora Park.

DIET
On Navarino Island, it feeds on seeds and grains on the ground. It also eats fruits of the Winter's Bark, Magellanic Currant, invertebrates such as coleopterans, dipterans, lepidopterans, hymenopterans, spiders, and mites.

SECTION B

LARGER-SIZED BIRDS INCIDENTALLY CAPTURED WITH MIST-NETS IN THE SUB-ANTARCTIC FORESTS AT OMORA PARK AND OTHER SITES ON NAVARINO ISLAND

STRIGIDAE

Austral Pygmy-Owl
Chuncho
Glaucidium nanum

MIST-NET DATA
Number of captures: 16
% of total captures: 0.18
Different individuals captured: 11
% individuals recaptured: 9.1
Maximum recorded longevity (years): 0.98

MORPHOMETRICAL MEASUREMENTS	MEAN ± SD	MIN-MAX	N
Total length (mm)	163.0 ± 4.6	159 - 168	3
Tail length (mm)	62.5 ± 0.6	62 - 63	4
Tarsus length (mm)	28.1 ± 1.0	26.5 - 29.3	10
Wing length (mm)	95.8 ± 4.7	89.5 - 103.0	13
Head and bill length (mm)	38.0 ± 0.4	37.4 - 38.6	8
Bill length (mm)	13.1 ± 1.2	10.6 - 15.0	13
Bill depth (mm)	9.6 ± 0.8	8.5 - 10.8	8
Bill width (mm)	6.4 ± 1.2	5.1 - 8.3	8
Weight (g)	69.9 ± 10.6	56.7 - 86.0	13

MORPHOLOGY

The Austral Pygmy-Owl is the smallest of the owls of the temperate forest biome of South America. In the Omora Park, its size (total length 163 mm and weight 70 g) is similar to records from populations further north. In the CHBR we have recorded the three coloration morphotypes described for *G. nanum*: 1) brown plumage with numerous light and narrow caudal bars; ii) grey with few light and wide caudal bars; iii) reddish and caudal bars may be absent.

DIET

Fierce hunter. Captures a wide range of birds, rodents, spiders, insects, and other invertebrates. It often only eats part of the bird; consequently, the wings and intact heads of birds that the Austral Pygmy-Owl has killed are often found on the ground. In Omora Park, it captures birds such as Sierra-Finches and Thorn-tailed Rayaditos, and rodents such as the Yellow-nosed Mouse (*Abrothrix xanthorhinus*).

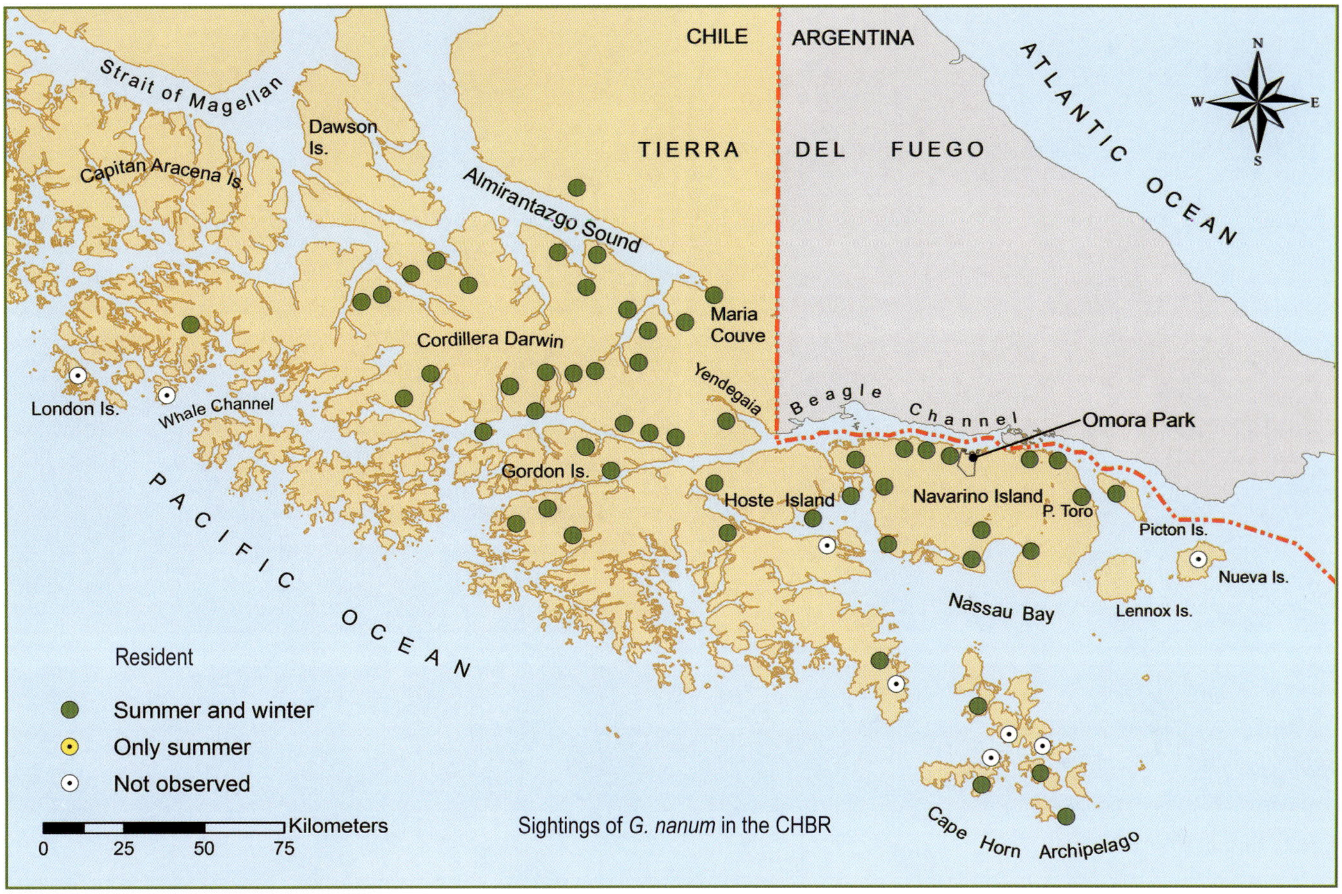

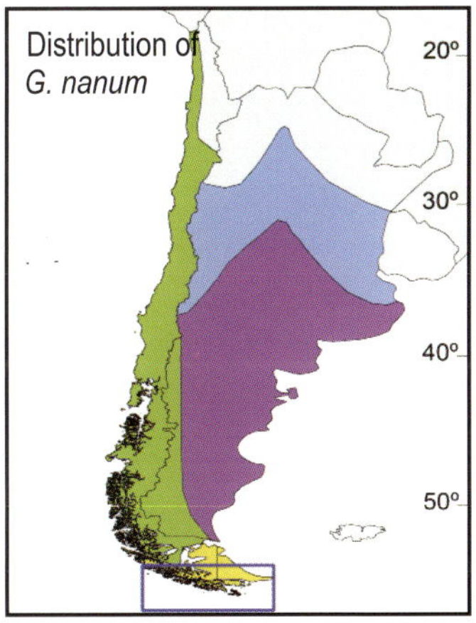

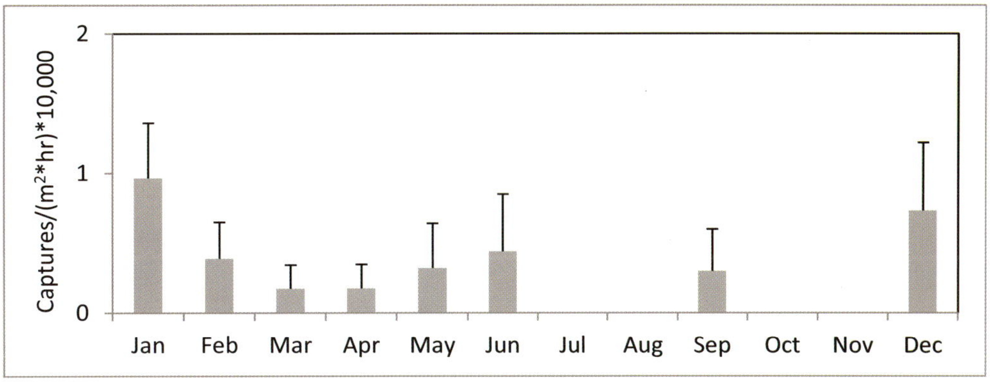

Mean number of captures of the Austral Pygmy-Owl standardized by monthly net surface area and hours of capture effort of mist-nets between 2003 and 2010 in the Omora Park.

DISTRIBUTION AND HABITAT

It is the most common owl in Chile. It is found from shrubby semiarid environments of Atacama to the sub-Antarctic forest of Cape Horn. It occupies a wide variety of habitats including old-growth and secondary forests, scrublands, as well as rural and urban areas. It has a wide distribution in all of the CHBR, where it even reaches Horn Island.

HABITS AND ECOLOGY

Although there were no captures of Austral Pygmy-Owls in the nets during some winter months, the records from Omora Park indicate that this is a resident species and that it hunts at all hours. It is not afraid of human presence and it is possible to observe it during the day at short distances perched on exposed branches. It is also detected through alarm calls and behaviors that passerines utter to warn of its presence. We have recorded nesting in hollow trunks and cavities made by Magellanic Woodpeckers.

CONSERVATION

The IUCN has classified it as of Least Concern.

PICIDAE

Magellanic Woodpecker
Carpintero
Campephilus magellanicus

MIST-NET DATA
Number of captures: 11
% of total captures: 0.12
Different individuals captured: 8
% individuals recaptured: 25.0
Maximum recorded longevity (years): 1.48

	MALE			FEMALE		
MORPHOMETRICAL MEASUREMENTS	MEAN ± SD	MIN-MAX	N	MEAN ± SD	MIN-MAX	N
Total length (mm)	-	-	-	-	-	-
Tail length (mm)	158	158	1	164.3 ± 6.1	158 - 170	4
Tarsus length (mm)	44.8 ± 6.8	40.0 - 49.6	2	43.6 ± 4.7	37.9 - 48.9	4
Wing length (mm)	202.5 ± 31.8	180 - 225	2	217.3 ± 8.9	210 - 230	4
Head and bill length (mm)	100.7	100.7	1	96.9 ± 1.8	95.6 - 98.9	3
Bill length (mm)	56.1 ± 4.5	52.9 - 59.2	2	55.7 ± 5.1	49.8 - 60.9	4
Bill depth (mm)	-	-	-	14.1 ± 0.4	13.6 - 14.6	4
Bill width (mm)	-	-	-	14.6 ± 0.5	14.0 - 15.3	4
Weight (g)	319.5 ± 0.7	319 - 320	2	303.7 ± 17.0	286 - 320	3

MORPHOLOGY
The Magellanic or Giant Woodpecker is the largest of the woodpeckers of South America. On Navarino Island, it reaches a weight over 319.5 g. It has a prominent sexual dimorphism. Although males and females have similar bill lenghts, males have a scarlet head and neck, whereas the female's head and neck are black. The tuft of the female is larger than that of the male and curls forward.

DIET
In Omora Park, it feeds primarily on the larvae of long-horned beetles (*Microplophorus magellanicus*) that it extracts with its tongue after excavating the trunks with its sharp and powerful bill. We have also observed it eating earthworms and other soli invertebrates, fruits of Winter's Bark, and occasionally sap. When nesting, it occasionally feeds its young with the chicks of passerines.

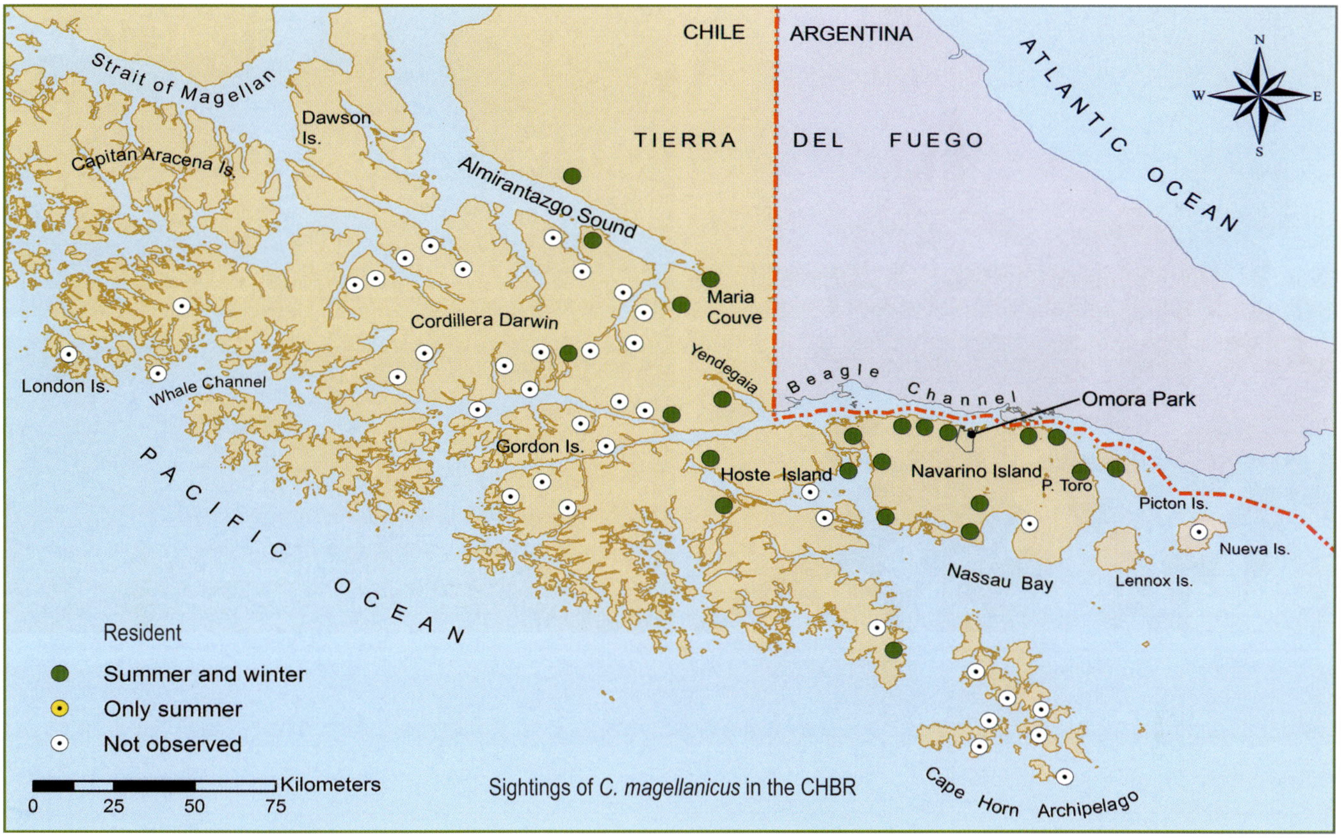

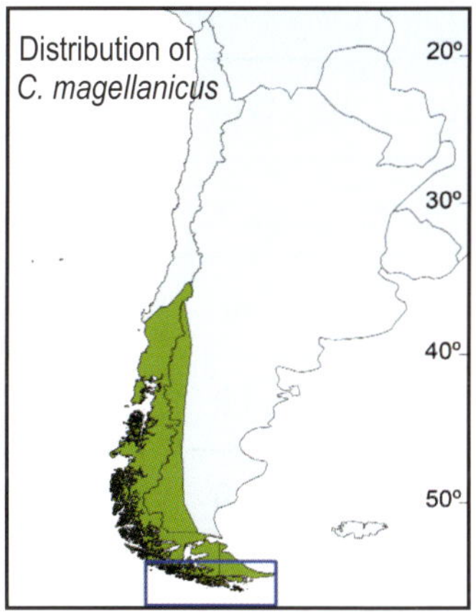

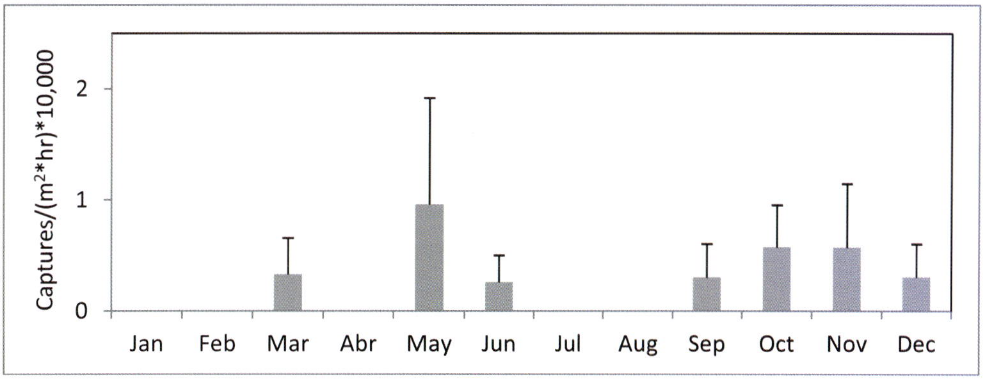

Mean number of captures of the Magellanic Woodpecker standardized by monthly net surface area and hours of capture effort of mist-nets between 2003 and 2010 in the Omora Park.

DISTRIBUTION AND HABITAT

It is endemic to the *Nothofagus* forests of southern Chile and Argentina. Its distribution extends the furthest south of any of the world's woodpeckers. Our southernmost records correspond to the forests of Magellanic Coigüe in Schapeham Bay, near the False Cape Horn, in Hoste Island. Interestingly, it is also possible to observe the Magellanic Woodpecker in Puerto Williams foraging on trunks of the old trees along the streets in the town.

HABITS AND ECOLOGY

It is a resident in the CHBR. It is very specialized in its habitat requirements and only nests in trees of the genus *Nothofagus*. In the Omora Park, we have found cavities where it nests in Magellanic Coigüe, Lenga, and Ñirre. The Magellanic Woodpecker is considered an ecological keystone species that provides cavities where other species nest, such as the Austral Parakeet, Austral Pygmy-owl, and Chilean Swallow. It patrols its territories in multi-aged family groups, communicates with strong loud screams, and drumming on dead trees.

CONSERVATION

According to the Chilean Hunting Law, the Magellanic Woodpecker is Endangered in the Regions VI and VII, and Vulnerable between the Regions VIII and XII. On Navarino Island, it seems to be more abundant than in more northern forests. In Omora Park we have identified at least seven family groups protected in this reserve.

PSITTACIDAE

Austral Parakeet
Cachaña
Enicognathus ferrugineus

MIST-NET DATA
Number of captures: 6
% of total captures: 0.07
Different individuals captured: 4
% individuals recaptured: 0
Maximum recorded longevity (years): No recaptures

MORPHOMETRICAL MEASUREMENTS	MEAN ± SD	MIN-MAX	N
Total length (mm)	-	-	-
Tail length (mm)	146.0 ± 12.7	137 - 155	3
Tarsus length (mm)	22.9 ± 1.3	21.4 - 24.5	4
Wing length (mm)	191.0 ± 5.7	184 - 197	4
Head and bill length (mm)	49.8 ± 2.9	46.5 - 51.6	3
Bill length (mm)	23.6 ± 3.1	20.6 - 26.7	3
Bill depth (mm)	23.6 ± 3.1	21.9 - 22.5	3
Bill width (mm)	12.1 ± 1.0	11.0 - 13.0	3
Weight (g)	188.8 ± 7.2	179 - 198	5

MORPHOLOGY
The Austral Parakeet is a medium sized parrot reaching 198 g on Navarino Island. Our measurements confirm that the southern subspecies, *ferrugineus*, is larger than the subspecies *minor*. The wing lenght of *E. ferrugineus* extends 191 mm. The body feathers are green with reddish feathers in its forehead, a rufous belly and a long rusty-colored tail (from where it gets its specific name *ferrugineus*). The subspecies *ferrugineus* is more yellowish than that of the north. It has a short hooked bill that in the individuals captured in the Omora Park reaches up to 26.7 mm long.

DIET
It has a broad plant diet that varies with the seasons of the year. In Omora Park, it feeds in October on tender leaf buds and flowers (which have stamens loaded with protein-rich pollen) of the Lenga (*Nothofagus pumilio*). In November it feeds on flowers of the Firebush (*Embothrium coccineum*), rich with nectar and pollen. In the summer, it eat seeds of the lipid-rich Lenga and Ñirre, and in March fruits of the Winter's Bark (*Drimys winteri*). During the winter, it eats the flower buds and leaves of the epiphytic mistletoe (*Misodendrum* spp.).

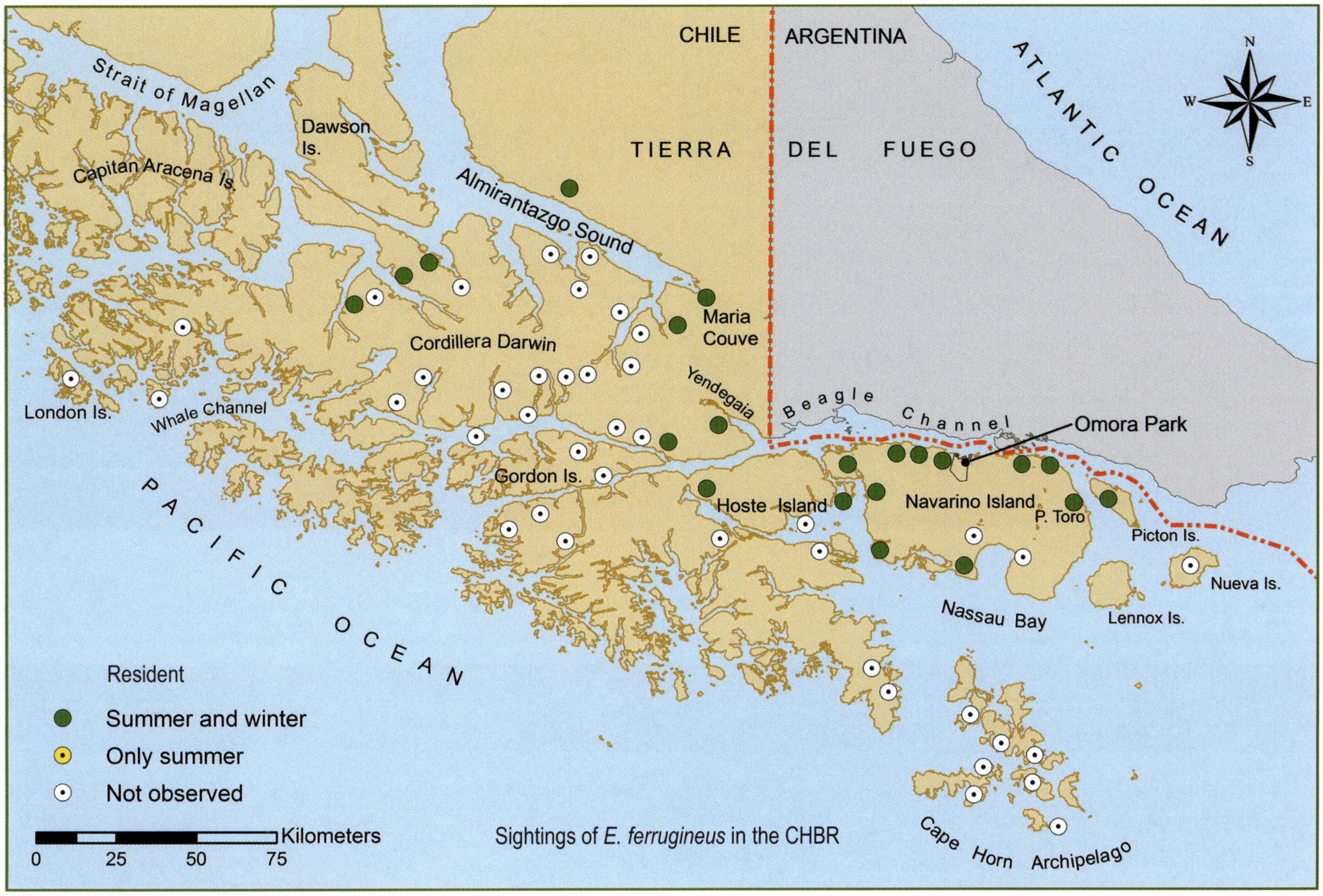

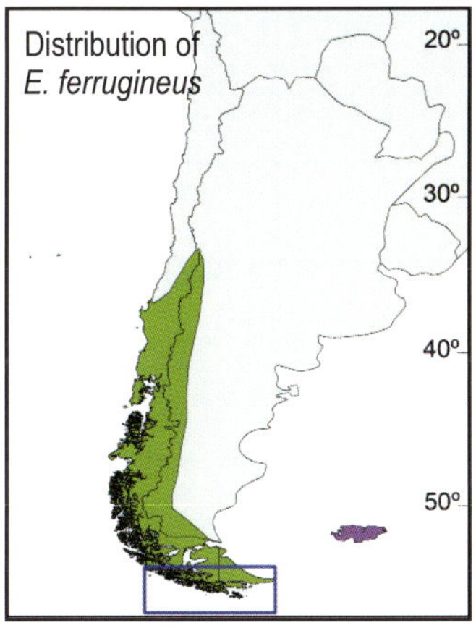

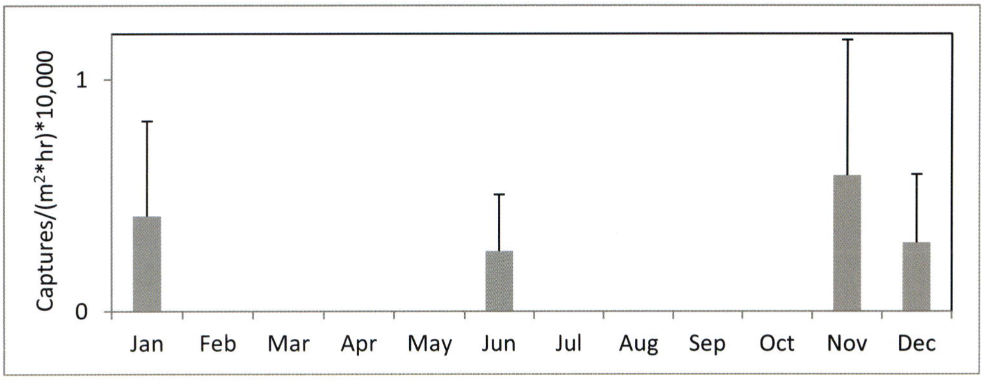

Mean number of captures of the Austral Parakeet standardized by monthly net surface area and hours of capture effort of mist-nets between 2003 and 2010 in the Omora Park.

DISTRIBUTION AND HABITAT

It is the southernmost parrot of the planet and the only one that inhabits the sub-Antarctic forests. The Austral Parakeet is endemic to the temperate forests of South America, and is distributed from the Metropolitan Region to Navarino and Hoste islands. In the CHBR it mainly inhabits the areas of old-growth forests, and is more abundant in the Lenga forests.

HABITS AND ECOLOGY

It is a resident species in Omora Park. It nests in natural or constructed cavities excavated by the Magellanic Woodpecker (*Campephilus magellanicus*) in trunks of large Magellanic Coigüe (*Nothofagus betuloides*) or Lenga trees. They are gregarious birds whose noisy flocks are observed fliyng overhead or perching on emerging treetops.

CONSERVATION

Enicognathus is an endemic genus to the temperate forests biome of South America that has only two species: *E. ferrugineus* (33-55ºS) and the Slender-billed Parakeet *E. lepthorhynchus*, which has a more restricted distribution (37-43ºS). Although in some locations the chicks of these parrots are poached from their nests for the pet trade, they are considered of Least Concern (IUCN).

FALCONIDAE

Chimango Caracara
Tiuque
Milvago chimango

MIST-NET DATA
Number of captures: 4
% of total captures: 0.05
Different individuals captured: 4
% individuals recaptured: 0.0
Maximum recorded longevity (years): No recaptures

MORPHOMETRICAL MEASUREMENTS	MEAN ± SD	MIN-MAX	N
Total length (mm)	-	-	-
Tail length (mm)	187.5 ± 7.8	182 - 193	2
Tarsus length (mm)	60.3 ± 8.5	52.0 - 69.0	3
Wing length (mm)	301.7 ± 6.0	296 - 308	3
Head and bill length (mm)	-	-	-
Bill length (mm)	22.2 ± 5.0	19.3 - 28.0	3
Bill depth (mm)	13.4 ± 1.0	12.6 - 14.5	3
Bill width (mm)	10.6 ± 1.7	9.3 - 12.5	3
Weight (g)	392.0 ± 37.4	367 - 435	3

MORPHOLOGY

The Chimango Caracara is a raptor that on Navarino Island reaches an average a weight of 392 g. Our measurements indicate that the subspecies *fuegiensis* is of larger size than *temucoensis* and *chimango*; its wing length exceeds 300 mm. This subspecies also has a darker brown color, but it retains the white rump. In flight it displays white "windows" at the base of the primaries. Its tail has a dark brown subterminal band with a white edge. The legs can be light blue or yellow.

DIET

Despite being a falcon taxonomically, its diet is very broad and includes carrion, larvae, maggots, worms, and rodents. On Navarino Island it is usually seen feeding in the intertidal zone from where it carries mollusks to feed on in the forest. This is why the shells of bivalves and gastropods are found on the ground in the forests of the Omora Park.

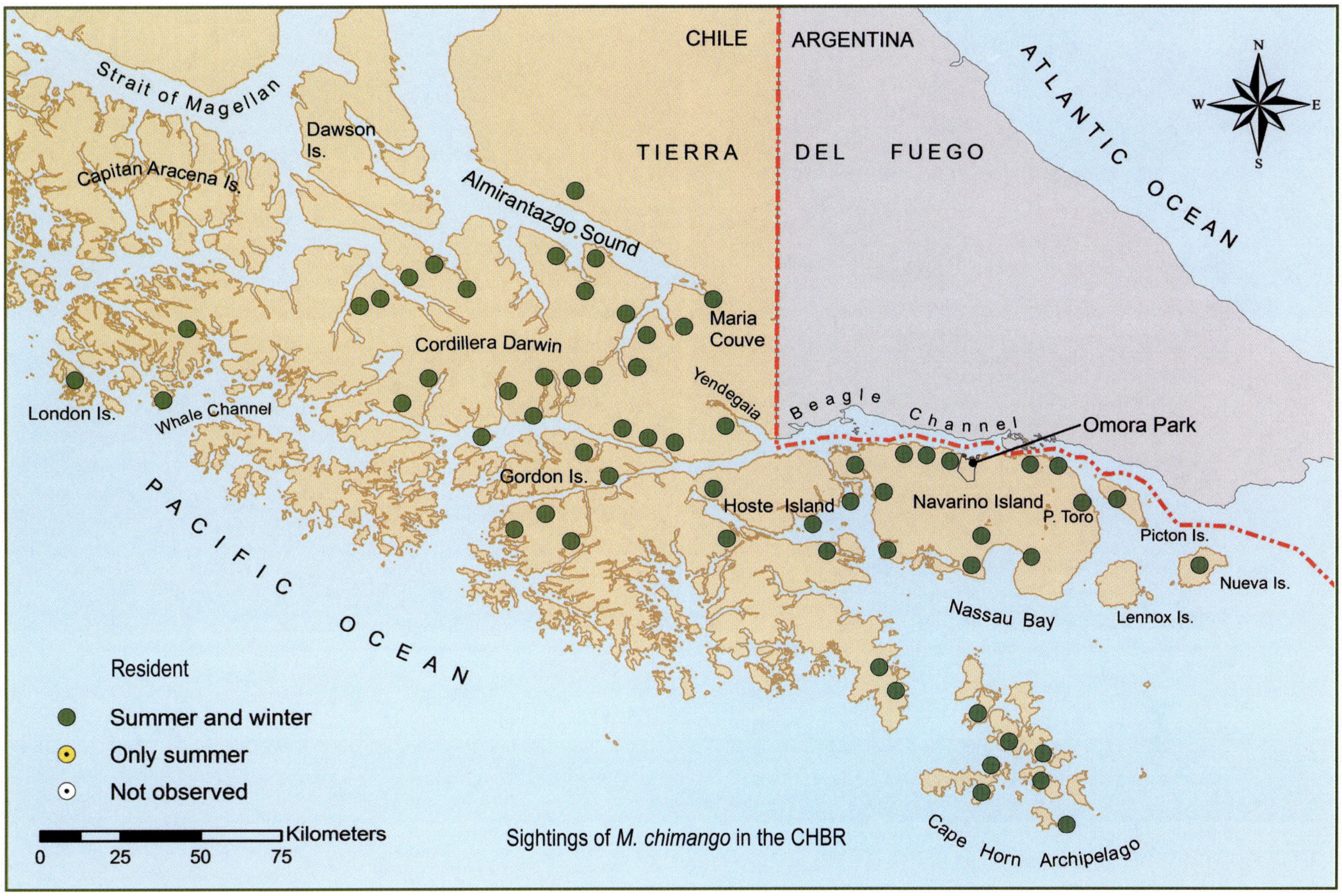

Sightings of *M. chimango* in the CHBR

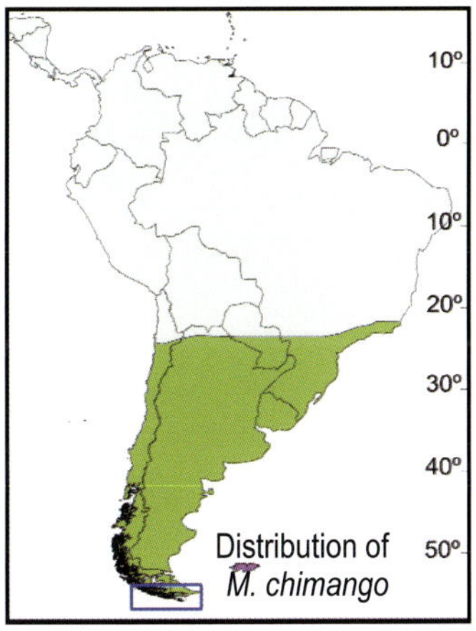

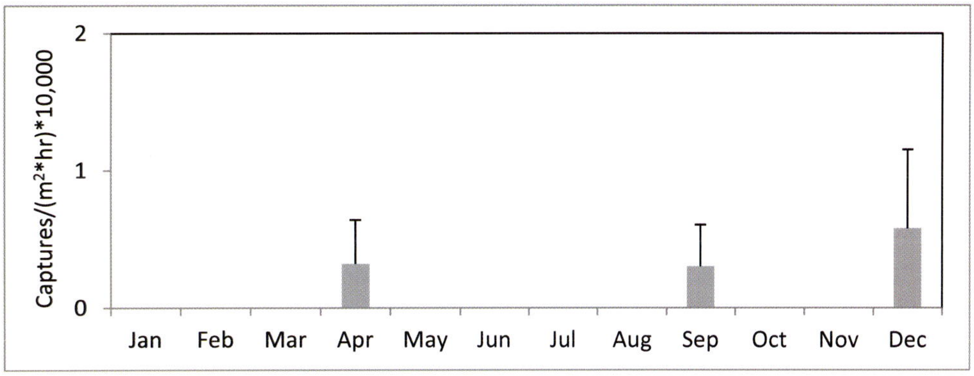

Mean number of captures of the Chimango Caracara standardized by monthly net surface area and hours of capture effort of mist-nets between 2003 and 2010 in the Omora Park.

DISTRIBUTION AND HABITAT

It is the most common raptor of southern Chile and occupies a great variety of environments: from the forest interior to fragmented landscapes, coastal, and urban areas. It ranges from Chañaral to Cape Horn. Of the three recognized subspecies *M. chimango fuegiensis* is the one that is found south of the Strait of Magellan. It has a broad distribution throughout the CHBR, and is abundant in the town and landfill of Puerto Williams.

HABITS AND ECOLOGY

It is resident in the CHBR. During the reproductive season, it is territorial and makes frequent aerial displays. During the rest of the year, the Chimango Caracaras form loose flocks. In the Omora Park, it uses large trees to nest and as a roost from which it calls loudly. This bird moves easily on the ground, where it finds its food.

CONSERVATION

Species of Least Concern (IUCN).

STRIGIDAE

Rufous-legged Owl
Concón
Strix rufipes

MIST-NET DATA
Number of captures: 2
% of total captures: 0.02
Different individuals captured: 2
% individuals recaptured: 0
Maximum recorded longevity (years): No recaptures

MORPHOMETRICAL MEASUREMENTS	MEAN ± SD	MIN-MAX	N
Total length (mm)	-	-	-
Tail length (mm)	157.5 ± 3.5	155 - 160	2
Tarsus length (mm)	59.8 ± 5.6	55.8 - 63.7	2
Wing length (mm)	277.0 ± 9.9	270 - 284	2
Head and bill length (mm)	72.8 ± 1.4	71.8 - 73.8	2
Bill length (mm)	32.2 ± 4.6	28.9 - 35.4	2
Bill depth (mm)	14.4 ± 0.7	13.9 - 14.9	2
Bill width (mm)	9.8 ± 0.1	9.8 - 0.1	2
Weight (g)	447.0 ± 12.7	438 - 456	2

MORPHOLOGY
The Rufous-legged Owl is a robust owl that on Navarino Island reaches 447 g of weight. It has a large head, lacks ear tufts, has a facial disk, and brown bars on its chest, abdomen, and sides. Our data are consistent with the fact that the local subspecies *rufipes* is larger than the northerly *sanborni*; its wings are longer than 277 mm. It has a hooked bill that in individuals captured in the Omora Park, reaches 35.4 mm in length.

DIET
On Navarino Island, we documented that this owl not only feed on rodents, but also on birds. In other locations it has also been recorded that its diet includes insects and other invertebrates.

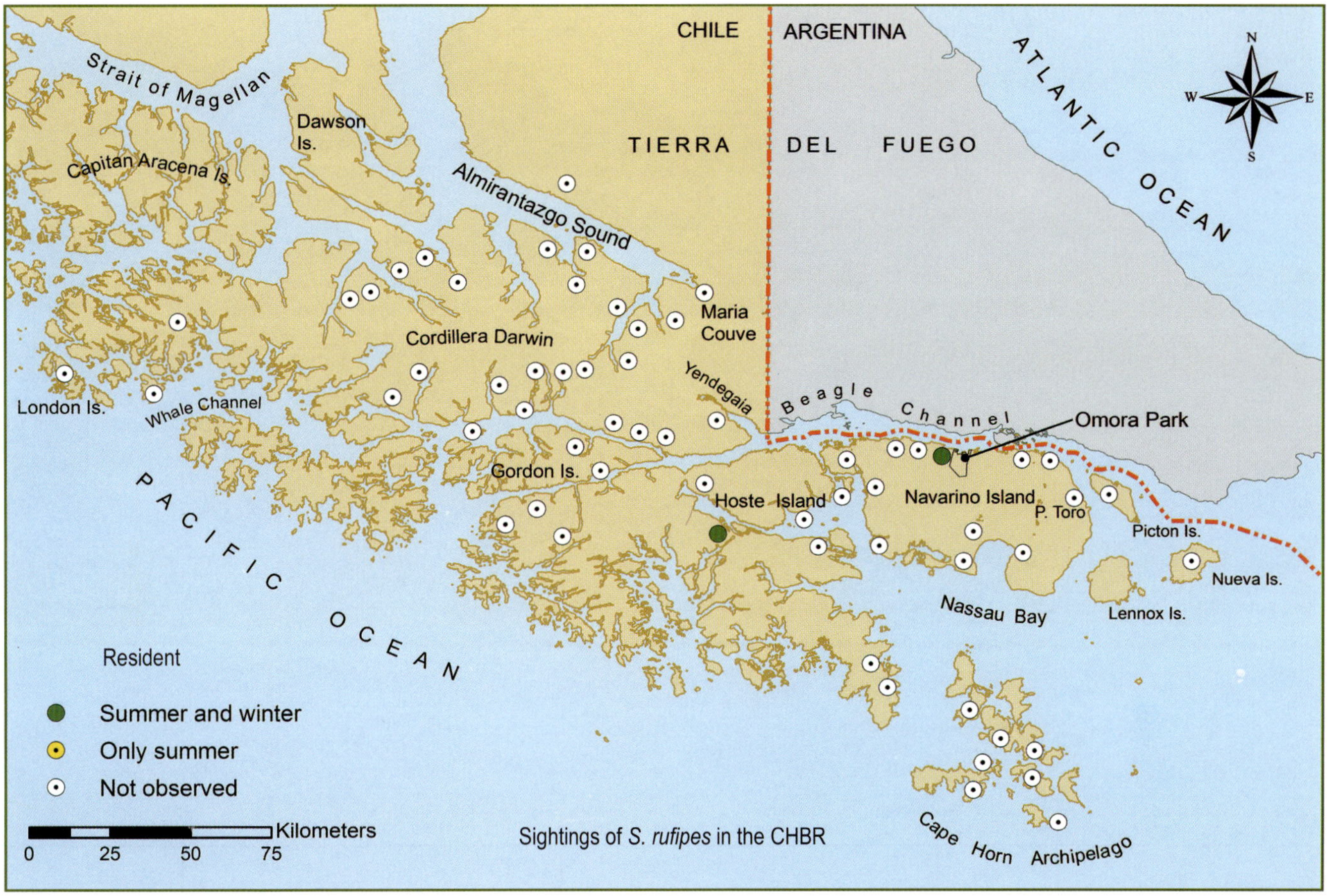

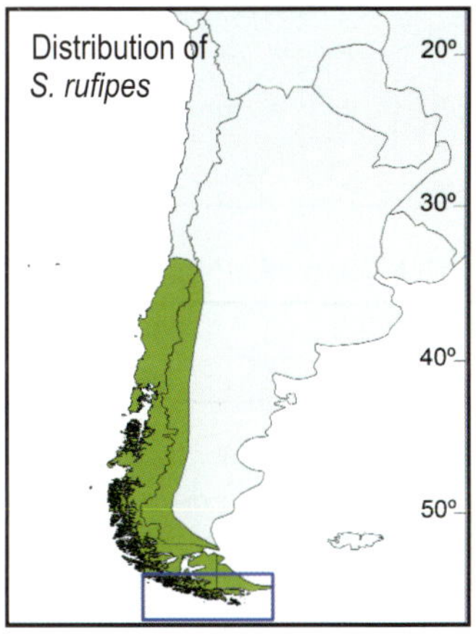

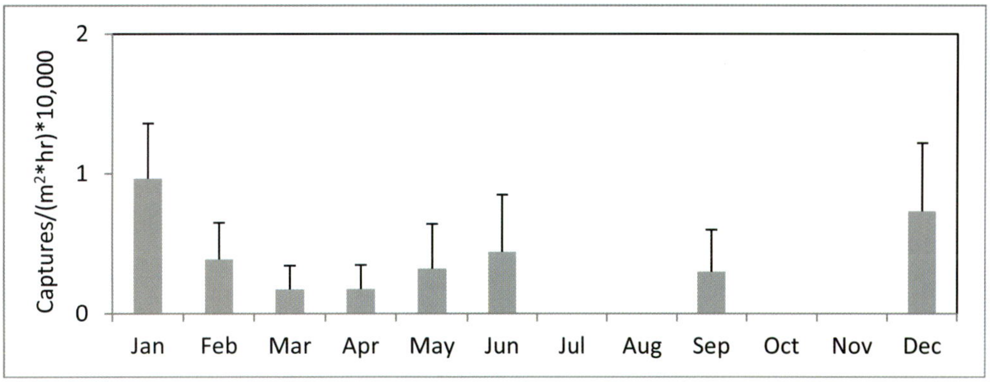

Mean number of captures of the Rufous-legged Owl standardized by monthly surface and hours of operation of mist-nets between 2003 and 2010 in the Omora Park.

DISTRIBUTION AND HABITAT

The subspecies *S. rufipes sanborni* is restricted to the Isla Grande of Chiloé, but *S. r. rufipes* is distributed from central Chile to south of the Beagle Channel. In the CHBR we have recently recorded it on the Hoste and Navarino islands, in old-growth forests of Magallanic Coigüe and Winter's Bark.

HABITS AND ECOLOGY

It is resident in Omora Park where it is active during both the day and night. Its population is small. They announce their presence on their territories with loud calls.

CONSERVATION

According to the Chilean Hunting Law, this is an Inadequately Known species. Among the Chilean raptors, this is one with the highest conservation priority. It should be considered a charismatic species for the conservation of the old-growth forests. In North America, its congeneric, the Spotted Owl (*Strix occidentalis*) was listed under the Endangered Species Act, providing a policy tool for the protection of old-growth forests.

C

SECTION

PASSERINES INHABITING THE HIGH-ANDEAN HABITATS ABOVE THE TREE-LINE CAPTURED WITH MIST-NETS AT ROBALO MOUNTAIN, OMORA PARK, NAVARINO ISLAND

TYRANNIDAE

Dark-faced Ground-Tyrant
Dormilona Tontita
Muscisaxicola macloviana

MIST-NET DATA
Number of captures: 85
% of total captures: 0.96
Different individuals captured: 84
% individual recaptured: 1.2
Maximum recorded longevity (years): 0.003

MORPHOMETRICAL MEASUREMENTS	MEAN ± SD	MIN-MAX	N
Total length (mm)	148.9 ± 5.4	137 - 161	29
Tail length (mm)	65.2 ± 3.76	58 - 73	30
Tarsus length (mm)	30.7 ± 1.2	28.1 - 32.7	30
Wing length (mm)	99.0 ± 3.3	93 - 106	30
Head and bill length (mm)	-	-	-
Bill length (mm)	9.7 ± 0.7	8.3 - 11.4	28
Bill depth (mm)	3.4 ± 0.2	3.1 - 3.7	28
Bill width (mm)	3.8 ± 0.3	3.3 - 4.3	29
Weight (g)	22.8 ± 1.4	20.0 - 25.7	30

MORPHOLOGY
The Dark-faced Ground-Tyrant is a bird with a grey back, whitish underparts and distinctive dark face (hence the name in English) that reaches an average weight of 22.8 g in the Omora Park. Its beak and tarsi are black and measure 10 and 30 mm, respectively. The tail measures 65.2 mm and is characterized by its white outer edge.

DIET
In the lakes and watercourses of the Robalo River watershed we have observed it to consume aquatic and flying insects. In Tierra del Fuego it feeds on larvae and small beetles.

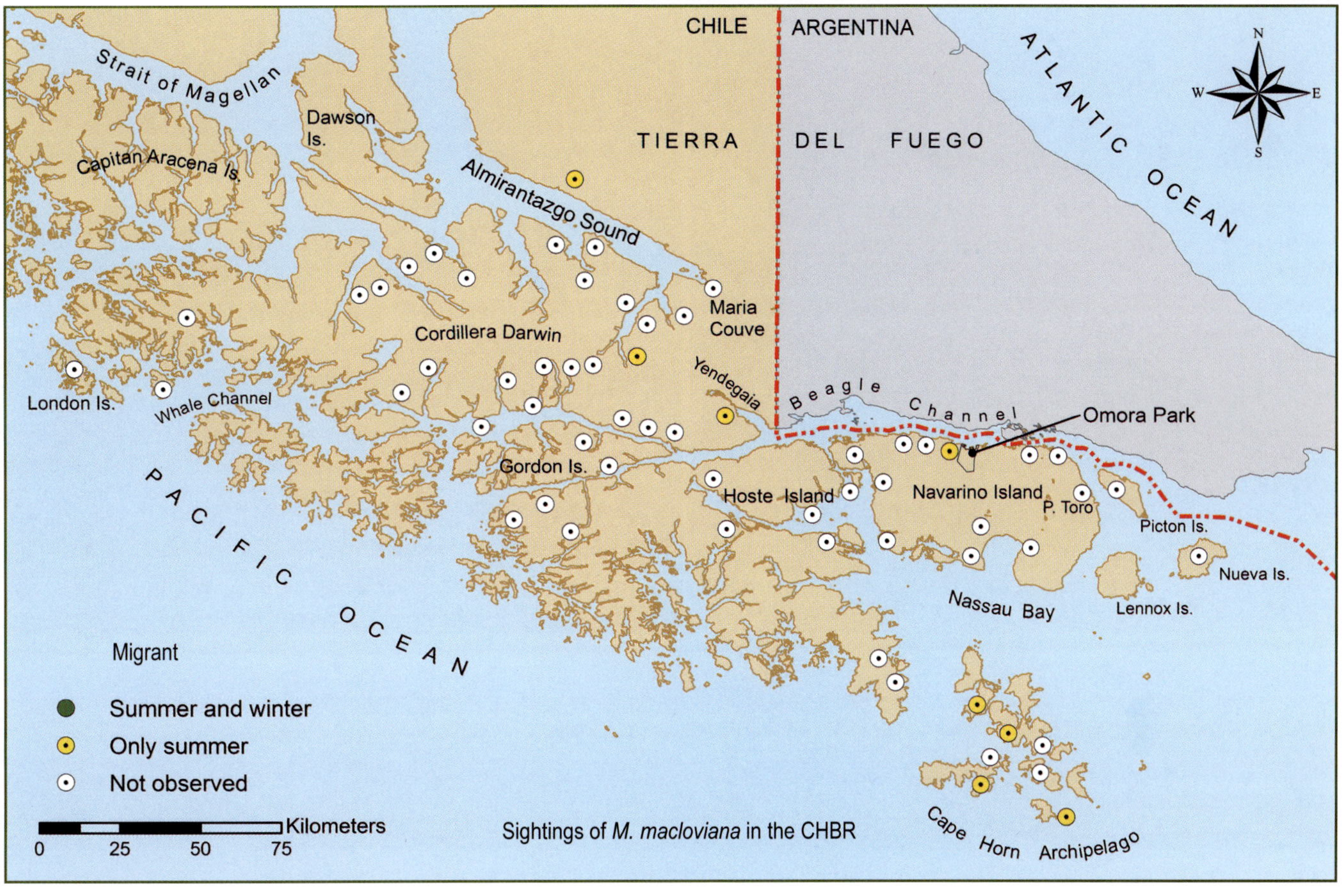

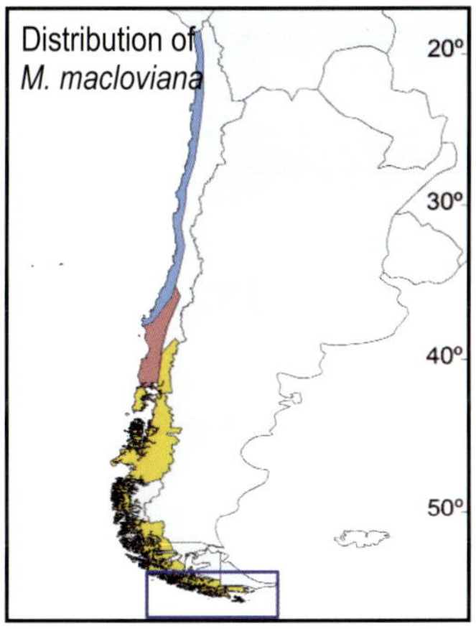

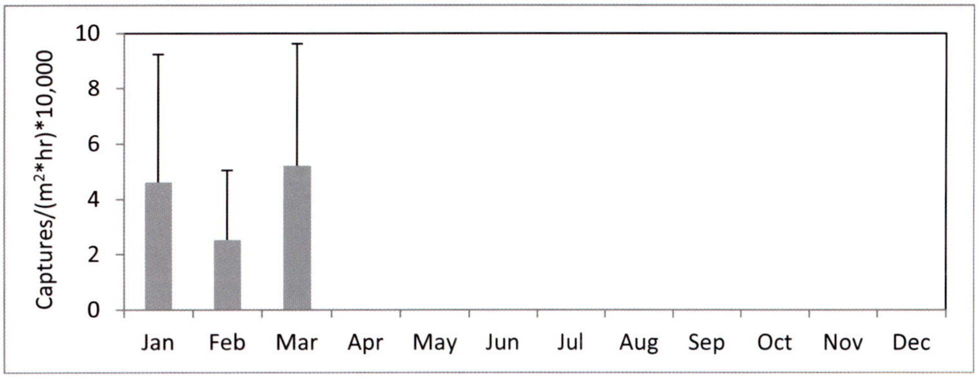

Mean number of captures of the Dark-faced Ground-Tyrant standardized by monthly net surface area and hours of capture effort of mist-nets between 2003 and 2010 in the Omora Park.

DISTRIBUTION AND HABITAT

The subspecies *M. macloviana mentalis* is endemic to southern South America. It breeds from Llanquihue to Cape Horn, migrating in winter along the Andes up to Peru. In the summer, we have observed it nesting along the banks of the Robalo Lake and other water bodies in the higher areas of the Robalo Mountain. It is typical to observe it resting on cushion plants and moving around in open Alpine habitats searching for its food. Before it migrates, it can be seen on the coast of the Beagle Channel.

HABITS AND ECOLOGY

It is a migrant that nests in ground cavities among rocks near water bodies at high elevations, near the tree line on Navarino Island. During the breeding season, scattered individuals are observed, but at the end of the warm season, they migrate in flocks. It moves on the ground by short flights, moving nervously its wings and tail.

CONSERVATION

According to the Chilean Hunting Law, it has no conservation issues and is a beneficial species, as it contributes to controlling insect populations.

EMBERIZIDAE

Yellow-bridled Finch
Yal Cordillerano
Melanodera xanthogramma

MIST-NET DATA
Number of captures: 20
% of total captures: 0.23
Different individuals captured: 19
% individual recaptured: 5.3
Maximum recorded longevity (years): 1.04

	MALE			FEMALE			TOTAL		
MORPHOMETRICAL MEASUREMENTS	MEAN ± SD	MIN-MAX	N	MEAN ± SD	MIN-MAX	N	MEAN ± SD	MIN-MAX	N
Total length (mm)	154.5 ± 3.5	152 - 157	3	153.3 ± 8.3	144 - 160	3	153.9 ± 5.9	144.0 - 160.0	6
Tail length (mm)	63.5 ± 0.7	63 - 64	3	69.0 ± 1.0	68 - 70	3	66.2 ± 0.8	63.0 - 70.0	6
Tarsus length (mm)	23.9 ± 1.0	22.7 - 24.7	3	23.7 ± 1.5	22.0 - 24.9	3	23.8 ± 1.2	22.0 - 24.9	6
Wing length (mm)	99.3 ± 4.5	95 - 104	3	100.5 ± 1.3	99 - 102	4	99.9 ± 2.9	95.0 - 104.0	7
Head and bill length (mm)	-	-	-	-	-	-	-	-	-
Bill length (mm)	10.7 ± 0.3	10.5 - 11.0	3	10.2 ± 0.4	9.8 - 10.6	3	10.4 ± 0.3	9.8 - 11.0	6
Bill depth (mm)	8.8 ± 0.4	8.4 - 9.1	3	8.1 ± 0.6	7.5 - 8.6	3	8.7 ± 0.5	7.5 - 9.1	6
Bill width (mm)	7.8 ± 0.4	7.3 - 8.1	3	6.5 ± 0.6	5.8 - 7.0	3	7.1 ± 0.5	5.8 - 8.1	6
Weight (g)	37.7 ± 1.8	36.1 - 40.2	4	33.1 ± 1.0	32.0 - 33.9	3	35.4 ± 1.4	32.0 - 40.2	7

MORPHOLOGY

The Yellow-bridled Finch is a granivorous bird with a thick bill that in the Omora Park reaches 8.8 mm high and 7.8 mm wide. It has a pronounced sexual dimorphism. The males have a throat and an area between the bill and the eyes that is black. The eyes have yellow around them. The rest of the body is ashy grey. In the subspecies that inhabits Tierra del Fuego and Cape Horn, the external tail feathers are yellow. The female is light brown with dark streaks through the body.

DIET

Little is known about its diet. Because of its morphology and behavior, it is assumed that it eats seeds that it find on the ground, as has been described in Tierra del Fuego.

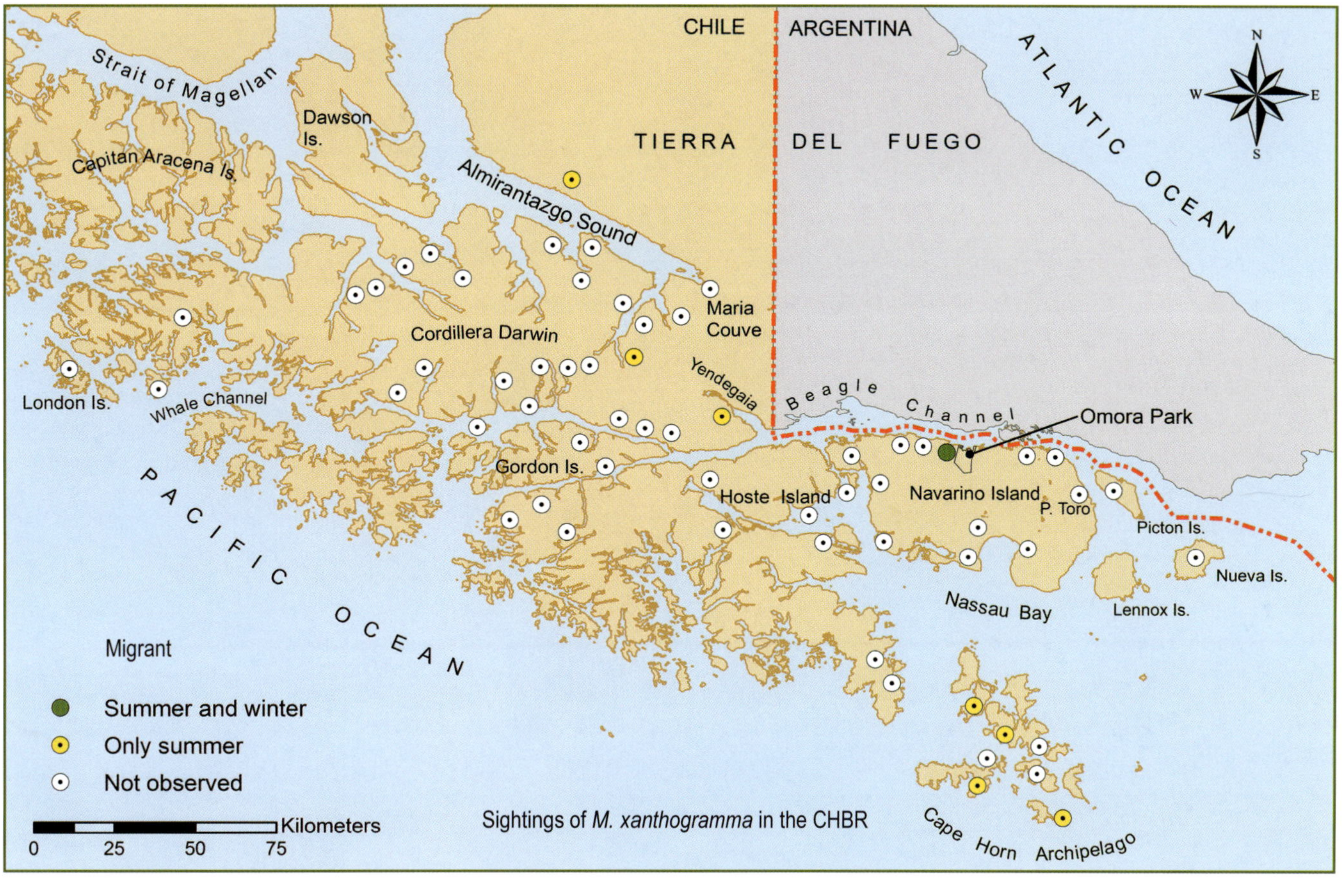

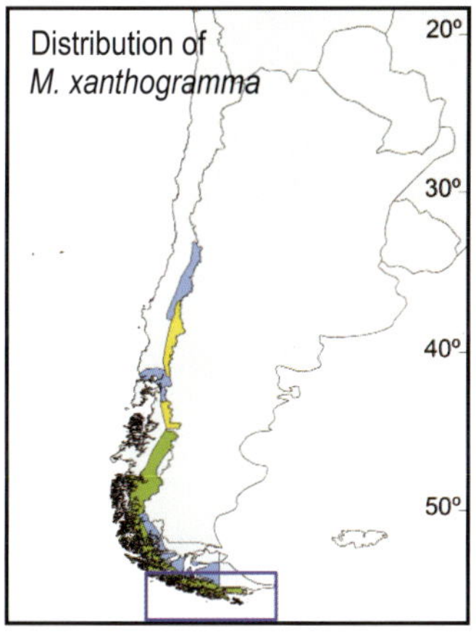

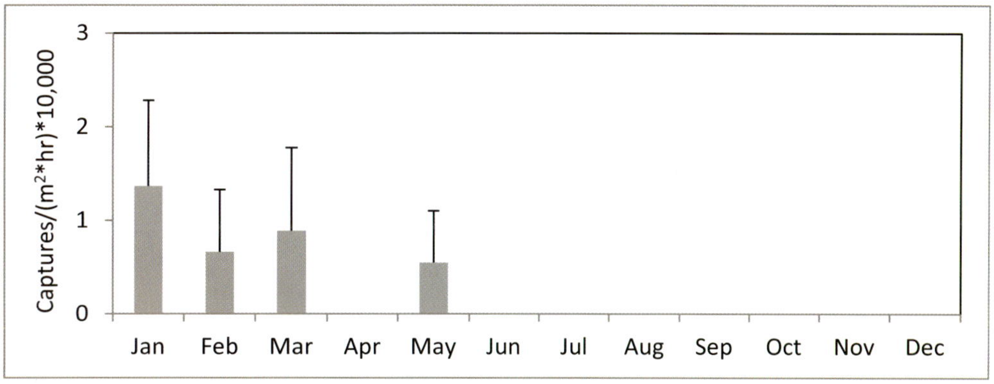

Mean number of captures of the Yellow-bridled Finch standardized by monthly net surface area and hours of capture effort of mist-nets between 2003 and 2010 in the Omora Park.

DISTRIBUTION AND HABITAT

It inhabits high Andean environments, both rocky with cushion plants and low shrubs, and meadows. On Navarino Island, it goes down to the coast in winter. It is distributed along the Andes Mountain Range from the Metropolitan Region to Cape Horn. Two subspecies are recognized; *M. xanthogramma xanthogramma* is found in the CHBR.

HABITS AND ECOLOGY

It is resident on Navarino Island. During the winter the populations of the Cape Horn Archipelago move toward the more protected areas of the CHBR. It is a ground dweller and is silent, lives solitary or in pairs, and occasionally forms small flocks. Its ecology is little known.

CONSERVATION

Rare species with restricted distributional range, but considered by the IUCN as of Least Concern.

D

SECTION

EXOTIC PASSERINES CAPTURED WITH MIST-NETS IN THE SUB-ANTARCTIC FORESTS AT OMORA PARK AND PUERTO WILLIAMS ON NAVARINO ISLAND

PLOCEIDAE

House Sparrow
Gorrión
Passer domesticus

MIST-NET DATA
Number of captures: 5
% of total captures: 0.06
Different individuals captured: 5
% individuals recaptured: 0.0
Maximum recorded longevity (years): No recaptures

MORPHOMETRICAL MEASUREMENTS	MEAN ± 1SD	MINIMUM– MAXIMUM	N
Total length (mm)	-	-	-
Tail length (mm)	56	56	1
Tarsus length (mm)	21.4 ± 1.7	18.9 - 23.1	5
Wing length (mm)	76.8 ± 3.1	72.4 - 80.0	5
Head and bill length (mm)	27.5 ± 3.0	24.9 - 30.8	3
Bill length (mm)	9.5 ± 1.4	7.9 - 11.4	4
Bill depth (mm)	5.8 ± 0.4	5.50 - 6.2	3
Bill width (mm)	5.0 ± 0.5	4.4 - 5.3	3
Weight (g)	23.9 ± 4.9	20.0 - 31.6	5

MORPHOLOGY
The House Sparrow is a bird with a granivorous bill and robust body. On Navarino Island its average weight is 23.9 g and its beak is 9.5 mm long, 6 mm high, and 5 mm wide. The males have a grey crown, a black bib, and a reddish back. The females are light brown and duller.

DIET
It is an omnivorous and opportunistic species that feeds on leftover food around houses, seeds, and insects.

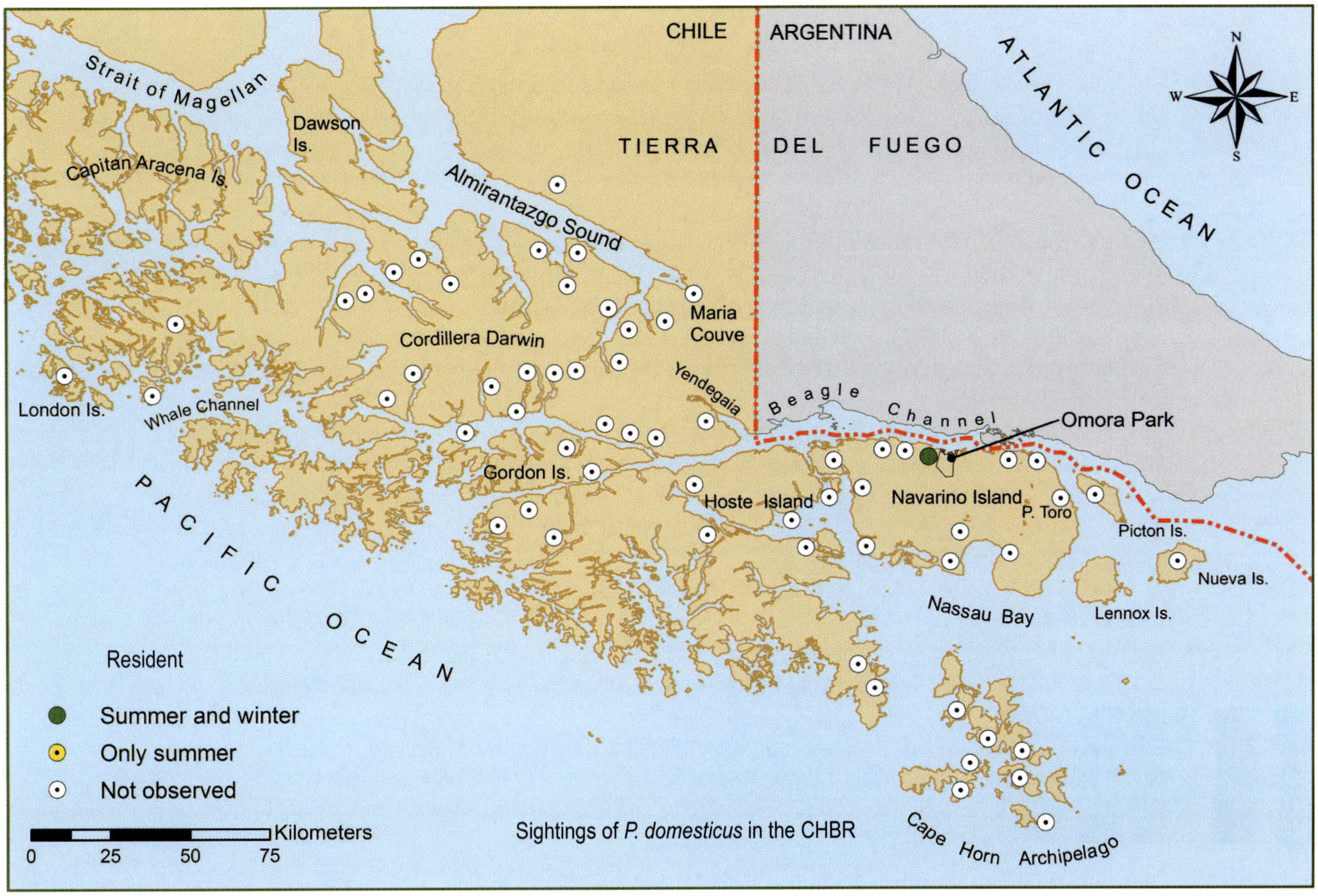

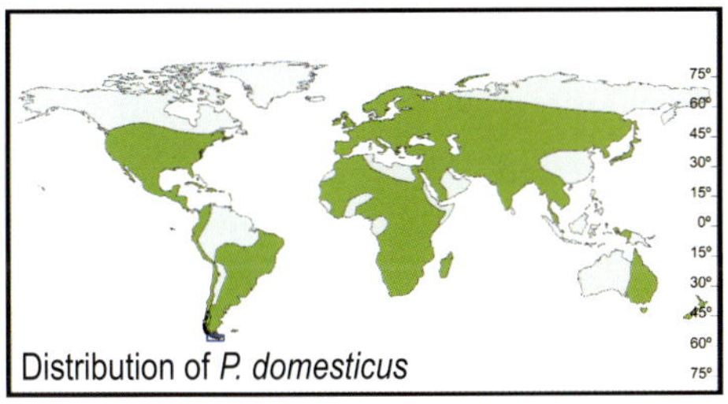
Distribution of *P. domesticus*

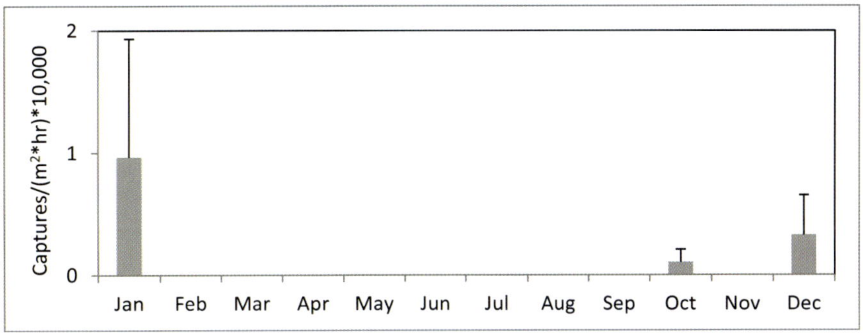

Mean number of captures of the House Sparrow standardized by monthly net surface area and hours of capture effort of mist-nets between 2003 and 2010 in the Omora Park.

DISTRIBUTION AND HABITAT

It is a cosmopolitan species, introduced to Chile from Europe in 1904. The subspecies *P. domesticus domesticus* is distributed from Arica to Navarino Island. In the CHBR, we have not detected it in the Cape Horn Archipelago, nor on the islands toward the Pacific Ocean. However, we have occasionally captured it in disturbed habitats in the Omora Park. It is abundant in Puerto Williams.

HABITS AND ECOLOGY

It is a resident species on Navarino Island. It is aggressive with other species, driving the Chilean Swallow from its nests when they are located in human structures, and displacing the Rufous-collared Sparrow from food. They are very successful in urban environments.

CONSERVATION

It is the only introduced bird species recorded in Omora Park.

REFERENCES

ALLEN, T.B. (ED.). 1983. *Field Guide to the Birds of North America*. National Geographic Society, Washington.

ARAYA, B., G. MILLIE & M. BERNAL. 1986. *Guía de Campo de las Aves de Chile*. Editorial Universitaria, Santiago, Chile.

ARAYA, B. & S. CHESTER. 1993. *The Birds of Chile*. Latour, Santiago, Chile.

BARROS, A. 1971. Aves observadas en las Islas Picton, Nueva, Lennox y Navarino Oriental. *Anales del Instituto de la Patagonia* 2: 166-180.

BARROS, V.A. 1976. Nuevas aves observadas en las Islas Picton, Nueva, Lennox y Navarino oriental. *Anales del Instituto de la Patagonia* 7: 190-193.

BLAKE, E.R. 1977. *Manual of Neotropical Birds*. University of Chicago Press, Chicago.

BROWN, C.E., C.B. ANDERSON, S. IPPI, M.F. SHERRIFFS & R. ROZZI. 2007. The autecology of the Fío-fío (*Elaenia albiceps* Lafresnaye & d'Orbigny) in Subantarctic forests of the Cape Horn Biosphere Reserve, Chile. *Anales del Instituto de la Patagonia* 35: 29-40.

BUTLER, T.Y. 1979. *The Birds of Ecuador and the Galapagos Archipelago: A Checklist of all the Birds Known in Ecuador and the Galapagos Archipelago and a Guide to Help Locate and See Them*. Ramphastos Agency, Portsmouth, N.H.

CLARK, R. 1986. *Aves de Tierra del Fuego y Cabo de Hornos. Guía de Campo.* L.O.L.A., Buenos Aires, Argentina.

CORNELIUS, C., H. COFRÉ & P.A. MARQUET. 2000. Effects of habitat fragmentation on bird species in a relict temperate forest in semiarid Chile. *Conservation Biology* 14: 534-543.

CORREA, A., J.J. ARMESTO, R. SCHLATTER, R. ROZZI & J.C. TORRES-MURA. 1990. La dieta del chucao (*Scelorchilus rubecula*), un paseriforme terrícola del bosque templado húmedo de Sudamérica Austral. *Revista Chilena de Historia Natural* 63: 197-202.

COUVE, E. & C. VIDAL. 2000. *Aves del Canal Beagle y Cabo de Hornos*. Editorial Fantástico Sur Birding LTDA, Punta Arenas, Chile.

COUVE, E. & C. VIDAL. 2003. *Birds of Patagonia, Tierra del Fuego & Antarctic Peninsula*. Editorial Fantástico Sur Birding Ltda., Punta Arenas, Chile.

DAVIS, L.I. 1972. *A Field Guide to the Birds of Mexico and Central America*. University of Texas Press, Austin.

DE LA PEÑA, M. 1988. *Guía de Aves Argentinas. Columbiformes a Piciformes*. Facultad de Agronomía y Veterinaria, U.N.L.

DE LA PEÑA, R. & M. RUMBOLL. 1998. *Birds of Southern South America and Antarctica*. Collins Illustrated Checklist. HarperCollins Publishers, Rotolito Lombarda, Italy.

DE SANTO, T.L., M.F. WILLSON, K.E. SIEVING & J.J. ARMESTO. 2002. Nesting biology of tapaculos (Rhinocryptidae) in fragmented South-Temperate rainforests of Chile. *Condor* 104: 482-495.

DE SCHAUENSEE, R.M. 1964. *The Birds of Colombia, and Adjacent Areas of South and Central America*. Livingston Pub. Co. Narberth, PA.

DE SCHAUENSEE, R.M. 1966. *The Species of Birds of South America and their Distribution*. Academy of Natural Sciences; dist. by Livingston, Narberth, PA., Philadelphia.

DE SCHAUENSEE, R. 1970. *A Guide to the Birds of South America*. Academy of Natural Sciences of Philadelphia by Livingston Pub. Co., Wynnewood, PA.

DE SCHAUENSEE, R.M. & W.H. PHELPS, JR. 1978. *A Guide to the Birds of Venezuela*. Princeton University Press, Princeton, N.J.

DEL HOYO, J., A. ELLIOTT & J. SARGATAL (EDS.). 1992. *Handbook of the Birds of the World*. Volume 1: Ostrichs to Ducks. Lynx Editions, Barcelona.

DEL HOYO, J., A. ELLIOTT & J. SARGATAL (EDS.). 1994. *Handbook of the Birds of the World*. Volume 2: New World Vultures to Guineafowls. Lynx Editions, Barcelona.

DEL HOYO, J., A. ELLIOTT & J. SARGATAL (EDS.). 1996. *Handbook of the Birds of the World*. Volume 3: Hoatzin to Auks. Lynx Edicions, Barcelona.

DEL HOYO, J., A. ELLIOTT & J. SARGATAL (EDS.). 1997. *Handbook of the Birds of the World*. Volume 4: Sandgrouse to Cukoos. Lynx Editions, Barcelona.

DEL HOYO, J., A. ELLIOTT & J. SARGATAL (EDS.). 1999. *Handbook of the Birds of the World*. Volume 5: Barn Owls to Hummingbirds. Lynx Editions, Barcelona.

DEL HOYO, J., A. ELLIOTT & J. SARGATAL (EDS.). 2001. *Handbook of the Birds of the World*. Volume 6: Mousebirds to Hornbills. Lynx Editions, Barcelona.

DÍAZ, I.A., J.J. ARMESTO, S. REID, K.E. SIEVING & M.F. WILLSON. 2005. Linking forest structure and composition: avian diversity in successional forests of Chiloé Island, Chile. *Biological Conservation* 123: 91-101.

DUNNING, J.S. 1982. *South American Land Birds*. Harrowood Book, Newtown Square, Pennsylvania.

EGLI, G. 1987. Anillamiento, captura y recaptura de aves en una comunidad ornitológica del matorral centro chileno. *Boletín Informativo Unión de Ornitólogos de Chile* (UNORCH) (Chile) (3): 8-15.

EGLI, G. 1996. Biomorfología de algunas aves de Chile central. *Boletín Chileno de Ornitología* 3: 2-9.

EGLI, G. & J. Aguirre. 2000. *Aves de Santiago*. UNORCH, Z&D Servicios Gráficos, Chile.

ELPHICK, C., J.B. DUNNING JR. & D. A. SIBLEY (EDS.). 2001. *The Sibley Guide to Bird Life & Behavior*. National Audubon Society, Alfred A. Knopf, New York.

ERAZO, S. 1984. Análisis de censos de avifauna realizados en un rodal boscoso de olivillo, Valdivia, Chile. *Revista Geográfica de Valparaíso (Chile)* 15: 49-71.

FJELDSÅ, J. & N. KRABBE. 1990. *Birds of the High Andes*. Zoological Museum and Apollo Books, Copenhagen and Stenstrup, Denmark.

GANTZ, A. & J. RAU. 1999. Relación entre el tamaño mínimo de fragmentos boscosos y su riqueza de especies de aves en el sur de Chile. *Anales del Museo de Historia Natural de Valparaíso* 24: 85-90.

GARCÍA, J.A. 1982. *Comunidad Avifaunística del Delta Río Gol-Gol. Una Necesidad de Conservación.* Tesis de Ingeniería Forestal, Universidad Austral de Chile, Valdivia. 126 pp.

GARRIDO, O.H. & F. GARCÍA-MONTAÑA. 1975. *Catálogo de las Aves de Cuba.* La Habana, Cuba.

GLADE, A. 1993. *Red book of Chilean Terrestrial Vertebrates.* 2nd ed. Corporación Nacional Forestal, Santiago, Chile.

GOODALL, J.D., A.W. JOHNSON & R.A. PHILIPPI. 1946. *Las Aves de Chile.* Volumen I. Platt Establecimientos Gráficos S.A., Buenos Aires, Argentina.

GOODALL, J.D., A.W. JOHNSON & R.A. PHILIPPI. 1951. *Las Aves de Chile.* Volumen II. Platt Establecimientos Gráficos S.A., Buenos Aires, Argentina.

HEINTZELMAN, D.S. 1979. *Hawks and Owls of North America: A Complete Guide to North American Birds of Prey.* Universe Books, N.Y.

HELLMAYR, C.E. 1932. *The Birds of Chile.* Publication 308, Field Museum of Natural History, Zoological Series 29: 1-472.

HOFFMANN, A. & I. LAZO. 2000. *Aves de Chile. Un Libro También para Niños.* Universidad Andrés Bello, Ril Editores, Santiago.

HOUSSE, R.E. 1945. *Las Aves de Chile en su Clasificación Moderna. Su Vida y Costumbres.* Ediciones Universidad de Chile, Santiago.

HUMPHREY, P.S., D. BRIDGE, P.W. REYNOLDS & R.T. PETERSON. 1970. *Birds of Isla Grande (Tierra del Fuego).* Preliminary Smithsonian Manual, Museum of Natural History, University of Kansas, Lawrence, Kansas, USA. 411 pp.

IBARRA, J.T., T. ALTAMIRANO, N. GÁLVEZ, I. ROJAS, J. LAKER & C. BONACIC. 2010. Avifauna de los bosques templados de Araucaria araucana del sur de Chile. *Ecología Austral* 20: 33-45.

IBARRA, J.T., A. BARREAU, F. MASSARDO & R. ROZZI. 2012. El cóndor andino: una especie biocultural clave del paisaje sudamericano. *Boletín Chileno de Ornitología* 18(1-2): 1-22.

IPPI, S., C.B. ANDERSON, R. ROZZI & C.S. ELPHICK. 2009. Variación anual de la abundancia y la composición de los ensambles de aves de bosque en isla Navarino, Reserva de Biosfera Cabo de Hornos, Chile. *Ornitología Neotropical* 20: 231-245.

IPPI, S., R.A. VÁSQUEZ, J. MORENO, S. MERINO & C.P. VILLAVICENCIO. 2012. Breeding biology of the Southern house wren on Chiloé island, southern Chile. *Wilson Journal of Ornithology* 124: 531–537.

IUCN. 2013. *IUCN Red List of Threatened Species. Version 2013.* 2. <www.iucnredlist.org>. Downloaded on 23 November 2013.

JAKSIC, F.M. & P. FEINSINGER. 1991. Bird assemblages in temperate forests of North and South America: a comparison of diversity, dynamics, guild structure and resource use. *Revista Chilena de Historia Natural* 64: 491-510.

JARAMILLO, A. 2003. *Birds of Chile.* Princeton Univ. Press, Princeton, New Jersey. 240 pp.

JARAMILLO, A. 2005. *Aves de Chile.* Lynx, Barcelona, España.

JARVINEN, O. & R.A. VAISÄNEN. 1977. Long term changes of North European land bird fauna. *Oikos* 29: 225-228.

JIMÉNEZ, J.E. & F.M. JAKSIC. 1989. Biology of the Austral Pygmy-Owl. *Wilson Bulletin* 101: 377 389.

JIMÉNEZ, J.E., R. D. CREGO, G.E. SOTO, I. ROMÁN, R. ROZZI & P.M. VERGARA. 2013. Potential impact of the alien American mink (*Neovison vison*) on Magellanic woodpeckers (*Campephilus magallanicus*) in Navarino Island, southern Chile. *Biological Invasions*: DOI 10.1007/s10530-013-0549-1.

JOHNSGARD, P.A. 1990. *Hawks, Eagles and Falcons of North America.* The Smithsonian Institute Press, Washington D.C.

JOHNSON, A.W. 1965. *Birds of Chile and Adjacent Regions of Argentina, Bolivia and Peru.* Volume I. Platt Establecimientos Gráficos, Buenos Aires, Argentina. 398 pp.

JOHNSON, A.W. 1967. *Birds of Chile and Adjacent Regions of Argentina, Bolivia and Peru.* Volume II. Platt Establecimientos Gráficos, Buenos Aires, Argentina. 447 pp.

KELT, D.A. 2001. Differential effects of habitat fragmentation on birds and mammals in Valdivian temperate rainforests. *Revista Chilena de Historia Natural* 74: 769-777.

LAND, H. 1970. *Birds of Guatemala.* Published for the International Committee for Bird Preservation, Pan-American Section by Livingston Pub. Co., Wynnewood, Pa.

MARTÍNEZ, D. & G. GONZÁLEZ. 2004. *Las Aves de Chile. Nueva Guía de Campo.* Ediciones del Naturalista, Santiago, Chile.

MCGEHEE, S.M. 2007. Consumption of Leñadura (*Maytenus magellanica*) seeds by three primarily insectivorous bird species. *Anales del Instituto de la Patagonia* 35: 71-73.

MCGEHEE, S.M., R. ROZZI, C.B. ANDERSON, S. IPPI, R.A. VÁSQUEZ & S. WOODLAND. 2004. Late-summer presence of the Patagonian Tyrant, *Colorhamphus parvirostris* (Darwin) in Navarino island, Cape Horn County, Chile. *Anales del Instituto de la Patagonia* 32: 25-33.

MCGEHEE, S.M. & J.C. EITNIEAR. 2007. Diet of the Patagonian Sierra-Finch (*Phrygilus patagonicus*) on Navarino island, Chile. *Ornitología Neotropical* 18: 449-452.

MONROE, B.L. JR. 1968. *A Distributional Survey of the Birds of Honduras.* Ornithological Monographs N°7. American Ornithologists' Union, N.Y.

MOREIRA, A. 1999. *Guía de Campo. Caleu y Cerro El Roble*. Asociación de Comuneros La Capilla de Caleu, Chile.
MORENO, J., S. MERINO, E. LOBATO, M.A. RODRÍGUEZ-GIRONÉS & R.A. VÁSQUEZ. 2007. Sexual dimorphism and parental roles in the Thorn-tailed Rayadito (Furnariidae). *Condor* 109: 312-320.
NAROSKY, T. & D. YZURIETA. 1987. *Guía para la Identificación de las Aves de Argentina y Uruguay*. Vazquez Mazzini Editores, Buenos Aires.
NAROSKY, T. & A. BOSSO. 1995. *Manual del Observador de Aves*. Editorial Albatros Saci, Buenos Aires.
NAROSKY, T. & M. BABARSKAS. 2000. *Aves de la Patagonia. Guía para su Reconocimiento*. Vásquez Mazzini Editores, Buenos Aires.
NAROSKY, T. & M. BABARSKAS. 2000. *Guía de Aves de Patagonia y Tierra del Fuego*. Vázquez Mazzini Editores, Buenos Aires, Argentina.
OJEDA, V. 2004. Breeding biology and social behaviour of Magellanic woodpeckers (*Campephilus magellanicus*) in Argentine Patagonia. *European Journal of Wildlife Research* 50: 18–24
OLROG, C.C. 1963. *Lista y Distribución de las Aves Argentinas*. Tucumán, Argentina.
ORTIZ, D. & P. CAPLLONCH. 2011. La migración del chingolo (*Zonotrichia capensis*) en Argentina. *Historia Natural, tercera serie* 1: 105-109.
PARKER, T.A. III, D.F. STOTZ & J.W. FITZPATRICK. 1996. *Neotropical Birds: Ecology and Conservation*. University of Chicago Press, Chicago.
PAYNTER, R.A. JR. 1982. *Ornithological Gazetteer of Venezuela*. Bird Dept., Museum of Comparative Zoology, Harvard University, Cambridge, Mass.
PAYNTER, R.A. Jr. 1989. *Ornithological Gazetteer of Paraguay*. Bird Dept., Museum of Comparative Zoology, Harvard University, Cambridge, Mass.
PAYNTER, R.A., JR., M.A. TRAYLOR, JR. & B. WINTER. 1975. *Ornithological Gazetteer of Bolivia*. Bird Dept., Museum of Comparative Zoology, Harvard University, Cambridge, Mass.
PAYNTER, R.A. & M.A. TRAYLOR, JR. 1977. *Ornithological Gazetteer of Ecuador*. Bird Dept., Museum of Comparative Zoology, Harvard University, Cambridge, Mass.
PAYNTER, R.A. & M.A. TRAYLOR, JR. 1981. *Ornithological Gazetteer of Colombia*. Bird Dept., Museum of Comparative Zoology, Harvard University, Cambridge, Mass.
PERRINS, C.M. 1990. *The Illustrated Encyclopedia of Birds: The Definitive Reference to Birds of the World*. Prentice Hall Editions, N.Y.
PINTO, O.M.O. 1960. *Beija-flores do Brasil*. Livraria São José, Rio de Janeiro.
RAND, D.M. & R.A. PAYNTER, JR. 1981. *Ornithological Gazetteer of Uruguay*. Bird Dept., Museum of Comparative Zoology, Harvard University, Cambridge, Mass.
REID, S., C. CORNELIUS, O. BARBOSA, C. MEYNARD, C. SILVA-GARCIA & P.A. MARQUET. 2002. Conservation of temperate forest birds in Chile: Implications from the study of an isolated forest relict. *Biodiversity and Conservation* 11: 1975-1990.
REMSEN, JR, J.V. & M.A. TRAYLOR, JR. 1989. *An Annotated List of the Birds of Bolivia*. Buteo Books, Vermillion, South Dakota.
RIDGELY, R.S. & G. TUDOR. 1989. *The Birds of South America*. Volume 1, The Oscine Passerines. University of Texas Press, Austin.
RIDGELY, R.S. & J.A. GWYNNE, JR. 1989. *A Guide to the Birds of Panama, with Costa Rica, Nicaragua, and Honduras*. 2nd ed. Princeton University Press, Princeton, N.J.
RIDGELY, R.S. & G. TUDOR. 1994. *The Birds of South America*. Volume 2. University of Texas Press, Austin, Texas.
ROBBINS, C.S., B. BRUUN & H.S. ZIM. 1983. *Birds of North America: A Guide to Field Identification*. Golden Press, N.Y.
ROZZI, R., D. MARTÍNEZ, M.F. WILLSON & C. SABAG. 1996. Avifauna de los Bosques Templados de Sudamérica. In *Ecología de los Bosques Nativos de Chile* (J.J. Armesto, C. Villagrán & M.T. Kalin, eds.), pp: 135-152. Editorial Universitaria, Santiago, Chile.
ROZZI, R., J.J. ARMESTO, A. CORREA, J.C. TORRES-MURA & M. SALLABERRY. 1996. Avifauna de bosques primarios templados en islas deshabitadas del archipiélago de Chiloé. *Revista Chilena de Historia Natural* 69: 125-139.
ROZZI, R. & M. SHERRIFFS. 2003. El visón (*Mustela vison*, Schreber) un nuevo mamífero exótico para la isla Navarino. *Anales del Instituto de la Patagonia (Serie Ciencias Naturales)* 31: 97-104.
ROZZI, R., F. MASSARDO, C.B. ANDERSON, G. CLARK, G. EGLI, S. MCGEHEE & E. RAMILO. 2010. *Multi-ethnic Bird Guide of the Sub-Antarctic Forests of South America*. University fo North Texas Press y Universidad de Magallanes. 235 pp.
SCHLATTER, R. & P. VERGARA. 2005. Magellanic Woodpecker (*Campephilus magellanicus*) sap feeding and its role in the Tierra del Fuego forest bird assemblage. *Journal of Ornithology* 146: 188-190.
SERVICIO AGRÍCOLA Y GANADERO (SAG). 2009. *La Ley de Caza y su Reglamento*. Ministerio de Agricultura, República de Chile.
SHORT , L. L. 1970. The habits and relationships of the Magellanic woodpecker. *Wilson Bulletin* 82: 115–129.
SIBLEY, C.G. & B.L. MONROE, JR. 1990. *Distribution and Taxonomy of Birds of the World*. Yale University Press, N.H.
SNYDER, D.E. 1966. *The Birds of Guyana (formerly British Guiana); A Checklist of 720 Species, with Brief Descriptions, Voice and Distribution*. Peabody Museum, Salem.
STEPHENS, L. & M.A. TRAYLOR, JR. 1985. *Ornithological Gazetteer of the Guianas*. Bird Dept., Museum of Comparative Zoology, Harvard University, Cambridge, Mass.
STILES, F.G. & A.F. SKUTCH. 1989. *A Guide to the Birds of Costa Rica*. Comstock, Ithaca.
TEXERA, W. 1972. Distribución y diversidad de mamíferos y aves en la Provincia de Magallanes. Análisis preliminar de la diversidad ecológica y variación taxonómica. *Anales del Instituto de la Patagonia (Chile)* 3: 171-200.

TEXERA, W. 1973. Distribución y diversidad de mamíferos y aves en la Provincia de Magallanes. II: Algunas notas ecológicas sobre los canales patagónicos. *Anales del Instituto de la Patagonia (Chile)* 4(1-3): 291-305.

VENEGAS, C. 1981. Aves de las Islas Wollaston y Bayly, Archipiélago del Cabo de Hornos. *Anales del Instituto de la Patagonia* 12: 213-219.

VENEGAS, C. 1991. Ensambles avifaunísticos estivales del Archipiélago Cabo de Hornos. *Anales del Instituto de la Patagonia* 20: 69-81.

VENEGAS, C. 1994. Aves de Magallanes. Ediciones de la Universidad de Magallanes, Punta Arenas, Chile.

VERGARA, P. & R. P., SCHLATTER. 2004. Magellanic Woodpecker (*Campephilus magellanicus*) abundance and foraging in Tierra del Fuego, Chile. *Journal of Ornithology* 145: 343- 351.

VERGARA, P.M. & J.J. ARMESTO. 2009. Responses of Chilean forest birds to anthropogenic habitat fragmentation across spatial scales. *Landscape Ecology* 24: 25-38.

VERGARA, P.M., J. E. JIMÉNEZ & R. P. SCHLATTER. 2010. Effective point-count duration for estimating bird species' richness in Chilean forests. *Zoological Studies* 49: 381-391.

VIGIL, C. 1973. Aves Argentinas y Sudamericanas. Editorial Atlántida, Buenos Aires.

VUILLEUMIER, F. 1985. Forest birds of Patagonia: ecological geography, speciation, endemism, and faunal history. *Neotropical Ornithology, Ornithological Monographs* 36: 255-304.

WILLSON, M.F. 1991. Dispersal of seeds by frugivorous animals in temperate forests. Revista Chilena de Historia Natural 64: 537-554.

WILLSON, M.F., T.L. DE SANTO, C. SABAG & J.J. ARMESTO. 1994. Avian communities of fragmented south-temperate rainforests in Chile. *Conservation Biology* 8:508-520.

WILLSON, M., T.L. DE SANTO, C. SABAG & J.J. ARMESTO. 1996. Avian communities in temperate rainforests of North and South America. Pp. 228-246 in *High-Latitude Rainforests and Associated Ecosystems of the West Coasts of the Americas. Climate, Hydrology, Ecology and Conservation*. Lawford, R.G., P.B. Alaback & E. Fuentes (eds.), Springer, New York.

WOODS, R.W. 1975. *The Birds of the Falkland Islands*. Anthony Nelson, Oswestry.

ZALLES, J.I. & K.L. BILDSTEIN (EDS.). 2000. *Raptor Watch: A Global Directory of Raptor Migration Sites*. Birdlife International, Cambridge.

WEBPAGES

Aves de Chile:
http://www.avesdechile.cl/

UNEP-WCMC Species Database:
http://www.unep-wcmc.org/isdb/Taxonomy/index.cfm?displaylanguage=ENG

The IUCN Red List of Threatened Species:
http://www.iucnredlist.org/apps/redlist/details/141778/0

Birdlife International:
http://www.birdlife.org/

Long-Term Socio-Ecological Research Sites (LTERS-Chile):
http://www.ieb-chile.cl/index.php/main/seccion/132

BOSQUES EN MINIATURA DEL CABO DE HORNOS
MINIATURE FORESTS OF CAPE HORN

3

A LONG-TERM PROGRAM OF BIOCULTURAL RESEARCH, EDUCATION, AND CONSERVATION AT THE SOUTHERN END OF THE AMERICAS

Ricardo Rozzi[1,2,3], Francisca Massardo[1,2], and Orlando Dollenz[1]

[1]Universidad de Magallanes, Punta Arenas, Chile
[2]Omora NGO, Puerto Williams, Chile
[3]Department of Philosophy and Religion Studies, University of North Texas, Denton, Texas, USA

Austro Universitaria 14: 50-59 (2003)

Originally published in Spanish as **Un programa de investigación, educación y conservación biocultural a largo plazo en el extremo austral de América.**

I. Introduction

Since 1998, a group of researchers from the University of Magallanes (UMAG) and other associated research institutions have developed a program called Ecology and Conservation of the Magellan Archipelago (ECOBAMAG). This program, among other things, has contributed to: 1) identifying the Magellanic sub-Antarctic or sub-polar ecoregion forests, as one of the 37 most pristine ecoregions of the world[1]; 2) identifying the Magellanic sub-Antarctic ecoregion as a "Biodiversity Hotspot" for non-vascular flora at the national and worldwide level[2]; and 3) creating the Omora Ethnobotanical Park.[3] This park is a biological reserve near Puerto Williams, Navarino Island, whose objective is to contribute to social and ecological well-being and the conservation of biocultural diversity at the southern tip of the Americas.

To concisely introduce the current scenario of the Cape Horn County and to present the motivations that drive Omora Ethnobotanical Park initiative, we begin this article with the speech by UMAG researcher, Dr. Ricardo Rozzi, at the ceremony opening the international passage between Chile and Argentina through the Beagle Channel. This ceremony took place on December 5, 2001, in Puerto Navarino, and was attended by senior government officials from Chile and Argentina, including local government, Navy authorities, Ministry of Foreign Affairs (*e.g.*, DIFROL), Customs Service, the Chilean National Tourism Service (SERNATUR), and other distinguished representatives from both countries to sign the agreement that opened the navigation path crossing the Beagle Channel between the towns of Puerto Navarino (Chile) and Ushuaia (Argentina). This agreement, without a doubt, will influence the historical destiny of Navarino Island and Cape Horn County. On one hand, it opens considerable opportunities for development. On the other hand, it generates social and environmental risks for one of the few regions of the world that still contains non-fragmented temperate forests in an almost pristine condition.[4]

In his speech, on behalf of the Omora Ethnobotanical Park, Dr. Rozzi synthesized the significance of this opening and the historical moment faced by the Cape Horn County as follows:

I invite you to look at this rock to see the diversity of mosses including the new species, *Orthotrichum omorae*, which was discovered a few weeks ago in the coastal rocks of the Navarino Island.

This small species may not surprise you as much as a tall tree, but it is precisely the unique biological and cultural character of Navarino Island and its people that inspired the naturalist Charles Darwin to think about human evolution. Navarino Island is just as important as the Galapagos Islands in inspiring the theory of evolution, and it is our responsibility to communicate this fact for education and tourism purposes.

It was also here, in Navarino, where the archaeologist Junius Bird, in the 1930s, began the work that changed the view on the native people of America from his findings in southern middens. Moreover, it was here in the far south where the Vienna School of Anthropology through the work of Martin Gusinde on the sophisticated Yahgan culture transformed the notion of linear progress in human history.

[1] The areas identified at world level have: (a) 70% or more of the intact original vegetation, (b) more than 10.000 km^2 of extension, and (c) have population density less than 5 inhabitant/km^2. See Mittermeier, R.A. et al. 2002. *Wilderness: Earth's Last Wild Places.* CEMEX – Conservation International, Washington DC; and Sanderson, E.W. *et al.* (2002), The human footprint and the last of the wild. *BioScience* 52: 891-904.

[2] See Rozzi, R. 2002. *Biological and Cultural Conservation in the Archipelago Forest Ecosystems of Southern Chile.* Ph.D. Dissertation, Department of Ecology and Evolutionary Biology, University of Connecticut, USA.

[3] See Rozzi R., F. Massardo, J. Silander Jr., C. Anderson, O. Dollenz & A. Marin. 2003. El Parque Etnobotánico Omora: una alianza público–privada para la conservación biocultural en el confín del mundo. *Ambiente & Desarrollo* XIX (1): 43-55.

[4] Silander, J.A., Jr. 2000. Temperate forests: plant species biodiversity and conservation, *in Encyclopedia of Biodiversity*, S.A. Levin, ed., Academic Press, New York, pp.: 607-626.

4 THE OMORA BIRD BANDING PROGRAM IN THE SUB-ANTARCTIC FORESTS: STANDARDIZATION OF THE APPROPRIATE BAND SIZES FOR THE BIRDS OF THE MAGELLANIC REGION

CHRISTOPHER ANDERSON[1,2], RICARDO ROZZI[1,3,4], CHRISTOPHER ELPHICK[4], AND STEVEN MCGEHEE[1]

[1]ONG Omora, Puerto Williams, Isla Navarino, Región de Magallanes, Chile www.geocities.com/omorapark/home
[2]Institute of Ecology, University of Georgia, Athens, GA 30602 USA, cba@uga.edu
[3]Departamento de Recursos y Ciencias Naturales, Universidad de Magallanes, Casilla 113–D, Punta Arenas, Región de Magallanes, Chile
[4]Department of Ecology and Evolutionary Biology, University of Connecticut, Storrs, CT 06269–30443 USA

Boletín Chileno de Ornitología 9: 2-11 (2002) (Originally published in Spanish)

ABSTRACT. The Omora bird banding program in the sub-Antarctic forests: standardization of the appropriate band sizes for the birds of the Magellanic Region. In January 2000 the Omora NGO began a bird banding program in the sub–Antarctic forests of the Magallanes region. In the context of this long-term study, concerning the morphology, composition, territoriality, migratory routes, nesting, diet and other ecological characteristics of the forest avifauna, we detected that the scarce information regarding the appropriate size of bands for each species was a serious methodological limitation for banding studies in Chile. Here, we define the band sizes (the internal diameter and height) for the most characteristic birds of the Magellanic forests, based on captures conducted at isla Navarino and península Antonio Varas and from museum specimens. Additionally, we provide information on captures and recaptures, and morphological and body mass data for 18 and 16 species, respectively.

Key words: Banding, band size, Magallanes, morphology.

RESUMEN. El Programa Omora de anillamiento de aves en los bosques subantárticos: estandarizacion del tamaño de anillos apropiados para las aves de la región de Magallanes. En enero del año 2000 la ONG Omora inició un programa de anillamiento de aves de los bosques subantárticos en la Región de Magallanes. En el marco de este estudio a largo plazo sobre la morfología, composición, territorialidad, patrones migratorios, nidificación, dieta y otras características ecológicas de la avifauna forestal, detectamos que una seria limitación metodológica para los estudios de anillamiento de aves, en la actualidad en Chile, es la escasa información acerca del tamaño de anillos apropiados para cada una de las especies de aves. Este trabajo define tamaños de anillos de acuerdo a su diámetro interno y altura, para las aves más características de los bosques magallánicos, basándose en capturas realizadas en isla Navarino y península Antonio Varas e información de museos. Además se proporciona información sobre capturas y recapturas y datos morfológicos y de masa corporal para un total de 18 y 16 especies, respectivamente.

Palabras clave: Anillamiento, Magallanes, medidas morfológicas, tamaño de anillos.

INTRODUCTION

The composition of bird communities in temperate forests of Chile has been described for numerous areas, including the extreme South (Anderson & Rozzi 2000, Barros 1971, 1976, Humphrey *et al.* 1970, Texera 1972, Siefield 1977, Venegas 1981, 1991, 1994, Venegas & Schlatter 1999). However, morphological studies have focused only on populations of birds in shrubland and sclerophilic forests of central Chile (Elgi 1996, López-Calleja 1990), Valdivian forest (Erazo 1984, García 1987) and Olivillo forest of insular Chiloé (Espinosa & Egli 1997, Sabag 1993, Rozzi *et al.* 1996a). Morphological and body mass data provide essential information for studies of energy, trophic interactions or ecological interactions in general, as well as for the analysis of latitudinal size patterns (Atkinson & Sibly 1997).

Morphological studies of birds in the field involve captures which, if accompanied by banding associated with studies of capture and recapture, permit the investigation of essential life history variables (longevity, territoriality, etc.).

One methodological limitation of banding birds in Chile is the lack of information concerning appropriate band sizes for the distinct bird species found in the country. With the goal of improving its own banding program, in January of 2000 the Omora NGO started a study of the appropriate band size for bird species associated with Magellanic forests. The determination of these dimensions represents a central objective of this study. Additionally, we report recapture data and morphological measurements of birds obtained during the first two years of the banding program.

STUDY SITES

The banding program took place in two areas in the ecoregion of sub-Antarctic (also called sub-polar) forests of Magallanes (see Wilcox 1995, Veblen 1996, Mittermeier *et al.* 2002): the Antonio Varas Peninsula (52°S) and Navarino Island (55°S) (Figure 1). These sites include forests dominated by Magellan's Beech (*Nothofagus betuloides*) near the coast, in a mosaic of primary forest and areas of anthropogenic intervention with scrubland formations, and areas dominated by Lenga Beech (*Nothofagus pumilio*) (for a detailed description of the area see Anderson & Rozzi 2000, Rozzi 2002).

The Antonio Varas Peninsula (AV) is located south of the southern Patagonian ice field, at the head of the channels of Puerto Natales. The investigation sites are located on the coast of Talcahuano Bay (51°44'S, 72°53'W) in an area that receives between 1,000 and 1,500 mm of precipitation per year and has an annual average temperature of 7°C (McCulloch *et al.* 1997).

Navarino Island (NI) is located south of Tierra del Fuego, between the Cape Horn archipelago and Beagle Channel. The banding sites are on the northern coast of NI (Omora Ethnobotanical Park [54°57'S, 67°39'W], Guerrico Point [54°55'S, 67°54'W], Mejillones Bay [54°57'S, 67°39'W]) and the western coast of the island (Puerto Inútil [54°59'S, 68°13'W] and Wulaia [55°03'S, 68°09'W]). The sector of Williams Port on the northern coast of NI receives an average of 650 mm of precipitation annually and has an average temperature of 6°C (Rozzi 2002).

METHODS

Between January of 2000 and November of 2001, we arranged between one and five mist-nets 6 or 12 m long x 2.6 m high, with a crosslink of 30 mm in open habitats, at the edge or interior of forests. In total, we opened the nets during 1019 hours. To standardize the sampling, we calculated the area of exposed net in m^2 per unit of time. In total, we opened the mist-nets for 16,604.9 hm^2 and 24, 124.6 hm^2 (hours of net use x $meter^2$ of open net) in AV and NI, respectively. We revised the nets every 15 to 30 minutes in search of captured birds.

Each captured bird was identified, banded, and subject to five measurements (cfr. Engli 1996):

1) total length, defined as the length between the tip of the tail and the tip of the beak with the bird extended and supported dorsally over an ornithological ruler with 1 mm precision,

2) wing length, defined as the length of the extended wing from the wrist to the tip of the longest primary feather, with the wing extended over an ornithological ruler with 1 mm precision,

3) beak length, defined as the length of the culmen measured with a caliper with 0.1 mm precision,

4) tarsus length, measured with a caliper with 0.1 mm precision,

5) body mass, measured with a pesola scale with a capacity of 30 g and 0.5 g sensibility, in the case of the majority of the species. For larger species we used a balance with a 100 g capacity and 0.5 g sensibility, and a 10 g balance with 0.1 g sensibility for hummingbirds.

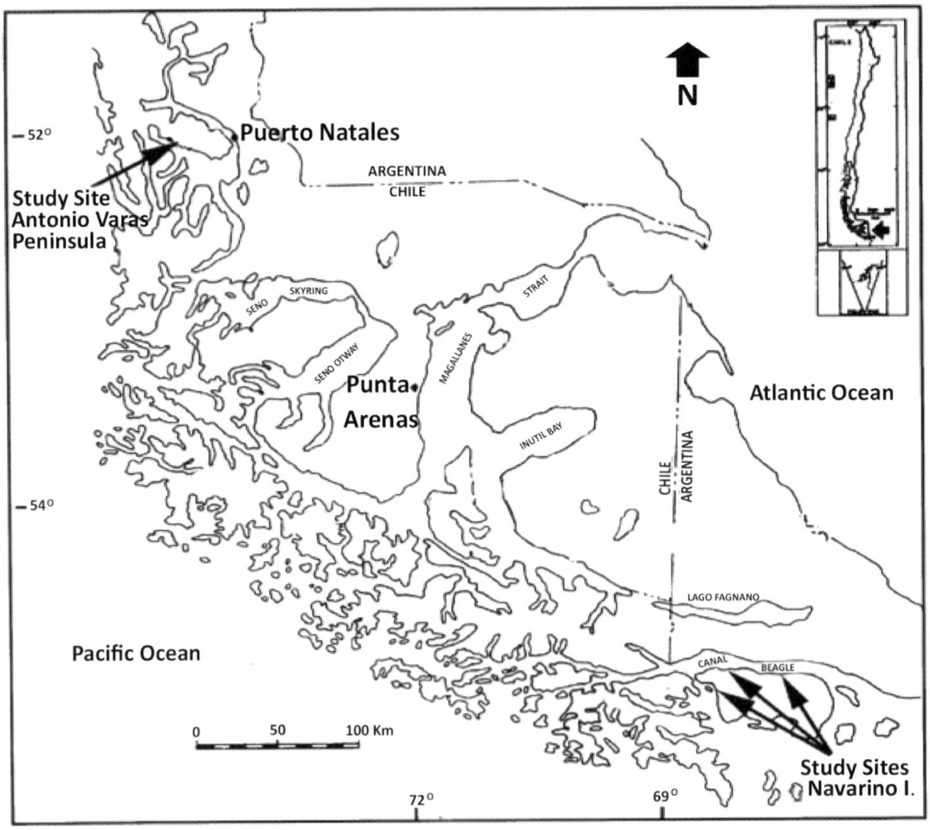

Fig. 1. Study area where mist-netting was conducted.

At the start of the study we used aluminum bands of uniform size with a 2.78 mm internal diameter, modifiable through folding one end of the band over the other upon placement on the bird's tarsus. These bands were numbered 1-500 and followed by the letters RR00 (Ricardo Rozzi year 2000), to facilitate the distinction of recaptured birds that could have been banded by other studies.

In November of 2000, we decided to use a variable-sized band set to prevent folding to obtain appropriately sized bands for each species of bird. To determine the size of the bands we used a leg gauge designed by the North American Banding Program (http://www.pwrc.usgs.gov/bbl/), of the United States Geological Survey (USGS) and the Canadian Wildlife Service (CWS). The tarsus of at least three individuals per species was measured with this instrument to determine the appropriate band size for each species. Specimens of the bird species mainly came from the collection of the Patagonian Institute, complimented with specimens of Green-backed Firecrown (*Sephanoides sephaniodes*) and Magellanic Tapaculo (*Scytalopus magellanicus*) from the National Natural History Museum collection, and measurements of birds captured in the field with mist-nets in NI and AV in November of 2000.

From January of 2001, we used bands of specific diameter for each species of bird. For each size category we obtained aluminum bands with numbers from 0000 to 1000. Unfortunately during that year it was not possible to include the abbreviation RR on the bands to facilitate the distinction of birds banded by our program. It is worth mentioning that hummingbirds require special care (Russell & Russell 2000). For each species a small band is prepared, cut from a sheet of aluminum. Each band is marked with successive numbers using an engraving instrument. Then the bands are neatly sanded to remove any sharp edges. These bands are placed on the legs of hummingbirds by using special pliers.

RESULTS AND DISCUSSION

Band Sizes

The band sizes for the 34 species studied range from USGS 8, with an internal diameter of 17.5 mm and height of 17 mm to USGS 0, with an internal diameter of 2.11 mm and a height of 5.5 mm, and also included are bands <0 necessary for *Sephanoides sephaniodes* (Table 1). The largest-sized band is required for the Rufous-legged Owl (*Strix rufipes*), a species associated with forest habitats. The small bands, size USGS 0, are appropriate for small birds like the Tufted Tit-Tyrant (*Anairetes parulus*) and the Plain-mantled Tit-Spinetail (*Leptasthenura aegithaloides*), and the appropriate measurements for the bands made for the Green-backed Firecrown (*S. sephaniodes*) are: 1.27-1.52 mm internal diameter and a height of 1.6 mm.

The majority of species require bands of size categories USGS 0 (6 species), 1 (6 species) or 1B (5 species) (Figure 2). These sizes serve for all passeriform birds in the area, with the exception of the Austral Thrush (*Turdus falcklandii*) that requires size USGS 3. The three species of woodpeckers, the Ringed Kingfisher (*Ceryle torquata*) and the Austral Parakeet (*Enicognathus ferrugineus*) fell in a range of band sizes between USGS 2 and 4. This range also includes the Austral Pygmy-Owl (*Glaucidum nanum*), whereas the other two species of owl studied require larger-sized bands.

Prior to the study, very little information existed about appropriately sized bands for the diverse avifauna of South American temperate forests. Table 1 shows the appropriate band sizes for 34 species common in the study sites of the sub-Antarctic forest ecoregion of Magallanes. We suggest the use of these band sizes in future studies in southern Chile and at the same time we recommend evaluating the necessary band sizes on the field for the listed species if in different study sites.

The body size of species of birds varies between populations, especially at significant altitudinal or latitudinal gradients (Graves 1991). At high latitudes and altitudes temperature decreases markedly (Rozzi *et al.*1989), and according to Bergmann's rule (1847), at low temperatures there should be a selection for larger body sizes due to lower heat loss. In accordance with Bergmann's rule, it has been found that 83% of endothermic animals reach larger sizes in areas with lower temperatures (Atkinson 1994). This type of morphological variation between populations makes it recommendable to measure the appropriate band sizes for studies of avifauna that involve new sites, populations, or species. It is also worth pointing out that tarsus measurements of museum specimens can differ slightly from live individuals due to desiccation. Another important aspect includes intraspecific size differences due to sexual dimorphism, age, or food intake (Rozzi *et al.* 1996b).

Finally, we find the development of a coordinated banding program in the Neotropical region and American continent in general highly recommendable. Such a program should consider: a) the determination of appropriate band sizes for each species; b) the development of a numbering system for the bands that would permit the identification of the area and date of banding of birds recaptured by any investigator; c) accessible databases and the exchange of information and experiences at the regional and international level because the migratory routes of a considerable number of birds cross political and administrative boundaries.

Recaptures

In total during the study period we recaptured 62 banded individuals. This represents 8.7% of the banded individuals (706 individuals). The percentages of recapture were similar in AV (8.3%) and NI (9.1%). The recaptured birds consisted of 11 species, 69% of the 18 species banded (Table 2). In terms of numbers of individuals, the species most frequently recaptured were the White-crested Elaenia (*Elaenia albiceps*, 17 individuals), the Thorn-tailed Rayadito (*Aphrastura spinicauda*, 14 individuals) and the House Wren (*Troglodytes aedon*, 9 individuals). However, the percentages of recapture varied greatly between sites and species. Considering only the species with more than 10 individuals in each study site, *T. aedon* and *A. spinicauda* present the largest recapture percentages, which could be associated with smaller territories during the nesting period (Rozzi *et al.*, in preparation). In various instances we obtained a recapture rate greater than 100% because the number of captures was low and the same individual was captured various times.

Two individuals of migratory birds, one *E. albiceps* (band #419RR00) and one Chilean Swallow (*Tachycineta meyeni*, band #29RR00), were recaptured in successive years. Both individuals were banded and recaptured in AV. The White-crested Elaenia was

5 FIELD VERIFICATION OF ZIMMER'S WING-FORMULA FOR IDENTIFICATION OF *ELAENIA ALBICEPS CHILENSIS*

Steven M. McGehee[1,2] and Jack C. Eitniear[2]

[1]Parque Etnobotánico Omora, Puerto Williams, Magallanes, Chile. whitethroatedcaracara@yahoo.com
[2]Center for the Study of Tropical Birds, Inc., 218 Conway Drive, San Antonio, Texas 78209, U.S.A. jce@cstbinc.org

Boletín SAO 16: 58-63 (2006)

ABSTRACT. Field verification of Zimmer's wing-formula for identification of *Elaenia albiceps chilensis*. Based on examination of museum specimens, Zimmer proposed that *Elaenia albiceps chilensis* could be differentiated from all other *Elaenia albiceps* spp. by the longer length of the 10th primary (compared to that of the 5th). We confirm the theory, with limitations, using wild caught birds.

Key words: Chile, *Elaenia albiceps chilensis*, identification, wing-formula.

RESUMEN. Verificación de campo de la fórmula del ala de Zimmer para la identificación de *Elaenia albiceps chilensis*. Basado en el estudio de especímenes, Zimmer propuso que *Elaenia albiceps chilensis* se podría diferenciar de las demás subespecies de *Elaenia albiceps* por poseer la décima primaria más larga (comparada con la quinta). Confirmamos esta teoría, con limitaciones, usando individuos capturados en la naturaleza.

Palabras clave: Chile, *Elaenia albiceps chilensis*, fórmula del ala, identificación.

The White-crested Elaenia, *Elaenia albiceps* (Tyrannidae), is a small flycatcher distributed from southern Colombia and eastern Brazil to Cape Horn, Chile and the Diego Ramirez Islands (Ridgley & Tudor 1994, Couve & Vidal 2003, Schlatter & Riveros 1997). It is usually divided into five (Zimmer 1941) or six (Fjeldså & Krabbe 1990, del Hoyo *et al.* 2004) subspecies. Disagreement remains as to its taxonomic status with several authors supporting the elevation of *E. a. modesta* to full species status (Fjeldså & Krabbe 1990, Ridgley & Tudor 1994, Jaramillo 2003, Zimmer 1941).

Of the five subspecies of *E. albiceps* most are considered resident with only a single race (*chilensis*) classified as migratory (Ridgley & Tudor 1994, del Hoyo *et al.*2004). Currently subspecies are differentiated by geographic location, strength of wing bars, paleness of under parts, amount of white in eye ring, dullness and shape of crown patch and thickness of bill (Fjeldså & Krabbe 1990, del Hoyo *et al.*2004). All of these characteristics vary between individuals, age and often time of year, making subspecies identification difficult even with live specimens. Zimmer (1941) proposed that *E. a. chilensis* could be differentiated from all other *E. a.* spp. by the longer length of the 10th (outermost) primary (compared to that of the 5th) (Fig. 1). Zimmer speculated that this was to aid *chilensis* in migration. Other authors have also used this wing formula for separating *chilensis* from the other subspecies (Fjeldså & Krabbe 1990, Kratter *et al.* 1993, Sick 1993, Jaramillo 2003, Brumfield *et al.* 2004). While Zimmer (1941) examined 89 museum specimens of *chilensis* (52 males, 31 females, 6 unsexed) one can only assume that all specimens examined supported the wing formula, as he provided no detailed statistical analysis. We felt that since there are no published data on wild caught *E. a. chilensis*, verification of the validity of the wing formula with wild caught birds was justified. Below we report on data collected from *E. a. chilensis* caught in mist-nets in southern Chile.

Elaenia were mist-netted at Omora Ethnobotanical Park (54°57'S, 67°39'W) and other localities (within 25 kilometers of the park) on Navarino Island from 1999-2005 (Anderson & Rozzi 2000, Anderson *et al.* 2002). Each captured bird had its right wing visually inspected (Fig. 1). *E. a. chilensis* was the only race documented to occur at the capture site. All birds captured had two white wing bars, strong eye ring, pure white crown patch and white edges on secondaries characteristic of *E. a. chilensis* (Fjeldså & Krabbe 1990, del Hoyo *et al.* 2004). The only vocalizations heard in the forests of Navarino Island were ones associated with *chilensis* (Fjeldså & Krabbe 1990). We examined 330 *E. a. chilensis* with the results indicating that 88% of the birds caught complied with Zimmer's wing formula (Table 1). There was a slight difference between male and female and a greater difference between unknown adults and known sexed adults. 82% of juveniles fit Zimmer's formula. There were differences in months with January having the most birds with the shorter wing. *E. albiceps* arrives on Navarino the third week in October and departs by the fourth week in March with a few juveniles remaining until the end of April (McGehee *et al.* in prep.). The 39 birds with a shorter tenth primary did not have any worn

Fig. 1. Adult White-crested Elaenia (*Elaenia albiceps chilensis*) showing 10th (outer primary) longer than 5th (Photo: Steven McGehee).

feathers. Molting was not noted in any bird (McGehee & Elphick in prep.).

Molting in this species is said to be between March and October (Zimmer 1941) presumably on the wintering grounds. Ten adults were subsequently recaptured from one month to two years later. Six of the 10 (60%) had the 10th primary the longest each time. Four had the 10th primary longer (40%) in the first capture and shorter in the next capture (one year later). One adult that was first captured with a 5th primary longer was recaptured one month later and had the 10th primary longer. It appears that some birds arrive from migration still growing in their 10th primary. Juveniles were identified by their lack of white crest and buffy wing bars (Jaramillo 2003). Adults were identified as female if they had a brood patch and as male if they had a large cloacal protuberance. These sexual characters are fairly reliable for use as sexing in the breeding season for Tyrannidae (Pyle 1997). However in this study 56 adults were unable to be sexed having neither character. Of these unsexed adults 79% had their 10th primary longer than the fifth. Many of these could be second year birds. Second year males are considered less likely to obtain mates or prime breeding territories (Greenwood *et al.* 1983, Heise & Rimmer 2000). Evolution could favor a more prolonged molt in second year birds that have a long migration to reach a summer grounds with limited availability of mates and breeding sites. Less energy expended in a rapid molt means first year birds do not have to find as much food on their wintering grounds. This hypothesis awaits further study.

To enhance probability of an accurate *E. a. chilensis* identification we encourage Zimmer's wing formula be used in association with plumage and morphological characteristics.

ACKNOWLEDGMENTS

We thank the many volunteers who have assisted in Omora Parks mist-netting program. A special appreciation is due to Chris Anderson, Chris Elphick and Ricardo Rozzi for implementing the netting and banding project and for always being available to advise. Thanks to Tomás Ibarra and Ximena Arango for assistance in the field. Thanks to a couple of anonymous reviewers for their helpful suggestions on earlier drafts of the manuscript. This note is part of the ongoing research and conservation programs at the Omora Ethnobotanical Park (www.omora.org).

REFERENCES

ANDERSON, C.B. & R. ROZZI. 2000. Bird assemblages in the southernmost forests in the world: Methodological variations for determining species composition. *Anales del Instituto de la Patagonia, Serie Ciencias Naturales* 28: 89-100.

ANDERSON, C., R. ROZZI, C. ELPHICK & S. MCGEHEE. 2002. El programa Omora de anillamiento de aves en los bosques subantárticos: La estandarización del tamaño de los anillos apropiados para las aves de la Región de Magallanes. *Boletín Chileno de Ornitología* 9: 2-11.

BRUMFIELD, R.T., Z.O. MAILLARD, R.C. FAUCETT, G. SÁNCHEZ, V.G. ROHWER, J.C. CATARI, C.G. SCHMITT, D.C. SCHMITT, R. STREM & A.M. MAMANI. 2004. Birds of the Laguna Kaucaya area, a semi-humid valley in the Andean foothills of Departamento Santa Cruz, Bolivia. *Ornitologia Neotropical* 15: 381-398.

TABLE 1. White-crested Elaenia (*Elaenia albiceps chilensis*) with P10 primary longer then P5. ( ) Indicate number of birds where P5 is longer than P10.

Month	Adult Male	Adult Female	Unsexed Adults	Juvenile	N
Nov	14 (2)	17 (3)	11 (2)	0 (0)	49
Dec	58 (1)	45 (6)	22 (3)	0 (0)	135
Jan	18 (1)	13 (1)	8 (5)	7 (3)	56
Feb	5 (0)	20 (1)	3 (1)	36 (7)	72
Mar	0 (0)	0 (0)	0 (2)	12 (2)	16
Apr	0 (0)	0 (0)	0 (0)	2 (0)	2
Totals	95 (4)	95 (11)	44 (12)	57 (12)	330

COUVE, E. & C. VIDAL. 2003. *Birds of Patagonia, Tierra del Fuego and Antarctic Peninsula, the Falkland Islands and South Georgia*. Editorial Fantástico Sur Birding Ltda., Punta Arenas, Chile.

DEL HOYO, J., A. ELLIOT & D.A. CHRISTIE. 2004. *Handbook of the Birds of the World*. Volume 6. Lynx Edicions, Barcelona.

FJELDSÅ, J. & N. KRABBE. 1990. *Birds of the High Andes*. Zoology Museum, University of Copenhagen & Svendborg.

GREENWOOD, H., P.J. WEATHERHEAD & R.D. TITMAN. 1983. A new ageand sex-specific molt scheme for the Red-winged Blackbird. *Condor* 85: 104-105.

HEISE, C.D. & C.C. RIMMER 2000. Definitive prebasic molt of Gray Catbirds at two sites in New England. *Condor* 102: 894-904.

JARAMILLO, A. 2003. *Birds of Chile*. Princeton University Press, Princeton.

KRATTER, A.W., T.S. SILLETT, R.T. CHESSER, J.P. O'NEIL, T.A. PARKER, III & A. CASTILLO. 1993. Avifauna of a Chaco locality in Bolivia. *Wilson Bulletin* 105: 114-141.

PYLE, P. 1997. *Identification Guide to North American Birds*, Part 1. Slate Creek Press, Bolinas, CA.

RIDGELY, R. & G. TUDOR. 1994. *The Birds of South America*. Volume II. The Suboscine Passerines. Oxford University. Press, UK.

SCHLATTER, R.P. & G.M. RIVEROS. 1997. Historia natural del Archipiélago Diego Ramírez, Chile. *Serie Científica INACH* 47: 87-112.

SICK, H. 1993. *Birds in Brazil*. Princeton U. Press. Princeton.

ZIMMER, J.T. 1941. Studies of Peruvian birds 36. *American Museum Novitates*1108: 1-23.

PART II

SINGULARITIES OF THE MAGELLANIC SUB-ANTARCTIC AVIFAUNA

6 INTRODUCTION TO THE SINGULARITIES OF THE MAGELLANIC SUB-ANTARCTIC AVIFAUNA

SEBASTIÁN DARDANELLI, JAIME E. JIMÉNEZ, RODRIGO A. VÁSQUEZ, AND CHRISTOPHER ELPHICK

The study of aspects of the basic biology of the species provides valuable information for the development of strategies for their conservation and their habitats, and to build links between humans and these species. Several authors (Soule & Kohn 1988, Primack 1993, Wiens 2007) point out that the cornerstone of the development of any program of conservation of species or habitats is the knowledge of their basic biology. This aspect is crucial, since the responses of each species to processes such as habitat loss and climate change are always "species-specific" (Lindenmayer & Fischer 2006). In addition, some species in the southern hemisphere, particularly understudied passerine families inhabiting the neotropics that are found in sub-Antarctic areas (*i.e.*, high latitudes of the southern hemisphere), such as several species of the family Tyrannidae and Furnariidae, among others, have unique biological characteristics and are very different from passerine species inhabiting high latitudes of the northern hemisphere (see *e.g.* Moreno *et al.* 2005, Moreno *et al.* 2007).

The Cape Horn Biosphere Reserve (CHBR) houses the world's southernmost forests where the birds are the dominant group of vertebrates. This is why the Omora Park prioritized the long-term study of birds in the CHBR, developing since 2000, a program of monitoring land birds. The program consists of bird banding in forests and scrublands, as well as tracking the population dynamics through the implementation of fixed points of capture, recapture, and release. Through the analysis of the results of this banding program, we are starting to learn the basic ecology of the almost unknown species that inhabit the sub-Antarctic forests.

This section reprints seven articles published in various journals between 2004 and 2008, which deal with very different aspects of the basic biology of sub-Antarctic birds from autoecology, behavior and migration patterns, diets to the study of latitudinal gradients of their blood parasites. These varieties of articles have been collected in one section called "singularities."

The owls of the genus *Strix* inhabit mainly forests on all continents except Antarctica (del Hoyo *et al.* 1999). Their affinity for forest environments and their frequent sensitivity to alterations of forests makes them good indicators of environmental health of forests and prey availability in those habitats. The best known case is that of the Northern Spotted Owl (*Strix occidentalis caurina*), from the forests of northwestern United States (Hunter & Gibbs 2007). In the first paper of this section, Ippi and Rozzi (2004) provide relevant data on the Rufous-legged Owl (*Strix rufipes*), a medium-sized owl (38 cm body length) endemic to the temperate forests of Chile and Argentina. Despite being considered a crepuscular or nocturnal owl, the authors have revealed the diurnal activity of this species. The Rufous-legged Owl was recorded not only moving in the forest, but hunting during the day. Also notable are the observations on its feeding behavior, including the use of small birds as prey. These facts were confirmed after the publication through daytime observations of activity of several individuals on the island and the capture of two individuals who tried to take small birds caught in mist-nets during the monthly long-term monitoring in the Omora Park (S. Dardanelli, personal observation).

The second paper of this section discusses a migratory species, albeit of short distance, of limited distribution and abundance. McGehee *et al.* (2004) reported the records of the Patagonian Tyrant (*Colorhampus parvirostris*) on Navarino Island. Additionally, the authors analyze the presence of this species in these high latitudes between the months of March and April, corresponding to the post-breeding period of the species. The authors propose an interesting discussion on the presence of *C. parvirostris* in the Cape Horn region, suggesting that this species may have expanded its range to the south in recent years, which could be associated with the phenomenon of climate change. Moreover, considering that they were sighted after the reproductive period, suggests that their presence could be due to southern movements toward high latitudes shortly before starting again northward migration, which is known as "reverse post-reproductive migration." This would be the first reported case of this behavior in the sub-Antarctic forests. However, it is noteworthy that after the publication of this study, we were able to observe the presence of *C. parvirostris* for most of the year except between October and December (Ippi *et al.* 2009, S. Dardanelli, personal observation). This suggests that some individuals may behave as residents and others as migrants. Based on this evidence, *C. parvirostris* seems to have a partially migratory population. This could be a recent strategy acquired by the population in Navarino Island. Alternatively, given the rarity of this species throughout its range, it has non been detected in Navarino in the Fall before 2004. These new questions encourage the development of further studies to confirm, for example, the nesting of the species in the southernmost forests.

Studies on the autoecology of bird species are essential to know the species and preserve them, and to conserve the ecosystems they inhabit (Lindenmayer & Fischer 2006, Wiens 2007). The third article in this section (Brown *et al.* 2007) deals with the study of the autoecology of *E. albiceps*. The authors help us understand more about a migratory species that breeds in subtropical and temperate southern South American forests and thickets, and winters in tropical forests of northern South America. This paper contains valuable information about the longevity of *Elaenia albiceps*, information which is only possible to obtain from long-term studies. There is a widespread ignorance about the longevity of birds in the wild, particularly in the southern hemisphere; this information is relevant to know the life history of a species and to model the dynamics of their populations over time. Thus, it is possible to predict changes in population behavior as compared to habitat degradation or climate change. Besides contributing to the knowledge of the autoecology of the species, the recording of the dates of arrival and departure of individuals of *E. albiceps* to and from Navarino Island is useful to detect marked changes in the phenology of migration, which can be used as an indicator of changes associated with climate change. For example, some migrants advance or delay their arrivals and departures as a response to changes in rainfall and temperature patterns (Jenni & Kéry 2003, Both 2010).

Unraveling the trophic relationships of species in an ecosystem is crucial to understand the distributions and abundances of organisms. These also determine the flows of energy and nutrients, and ultimately the ecosystem functioning. The diet of the sub-Antarctic birds has hardly been studied. The following two papers describe the diet of the Patagonian Sierra-Finch (*Phrygilus patagonicus*) and of three other bird species that feed on the Leñadura tree (*Maytenus magellanica*). In this way, McGeehe & Eitnier (2007), based on behavioral observations, describe that the Patagonian Sierra-Finch feeds on seeds of eight plant species, on the nectar of three, on floral parts of five, and on fruits of four, in addition to the Old-man's Beard lichen (*Usnea* spp.). This wide trophic spectrum of the most abundant bird of the southern forests likely relates closely with trees (three species), shrubs (five species), and grasses of small plants (six species), fulfilling complex roles as a predator of fruits, seeds, and flowers, and as a pollinator of forest species. It is interesting to note that the Sierra-Finch is not a seed disperser, as the seeds eaten are destroyed in its passage during digestive tract. It would also be interesting to know what the Patagonian Sierra-Finch feed on during the winter, as it is a resident bird and no dietary observations during the cold period have been reported.

In the fifth paper presented, McGeehe (2007) documented interesting observations in which three essentially insectivorous bird species (two of which are of the flycatchers family) eat seeds and fruits of a shrub, the Leñadura that has been decimated by cattle grazing. Despite the fact that these observations seem unique, it is likely that this behavior of insectivorous birds could be more widespread than believed, and that it will be better documented with more detailed observations of sub-Antarctic birds. The author suggests that these unexpected trophic behaviors may result from a decline in the environmental availability of insects due to adverse weather conditions in the fall, when these observations were made. However, it remains to be seen if these birds play a role in dispersing seeds of this commomn shrub of the southern region, and whether these behaviors are repeated in more northern populations of these bird species.

Since 2004, a program to monitor the reproductive biology of birds that nest in cavities began. This study was conducted by installing nest boxes in the Omora Park and surrounding areas, and was linked to investigations initiated earlier in Chiloé (see Moreno *et al.* 2005). Species that have predominantly used nest boxes in Navarino are the House Wren (*Troglodytes aedon*) and the Thorn-tailed Rayadito (*Aphrastura spinicauda*). The latter species, a member of the Furnariidae family, has proven to be a particularly interesting species by being abundant in temperate forests and by possessing unique characteristics related to sexual dimorphism and reproductive behavior. The sixth paper in this collection corresponds to Moreno *et al.* (2007), who studied the sexual dimorphism and parental care of the Thorn-tailed Rayadito in Chiloé and in Navarino. The authors show that males and females are morphologically similar, differing only in body size, with males being 2 to 10% larger than females. In addition, the parental care of eggs and chicks is generally shared by adult males and females, collaborating both in incubation, feeding the chicks, and cleaning the nest. This type of bi-parental care, associated with other unique features of the species at this latitude, such as having low annual fecundity, large egg size relative to other passerines, and an extended growth period (Moreno *et al.* 2005), suggest that the Thorn-tailed Rayadito is a singular species that deserves further research on various aspects of its biology.

There is a marked asymmetry between the knowledge of the existence of latitudinal gradients in species diversity among free-living organisms and their parasites, with many studies for the former and lack of studies on the latter (Hillebrand *et al.* 2001). This lack of studies on latitudinal gradients of parasites is even more marked in the southern hemisphere (Merino *et al.* 2008). In the seventh paper of this section, Merino *et al.* (2008) provide valuable information on the latitudinal gradients of richness and abundance of lineages of bird blood parasites. The authors describe latitudinal gradients for the prevalence of the three parasite genera (*Leucocytozoon*, *Haemoproteus*, and *Plasmodium*) as well as for mixed infections. They also suggest that some lineages have evolved in isolation in some localities and/or species. It is remarkable that the genus *Haemoproteus* always had hosts of the family Emberizidae. Individuals of White-crowned Elaenia (*E. albiceps*), a long-distance migratory species, were found infected by the same parasite lineages in locations separated by 20° of latitude. Infections by these lineages were recorded in other non-migratory species, so the authors suggest that long distance migratory birds could prevent latitudinal gradients in parasites due to the potential role of these migratory birds in the spread of diseases to other species.

In short, the study of these singularities of sub-Antarctic avifauna provides valuable information that helps understand the basic biology of almost unknown species. The value of these data becomes even more important when considering that, in analyzing large databases and comparing different taxa, it was found that the birds, involved in the lowest monitoring cost, was the taxonomic group that performed the best as a biological indicator (Gardner *et al.* 2008). The knowledge of the biology of sub-Antarctic birds is critical for the development of conservation strategies for the birds and for their habitats. Furthermore, the communication to society of its most notable features will enhance the relationship of the human inhabitants with the birds and their habitats. The validity and duration of the long-term banding program, and more recently, the tracking of the reproductive biology using nest boxes in the Omora Park, ensures continuity in the study of these fundamental aspects of the ecology of the avifauna of the southernmost forests of the planet.

Looking into the future there are infinite possibilities and needs to further explore the "uniqueness" of sub-Antarctic avifauna. Thus, the ongoing studies on the Magellanic Woodpecker (*Campephilus magellanicus*), the most charismatic species in the forest of the CHBR (Arango *et al.* 2007), promise to continue revealing singularities of fundamental importance to help us understand, appreciate, enjoy, and conserve the sub-Antarctic avifauna.

REFERENCES

ARANGO, X., R. ROZZI, F. MASSARDO, C.B. ANDERSON & T. IBARRA. 2007. Descubrimiento e implementación del pájaro carpintero gigante (*Campephilus magellanicus*) como especie carismática: una aproximación biocultural para la conservación en la Reserva de Biosfera Cabo de Hornos. *Magallania* 35: 71-88.

BOTH, C. 2010. Flexibility of timing of avian migration to climate change masked by environmental constraints en route. *Current Biology* 20: 243-248.

BROWN, C.E., C.B. ANDERSON, S. IPPI, M.F. SHERRIFFS & R. ROZZI. 2007. The autecology of the Fio-fio (*Elaenia albiceps*). *Anales del Instituto de la Patagonia* 35: 29-40.

DEL HOYO, J., A. ELLIOTT & J. SARGATAL. 1999. *Handbook of the Birds of the World.* Volume 5. Barn-Owls to Hummingbirds. Lynx Edicions, Barcelona, España.

GARDNER, T.A., J. BARLOW, I.S. ARAUJO, T.C. AVILA-PIRES, A.B. BONALDO, J.E. COSTA, M.C. ESPOSITO, L.V. FERREIRA, J. HAWES, M.I. HERNANDEZ, M.S. HOOGMOED, R.N. LEITE, N.F. LO-MAN-HUNG, J.R. MALCOLM, M.B. MARTINS, L.A. MESTRE, R. MIRANDA-SANTOS, W.L. OVERAL, L. PARRY, S.L. PETERS, M.A. RIBEIRO-JUNIOR, M.N. DA SILVA, C. DA SILVA MOTTA & C.A. PERES. 2008. The cost-effectiveness of biodiversity surveys in tropical forests. *Ecology Letters* 11: 139-150.

HILLEBRAND, H., F. WATERMANN, R. KAREZ & U.G. BERNINGER. 2001. Differences in species richness patterns between unicellular and multicellular organisms. *Oecologia* 126: 114-124.

HUNTER, M.L. JR. & J.P. GIBBS. 2007. *Fundamentals of Conservation Biology*. Third Ed. Blackwell Publishing Ltd. Malden, Massachusetts, EE.UU.

IPPI, S. & R. ROZZI. 2004. Actividad diurna y nocturna del concón (*Strix rufipes*) en los bosques del Cabo de Hornos. *Boletín Chileno de Ornitología* 10: 9-12.

IPPI, S., C.B. ANDERSON, R. ROZZI & C.S. ELPHICK. 2009. Intra-annual variation in subantarctic forest bird communities. *Ornitología Neotropical* 20: 231-245.

JENNI, L. & M. KÉRY. 2003. Timing of autumn bird migration under climate change: advances in long-distance migrants, delays in short-distance migrants. *Proceedings of the Royal Society of London* B 270: 1467-1471.

LINDENMAYER, D.B. & J. FISCHER. 2006. *Habitat Fragmentation and Landscape Change: An Ecological and Conservation Synthesis.* Island Press, Washington, D.C.

MCGEHEE, S., ROZZI, R., S. IPPI, C. ANDERSON, R.A. VÁSQUEZ & S. WOODLAND. 2004. Late-summer presence of the Patagonian Tyrant, *Colorhamphus parvirostris* (Darwin) in Navarino Island, Cape Horn County. *Anales del Instituto de la Patagonia, Serie Ciencias Naturales* 32: 25-33.

MCGEHEE, S.M. & J.C. EITNIEAR. 2007. Diet of the Patagonian Sierra-Finch (*Phrygilus patagonicus*) on Navarino Island, Chile. *Ornitología Neotropical* 18: 1-4.

MCGEHEE, S.M. 2007. Consumption of Leñadura (*Maytenus magallanica*) seeds by three primarily insectivorous birds species. *Anales del Instituto de la Patagonia* 35: 71-73.

MERINO, S., J. MORENO, R.A. VÁSQUEZ, J. MARTÍNEZ, I. SÁNCHEZ-MONSÁLVEZ, C.F. ESTADES, S. IPPI, P. SABAT, R. ROZZI & S. MCGEHEE. 2008. Haematozoa in forest birds from southern Chile: Looking for latitudinal gradients in prevalence and parasite lineage richness. *Austral Ecology* 33: 329-340.

MORENO J., S. MERINO, R.A. VÁSQUEZ, J.J. ARMESTO. 2005. Breeding biology of the Thorn-tailed Rayadito (Furnariidae) in south-temperate rainforests of Chile. *Condor* 107: 69-77.

MORENO, J., S. MERINO, E. LOBATO, M.A. RODRÍGUEZ-GIRONÉS & R.A. VÁSQUEZ. 2007. Sexual dimorphism and parental roles in the Thorn-tailed Rayadito (Furnariidae). *Condor* 109: 312-320.

PRIMACK, R.B. 1993. *Essentials of Conservation Biology.* Sinauer Associates, Inc. Sunderland, Massachusetts, USA.

SOULE, M.E. & K.A. KOHN. 1989. *Research Priorities for Conservation Biology*. Island Press. Washington D.C., EE.UU.

WIENS, J.A. 2007. Does conservation need landscape ecology? A perspective from both sides of the divide. Pp. 479-493 *in Managing and Designing Landscapes for Conservation: Moving from Perspectives to Principles*, D.B. Lindenmayer & R.J. Hobbs (eds.). Blackwell Publishing Ltd. Malden, Massachusetts, EE.UU.

7 DIURNAL AND NOCTURNAL ACTIVITY OF THE RUFOUS-LEGGED OWL (*STRIX RUFIPES*) IN THE FORESTS OF CAPE HORN

SILVINA IPPI[1,2] AND RICARDO ROZZI[1, 3]

[1]Parque Etnobotánico Omora, Universidad de Magallanes-Fundación Omora
Puerto Williams, Isla Navarino, Magallanes, Chile, silvippi@yahoo.com
[2]Dept. de Ecología, Facultad de Ciencias, Universidad de Chile, Santiago
[3]Department of Philosophy & Religion Studies, University of North Texas Denton, TX 76203-0920

Boletín Chileno de Ornitología 10: 9-12 (2004) (Originally published in Spanish)

ABSTRACT. Diurnal and nocturnal activity of the Rufous-legged Owl (*Strix rufipes*) in the forests of Cape Horn. Observations on daily activity patterns and habitat use by Rufous-legged Owls (*Strix rufipes*) on Navarino Island, Southern Chile.

RESUMEN. Actividad diurna y nocturna del concón (*Strix rufipes*) en los bosques del Cabo de Hornos. Se presentan observaciones de patrones de actividad y uso de habitat por el concón (*Strix rufipes*) en la isla Navarino, Chile.

The Rufous-legged Owl (*Strix rufipes*, King 1828) is a nocturnal strigiform endemic to the temperate forests of southern Chile and Argentina (Hoyo *et al.* 1999). Recently, it has been separated from its congener, *S. chacoensis*, that inhabits the Chaco ecoregion in northern Argentina, southern Bolivia, and western Paraguay (del Hoyo *et al.* 1999). The known latitudinal range and distribution of the Rufous-legged Owl extends from the sclerophilous forests of central Chile (35°S, Díaz 1999) south toward the forests of the insular region of Cape Horn (55°S) (Sielfeld 1977).

The Red Book of Chilean Vertebrates (Glade 1987) classifies *S. rufipes* as Inadequately Known. However, recently it has been suggested that its populations are declining due to fragmentation, degradation, and habitat loss (Martínez & Jaksic 1996, Jaksic *et al.* 2001) because it is highly sensitive to landscape modifications (Vásquez & Simonetti 1999) and requires old-growth or primary forests (Martínez & Jaksic 1996, Rozzi *et al.* 1996a,b). Both in the Northern and Southern Hemispheres, it is thought that owls of the genus *Strix* preferably inhabit extensive, old-growth temperate forests, with dense canopies and complex vegetation structure (Murphy & Noon 1992, Martínez & Jaksic 1996). However, the degree of dependence of old-growth forests by *S. rufipes* has been questioned by Estades *et al.* (1998), based on its presence and nesting in the interior of Monterey Pine plantations (*Pinus radiata*), an exotic species that covers large areas in central and southern Chile.

It has been suggested that *S. rufipes* presents generalist dietary habits. Regarding this, Martínez (1993) and Martínez & Jaksic (1996) found that small, nocturnal mammals with arboreal and scansorial habits were the most frequent prey of *S. rufipes* in Valdivian forest. In the northern limit of its distribution, Díaz (1999) revealed a diet primarily composed of rodents, but that also included insects and reptiles. None of these studies mention remains of birds in its pellets, suggesting that these are not part of its normal diet. Regarding its activity pattern, the Rufous-legged Owl is considered a strictly nocturnal strigiform (Couve & Vidal 2003).

In this note, we report the first observations of habitat use, activity patterns, and diet of *S. rufipes* in the forests of the extreme south of the Americas on Navarino Island (55°S). The study area spans the northern and western coasts of Navarino Island, and the principal site corresponds to the Omora Ethnobotanical Park (54°56'S, 67°39'W), 3 km west of the city of Puerto Williams. The landscape includes large patches of broadleaf evergreen forests dominated by Magellan's Beech (*Nothofagus betuloides*) and Winter's Bark (*Drimys winteri*), deciduous forests dominated by Antarctic Beech (*N. antarctica*) and/or Lenga Beech (*N. pumilio*), and mixed forests dominated by Magellan's and Lenga Beech (Rozzi 2001). Navarino Island has an oceanic climate with an annual mean of maximum temperature of 10.6°C and a minimum temperature of 1.6°C (between November 1999 and October 2000). Annual average precipitation in this period was 429 mm (Rozzi 2001). Observations of the Rufous-legged Owl were made by point counts and captures using mist-nets in the Omora Park between January of 2000 and September of 2003 (see Anderson & Rozzi 2000, Anderson *et al.* 2003). Diurnal observations totaled 540 hrs. During January and February of 2001 and 2002, we conducted point counts between 18:00 and 24:00, totaling 140 hrs. Further, since January of 2000, a record of monthly checklists and occasional sightings in the area has been kept. This note considers records derived from systematic and opportunistic sampling. For the analysis, we defined three time intervals: morning hours (6:00-11:59), afternoon (12:00-17:59), and twilight-nighttime (18:00-23:59).

Our observations of *S. rufipes* represent the first records of this species on Navarino Island, including its most southern detection (55°05'S, 68°13'W) in Douglas Bay (Table 1). Most of the records

(66%) occurred in old-growth forests of Magellan's Beech-Winter's Bark, mainly in the interior (N = 6). Other observations were associated with new-growth Magellan's Beech forests, an open space (>100 m in diameter) and a forest of Antarctic Beech, all adjacent to adult Magellan's Beech forests.

The number of sightings of *S. rufipes* was the same for morning hours, the afternoon, and twilight-nighttime: three records in each time interval (Table 1). In two instances, the Rufous-legged Owl was detected by its vocalizations, both cases during the twilight-nighttime hours (Table 1). Two records corresponded to prey searching. The first record, at 23:30 in January of 2002, we observed an individual that flew directly from a dead tree branch toward a wetland where it rapidly took off with a rodent in its talons and returned to the same perch. The second instance occurred in July of 2003; we spotted another individual during the day while we placed mist-nets in the Magellan's Beech-Winter's Bark evergreen forest. Afterward at 12:00, while cheking the nets for captured birds we found an individual on the ground eating a Thorn-tailed Rayadito (*Aphrastura spinicauda*, Furnariidae) that was entangled in the net and was killed by the owl. The individual immediately flew and distanced itself from the site. In the same net, an Austral Thrush (*Turdus falcklandii*, Turdidae) was also apparently killed by the Rufous-legged Owl, but at a height of 1.8 m, which had prevented it from being consumed. On two other occasions we detected individuals while they moved during the day, flying between forested areas of Omora Park. Finally, on three occasions we observed individuals perched during the day. On all occasions that *S. rufipes* was perched, its presence was detectable because of the warning behavior of species like *T. falcklandii*, *A. spinicauda*, and *Phrygilus patagonicus* (Patagonian Sierra-Finch), which consistently approached and sounded alarm calls.

The records of *S. rufipes* on Navarino Island suggest that:

1) This species of owl displays diurnal activity in the forests of the extreme south of the Americas, as it was observed flying and feeding during afternoon hours (approximately 4 hours after dawn). We should note that observations of diurnal movement occurred before the detection of our presence by the bird.

2) The diet of the Rufous-legged Owl in Magellanic forests could include birds. Not only was it observed consuming species of birds, but these same species of passeriforms responded to the presence of the owl with alarm calls. The presence of birds in its diet widens the spectrum of prey species for this strigiform, and is consistent with the generalist feeding habits described in other studies (Martínez 1993, Martínez & Jaksic 1997, Díaz 1999). This type of generalist behavior is also found in the diet of other non-raptor species on Navarino Island, and could be associated with the fluctuations in food availability and the rigorous climatic conditions at high latitudes (Ippi *et al.* 2003).

3) Regarding the habitat use of *S. rufipes*, it is worth pointing out that although the sampling effort was similar for the diverse types of forest, all observations occurred in the interior or in the vicinity of old-growth evergreen Magellan's Beech-Winter's Bark forest.

These behavior and feeding records of *S. rufipes* not only widen its known distribution, but also present new questions regarding its habitat requirements and feeding habits in the extreme South, where the availability of food varies markedly in relation to the seasonality of Magellanic forests. Studies of the reproductive and trophic ecology of the Rufous-legged Owl are necessary to adequately conserve this species and the forests it inhabits.

TABLE 1. Records of *S. rufipes* on Navarino Island, between January of 2000 and September of 2003.

Date	Time	Activity[1]	Forest Type[2]	Site[3]
2000				
23-Feb	23:30	Voc	OGF	DB
2001				
03-Dec	19:00	Voc	SFM	OP
2002				
08-Jan	23:30	H-R	FGM	OP
2003				
08-Jan	11:00	Fly	OGF	OP
20-Jan	13:00	Per	FMA	OP
10-Apr	13:30	Per	OGF	OP
25-Apr	9:00	Per	OGF	OP
21-Jun	12:50	Fly	OGF	OP
19-Jul	9:25 - 12:08	H-B	OGF	OP

[1] Voc: vocalizing, Fly: flying, H-R: hunting rodents, H-B: hunting birds, Per: perched

[2] OGF: old-growth forest dominated by or Magellan's Beech and Winter's Bark, SFM: secondary forest of Magellan's Beech, FGM: forest gap of Magellan's Beech, FMA: forest margin of a patch dominated by Antarctic Beech- Winter's Bark.

[3] DB: Douglas Bay, OP: Omora Park

ACKNOWLEDGMENTS

We thank the participants of project Earthwatch "The Owls of Cape Horn." This study forms part of the investigation program of the Omora Ethnobotanical Park, with support from the projects BIOKONCHIL of the German Ministry of Education and Research (BMBF, FKZ 01 LM 0208) and the Millenium Center for Advanced Studies in Ecology and Biodiversity (CMEB, P02-051-FICM).

REFERENCES

ANDERSON, C., R. ROZZI, C. ELPHICK & S. MCGEHEE. 2002. El programa Omora de anillamiento de aves en los bosques subantárticos: estandarización del tamaño de anillos apropiados para las aves de la Región de Magallanes. *Boletín Chileno de Ornitología* 9: 2-11.

ANDERSON, C. & R. ROZZI. 2000. Bird assemblages in the southernmost forests in the world: methodological variations for determining species composition. *Anales del Instituto de la Patagonia, Serie Ciencias Naturales* 28: 89-100.

ARMESTO, J.J., C. VILLAGRÁN & M.T. KALIN (EDS.). 1996. *Ecología de los Bosques Nativos de Chile*. Editorial Universitaria, Santiago, Chile.

COUVE, E. & C. VIDAL. 2003. *Birds of Patagonia, Tierra del Fuego and Antarctic Peninsula*. The Falkland Islands and South Georgia. Chile. Editorial Fantástico Sur.

GLADE, A. (ED.). 1993. *Libro Rojo de los Vertebrados Terrestres de Chile*. CONAF. Santiago de Chile.

DEL HOYO, J., A. ELLIOT & J. SARGATAL. 1999. *Handbook of the Birds of the World*. Volume 5. Barn-owls to Hummingbirds. Lynx editions.

DÍAZ, I. 1999. Food habits of the Rufous-egged Owl (*Strix rufipes*) in the Mediterranean sclerophyllous forest of Central Chile. *Journal of Raptor Research* 33: 260-264.

ESTADES, C.F., S.A. TEMPLE & A. GAJARDO. 1998. Unusual nesting of the Rufous-legged Owl? *Journal of Raptor Research* 33: 183.

IPPI, S., M. SHERRIFFS, S. MCGEHEE, C. ANDERSON & R. ROZZI. 2003. *Dieta del ensamble de paseriformes de los bosques subantárticos en la Isla Navarino (55ºS), Chile*. VII Congreso de Ornitología Neotropical, Puyehue.

JAKSIC, F.M., E. PAVEZ, J.E. JIMÉNEZ & J.C. TORRES-MURA. 2001. The conservation status of raptors in the metropolitan region, Chile. *Journal of Raptor Research* 35: 151-158.

MARTÍNEZ, D.R. 1993. Food habits of the Rufous-legged Owl (*Strix rufipes*) in temperate rainforests of southern Chile. *Journal of Raptor Research* 27: 214-216.

MARTÍNEZ, D.R. & F.M. JAKSIC. 1996. Habitat, relative abundance, and diet of Rufous-legged Owls (*Strix rufipes* King) in temperate forest remnants of southern Chile. *Ecoscience* 3: 259-263.

MARTÍNEZ, D.R. & F.M. JAKSIC. 1997. Selective predation on scansorial and arboreal mammals by Rufous-legged Owls (*Strix rufipes*) in southern Chilean rainforest. *Journal of Raptor Research* 31: 370-375.

MURPHY, D.D. & B.R. NOON. 1992. Integrating scientific methods with habitat conservation planning: reserve design for Northern Spotted Owls. *Ecological Applications* 2: 3-17.

ROZZI, R. 2001. *Biological and Cultural Conservation in the Archipelago Forest Ecosystems of Southern Chile*. Ph.D. Dissertation, Department of Ecology and Evolutionary Biology, University of Connecticut, USA.

ROZZI, R., D. MARTÍNEZ, M.F. WILLSON & C. SABAG. 1996. Avifauna de los bosques templados de Sudamérica. Pp. 135-152 *in Ecología de los Bosques Nativos de Chile*, J.J. Armesto, C. Villagrán & M.T. Kalin (eds.). Editorial Universitaria, Santiago, Chile.

SIELFELD, W.H. 1977. Reconocimiento macrofaunístico terrestre en el área de Seno Ponsonby (Isla Hoste). *Anales del Instituto de la Patagonia, Serie Ciencias Naturales* 8: 275-296.

VÁSQUEZ, R.A. & J.A. SIMONETTI. 1999. Life history traits and sensitivity to landscape change: the case of birds and mammals of Mediterranean Chile. *Revista Chilena de Historia Natural* 72: 517-525.

8

LATE-SUMMER PRESENCE OF THE PATAGONIAN TYRANT, *COLORHAMPHUS PARVIROSTRIS* (DARWIN) ON NAVARINO ISLAND, CAPE HORN COUNTY, CHILE

Steven M. McGehee[1], Ricardo Rozzi[1, 2, 3], Christopher Anderson[1, 4], Silvina Ippi[1, 5], Rodrigo Vásquez[5], and Sue Woodland

[1]Parque Etnobotánico Omora, Puerto Williams, Magallanes, Chile. fundacion@omora.org
[2]Universidad de Magallanes, Casilla 113-D, Punta Arenas, Chile
[3]Department of Philosophy & Religion Studies, University of North Texas, P.O. Box 310980, Denton Texas 76203-0980, Texas, USA
[4]Institute of Ecology, University of Georgia, Athens, GA 30602, USA
[5]Facultad de Ciencias, Universidad de Chile, Las Palmeras 3425, Santiago, Chile

Anales del Instituto de la Patagonia (Chile) 32: 25-33 (2004)

ABSTRACT. Late-summer presence of the Patagonian Tyrant (*Colorhamphus parvirostris*, Darwin) on Navarino Island, Cape Horn County, Chile. The southernmost representative of the Tyrannidae, a diversified family with approximately 420 species endemic to the Americas, belongs to a monotypic genus *Colorhamphus* with only one species: *C. parvirostris*. Little is known about the Patagonian Tyrant, which is considered rare throughout its breeding range and vulnerable to forest disturbances and habitat fragmentation. Even its distribution and life history are poorly known. During the breeding season it migrates to higher latitudes (>40°S) in southern Chile and Argentina, but small numbers remain within central-southern latitudes year-round (30-40°S). In the high latitudes of Tierra del Fuego it is reported to be an accidental visitor, but records of *C. parvirostris* are extremely rare.

As a part of the long-term mist-netting and bird census program at the Omora Ethnobotanical Park on Navarino Island, Cape Horn County, Chile (55°S) a total of 1,542 birds belonging to 17 species have been captured with mist-nets on Navarino Island. Only 12 were *C. parvirostris* (<1%). Therefore, this species is indeed rare. However, from 2000 to 2003 all of the few Patagonian Tyrants have been captured from late March and April. Its presence on Navarino Island may have become more frequent due to an extension of its latitudinal range associated with global climate change. However, additional observations made on successive years during late summer and fall suggest an alternative, and perhaps complementary, explanation: some southward movement by some individuals prior to their northward migration. This would be the first time that this phenomenon is reported for the Southern Hemisphere. Taken together these two hypotheses could reinforce one another, whereby the southern limit of post-reproductive southerly movements could be extending due to global warming.

Key words: *Colorhamphus parvirostris*, mist-netting, Navarino Island, reverse migration, Sub-Antarctic forests, Tyrannidae.

RESUMEN. Presencia de la viudita (*Colorhamphus parvirostris*, Darwin) a fines de verano en isla Navarino, comuna Cabo de Hornos, Chile. El representante más austral de los Tyrannidae, familia endémica de América que posee unas 420 especies, pertenece al género monotípico *Colorhamphus*, representado por una única especie: *C. parvirostris*. Poco se conoce acerca de la viudita, que es considerada como rara en su área de reproducción y vulnerable a la degradación y fragmentación del hábitat. Incluso su distribución e historia de vida son poco conocidas. Durante la estación reproductiva las poblaciones migran a latitudes australes (>40ºS) en el sur de Chile y Argentina, pero algunos individuos permanecen en latitudes del centro sur (30-40ºS). En el extremo sur, en Tierra del Fuego, los registros de *C. parvirostris* son extremadamente escasos y se la ha considerado como un visitante accidental.

Como parte del programa a largo plazo de censo y anillamiento de aves en el Parque Etnobotánico Omora, isla Navarino, Comuna Cabo de Hornos, Chile (55ºS) se capturó un total de 1.542 aves correspondientes a 17 especies en la isla Navarino. Sólo 12 individuos (<1%) eran de *C. parvirostris*. En consecuencia, ésta puede ser considerada una especie rara. Primero, especulamos que la mayor presencia de esta especie en la isla Navarino podría estar relacionada con una extensión de su ámbito de distribución asociada al cambio climático global. Sin embargo, observaciones adicionales de *C. parvirostris* en años sucesivos que se restringían al verano tardío y otoño nos sugirieron una explicación alternativa: una corta migración al sur efectuada por algunos individuos antes de emprender su migración hacia el norte, que constituiría el primer caso de esta conducta conocida para el hemisferio sur. Tomadas en conjunto estas dos hipótesis podrían reforzarse, lo que significaría que el límite austral de la migración reversa post-reproductiva podría extenderse debido al calentamiento global.

Palabras clave: anillamiento de aves, bosques subantárticos, *Colorhamphus parvirostris,* isla Navarino, migración reversa, Tyrannidae.

INTRODUCTION

Tyrannidae (tyrant flycatchers) is a family endemic to the Americas (Welty 1975), which includes about 420 species that inhabit ecosystems from the Canadian arctic tree-line to Patagonia (Clark 2001). *Colorhamphus* is a monotypic genus with only one species: *C. parvirostris* (Darwin), or the Patagonian Tyrant (Araya *et al.* 1996, Fjeldsa & Krabbe 1990). In large part its distribution and life history are poorly known (Marin *et al.* 1989, Chesser & Marin 1994). It is a small insect-eating bird (Rozzi *et al.* 1996a,b), and it migrates to higher latitudes (>42°S) in southern Chile and Argentina during the austral spring and summer (October-March), to later return to central-southern latitudes (Fig. 1). It is uncommon throughout all of its breeding range (Chesser & Marin 1994, Ridgely & Tudor 1994) and has been considered vulnerable to forest disturbances and habitat fragmentation (Cofre 1999, Willson *et al.* 1994). Most breeding populations exhibit a southward migration in the spring, but small numbers remain within the northern part of the breeding range all year long (Chesser & Marin 1994). In the high latitudes of Tierra del Fuego and Cape Horn, it is said to be an accidental visitor. Humphrey *et at.* (1970) comment that:

"Philippi *et al.* (1954) stated that from comments by Reynolds and others they had hoped to be able to study the nesting of this little flycatcher during their trips to the southern regions, but it was so rare there that they were disillusioned. They never saw or heard it anywhere on Isla Grande (Tierra del Fuego)."

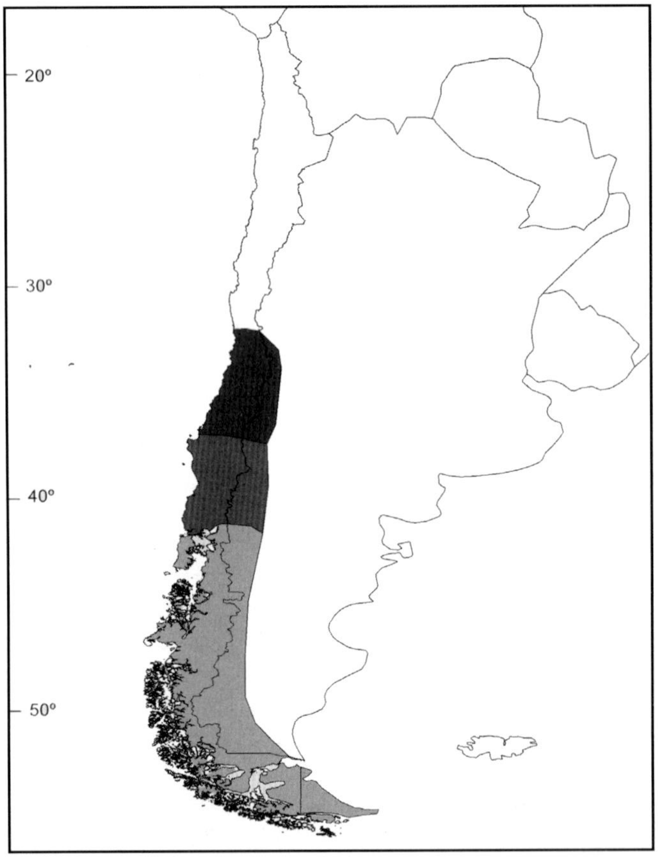

Fig. 1. Distribution map of *Colorhamphus parvirostris*; the black line delimits the restriction of its range to the southern temperate forests. The darkest grey area shows winter distribution, the intermediate shading represents year-round range and the white zone demonstrates summer distribution all the way south to Isla Navarino (Figure modified from Rozzi *et al.* 2003).

Indeed, south of the Strait of Magellan records of *C. parvirostris* are very scarce. Previously published records for this species south of the strait are: 1) a male collected in Lapataia, Tierra del Fuego in 1895 (Schalow 1898); 2) a specimen collected at False Cape Hom, southern Isla Hoste in 1895 (Schalow 1898); 3) a pair seen in January 1977 near Seno Ponsonby, Isla Hoste (Sielfeld 1977); and 4) an individual captured by mist-net in March 2000 on the northern coast of Isla Navarino (Anderson & Rozzi 2000). The latter authors suggested that the presence of *C. parvirostris* on Navarino Island might become more frequent in relation to an extension of its latitudinal range associated with global climate change. Such southward expansions of latitudinal ranges have been already reported and discussed for other species in southern Chile (McGehee *et al.* 1999).

In this article, we report additional observations of Patagonian Tyrants made at the southern tip of their range and discuss an alternative, and perhaps complementary, explanation to its presence on Navarino Island: the possibility of short post-breeding movements performed by some individuals prior to their northward migration. This pattern is common in other migratory passerine birds and can occur in both autumn (post-reproductive) and spring (pre-reproductive). Orientation errors, effect of wind drift and response to unsuitable weather have been suggested as possible explanations for reoriented movements (Akesson *et al.* 1996, Akesson 1999, Akesson & Sandberg 1999), but it is also possible that reverse migration is an adaptive strategy when species encounter ecological barriers that must be crossed (Akesson 1997, Bruderer & Liechti 1998). If such patterns are exhibited by *C. parvirostris*, then we should expect the presence of Patagonian Tyrants at high latitudes on Navarino Island (55°S) in late summer or early fall (post-reproductive season), before their northward migrations. To evaluate this pattern we analyze the abundance of *C. parvirostris* on the northern coast of Navarino Island at different months of the breeding and post-breeding seasons.

Study Area

The study was conducted on Navarino Island (55°S), which is situated between the Cape Horn Archipelago (56°S) and Tierra del Fuego (54°S). Mist-netting was conducted primarily at two sites on the northern coast of the island: the Omora Ethnobotanical Park (54°57'S; 67°39'W), near the town of Puerto Williams, and on Guerrico Hill (54°55'S, 67°54'W) (Fig. 2). To complement and expand the study done on the northern coast of Navarino Island, mist-netting was also done at three sites on the western coast of Navarino Island: Puerto Inutil (54°59'S, 68°13'W), Wulaia (55°03'S, 68°09'W) and Douglas Bay (55°05'S, 68°09'W), during January and February 2000. All sites are relatively low elevation, ranging from sea-level to 100 m.

The Omora Ethnobotanical Park comprises a mosaic of diverse forest types, which include old growth forests dominated by *Nothofagus* species, as well as naturally- and anthropogenically-perturbed forests. Among the latter is a shrubland area dominated by the Firebush (*Embothrium coccineum*), adjacent to evergreen forests of *Nothofagus betuloides* on a north-facing slope of Robalo

Mountain. The other site on Navarino Island is a gentle, north facing slope on Guerrico Hill (54°55'S, 67°54'W), which is also dominated by *E. coccineum*. This area is a post fire regenerated forest of *Nothofagus betuloides* and *N. pumilio* adjacent to the main dirt road along the northern coast of Navarino Island. As in Omora Park, the landscape near Guerrico comprises forests subject to natural and anthropogenic perturbations, including forest burning, wood cutting, livestock ranching, and disturbance effects of the introduced beaver (*Castor canadensis*) since the 1960s (Sielfeld & Venegas 1980).

Navarino Island is characterized by an oceanic climate type with a mean annual temperature of 5.6°C and an annual temperature fluctuation of less than 10°C (Rozzi 2002). Moreover, minimum temperatures below 0°C prevail only during the winter months of July and August, and rarely are lower than 5°C. In 2000, minimum mean temperature was 1.6°C and maximum mean temperature was 10.2°C (Rozzi 2002). The average annual rainfall in Puerto Williams is less than 500 mm, and in the year between November 1999 and October 2000, total precipitation was 429 mm (Rozzi 2002).

MATERIALS AND METHODS

We studied the presence and abundance of forest bird species using mist-netting (permits granted by the *Servicio Agricola y Ganadero* to S. McGehee and R. Rozzi) (for further details on mist-netting program see Anderson & Rozzi 2000, Anderson *et al.* 2002). A total of 2,023.54 hours of mist-netting were done on Navarino Island between January 2000 and May 2003 (Table 1). Mist-nets had a grid of squares with a diagonal of 30 mm. The nets were checked every 15 to 30 min. Each sampling used a number of mist-nets of 6 m long by 2.6 m height and/or 12 x 2.6 m. In order to standardize the sampling effort for each site, we considered the size of the net and the time of display. The products of the net-area (m^2) and the number of hours of net sampling (based on the time of installation and removal of each net) were calculated for each net and sampling day and summed up for each sampled month as h x m^2 (Table 1).

To compare presence/absence of *Colorhamphus parvirostris* with four other species, two residents (*Aphrastura spinicauda* and *Phrygilus patagonicus*), and two migratory (*Elaenia albiceps* and *Troglodytes aedon*), during the post-reproductive season and the rest of year, we used Fisher exact test. We defined the reproductive season as the period November to February, post- reproductive season from March to May, and pre-reproductive season from June to October.

RESULTS

In total, 1,542 individuals belonging to 17 forest bird species were caught with mist-nets on Navarino Island (Table 2). The most abundant species were *Aphrastura spinicauda*, *Phrygilus patagonicus*, *Carduelis barbata*, *Troglodytes aedon*, and *Elaenia albiceps*. In total, only 12 *Colorhamphus parvirostris* individuals were captured over 4 years, which is 0.7% of all birds caught. Therefore, *C. parvirostris* indeed can be considered a rare species for the avifauna of the forests of Navarino Island.

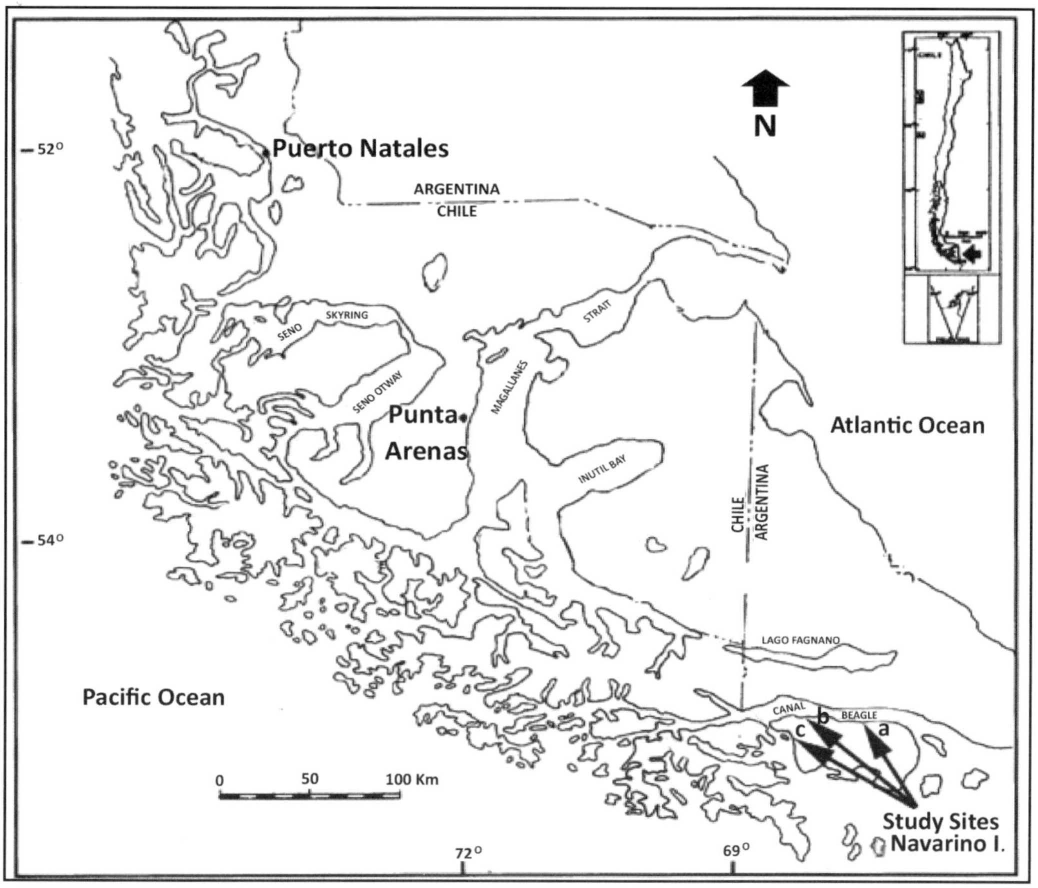

Fig. 2. Study sites on Navarino Island (55°S): a = Omora Ethnobotanical Park; b = Guerrico; c = West coast of Navarino (Puerto Inutil, Wulaia, Douglas Bay).

When analyzing the data for each month, an interesting pattern emerges: all *C. parvirostris* individuals were captured during the months of March or April. No Patagonian Tyrants were caught during other months, even in months when bird abundances were extremely high with totals of more than one hundred birds captured per day (*e.g.,* November 2000, January 2001, January and February 2002). On the other hand, during the months of March and April, captured *C. parvirostris* represented more than 2% of the total birds captured with mist-nets. Therefore, the recorded presence of the Patagonian Tyrant on Navarino Island has been restricted to the post-reproductive season.

To test and contrast the temporal pattern of *C. parvirostris* presence on Navarino Island, we compared its frequency of presence/absence during pre-reproductive, reproductive, and post-reproductive months, and we performed the same comparisons for four other species, two resident and two migratory (Table 3). A Fisher exact test, shows that the temporal distribution of Patagonian Tyrant restricted to the post-reproductive season (March and April) is significant (Fisher exact test, p=0.012). For the other four species, no significant difference was found regarding their pre-reproductive, reproductive, and post reproductive seasonal presence on Navarino Island: *T. aedon* (Fisher exact test, p>0.5), *E. albiceps* (p>0.5), *A. spinicauda* (p>0.9) and *P. patagonicus* (p>0.9). Therefore, *C. parvirostris* presents a distinct pattern of post-reproductive presence on Navarino Island, compared to other resident and migratory passerines.

DISCUSSION

Other than one sighting of a pair of *Colorhamphus parvirostris* on Isla Hoste (Sielfeld 1977), there was no previous evidence of breeding on the islands south of Tierra del Fuego (Johnson 1967). Most Patagonian Tyrants migrate further north in Chile and Argentina from their nest sites, and only one bird has been recorded south of 43 degrees during the austral winter (Chesser & Marin 1994). Our long-term mist-netting and banding program at the Omora Ethnobotanical Park first detected this species on Navarino Island in March 2000 (Anderson & Rozzi 2000). This study confirms the presence of *C. parvirostris* on Navarino Island to be more than a one time event. However, it is intriguing that this species was not reported previously for Navarino Island, and that the Patagonian Tyrant is the only forest bird species that lacks a Yaghan name (Massardo & Rozzi 2004).

The Yaghan culture has a rich ornithological knowledge (Rozzi *et al.* 2003, Rozzi 2004), and the lack of name for this species, which has a characteristic and conspicuous call, suggests that the Patagonian Tyrant may have not been present on Navarino Island and the Cape Hom archipelago region in earlier times. Consequently, Anderson & Rozzi (2000) suggest that the presence of *C. parvirostris* on Navarino Island today, might correspond to a recent latitudinal extension of its distribution range. Recent extensions for distribution ranges detected for bird species at high latitudes have been associated with global change (see McGehee

TABLE 1. Mist-net effort per month during the study period on Navarino Island. Mist-net effort was calculated as the number of hours each mist-net was open multiplied by the area of each mist-net (length x height of the mist-net). Percentage of *Colorhamphus parvirostris* is calculated as (number of caught *C. parvirostris*/number of caught birds) x 100.

Year	Month	Mist-net hours	Mist-net effort (h x m^2)	Number of birds caught	Number of caught *C. parvirostris*	Percent *C. parvirostris* (%)
2000	January	59.3	1796.2	43	0	0
2000	March	63.3	2258.1	40	1	2.5
2000	May	59.1	2111.3	52	0	0
2000	September	19.1	711.3	10	0	0
2000	November	108.3	3988.4	112	0	0
2000	December	144.4	4348.5	85	0	0
2001	January	156.2	5517.7	165	0	0
2001	February	41.3	1287.5	32	0	0
2001	December	15.1	545.3	12	0	0
2002	January	363.1	11328.2	129	0	0
2002	February	166.4	5190.1	165	0	0
2002	March	193.5	6035.6	150	3	2.0
2002	April	181.3	6015.1	183	4	2.2
2003	January	67.8	2035.8	24	0	0
2003	February	137.1	5544.2	90	0	0
2003	March	144.1	7284.4	60	2	3.3
2003	April	70.8	6290.4	99	2	2.0
2003	May	33.9	3351.4	91	0	0
Total		2023.5	75639.7	1542	11	0.7

TABLE 2. Bird species captured with mist-nets from January 2000 to May 2003. Shown in the right column is each species' relative abundance, as a percentage of total captures (N=1542).

Scientific name	Common name	Relative percentage (%)
Anairetes parulus	Tufted Tit-Tyrant	2.2
Aphrastura spinicauda	Thorn-tailed Rayadito	20.2
Carduelis barbata	Black-chinned Siskin	6.0
Cinclodes fuscus	Bar-winged Cinclodes	0.3
Cinclodes patagonicus	Dark-bellied Cinclodes	0.2
Colorhamphus parvirostris	**Patagonian Tyrant**	**0.7**
Elaenia albiceps	White-crested Elaenia	17.8
Enicognathus ferrugineus	Austral Parakeet	0.4
Glaucidium nanum	Austral Pygym-Owl	0.4
Phrygilus patagonicus	Patagonian Sierra-Finch	35.2
Pygarrhichas albogularis	White-throated Treerunner	0.7
Sephanoides galeritus	Green-backed Firecrown	0.7
Tachycineta meyeni	Chilean Swallow	1.6
Troglodytes aedon	Magellanic Wren	8.0
Turdus falcklandii	Austral Thrush	1.2
Xolmis pyrope	Fire-eyed Diucon	0.7
Zonotrichia capensis	Rufous-collared Sparrow	3.7

TABLE 3. Presence/absence on Navarino Island during pre-reproductive, reproductive, and post-reproductive seasons, of *Colorhamphus parvirostris*, and four other common species: two residents (*Aphrastura spinicauda* and *Phrygilus patagonicus*), and two migrants (*Elaenia albiceps* and *Troglodytes aedon*).

Season	Resident species		Migratory species		
	Aphrastura spinicauda	*Phrygilus patagonicus*	*Troglodytes aedon*	*Elaenia albiceps*	*Colorhamphus parvirostris*
Reproductive 2000 (January)	1	1	1	1	0
Post-reproductive 2000 (March)	1	1	0	1	1
Pre-reproductive 2000 (May and September)	1	1	0	0	0
Reproductive 2000-2001 (November-February)	1	1	1	1	0
Reproductive 2001-2002 (December-January)	1	1	1	1	0
Post-reproductive 2002 (March-April)	1	1	1	1	1
Reproductive 2003 (January-February)	1	1	1	1	0
Post-reproductive 2003 (March-April)	1	1	0	1	1
Pre-reproductive 2003 (May)	1	1	0	0	0

et al. 1999), and the presence of the Patagonian Tyrant might be another case.

However, if this were the only reason for the presence of Patagonian Tyrant in Cape Horn County, then one could expect to find them at any time of the Spring or Summer. The fact that the presence of *C. parvirostris* is restricted to the months of March and April seems to confirm our hypothesis of a reverse migration in autumn, because in this season the normal migration route would be toward the north. It is important to note that global warming could also be a contributing factor that has extended this reverse migration further southwards. A short southward migration performed by some Patagonian Tyrants prior to their northward migration might represent the first reported case of reverse post- reproductive migration for the Southern Hemisphere. Post-reproductive reverse migration, or pre-migratory vagrancy, seems to be common among some migratory passerine birds in the Northern Hemisphere (Akesson *et al.* 1996, Akesson 1997, Bruderer & Liechti 1998, Akesson 1999, Akesson & Sandberg 1999), but this phenomenon previously has not been observed in southern temperate latitudes.

ACKNOWLEDGMENTS

The authors thank the support of the Millennium Center for Advanced Studies in Ecology and Research on Biodiversity at the University of Chile (CMEB, Grant No. P02-051-F ICM). This article is part of the BIOKONCHIL biodiversity assessment project (FKZ 01 LM 0208, German Ministry of Education and Research, BMBF) and the ongoing research and conservation programs conducted by the Omora Ethnobotanical Park, Puerto Williams, Chile (www.omora.org). The capture of birds is authorized by the Chilean *Servicio Agrícola y Ganadero.*

REFERENCES

ÅKESSON, S. 1997. *Bimodal Orientation and the Adaptive Significance of Temporary Reverse Bird Migration in Autumn.* The Royal Institute of Navigation. Oxford.

ÅKESSON, S. 1999. Do Passerine migrants captured at an inland site perform temporary reverse migration in autumn? *Ardea* 87: 129-137.

ÅKESSON, S., L. KARLSSON, G. WALINDER & T. ALERSTAM. 1996. Bimodal orientation and the occurrence of temporary reverse bird migration during autumn in south Scandinavia. *Behavioral Ecology and Sociobiology* 38: 293-302.

ÅKESSON, S. & R. SANDBERG. 1999. Ecological barriers and the flexible orientation system of migrating birds. Pp. 1005-1016 *in Proc. 22 International Ornithological Congress.* N.J. Adams & R.H. Slotow (eds.), Durban. Johannesburg: BirdLife South Africa.

ANDERSON, C. & R. ROZZI. 2000. Bird assemblages in the southernmost forests of the world: Methodological variations for determining species composition. *Anales Instituto Patagonia, Serie Ciencias Naturales* (Chile) 28: 89-100.

ANDERSON, C., R. ROZZI, C. ELPHICK & S. MCGEHEE. 2002. El programa Omora de anillamiento de aves en los bosques subantárticos: estandarización del tamaño de anillos apropiados para las aves de la Región de Magallanes. *Boletín Chileno de Ornitología* 9: 2-11.

ARAYA, B., G. MILLIE & M. BERNAL. 1996. *Guía de Campo de las Aves de Chile.* 7ª edición, Editorial Universitaria, Santiago, Chile.

BRUDERER, B. & F. LIECHTI. 1998. Flight behaviour of nocturnally migrating birds in coastal areas crossing or coasting. *Journal of Avian Biology* 29: 499-507.

CHESSER, R.T. & M. MARIN. 1994. Seasonal distribution and natural history of the Patagonian Tyrant *(Colorhamphus parvirostris). Wilson Bulletin* 106: 649-667.

CLARK, G. 2001. Tyrant Flycatchers. Pp. 384-398. In *The Sibley Guide to Bird Life & Behavior,* C. Elphick, J. Dunning & D. Sibley, eds. Alfred A. Knopf, New York.

COFRÉ, H. 1999. Patrones de rareza de las aves del bosque templado de Chile: implicancias para su conservación. *Boletín Chileno de Ornitología* 6: 8-16.

FJELDSA, J. & N. KRABBE 1990. *Birds of the High Andes.* Zoological Museum, University of Copenhagen, Svendberg, Denmark.

HUMPHREY, P.S., D. BRIDGE, P.W. REYNOLDS & R.T. PETERSON. 1970. *Birds of Isla Grande* (Tierra del Fuego). Smithsonian Institution. Washington, DC.

JOHNSON, A.W. 1967. *The Birds of Chile and Adjacent Regions of Argentina, Bolivia and Peru.* Vol 2, Platt Establ. Gráficos, Buenos Aires, Argentina.

MARIN, M., L. KIFF & L. PEÑA. 1989. Notes on Chilean birds, with descriptions of two new subspecies. *Bulletin of the British Ornithological Club* 109: 66-82.

MASSARDO, F. & R. ROZZI. 2004. Etno-ornitología yagán y lafkenche en los bosques templados de Sudamérica austral. *Ornitología Neotropical* 15: 395-407.

MCGEHEE, S., R. ROZZI, J.C. TORRES-MURA & M. WILLSON. 1999. Observaciones del bailarín *(Elanus leucurus)* en Chiloé (X Región). *Boletín Ornitológico de Chile* 6: 22-24.

MOORE, D.M. 1983. *Flora of Tierra del Fuego.* Anthony Nelson, England.

PHILIPPI, R.A., A.W. JOHNSON, J.D. GOODALL & F. BEHN. 1954. Notas sobre aves de Magallanes y Tierra del Fuego. *Boletín del Museo Nacional de Historia Natural* 26: 1-65.

RIDGELY, R.S. & G. TUDOR. 1994. *The Birds of South America.* Vol 2, University of Texas Press, Austin.

ROZZI, R. 2002. Biological and Cultural Conservation in the Archipelago Forest Ecosystems of Southem Chile. Ph.D. Dissertation, Department of Ecology and Evolutionary Biology, University of Connecticut, USA.

ROZZI, R. 2004. Implicaciones éticas de narrativas yaganes y mapuches sobre las aves de los bosques templados de Sudamérica austral. *Ornitología Neotropical* 15: 435-444.

ROZZI, R., D. MARTÍNEZ, M.F. WILLSON & C. SABAG. 1996a. Avifauna de los bosques templados de Sudamérica. Pp. 135-152 *in Ecología de los Bosques Nativos de Chile,* J.J. Armesto, C. Villagrán & M.T. Kalin (eds.). Editorial Universitaria, Santiago, Chile.

ROZZI, R., J.J. ARMESTO, A. CORREA, J.C. TORRES-MURA & M. SALLABERRY. 1996b. Avifauna de bosques primarios templados en islas deshabitadas del archipiélago de Chiloé. *Revista Chilena de Historia Natural* 69: 125-139.

ROZZI, R., F. MASSARDO, J. SILANDER, JR., C. ANDERSON & A. MARIN. 2003. Conservación biocultural y ética ambiental en el extremo austral de América: oportunidades y dificultades para el bienestar ecosocial. Pp. 51-85 *in Biodiversidad y Globalización*, E. Figueroa & J. Simonetti (eds.). Editorial Universitaria, Santiago, Chile.

SCHALOW, H. 1898. Die Vogel der Sammlung Plate. *Zoologisches Jahrzeitung Suppl.* 4: 641-748.

SIELFELD, W. 1977. Reconocimiento macrofaunístico terrestre en el área del Seno Ponsonby (Isla Hoste). *Anales del Instituto de la Patagonia* (Chile) 8: 275-296.

WELTY, J. 1975. *The Life of Birds.* W.B. Saunders. Philadelphia, USA.

WILLSON, M.F., T.L. DE SANTO, C. SABAG & J.J. ARMESTO. 1994. Avian communities of fragmented South-temperate rainforests in Chile. *Conservation Biology* 8: 508-520.

9 THE AUTECOLOGY OF THE FÍO-FÍO (*ELAENIA ALBICEPS* LAFRESNAYE & D'ORBIGNY) IN SUBANTARCTIC FORESTS OF THE CAPE HORN BIOSPHERE RESERVE, CHILE

CLARE E. BROWN[1,2], CHRISTOPHER B. ANDERSON[1,2,3], SILVINA IPPI[1,4], MARGARET F. SHERRIFFS[1,5], RINA CHARLIN[1,3], STEVE MCGEHEE[6], AND RICARDO ROZZI[1,3,7]

[1]Parque Etnobotánico Omora, Universidad de Magallanes, Puerto Williams, Chile
[2]Omora Sub-Antarctic Research Alliance, Tacoma, Washington, USA
[3]Instituto Milenio de Ecología y Biodiversidad, Casilla 653, Santiago, Chile
[4]Departamento de Ecología Terrestre, Facultad de Ciencias, Universidad de Chile, Santiago, Chile
[5]Center for Population Biology, Section of Evolution and Ecology, University of California, Davis, California, USA
[6]Department of Biology, University of Victoria, Victoria, Canada
[7]Department of Philosophy & Religion Studies, University of North Texas, Denton, Texas, USA

Anales del Instituto de la Patagonia (Chile) 35: 29-40 (2007)

ABSTRACT. The autecology of the Fío-fío (*Elaenia albiceps* Lafresnaye & d'Orbigny) in Sub-Antarctic forests of the Cape Horn Biosphere Reserve, Chile. As the most diverse and abundant terrestrial vertebrates in the Cape Horn Archipelago, birds potentially play key roles in the ecology of subantarctic ecosystems. However, the lack of long-term ornithological studies in southern South America has previously hindered the identification of relevant aspects of the autecology of even the most common species. One abundant species in the austral summer is the Fio-fio (*Elaenia albiceps*), which migrates annually between the tropical regions of South America (Peru, Bolivia, Ecuador, Colombia, Paraguay, & Brasil) to as far south as the Cape Horn Biosphere Reserve (CHBR), Chile. Using the database accumulated during the 8 year mist-netting program of Passeriformes forest birds at the Omora Ethnobotancial Park, it was possible to describe various aspects of the Fio-fio's ecology in the CHBR by determining: a) longevity and site fidelity, b) arrival-departure dates, c) abundance of adults and juveniles, d) diet and e) morphology. A total 827 Fío-fíos were banded during the study period. Of the 67 recaptures, it was possible to determine that Fio-fios may live at least up to 7 years. Fio-fios arrived as early as 28 October and left as late as 15 April with juveniles fledging in January. Their abundance increased from October to reach approximately 35% of the forest bird assemblage by January. In the austral archipelago, the Fio-fios diet consisted of 10 invertebrate orders and at least five plant species. Males had significantly longer tails and wings than females, while females were significantly heavier in the pre-fledging season (Oct-Dec). This study provides new information on the autecology of Fio-fio in the world's southernmost forests.

Key words: avian community, ecological role, *Elaenia albiceps,* niche, *Nothofagus.*

RESUMEN. Autoecología del fío-fío (*Elaenia albiceps* Lafresnaye & d'Orbigny) en bosques subantárticos de la Reserva de Biosfera Cabo de Hornos, Chile. Las aves son los vertebrados terrestres más diversos en el archipiélago del Cabo de Hornos. Por esta razón, muchas de las especies de aves potencialmente pueden desempeñar un papel clave en las interacciones bióticas de los ecosistemas subantárticos. Sin embargo, la falta de estudios ornitológicos a largo plazo en el sur de Sudamérica ha limitado la identificación de aspectos relevantes de la autoecología de las aves, incluso de las especies más comunes. Entre ellas, una especie muy abundante en el verano austral es el fío-fío (*Elaenia albiceps*). Ésta migra anualmente desde regiones de bosques tropicales de Sudamérica (Perú, Bolivia, Ecuador, Colombia, Paraguay y Brasil) hasta los bosques templados del sur de Chile y Argentina, alcanzando incluso los bosques del extremo austral de América en la Reserva de la Biosfera Cabo de Hornos (RBCH). Utilizando la base de datos del programa de anillamiento de Passeriformes de bosque en el Parque Etnobotánico Omora iniciado hace 8 años, se describen cinco aspectos de la autoecología del fío-fío en la RBCH: a) longevidad y filopatría, b) fechas de arribo y partida de la región, c) abundancia de adultos y juveniles, d) dieta y e) morfología. Durante el período de estudio se han anillado 827 fío-fíos. De éstos, 67 fueron recapturados, lo que permitió determinar que los fío-fíos pueden vivir hasta al menos 7 años. El primer y último registro de fío-fíos en el área de estudio fueron el 28 de octubre y el 15 de abril, respectivamente. Los juveniles abandonaron el nido en enero. La abundancia aumentó a partir de octubre hasta enero cuando constituyó el 35% del ensamble de aves de bosque. La dieta del fío-fío se compuso de 10 órdenes de invertebrados y a lo menos cinco especies de plantas. Los machos tuvieron colas y alas significativamente más largas que las hembras, mientras éstas fueron significativamente más pesadas en la estación pre-reproductiva (oct-dic). Este estudio provee nueva información de la autoecología del fío-fío en los bosques más australes del mundo.

Palabras clave: comunidad de aves, *Elaenia albiceps,* nicho, *Nothofagus,* rol ecológico.

INTRODUCTION

The South American temperate forest biome occurs between 35° and 56°S (Arroyo *et al.* 1996, Armesto *et al.* 1996a) and is characterized by relatively low species richness and high levels of endemism, compared to Northern Hemisphere temperate forests (Armesto *et al.* 1995, Veblen *et al.* 1996). Close to 90% of the woody plant species found in this ecoregion are endemic to the temperate forests of South America, and among vertebrates, we find endemic species representing: 50% of fishes (26 spp.), 80% of amphibians (30 spp.), 36% of reptiles (16 spp.), 30% of land birds (44 spp.), and 33% of mammals (33 spp.) (Armesto *et al.* 1996b). At the southern limit of this biome, we find the world's southernmost forested ecosystems, encompassed in the Cape Horn Biosphere Reserve (CHBR) (approx. 55°S), declared by UNESCO on June 25, 2005, which is the largest biosphere reserve in the Southern Cone and the only one in Chile to include terrestrial as well as marine ecosystems and human populations (Rozzi *et al.* 2006, 2007).

Not only is the degree of endemism of the southern temperate forest region unusually high, but so is the frequency of plant-animal mutualisms, particularly related to seed dispersal and pollination (Aizen & Ezcurra 1998). A high percentage (67%) of plant species have fleshy fruits and avian seed dispersal, and more than 50% of the plant species require a biotic pollinator (Armesto *et al.* 1987, Armesto & Rozzi 1989, Riveros 1991, Rozzi *et al.* 1996a,b, Willson *et al.* 1996, Aizen & Ezcurra 1998). Most bird-plant pollination and dispersal studies, however, have been conducted in south-central Chile and the Valdivian rainforests (*e.g.*, Rozzi *et al.* 1996a).

South of the Gulf of Penas (49°S), the Magellanic subantarctic forest ecoregion extends over the archipelagoes reaching Cape Horn (56°S), forming one of the most pristine wilderness areas remaining on the planet (Mittermeier *et al.* 2001, Rozzi *et al.* 2006). Given its recent glacial history and the biogeographic barriers presented by the numerous channels and fjords, the terrestrial vertebrate communities in this region are relatively species poor, and birds constitute a diverse and ecologically important component of the archipelago's fauna (Venegas & Sielfeld 1998, Anderson *et al.* 2006).

Seminal research on the songbirds of the Magellanic subantarctic forests has permitted an initial understanding of the forest avian assemblage (Barros 1971, 1976, Sielfeld 1977, Venegas 1981, 1991, Anderson & Rozzi 2000). Nonetheless, the autecology of many species in this region remains poorly known (but see Schlatter & Vergara 2005). To fill this gap in our knowledge, the Omora Ethnobotanical Park, a Long-Term Ecological Research (LTER) site run by the University of Magallanes and Institute of Ecology and Biodiversity located near Puerto Williams (55°S) in the CHBR, began conducting a continuous mist-netting program of forest birds in January 2000. The program aims to provide answers to basic ecological questions about subantarctic avifauna, including life history, diet, morphology, and migration patterns.

Among forest birds, the Fio-fio (White-crested Elaenia or *Elaenia albiceps* Lafresnaye & d'Orbigny 1837) is a species of particular relevance due to its summer-time abundance and because it is the only member of the forest avian community that undertakes long-distance migrations between tropical and temperate forests of South America (Fjeldsa & Krabbe 1990, Rozzi *et al.* 1996a). It is known that Elaenia winters in the tropics, and based on observational studies conducted in its winter range, *E. albiceps* has been described as present in eastern Paraguay during the austral spring (October-November) and autumn (March-April) months (Hayes *et al.* 1994). Data from Chile in the Fio-fio's summer habitat indicate that *E. albiceps* migrates down the eastern edge of the Andes and crosses laterally into Chile from Argentina in an east-west direction, arriving as far south as the subantarctic archipelago (Goodall *et al.* 1951, Araya *et al.* 1995, Espinoza & Egli 1997, Venegas & Siefeld 1998).

As a migratory species inhabiting a very wide latitudinal range, the Fío-fío has shown wide feeding plasticity and thus occupies a broad and diverse ecological niche (Armesto *et al.* 1987, Rozzi *et al.* 1996a,b). On Chiloé Island, in spring it is largely a nectarivore (Smith-Ramírez & Armesto 1998, 2003), but in summer becomes principally frugivorous (Sabag 1993), despite pertaining to the family of "flycatchers." On the other hand, in Argentina's Nahuel Huapi National Park, the summer diet of the Fio-fio was constituted of 70% insects and 30% plant items (Grigera 1982). In Central Chile, López–Calleja (1990) recorded a diet composed mostly of seeds (41.67% summer and 1.22% spring) and insects (54.33% summer and 98.79% spring). Additionally, in its winter range, the genus *Elaenia* has been reported to be principally frugivorous in the Amazonian sector of northeastern Peru (Fitzpatrick 1980), and it is known to be an omnivorous fruit and insect eater in September and October in Paraguay (Foster 1987).

The goal of this study was to characterize five essential attributes of the autecology of *Elaenia albiceps* in the Magellanic subantarctic forests: a) longevity-site fidelity, b) arrival-departure dates, c) abundance of adult and juvenile individuals, d) diet, and e) morphology. To develop this analysis, we used the database of the long-term mist-netting program at the Omora Ethnobotancial Park on Navarino Island in the Cape Horn Biosphere Reserve, Chile, with over 6,000 captures since the year 2000.

MATERIALS AND METHODS

Study area

The study was conducted at sites on Navarino Island in the CHBR (Fig. 1). All surveyed locations were within 10 km of the coast on the northern and western slope of the island, which is characterized by mixed *Nothofagus* forests with a mosaic of bogs, high-Andean, and matorral ecosystems (Rozzi *et al.* 2006). The long-term mist-netting program is conducted within the Omora Ethnobotanical Park, which is approximately 4 km west of the town of Puerto Williams. Other sites outside the Omora Park were mist-netted only in summer and with point-count transects. These other study sites included: Mejillones and Guerrico on the island's north coast; Puerto Inútil and Wulaia on the west coast in the area of the Murray Channel; and Cerro La Bandera and Cerro Róbalo (adjacent to the Omora Park), which were the only high elevation sites studied (>350 m).

Mist-netting program

From 2000 to 2002, mist-netting was conducted only during spring, summer, and fall seasons. Systematic, year-round mist-netting began in January 2003, including 6 consecutive days each month with a minimum of 5 hours each day. Two permanent sites have been used: a) "Canelo Site," which has as dominant vegetation of Winter's Bark (Canelo in Spanish or *Drimys winteri*) in a mixed forest of Magellanic Coigue (*Nothofagus betuloides*) and Lenga (*N. pumilio*) and b) "Notro Site," which is an anthropogenically created shrubland with Firebush (Notro in Spanish or *Embothrium coccineum*), as well as Fashine (*Chilotrichium difussum*) and Box-leafed Barberry (*Berberis buxifolia*). The 6 days of netting are divided with 3 days each in the Canelo and Notro sites.

To date, the program has produced a working database of more than six thousand bird captures, 827 of which are Fio-fios (excluding recaptures). The protocol of the banding program is as per Anderson *et al.* (2002). Once banded, each bird is subjected to a set of measurements including: bill length, wing length, tarsal length, tail length, and mass. A caliper (0.1 mm) was used to measure bill length at the culmen and the tarsal bone. Wing and tail lengths were recorded with a ruler (1 mm), and mass was taken with a spring scale (1 g). When possible, sex and age were also determined. Additionally, all birds heard and seen during the mist-netting period were recorded.

During the years 2002-2003, all birds that defecated during handling had their feces collected for subsequent diet analysis. Samples were preserved and dried in paper envelopes, and later contents were identified using a dissecting microscope. In addition, the Fio-fios caught from December to January 2002 were sampled for nectar consumption by using Scotch tape to remove pollen grains from above and below the beak of the bird. The tape was then placed on a microscope slide to be observed under a compound microscope. Reference collections of common plant seeds and pollen grains were made to determine the species of remains in feces and the pollen on slides.

Data analysis

Recapture data were examined to determine minimum age of recaptured Fio-fios. Individuals were classified as at least 1 year old at the time of their first capture, unless they were a fledgling. In addition, we used these data to determine the percent of individuals that returned to the same nesting-feeding area each austral summer. Furthermore, arrival and departure dates were determined based on the first and last date in which Fio-fios were heard, seen or captured in the mist-nets.

Mist-netting capture data were standardized to account for sampling effort, whereby the number of Fio-fios captured was divided by the product of the area of the net used each sample day (m^2) multiplied by the time the net was open (h) to determine the abundance (number of individuals captured m^{-2} x h^{-1}), as per Anderson & Rozzi (2000). The abundance of Fío-fíos was then compared to the total abundance of songbirds over the eight breeding seasons to determine the mean relative abundance of *E. albiceps* in subantarctic forests on a monthly basis.

Dietary components found in the feces of the Fio-fio were tabulated as the percent of individuals having a particular food item in their diet and presented as annual and seasonal (spring and summer) values. Items were determined to the level of order for insects and species for plants. Morphological differences between

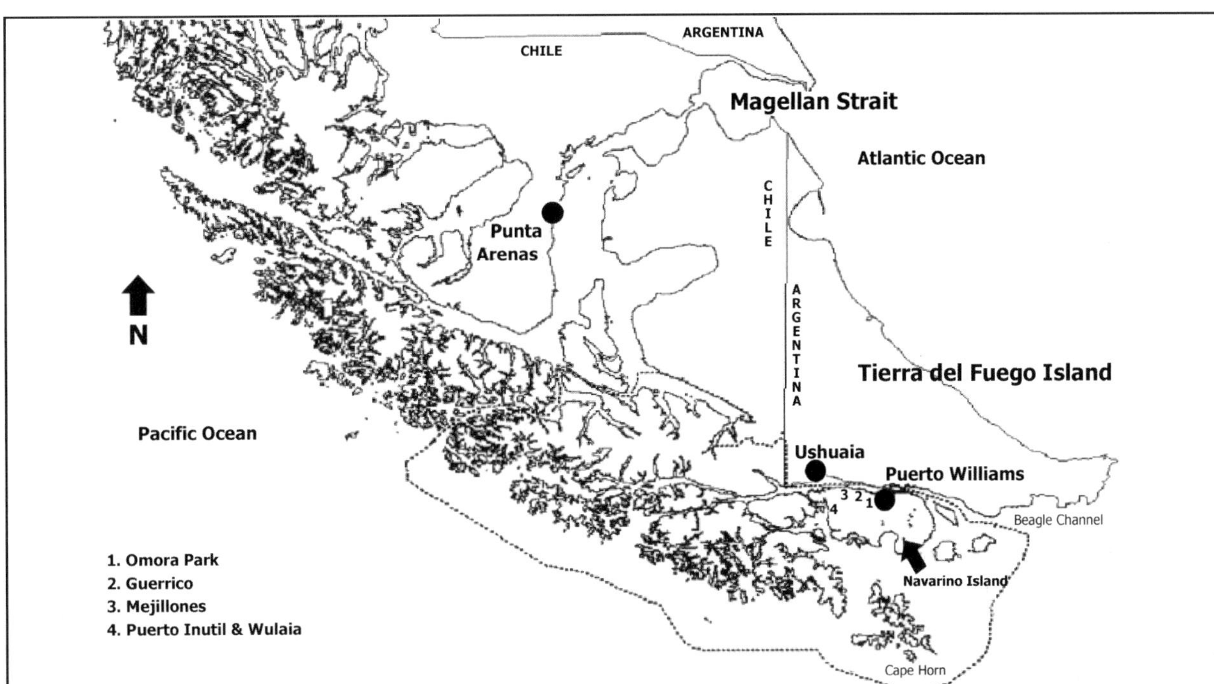

Fig. 1. Mist-netting was conducted systematically from 2000 to 2007 at the Omora Ethnobotanical Park (University of Magallanes – Institute of Ecology and Biodiversity) on Navarino Island in the Cape Horn Biosphere Reserve (dotted line). Summer surveys were conducted as well at Guerrico, Mejillones, Puerto Inutil, and Wulaia.

TABLE 5. Mean morphological measurements of the Fio-fio (*Elaenia albiceps*) (± 1S.D.) from the Cape Horn Biosphere Reserve. Significant differences between male and female characteristics were determined with ANOVA and Tukey HSD *post-hoc* test, p <0.05).

	Wing mm	Tarsus mm	Bill length mm	Tail mm	Mass g
Male	76.2 (2.7)	20.3 (1.6)	8.8 (0.8)	64.2 (3.3)	16.0 (1.1)
Range	66-86	16.5-23.4	5.3-11.2	50-76	12.3-22.4
N=272	N=249	N=196	N=194	N=167	N=256
Female	73.8 (2.4)	20.18 (1.72)	8.6 (0.7)	63.0 (2.5)	16.3 (1.3)
Range	63-81	12.8-24.4	7.1-11.0	56-71	13.7-21.1
N=222	N=210	N=163	N=160	N=137	N=208
F	101.15	0.21	2.67	14.02	9.53
p	<0.0001	0.65	0.1	0.0002	0.002

Society Group of the Institute of Ecology and Biodiversity (www.ieb-chile.cl). This study has been carried out by numerous assistants and volunteers and with the funding of the Millennium Scientific Initiative MIDEPLAN (ICM P05-002), the Earthwatch Institute and the BIOKONCHIL Project (FKZ 01 LM 0208 BMBF). The authors recognize the following institutional support: CEB: OSARA (www.osara.org); CBA: a postdoctoral fellowship from the Institute of Ecology and Biodiversity; SI: a doctoral scholarship from CONICYT; MFS: Gordon Grand Fellowship (Yale Univ.).

REFERENCES

AIZEN, M. & C. EZCURRA. 1998. High incidence of plant-animal mutualisms in the woody flora of the temperate forest of southern South America: biogeographical origin and present ecological significance. *Ecología Austral* 8: 217-236.

ANDERSON, C.B. & R. ROZZI. 2000. Avian communities in the southernmost forests of the world: Methodological variations for determining species composition. *Anales del Instituto de la Patagonia Serie Ciencias Naturales* (Chile) 28: 89-100.

ANDERSON, C.B., R. ROZZI, C. ELPHICK & S. MCGEHEE. 2002. El programa Omora de anillamiento de aves en los bosques subantárticos: La estandarización del tamaño de anillos apropiados para las aves de la Región de Magallanes. *Boletín Chileno de Ornitología* 9: 2-11.

ANDERSON, C.B., R. ROZZI, J.C. TORRES-MURA, S.M. MCGEHEE, M.F. SHERRIFFS, E. SCHUETTLER & A.D. ROSEMOND. 2006. Exotic vertebrate fauna in the remote and pristine sub-Antarctic Cape Horn Archipelago region of Chile. *Biodiversity and Conservation* 10: 3295-3313.

ARAYA, B., G. MILLIE & M. BERNAL. 1986. *Guía de Campo de las Aves de Chile*. Editorial Universitaria, Santiago, Chile.

ARMESTO, J.J., R. ROZZI, P. MIRANDA & C. SABAG. 1987. Plant/frugivore interaction in South American temperate forests. *Revista Chilena de Historia Natural* 60: 321-336.

ARMESTO, J.J. & R. ROZZI. 1989. Seed dispersal in the rain forests of Chiloé: a survey of syndromes in comparison with other temperate and tropical forests. *Journal of Biogeography* 16: 219-226.

ARMESTO, J.J., C. VILLAGRÁN & M.T. KALIN (EDS.). 1996a. *Ecología de los Bosques Nativos de Chile*. Editorial Universitaria, Santiago, Chile.

ARMESTO, J.J., R. ROZZI & P. LEÓN-LOBOS. 1996b. Ecología de los bosques chilenos: síntesis y proyecciones. Pp. 405-421 *in Ecología de los Bosques Nativos de Chile,* J.J. Armesto, C. Villagrán & M.T. Kalin (eds.). Editorial Universitaria, Santiago, Chile.

ARMESTO, J.J., P.L. LOBOS & M.T.K. ARROYO. 1996. Los bosques templados del sur de Chile y Argentina: una isla biogeográfica. Pp. 23-28 *in Ecología de los Bosques Nativos de Chile,* J.J. Armesto, C. Villagrán & M.T.K. Arroyo (eds.). Editorial Universitaria, Santiago, Chile.

ARROYO, M.T.K., M. RIVEROS, A. PEÑALOZA, L. CAVIERES & A.M. FAGGI. 1996. History and regional patterns of the cool temperate rainforest flora of Southern South America. Pp. 134-172 *in High Latitude Rain Forests and Associated Ecosystems of the West Coast of the Americas: Climate, Hydrology, Ecology and Conservation*, R.G. Lawford, P. Alaback & E.R. Fuentes (eds.). Springer, Berlin.

BARROS, A. 1971. Aves observadas en las islas Picton, Nueva, Lennox y Navarino oriental. *Anales del Instituto de la Patagonia* (Chile) 2: 166-180.

BARROS, A. 1976. Nuevas aves observadas en las islas Picton, Nueva, Lennox y Navarino oriental. *Anales del Instituto de la Patagonia* (Chile) 7: 190-193.

BERGMANN, C. 1847. Über die Verhältnisse der Wärmeökonomie der Thiere zu Ihrer Grösse. *Goettische Studien* 1: 595-708.

DEFERRARI, G., C. CAMILIONI, G. MARTÍNEZ PASTUR & P.L. PERI. 2001. Changes in *Nothofagus pumilio* forest biodiversity during the forest management cycle. 2. Birds. *Biodiversity and Conservation* 10: 2093-2108.

ESPINOZA, L.A. & G. EGLI. 1997. Nueva información biométrica y conductual del Fio-fio (*Elaenia albiceps chilensis*). *Boletín Chileno de Ornitología* 4: 9-13.

FITZPATRICK, J.W. 1980. Foraging behavior of Neotropical tyrant flycatchers. *Condor* 82: 43-57.

FJELDSA, J. & N. KRABBE. 1990. *Birds of the High Andes*. Zoological Museum, University of Copenhagen and Apollo Books, Svedbord, Denmark. 876 pp.

FOSTER, M.S. 1987. Feeding methods and efficiencies of selected frugivorous birds. *Condor* 89: 566-580.

GOODALL, J.D., A.W. JOHNSON & R.A. PHILIPPI. 1951. *Las Aves de Chile, su Conocimiento y sus Costumbres*. Volumen II. Platt Establecimientos Gráficos, Buenos Aires, Argentina. 442 pp.

GRIGERA, D.E. 1982. Ecología alimentaria de algunas passeriformes insectívoras frecuentes en los alrededores de S. C. de Bariloche. *Ecología Argentina* 7: 67-84.

HAYES, F.E., P.A. SCHARF & R.S. RIDGELY. 1994. Austral bird migrants in Paraguay. *Condor* 96: 83-97.

LENCINAS, M.V., G. MARTÍNEZ PASTUR, M. MEDINA & C. BUSSO. 2005. Richness and density of birds in timber *Nothofagus pumilio* forests and their unproductive associated environments. *Biodiversity and Conservation* 14: 2299-2320.

LÓPEZ-CALLEJA, M.V. 1990. *Variación Estacional en el Uso de los Recursos Alimenticios por Algunos Componentes de una Taxocenosis de Aves Paseriformes en Quebrada de la Plata, Chile Central*. Tesis de Magíster. Universidad de Chile.

MARANTZ, C.A. & J.V. REMSEN, JR. 1991. Seasonal distribution of the Slaty Elaenia, a little-known austral migrant of South America. *Journal of Field Ornithology* 62: 162-172.

MCNEIL, R. 1982. Winter resident repeats and returns of austral and boreal migrant birds banded in Venezuela. *Journal of Field Ornithology* 53: 125-132.

MITTERMEIER, R., C. MITTERMEIER, P. ROBLES-GIL, J. PILGRIM, G. FONSECA, J. BROOKS & J. KONSTANT. 2002. *Wilderness: Earth's Last Wild Places*. Cemex and Conservation International, Washington D.C. 573 pp.

RIVEROS, M. 1991. Aspectos sobre la Biología Reproductiva en dos Comunidades del Sur de Chile (40ºS). Ph.D. Dissertation, Facultad de Ciencias, Universidad de Chile. Santiago, Chile.

ROZZI, R. & P. FEINSINGER. 2006. Desafíos para la conservación biológica en Latinoamérica. Pp. 661-688 *in Fundamentos de Conservación Biológica: Perspectivas Latinoamericanas*, R. Primack, R. Rozzi, P. Feinsinger, R. Dirzo & F. Massardo. Second Edition. Fondo de Cultura Económica, México.

ROZZI, R., D. MARTÍNEZ, M.F. WILLSON & C. SABAG. 1996a. Avifauna de los bosques templados de Sudamérica. Pp. 135-152 *in Ecología de los Bosques Nativos de Chile*, J.J. Armesto, C. Villagrán & M.T. Kalin (eds.). Editorial Universitaria, Santiago, Chile.

ROZZI, R., J.J. ARMESTO, A. CORREA, J.C. TORRES-MURA & M. SALLABERRY. 1996b. Avifauna de bosques primarios templados en islas deshabitadas del archipiélago de Chiloé. *Revista Chilena de Historia Natural* 69: 125-139.

ROZZI, R., F. MASSARDO, A. BERGHOEFER, C.B. ANDERSON, A. MANSILLA, M. MANSILLA & J. PLANA (EDS.). 2006. *Reserva de Biosfera Cabo de Hornos*. Ediciones de la Universidad de Magallanes. Punta Arenas, Chile. 274 pp.

ROZZI, R., F. MASSARDO, A. MANSILLA, C.B. ANDERSON, A. BERGHÖFER, M. MANSILLA, M.R. GALLARDO, J. PLANA, U. BERGHÖFER, X. ARANGO, S. RUSSELL, P. ARAYA & E. BARROS. 2007. La Reserva de Biosfera Cabo de Hornos: un desafío para la conservación de la biodiversidad e implementación del desarrollo sustentable en el extremo austral de América. *Anales Instituto de la Patagonia* (Chile) 35: 55-70.

SABAG, C.E. 1993. *El Rol de las Aves en la Dispersión de Semillas en un Bosque Templado Secundario de Chiloé (42ºS)*. Tesis de Magíster. Universidad de Chile.

SCHLATTER, R.P. & P. VERGARA. 2005. Magellanic woodpecker *(Campephilus magellanicus)* sap feeding and its role in the Tierra del Fuego forest bird assemblage. *Journal of Ornithology* 146: 188-190.

SIELFELD, W.H. 1977. Reconocimiento macro-faunístico terrestre en el área de Seno Ponsonby (Isla Hoste). *Anales del Instituto de la Patagonia* (Chile) 8: 275-297.

SMITH-RAMÍREZ, C. & J. ARMESTO. 1998. Nectarivoría y polinización por aves en *Embothrium coccineum* (Proteaceae) en el bosque templado del sur de Chile. *Revista Chilena de Historia Natural* 71: 51-63.

SMITH-RAMÍREZ, C. & J. ARMESTO. 2003. Foraging behaviour of bird pollinators on *Embothrium coccineum* (Proteaceae) trees in forest fragments and pastures in southern Chile. *Austral Ecology* 28: 53-60.

VEBLEN, T.T., R.S. HILL & J. READ. 1996. *The Ecology and Biogeography of Nothofagus Forests*. Yale University Press, New Haven. 403 pp.

VENEGAS, C. 1981. Aves de las Islas Wollaston y Bayly, Archipiélago del Cabo de Hornos. *Anales del Instituto de la Patagonia* (Chile)12: 213-219.

VENEGAS, C. 1991. Ensambles avifaunísticos estivales del Archipiélago Cabo de Hornos. *Anales del Instituto de la Patagonia, Serie Ciencias Naturales* (Chile) 20: 69-81.

VENEGAS, C. & W. SIELFELD. 1998. *Catálogo de los Vertebrados de la Región de Magallanes y Antártica Chilena*. Ediciones de la Universidad de Magallanes, Punta Arenas. 122 pp.

VENEGAS, C. & R.P. SCHLATTER. 1999. Efecto de la intervención silvícola en bosques de *Nothofagus pumilio* sobre sus ensambles avifaunísticos estivales en Tierra del Fuego (Chile). *Anales del Instituto de la Patagonia, Serie Ciencias Naturales* (Chile) 27: 41-50.

VERGARA, P. & R.P. SCHLATTER. 2006. Aggregate retention in two Tierra del Fuego *Nothofagus* forests: short term effects on bird abundance. *Forest Ecology and Management* 225: 213-224.

WILLSON, M.F., C. SMITH-RAMÍREZ, C. SABAG & J.F. HERNÁNDEZ. 1996. Mutualismos entre plantas y animales en bosques templados de Chile. Pp. 251-264 *in Ecología de los Bosques Nativos de Chile,* J.J. Armesto, C. Villagrán & M.T. Kalin (eds.). Editorial Universitaria, Santiago, Chile.

YOUNG, B.E. 1994. Geographic and seasonal patterns of clutch-size variation in house wrens. *Auk* 111: 545-555.

Received 06/28/2007. Accepted 10/9/2007.

10 DIET OF THE PATAGONIAN SIERRA-FINCH (*PHRYGILUS PATAGONICUS*) ON NAVARINO ISLAND, CHILE

STEVEN M. McGEHEE[1,2] AND JACK CLINTON EITNIEAR[2]

[1]Parque Ethnobotánico Omora, Universidad de Magallanes, Fundación Omora, Isla Navarino, Magallanes, Chile.
[2]Center for the Study of Tropical Birds Inc., 218 Conway Drive, San Antonio, Texas 78109, USA. E-Mail: jce@cstbinc.org

Ornitología Neotropical 18: 499-452 (2007)

Diet of the Patagonian Sierra-Finch (*Phrygilus patagonicus*) on Navarino Island, Chile.

Key words: Chile, diet, Navarino Island, Patagonian Sierra-Finch, *Phrygilus patagonicus*.

Dieta del cometocino patagónico (*Phrygilus patagonicus*) en la Isla Navarino, Chile.

Palabras clave: Chile, cometocino patagónico, dieta, isla Navarino, *Phrygilus patagonicus*.

The Patagonian Sierra-Finch (*Phrygilus patagonicus*) is one of the most abundant species in the Subantarctic Magellanic evergreen forest ecoregion (Anderson & Rozzi 2000). It ranges from sea level to 1800 m and is most often seen in forests (Rozzi *et al.* 2003, Martínez & González 2005) and forest borders or shrubby cleared areas in southern Chile and Argentina (Ridgely & Tudor 1989, Vuilleumier 1991). In spite of its high abundance and potentially key ecological role, little is known about the diet of this species and Sierra-Finches (genus *Phrygilus* and subfamily Emberizinae) in general. Published records of Patagonian Sierra-Finch diet consist of anecdotal observations, stomach contents (Humphrey *et al.* 1970) and observations of nectar robbing (Traveset *et al.* 1998). We present here Patagonian Sierra-Finch dietary information at the southern extreme of its distribution on Navarino Island (55ºS), Cape Horn County, Chile.

From 25 September 2004 to 12 May 2006, Patagonian Sierra-Finches were observed in Omora Ethnobotanical Park (54°55´S, 67°39´W). The park is located at the northern end of Navarino Island (Anderson *et al.* 2002), 5 km east of the town of Puerto Williams. Navarino Island receives 650 mm of precipitation annually with a mean annual temperature of 6°C (di Castri & Hajek 1976). The northern part of the island is characterized by old growth Southern Beech (*Nothofagus betuloides*) and Lenga Beech (*N. pumilio*) forests along with areas of second growth, burned areas and areas converted to pasture land. Observations of foraging were made from sea level to above the tree line at 750 m with 8 x 10 binoculars.

The results of our observations are summarized in Table 1. We documented utilization of two tree species, four flowering plants, six grasses, and a single species of lichen that had not previously been recorded as food items. We also reaffirm the use of Chilean Firebush (*Embothrium coccineum*) and Winter's Bark (*Drimys winteri*) as food sources. Patagonian Sierra-Finches were observed in all high levels of the forest and clearings. While feeding observations were of birds on the ground or in the lower 4 m of trees and shrubs, Patagonian Sierra-Finches were often seen high up in beech trees and was noted to be a member of foraging flocks of the Thorn-tailed Rayadito (*Aphrastura spinicauda*) and the White-throated Treerunner (*Pygarrhichas albogularis*) and may have been consuming insects, but were too high in the canopy to observe. The only non-vegetative item noted in the diet was grit recovered in the stomachs of two individuals that died in the nets.

The species' preferred habitat is unclear. Observations of Sierra-Finches robbing nectar of Hardy Fuchsia (*Fuchsia magellanica*) were recorded as being principally in open habitat followed by forest interior (Traveset *et al.* 1998). Couve & Vidal (2003) state that it is "generally seen in the canopy or middle story where it feeds" while Jaramillo (2003) remarks that Patagonian Sierra-Finches are "terrestrial when foraging." In consideration of these contrasting statements no singular trophic level within the forest appears to be preferred. Rather, the species is likely inhabiting the level within the forest where food resources are most abundant.

Patagonian Sierra-Finch diet is described as being composed of seeds, buds, flowers, fruit, insects, and other invertebrates (Egli & Aguirre 2000). Specific documentation includes feeding on the nectar of Chilean Firebush (Smith-Ramirez & Armesto 2003), Chilean Needletree (*Rhaphithamnus cyanocarpus*) (Barros 1945), Austral Bellflower (*Philesia magellanica*) (Rozzi *et al.* 2003) and Hardy Fuchsia (Blumberg undated, Traveset *et al.* 1998). Patagonian Sierra-Finch has also been observed to consume the flowers and fruit of Winter's Bark (Chebez & Bertonatti 1994) and phloem-sap of Southern Beech (Schlatter & Vergara 2005).

Seeds have been noted in stomach samples (Humphrey *et al.* 1970, Markham 1970, Rozzi *et al.* 1996) as well as insects, pebbles, and vegetable matter (Schlatter *et al.* 1995). Finally, Crawshay (1907) reported that, around human settlements, it subsisted largely on refuse and he observed some individuals pecking at a raw sheep's head and beef bone.

Our observations add to the known list of items foraged on. Two plants noted as food items in other papers, Chilean Needletree and Hardy Fuchsia, do not grow on Navarino Island (Moore 1983), so could not be utilized by Sierra-Finches. The four exotic species of grasses they consumed were introduced in the last 150 years as forage for livestock in Tierra del Fuego. We found no evidence that Patagonian Sierra-Finches play a significant role as seed dispersers. The observed method of feeding consists of thoroughly grinding and crushing all food items before swallowing. On large fruits of species such as Winter's Bark and Magellanic Barberry (*Berberis buxifolia*), they took off small chunks of fruit, rarely getting any seeds in the center. A study of >100 Patagonian Sierra-Finch feces from birds captured in mist-nets in Omora Park revealed no whole seeds had passed through after digestion (S. Ippi pers. com.). This species has previously been noted as a destroyer of flowers with nectar (Traveset *et al.* 1998, Smith-Ramirez & Armesto 2003), a behavior we also observed. The Patagonian Sierra-Finch appears to act as a seed predator and a nectar robber. Its evolutionary relationship with the native plants inside its range and a more thorough study of this species' diet remains to be conducted. Our observations add to the known list of items in their diet.

ACKNOWLEGMENTS

We thank R. Charlin, D. Christie, O. Dollenz, A. Gutiérrez, and F. Massardo for identifying the plant species. Thanks to C. Anderson, R. Rozzi, and J.C. Torres-Mura for comments on earlier drafts. Special thanks to C. Anderson, X. Arango, S. Ippi, F. Massardo, R. Rozzi, and M. Sherriffs for their assistance in logistics. Support for this project came from the Omora Foundation and the Millennium Project, IEB, University of Chile.

REFERENCES

ANDERSON, C. & R. ROZZI. 2000. Bird assemblages in the southernmost forests in the world: Methodological variations for determining species composition. *Anales del Instituto de la Patagonia* 28: 89-100.

ANDERSON, C., R. ROZZI, C. ELPHICK & S. MCGEHEE. 2002. El programa Omora de anillamiento de aves en los bosques subantárticos: estandarización del tamaño de anillos apropiados para las aves de la region de Magallanes. *Boletin Chileno Ornitología* 9: 2-11.

BARROS, V. 1945. Aves polinizadoras y flores ornitófilas de Maullín. *Revista Universitaria* 30: 73-78.

BLUMBERG, C.A. *Sin fecha. Contemplando aves y mamíferos de Aisén*. Universidad de Los Lagos, Osorno, Chile.

COUVE, E. & C. VIDAL. 2003. *Birds of Patagonia, Tierra del Fuego and Antarctic Peninsula, the Falkland Islands and South Georgia*. Editorial Fantástico Sur, Punta Arenas, Chile.

CHEBEZ, J.C. & C.C. BERTONATTI. 1994. *La Avifauna de la Isla de Los Estados, Islas de Año Nuevo y Mar Circundante*. Monografia Especial No. 31, Literature of Latin America (L.O.L.A.), Buenos Aires, Argentina.

CRAWSHAY, R. 1907. *The Birds of Tierra del Fuego*. Bernard Quaritch, London, UK.

DI CASTRI, F. & E.R. HAJEK. 1976. *Bioclimatología de Chile*. Universidad Católica de Chile, Santiago, Chile.

TABLE 1. Fifteen plants consumed by the Patagonian Sierra-Finch (*Phrygilus patagonicus*) on Navarino, Island, Chile.

Plant species	Parts of plant consumed	Months
Nothofagus pumilio	Seed, buds, floral parts	Nov., Sep., Feb.
Embothrium coccineum	Seed, nectar	Oct., Dec., Jan., Mar., Apr.
Perezia magellanica	Seed	Mar.
Carex curta	Seed	Mar.
Rumex sp.*	Seed	Sep.
*Taraxacum officinale**	Seed	Nov., Dec.
*Poa pratensis**	Seed	Jan.
*Holcus lanatus**	Seed	Jan.
Berberis ilicifolia	Nectar, floral parts	Oct.
Berberis buxifolia	Fruit, nectar, floral parts	Oct., Nov., Feb.
Ribes magellanicum	Fruit	Feb.
Chiliotrichum diffusum	Floral parts	Jan.
Drimys winteri	Fruit, floral parts	Jan., Mar., Apr., May.
Maytenus magellanica	Fruit	Apr.
Usnea spp.	All	Mar., Apr.

*Non-native species.

EGLI, G. & J. AGUIRRE. 2000. *Aves de Santiago*. UNORCH, Santiago, Chile.

FJELDSA, J. & N. KRABBE. 1990. *The Birds of the High Andes*. Zoological Museum, University of Copenhagen, Copenhagen, Denmark.

HUMPHREY, P.S., D. BRIDGE, P.W. REYNOLDS & R.T. PETERSON. 1970. *Birds of Isla Grande (Tierra del Fuego)*. Preliminary Smithsonian Manual, Smithsonian Institution, Washington, D.C.

JARAMILLO, A. 2003. *Birds of Chile*. Princeton University Press, Princeton, New Jersey.

JOHNSON, A.W. 1967. *Birds of Chile and Adjacent Regions of Argentina, Bolivia and Peru*. Volume II. Platt Establecimientos Gráficos, Buenos Aires, Argentina.

MARKHAM, B.J. 1970. Reconocimiento faunístico del área de los fiordos Toro y Condor, Isla Riesco, Magallanes. *Anales del Instituto de la Patagonia* 1: 41-57.

MARTÍNEZ, D. & G. GONZÁLEZ. 2004. *Las Aves de Chile*. Ediciones del Naturalista, Santiago, Chile.

MOORE, D.M. 1983. *Flora of Tierra del Fuego*. Missouri Botanical Gardens, St. Louis, Missouri.

RIDGELY, R. & G. TUDOR. 1994. *The Birds of South America*. Volume 1. The Oscine Passerines. University of Texas Press, Austin, Texas.

ROZZI, R., J.J. ARMESTO, A. CORREA, J.C. TORRES-MURA & M. SALLABERRY. 1996. Avifauna de bosques primarios templados en islas deshabitadas del Archipielago de Chiloé, Chile. *Revista Chilena de Historia Natural* 69: 125-139.

ROZZI, R., F. MASSARDO, C.B. ANDERSON, S. MCGEHEE, G. CLARK, E. RAMILO, U. CALDERÓN, C. CALDERÓN, L. AILLAPAN & C. ZÁRRAGA. 2003. *Guía Multi-étnica de Aves de los Bosques Templados de Sudamérica Austral*. Editorial Fantástico Sur, Punta Arenas, Chile.

SCHLATTER, R. & P. VERGARA. 2005. Magellanic Woodpecker (*Campephilus magellanicus*) sap feeding and its role in the Tierra del Fuego forest bird assemblage. *Journal of Ornithology* 146: 188-190.

SCHLATTER, R.P., C. VENEGAS, C. BRAVO & J.C. TORRES-MURA. 1995. *Proyecto Río Cóndor Tierra del Fuego*. Aves. Informe científico, Universidad Austral de Chile, Valdivia, Chile.

SMITH-RAMIREZ, C. & J.J. ARMESTO. 2003. Foraging behavior of bird pollinators on *Embothrium coccineum* (Proteaceae) trees in forest fragments and pastures in southern Chile. *Animal Ecology* 28: 53-60.

TRAVESET, M.F., M.F. WILSON & C. SABAG. 1998. Effect of nectar-robbing birds on fruit set of *Fuchsia magellanica* in Tierra del Fuego: a disrupted mutualism. *Functional Ecology* 12: 459-464.

VUILLEUMIER, F. 1991. A quantitative survey of speciation phenomena in Patagonian birds. *Ornitología Neotropical* 2: 5-28.

Accepted 6 January 2007.

11 CONSUMPTION OF LEÑADURA (*MAYTENUS MAGELLANICA*) SEEDS BY THREE PRIMARILY INSECTIVOROUS BIRD SPECIES

STEVEN M. MCGEHEE*

Parque Etnobotánico Omora - Sede Puerto Williams, Universidad de Magallanes, Puerto Williams, Chile (www.umag.cl/williams)
*Current address: Center for the Study of Tropical Birds, 218 Conway Drive, San Antonio, Texas 78109 USA (whitethroatedcaracara@yahoo.com)

Anales del Instituto de la Patagonia (Chile) 35: 71-73 (2007)

Consumption of leñadura (*Maytenus magellanica*) seeds by three primarily insectivorous bird species.

Consumo de semillas de leñadura (*Maytenus magellanica*) por tres especies de aves principalmente insectívoras.

The temperate forests of southern Chile and Argentina are home to a number of endemic bird species (Rozzi *et al.* 2003). Many are wholly or principally insectivorous (Jaksic & Feinsinger 1991). Three of these species, a furnarid and two tyrannid species, are found as far south as Navarino Island in the Cape Horn Biosphere Reserve, Chile. The Patagonian Tyrant (*Colorhamphus parvirostris*) is considered an insectivorous species (Johnson 1967, del Hoyo *et al.* 2004), although it has been seen to take the seeds of Maiten (*Maytenus boaria*) in winter (Barros 1921). The Fire-eyed Diucon (*Xolmis pyrope*) is also known to be mainly insectivorous (Lazo & Anabalon 1992), but is reported to eat different types of fruits and seeds in certain parts of its range (Johnson 1967, Rozzi *et al.* 1996, Egli & Aguirre 2000). The Thorn-tailed Rayadito (*Aphrastura spinicauda*), long considered strictly an insect-eater, has recently been shown to consume some seeds of exotic pine trees (*Pinus radiata*) (Estades 2001) and peck at fruits of *Berberis, Ribes,* and *Gunnera* (del Hoyo *et al.* 2003). Here, three instances of observations of fruit-eating behavior in these bird species are reported for the subantarctic forests of the Cape Horn Biosphere Reserve.

Since 2000, the Omora Ethnobotanical Park has been conducting an ongoing mist-netting study of the forest birds on Navarino Island, Chile (54°55'S, 67°39'W) (Anderson & Rozzi 2000, Anderson *et al.* 2002, McGehee *et al.* 2004). A description of the program can be found in Anderson & Rozzi (2000) and Anderson *et al.* (2002). During mist-netting operations on March 24, 2004, a Patagonian Tyrant was seen actively perch hunting for insects in the Omora Park. It then flew into a Leñadura (*Maytenus magellanica*) tree and was seen to consume two fruits. This same Patagonian Tyrant then resumed its perch hunting for insects.

On April 10, 2005 two Patagonian Tyrants were seen in the same Leñadura tree. Each would sit on a branch of the tree or in a Firebush (*Embothrium coccineum*) intertwined with the Leñadura. The Patagonian Tyrants would fly from their perches, hover and pluck the fruit, then fly back to a branch and swallow the fruit. This was done on and off for over 6 hours. Twice they were seen to regurgitate the fruit, open their beaks, chew the seeds and then reswallow the food item. Usually two to three fruits were eaten in 10-15 seconds, followed by a 30 second pause that was followed by another series of fruit plucking and eating. This continued for approximately 40 minutes. The birds would then fly off and reappear approximately 30 minutes later and repeat the fruit eating off and on for the next 4 hours. The Patagonian Tyrants appeared to be swallowing the fruits whole.

During the time the Patagonian Tyrants were eating fruits, another species considered to be primarily insectivorous, the Diucon, was seen to consume these fruits as well. A Diucon flew in and ate a Leñadura fruit, plucking it while hovering and then flying to a nearby branch and swallowing it whole. The individual repeated this three times, then flew off. Throughout that day, flocks of 10-16 Rayaditos would pass though the trees as well. They would hang from a Leñadura branch, pluck a fruit and fly away. A few would land on a nearby branch, hold the fruit against the branch with their foot and proceed to tear off the fruit and swallow one or both seeds. The Rayaditos appeared to only take one or two fruits before moving on. It is not known whether the same flock was passing through all day or different flocks. In 2006, this tree produced very little fruit, but 3 times on April 16th, a Rayadito was seen flying off with a fruit in its beak. On April 15th and April 17th, 2005 in an area about 120 meters from the initial observation site, Rayaditos were seen eating fruit from another Leñadura tree. This tree was observed for 7 hours on the 15th and 8 hours on the 17th, and only Rayaditos were seen at this tree although Diucons were in the area. The Rayaditos were seen eating the fruit in this Leñadura tree during the full period of observation. This tree produced no fruit in 2006.

Maytenus is a tree genus restricted to Central and South America with two species recorded in southern Chile and Argentina (Dollenz 1995). Leñadura (*M. magellanica*) are found from Hoste and Navarino islands north through Chile and Argentina to 40°S (Moore 1983). In late summer, it produces a fruit capsule of 4-6 x

INTRODUCTION

Detailed information about sexual dimorphism and the role of each sex in parental care is lacking for the great majority of species in the large passerine family Furnariidae (Remsen 2003). Furnariids are characteristically sexually monochromatic, and in most species both sexes have brown coloration and complex plumage patterns, which are exhibited during the displays of some species. Sexual dimorphism in size is also slight. For the few species studied in detail, males are on average only slightly larger in linear measurements (Winker *et al.* 1994, Remsen 2003). This is in accordance with the presumed social monogamy of many species, and the apparent equality between the sexes in parental care: the few existing studies suggest that both members of the pair contribute to nest-building, incubation, and feeding of nestlings and fledglings (Kendeigh 1952, Fraga 1980, Nores & Nores 1994, Skutch 1996). Many furnariids have exceptionally long tails, in some cases among the longest tails with respect to body size for any passerine (Remsen 2003). These tails are often graduated, with rectrices ending in spines. The use of these tails for support on trunks or vertical branches has been proposed for several species in which the rachises are stiffened and the distal ends function as braces during climbing (Remsen 2003). However, a stiffened rachis is not always evidence of climbing behavior while foraging (Skutch 1967, Remsen 2003), and tails in some species are too long to be useful in foraging (*e.g.*, the extremely long tail of des Muir's Wiretail [*Sylviorthorhynchus desmursii*]). This suggests to us that sexual selection has emphasized tail development instead of body coloration in furnariids.

We studied sexual dimorphism, the role of each sex in parental care, and foraging behavior in a furnariid characteristic of southern temperate forests in Chile and Argentina, the Thorn-tailed Rayadito (*Aphrastura spinicauda*). Rayaditos are small, active, insectivorous birds that glean prey items from tree branches and trunks, and forage from the understory to the canopy (Vuilleumier 1967). They are resident cavity-nesters, with low fecundity and prolonged parental care of eggs and chicks (Moreno *et al.* 2005). In winter, Rayaditos travel in mixed-species flocks, in which they are the main species around which other birds congregate (Ippi & Trejo 2003). They show an apparent lack of sexual dichromatism and size dimorphism, although no quantitative morphological study on this species has been conducted to date. No detailed information about the role of each sex in parental care in this species has been published. The characteristic long and graduated spiny tail of Rayaditos suggests that it may function as a brace and support while foraging or as a sexually selected ornament in both sexes. Graduated tails incur an aerodynamic cost through increased drag (Thomas 1993), and are often sexually selected by acting as handicaps (Andersson 1982, Grafen 1990, Johnstone 1995). Using molecular sexing, morphological measurements, detailed spectral analyses of plumage coloration, and observations of parental provisioning at the nest and of foraging and habitat use, we aimed to establish: (1) if tails are normally used for support while foraging, and (2) the degree of sexual dimorphism in morphology, coloration, and parental care in this socially monogamous species.

METHODS

FORAGING BEHAVIOR

In December 2005, we conducted a study of habitat use by three local populations of Thorn-tailed Rayaditos foraging in mature *Nothofagus* forests on the island of Navarino (54°55'S, 67°40'W). Our aim was to elucidate the function of the tail while foraging. In each study site, we walked slowly through the forest along linear transects and stopped as soon as we observed a foraging Rayadito. After 10 sec of observation, we began to record on tape the following data: type of substrate (ground, trunk, thick branch directly protruding from trunk, thin outer branch growing from a higher order branch, or foliage), height of substrate above ground, Rayadito posture (climbing upward, climbing downward, hanging upside down, horizontally moving along a branch), and the position of the tail (in the air or leaning against the substrate). In cases where the tip of the tail leaned against the substrate, we noted if it barely touched the substrate or was pressed against it. Recordings were made every 5 sec. We made ≤10 records per observation (mean number of records per observation = 5.4 ± 0.4 SE, range = 2–10). To avoid repeating observations on the same individual, we left the immediate area and continued along the transect for at least 40–50 m before making further observations. For each individual observed, we calculated the proportion of records per observation on different substrates and in different postures, when the tail was in the air or touching the substrate, and the average height of the records. As sex was unknown in this study, we did not make any statistical comparison, and instead present average values for individuals to describe the use of the tail by foraging Rayaditos.

MORPHOLOGY AND COLOR MEASUREMENTS

The study of morphology and parental roles was conducted on Chiloé Island, Chile (41°52'S, 73°39'W), in the austral spring (November–December) of 3 years (2002–2004). Although Rayaditos on Chiloé are considered a different subspecies than birds in central Chile and Navarino (Johnson & Goodall 1967), they are morphologically very similar (RAV *et al.*, unpubl. data). Adults were captured in the context of a study of the breeding biology of Rayaditos using nest boxes (Moreno *et al.* 2005). For a description of the study area, see Moreno *et al.* (2005). Adults were captured with nest-box traps when chicks were 13 days old (hatching day = day 0). Birds were banded with individual combinations of color and metal bands (National Band and Tag Co., Newport, Kentucky, model 1242–3). Following Svensson (1984), we measured tarsus length and bill length (using digital calipers, to the nearest 0.1 mm), flattened wing chord (to the nearest mm with a stopped ruler), and the two central rectrices on the right side of the tail (to the nearest mm with a stopped ruler designed for tail measurements). The two central rectrices on each side of the graduated tail are the longest, while all rectrices have a spine (Fig. 1). We measured total rectrix length and length excluding the spine. All measurements were taken by the same person. There was no obvious difference between the sexes at this stage of the breeding cycle, as all birds showed signs of a brood patch and there was no detectable difference between mates in the size of the cloacal protuberance (differences may possibly be evident during earlier breeding stages). Mass was recorded with a Pesola (Baar, Switzerland) spring balance to the nearest 0.1 g. As all individuals were captured in the same nesting stage, it was not necessary to control for stage-dependent mass changes. A drop of blood was obtained by brachial venipuncture, and blood was stored in a buffer solution (100 mM of Trishydroxymethylaminomethane [TRIS] pH = 8.0, 100 mM of Ethylenediaminetetraacetic acid [EDTA] and 2% Sodium Dodecyl Sulfate [SDS]) for DNA preservation until analysis. The exact age was known for only two breeding adults, banded as chicks the previous year.

In 2004, we measured plumage color with a portable battery-driven Minolta spectrophotometer (model CM-2600d) that covered the range 360–700 nm. There is probably variation in color in the UV range below 360 nm that should be considered in future studies if furnariids are in the UV range (Ödeen & Håstad 2003). Tail feathers were analyzed in the laboratory with a spectrophotometer that included the entire reflectance range (see below). The target mask of the spectrophotometer, with a diameter of 1 cm, was placed directly on the breast, top of the head, and middle of the back and rump. Reflectance spectra for each part of the body were calculated as the mean of three sequential measurements obtained by the spectrophotometer. The apparatus was moved slightly between measurements. Reference calibrations against a white standard were performed periodically according to the manufacturer's specifications. We used brightness, chroma, and hue from the CIELAB color space (Commission Internationale de l'Eclairage 1976) to describe body color in the human visible spectrum. The L*a*b*, or CIELAB, color space was used to denote brightness (L*) and chromaticity (a*, b*). In the chromaticity diagram, a* goes from green (negative values) to red (positive values), and b* goes from blue (negative values) to yellow (positive values). The farther away from zero, the more saturated the color. Chroma in the L*C*h* (brightness, chroma, hue) color space is a composite measure of chromaticity and was calculated according to the formula: Chroma 5 $\sqrt{(a^*)^2 + (b^*)^2}$. Hue in this space is calculated as $\tan^{-1}(b^*/a^*)$. Geometric distances in the CIELAB color space approximate intuitive color differences.

In 2004, we collected the right central rectrix of each individual by cutting it at the base and stored it away from light until analysis. The dorsal side of rectrices was measured in the laboratory with a nonportable Ocean Optics S2000 fiber-optic spectrophotometer (Ocean Optics, Inc., Dunedin, Florida) using a 200 micron fiber-optic probe with a 2 mm diameter reading area held at a 90° angle to the feather surface. Reflectance spectra curves (300–700 nm) were generated with a Spectralon white standard (Labsphere, Inc., North Sutton, New Hampshire) as a reference. We did not use the CIELAB color space with respect to tail feathers because it disregards reflectance in the UV part of the reflectance spectrum. Three consecutive measurements were taken on each of the proximal, middle, and distal parts of the feather (not including the spine) to obtain repeatabilities of reflectance spectra. Principal components analyses (PCA) were performed on the three spectra obtained for each feather. Repeatabilities of PC1 and PC2 for all parts of the feather were significant (all $R > 0.47$, all $P < 0.001$), so PCA was performed on a single reflectance spectrum for each feather calculated as the average of the reflectance spectra for the three feather parts. To reduce the number of variables in relation to the number of individuals, we grouped the 36 10 nm intervals presented by the apparatus into 18 intervals of 20 nm each by taking the average of every two intervals (Cuthill *et al.* 1999).

Fig. 1. Thorn-tailed Rayadito with its characteristic graduated tail.

SEX DETERMINATION

DNA was isolated from blood samples using proteinase K digestion and ethanol precipitation by means of the "UltraClean DNA BloodSpin Kit" (Mo Bio Laboratories, Carlsbad, California). PCR reactions were performed in 20 µl of final volume using 0.5 U Taq Polymerase (Biotools, Madrid, Spain), 200 µM dNTP's, 1 mM Taq Buffer with $MgCl_2$, and 10 pmol of primers 2550F (5'-GTTACTGATTCGTCTACGAGA-3') and 2718R (5'-ATTGAAATGATC-CAGTGCTTG-3'). The forward and reverse primers amplified the CHD1 (chromo-helicase-DNA binding protein) sequence, present in both W and Z avian chromosomes (Fridolfsson & Ellegren 1999). The PCR reaction consisted of an initial denaturing step at 94°C for 90 sec, followed by 30 cycles of denaturation (94°C for 30 sec), annealing (48°C for 45 sec) and extension (72°C for 45 sec), and two final cycles of 48°C for 60 sec and 72°C for 5 min. PCR products were separated by electrophoresis at 90 V in 2% agarose gels in Trishydroxymethyl-aminomethane-acetate [TAE] buffer. The amplification products corresponded to CHD1W of 450 bp and CHD1Z of 600 bp. Birds were sexed as females (heterogametic, WZ) when both fragments were amplified and as males (homogametic, ZZ) when only the 600 bp band was visible.

PARENTAL CARE

Most observations of parental care by banded pairs were obtained on day 3 ($N = 5$) and days 14–17 after hatching ($N = 18$). Observations of banded pairs during the nestling stage were used to compare parental care between the sexes. Two observations of banded pairs and nine observations of unbanded pairs during incubation were also obtained. Observations of incubating pairs were used for descriptive purposes only, as sex of individuals could not be established in most cases. Activity at nest boxes was recorded with video cameras for periods of 4 hr in 2003 (all observations on day 3 and 12 observations on days 14–17) and 90 min in 2004. In total, 104 hr of observations were obtained in 2003 and 33 hr in 2004. The number of feeding visits and the number of times fecal sacs were removed were obtained from the videotapes. The time spent inside the nest box when chicks were 3 days old was also recorded.

STATISTICAL ANALYSIS

As several individuals were measured and observed in more than one year, we used data from only the last year in which an individual was studied. Similarly, for pairs that bred together in more than one year, only data for the last year were used in statistical comparisons. Different sample sizes for the two sexes in comparisons between sexes are due to elimination of some observations to avoid pseudoreplication. Comparisons between the sexes were conducted both independently of mates (using ANOVA) and within pairs (paired *t*-tests). All parental care variables are expressed per hour of observation. Sequential Bonferroni corrections for multiple tests were applied when two or more associated tests were conducted (Rice 1989). Means are presented $\pm$ SE.

RESULTS

FORAGING BEHAVIOR

In total, 46 individuals were observed for periods of less than one minute. Substrates were used in the following proportions of records per observation: 0.25 ± 0.05 foliage, 0.25 ± 0.05 thin outer branches, 0.22 ± 0.04 primary thick branches, 0.21 ± 0.04 trunk, and 0.04 ± 0.02 ground. The mean height per observation at which Rayaditos foraged was 7.7 ± 6.5 m. The proportions of records per observation in which foraging postures were used were: 0.53 ± 0.05 for horizontal movement along branches, 0.28 ± 0.04 for hanging upside down, 0.17 ± 0.04 for climbing upward along trunks or steep branches, and 0.02 ± 0.01 for climbing downward. Rayaditos held their tails in the air in 0.96 ± 0.01 of the records per observation, and they leaned their tails against the substrate in the remaining 0.03 ± 0.01 of the records per observation. All records of Rayaditos leaning their tails against the substrate were made when they climbed upward on trunks. In all these cases, the tip of the tail only briefly touched the trunk and appeared to be used for balance, not support.

MORPHOLOGY

Males were larger than females with respect to tarsus length, wing length, and mass (Table 1). However, no differences with respect to bill length or the length of the two longest rectrices and their protruding spines were found (Table 1). Within-pair comparisons returned similar results as the general comparisons between sexes, with differences in all morphological variables except rectrix and spine lengths (paired *t*-tests: all $t_{23-26} > 3$, all $P < 0.01$), for all variables except tail variables (all $t_{21-26} < 1.5$, all $P > 0.10$). Coefficients of variation were similar for all variables, except that the lengths of the central rectrices showed slightly higher values (Table 1). Tail spine lengths showed very high coefficients of variation, as is expected for parts of a structure compared with whole structures (Lande 1977). We took measurements from four females and four males in more than one year. Of these, three females and two males showed increases with time in the length of the longest rectrix, while one female showed no change and two males showed decreases. Significant within-pair correlations were found with respect to mass only ($r_{23} = 0.5$, $P = 0.009$; for other traits all $r_{20-25} < 0.3$, all $P > 0.10$), indicating that heavy males were paired with heavy females.

COLOR

Reflectance spectra of head, back, and rump plumage were typical of dark-brownish colored objects, while breast plumage showed reflectance spectra typical of whitish-brown objects (Fig. 2). No evidence of high reflectance close to the UV range was detected. There was no difference between the sexes in body plumage color as represented by brightness, chroma, and hue in the CIELAB color space (Table 2). Within-pair comparisons showed the same results as the independent comparisons between sexes (all $t_{5-8} < 1.8$, all $P > 0.10$). There were no within-pair associations with respect to body color after correcting for multiple tests (all $r_{4-7} < 0.4$, all $P > 0.33$).

Reflectance spectra of tail feathers were similar to those for body plumage, with reflectance in the UV range very low. Two principal components explained more than 92% of the variation in reflectance spectra for the longest tail feathers (PC1: 82%, eigenvalue = 29.3; PC2: 16%, eigenvalue = 5.8). PC1 was related to achromatic brightness and was flat with respect to wavelength (high values represent dark feathers and low values pale feathers), and PC2 was associated with hue and declined monotonically with wavelength, with the highest values at the UV end and the lowest

TABLE 1. Mass and morphological measurements (means ± SE, *N* in parentheses) of female and male Thorn-tailed Rayaditos on Chiloé Island, Chile, with coefficients of variation (CV). Significant differences between the sexes after sequential Bonferroni correction are denoted with an asterisk.

	Female	CV	Male	CV	*t*	*P*
Mass (g)	10.3 ± 0.1 (29)	4.9	11.3 ± 0.1 (26)	5.4	7.0	<0.001*
Tarsus length (mm)	9.5 ± 0.1 (29)	4.3	20.0 ± 0.1 (25)	2.3	2.8	0.008*
Wing length (mm)	55.0 ± 0.3 (29)	3.2	58.0 ± 0.3 (27)	3.0	6.4	0.001*
Bill length (mm)	13.9 ± 0.1 (29)	4.0	14.3 ± 0.2 (27)	6.8	2.3	0.03
Longest rectrix (mm)	71.6 ± 0.9 (29)	6.3	70.0 ± 0.9 (27)	7.2	1.2	0.22
Spine of longest rectrix (mm)	19.8 ± 0.8 (29)	20.9	18.3 ± 0.9 (27)	25.1	1.3	0.19
Second-longest rectrix (mm)	57.6 ± 0.7 (23)	6.8	58.1 ± 0.7 (24)	4.5	0.5	0.59
Spine of second-longest rectrix (mm)	8.9 ± 0.3 (23)	18.3	8.9 ± 0.3 (24)	19.7	0.01	0.99

values at the red end of the spectrum (Fig. 3). No difference was found between the sexes for PC1 (t_{20} = 1.9, *P* = 0.07) or PC2 (t_{20} = 0.7, *P* = 0.83) of the longest rectrix. The same result was obtained from within-pair comparisons (PC1: t_9 = 1.5, *P* = 0.17; PC2: t_9 = 1.2, *P* = 0.28). No within-pair correlations were found with respect to color of the longest rectrix (all r_8 < 0.01, all *P* > 0.90).

PARENTAL CARE

Parental attendance at the nest during incubation was very high, with an adult present in the nest box for 55.1 ± 2.9 min hr^{-1} (*N* = 11 nests). Most incubation sessions (52%) were terminated by the other member of the pair assuming incubation duty. Absences of both mates from the nest were short, with an average duration of 3.5 ± 1.2 min (*N* = 11). Incubation sessions had an average duration of 16.8 ± 3.1 min (*N* = 11).

No differences in parental care between the sexes were detected when chicks were three days old (Table 3). However, within-pair comparisons revealed a significant difference in duration of brooding sessions (t_4 = 5.6, *P* = 0.005), with males brooding longer per session than females. There were no within-pair differences for other variables (all t_4 < 2.1, all *P* > 0.11). Both sexes visited the nest at similar rates when chicks were approximately two weeks old (Table 3). However, females removed fecal sacs at rates twice as high as males at this stage of the nestling period (Table 3). Within-pair comparisons at this stage were similar (provisioning days 14–17: t_{16} = 0.4, *P* = 0.71; fecal sac removal days 14–17: t_{16} = 4.0, *P* < 0.001). No within-pair correlations with respect to parental care variables were significant after Bonferroni correction at day 3 (all r_3 < 0.93, all *P* > 0.03), although this was probably due to low sample size. On days 14–17, within-pair correlations were not significant either (all r_{15} < 0.06, all *P* > 0.85). There were no significant associations after Bonferroni correction between tail length and parental care variables (all r_{5-32} < 0.2, all *P* > 0.63), or between tail color (PC1 and PC2) and parental care variables (all r_{11} < 0.6, all *P* > 0.03) on days 14–17.

TABLE 2. Comparisons between body plumage color measures (means ± SE, number of individuals in parentheses) in the CIELAB color space of female and male Thorn-tailed Rayaditos on Chiloé Island, Chile. L* (lightness) ranges from 0 (extremely dark) to 100 (extremely light), C* (chroma) ranges from 0 (achromatic) to 60 (fully saturated color), and h* is an angular variable going from 0° (red) to 90° (yellow), 180° (green), 270° (blue), and full circle to red again (360°). No tests were significant after Bonferroni correction.

	Females	Males	*t*	*P*
L* Head	15.6 ± 1.9 (9)	17.9 ± 4.0 (10)	0.5	0.63
C* Head	6.2 ± 1.4 (9)	10.2 ± 3.2 (10)	1.1	0.28
h* Head	45.2 ± 7.0 (9)	53.8 ± 3.5 (10)	1.1	0.27
L* Breast	53.1 ± 2.8 (10)	52.2 ± 1.8 (10)	0.3	0.80
C* Breast	15.4 ± 2.3 (10)	15.7 ± 1.7 (10)	0.1	0.93
h* Breast	80.6 ± 1.3 (10)	80.9 ± 0.9 (10)	0.1	0.89
L* Back	25.4 ± 0.9 (10)	25.3 ± 1.4 (10)	0.1	0.93
C* Back	14.7 ± 1.4 (10)	15.1 ± 1.2 (10)	0.2	0.82
h* Back	71.0 ± 1.2 (10)	71.5 ± 0.8 (10)	0.3	0.75
L* Rump	26.1 ± 1.1 (10)	30.1 ± 1.1 (10)	2.5	0.02
C* Rump	23.1 ± 2.6 (10)	24.6 ± 3.1 (10)	0.4	0.71
h* Rump	64.4 ± 2.1 (10)	64.8 ± 1.5 (10)	0.1	0.88

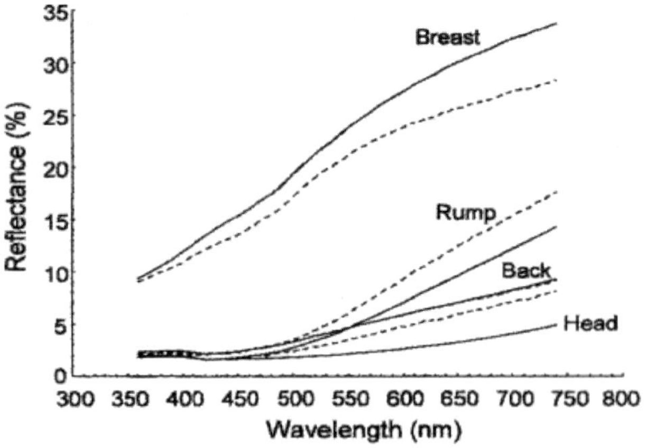

Fig. 2. Reflectance spectra of four body parts of female (solid lines) and male (dotted lines) Thorn-tailed Rayaditos on Chiloé Island, Chile. Only reflectance in the wavelengths visible to humans was measured. Data are means for all measured individuals (*N* = 9 for females, *N* = 10 for males).

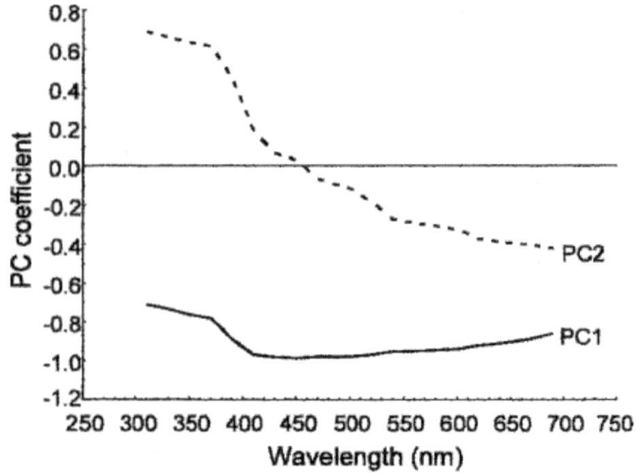

Fig. 3. Coefficients for the two main principal components derived from PCA in relation to wavelength for the right longest rectrices of Thorn-tailed Rayaditos on Chiloé Island, Chile. Multiplying these coefficients by reflectance at a certain wavelength interval gives the PCA scores.

DISCUSSION

Thorn-tailed Rayaditos exhibit an absence of obvious sexual dimorphism in color and morphology characteristic of furnariids (Remsen 2003). In addition, males exhibit brood patches and there is no detectable difference with respect to cloacal protuberance between the sexes during the nestling stage, making it necessary to use molecular methods to sex adults. However, despite overlap in measurements, males are between 2% and 10% larger than females in size and mass. Given that island populations may experience ecological release (Selander 1966), sexual dimorphism of mainland Rayaditos may be less than that reported here. Despite males being larger than females, there is no difference in length of the two longest rectrices or spines, as has been found for ornamental traits in other species (Boland *et al.* 2004, Kraaijeveld *et al.* 2004). Females are smaller and weigh less but have longer tails for their body size than males. Both sexes have very similar body plumage coloration, although we cannot confirm this for the UV part of the spectrum. The same is true for tail color, including UV reflectance. As has been reported for other furnariids (Kendeigh 1952, Fraga 1980, Nores & Nores 1994, Skutch 1996), males and females participate equally in incubating eggs and brooding and raising young, except for removal of fecal sacs, which is done predominantly by females. Monomorphism in body plumage and equality of the sexes in parental care suggest the absence of sexual selection for signaling traits in these socially monogamous birds.

All 12 tail feathers of Rayaditos end in thin spines, in which the distal portions of shafts are virtually without barbs. The lengths of both the rectrices and their spines decrease gradually from the inner feathers outwards. The two central tail feathers on each side have spines of 9 and 20 mm on average, respectively, with strengthened shafts. However, our observations indicate that Rayaditos seldom climb tree trunks by supporting themselves with their tails, and their foraging behavior consists mostly of horizontally moving along branches and hanging upside down while gleaning arthropods from bark and foliage. On the few occasions when we observed a bird leaning the tip of its tail against a trunk while climbing upward, it was during a quick balancing movement. Furnariids with similar tails to Rayaditos, like the Spotted Barbtail (*Premnoplex brunnescens*), sometimes move up vertical trunks with the body upright and the tail pressed against the bark, but they seem to depend little upon the strongly barbed tail feathers for support (Skutch 1967). Several genera exhibiting tail spines do not climb on trunks while foraging (Remsen 2003). Furnariid tree-climbing specialists like White-throated Treerunners (*Pygarrhichas albogularis*) have much shorter tails and stronger spines than Rayaditos (RAV and JM, pers. obs.). Thus, it is doubtful that tail spines in Rayaditos have any utility related to support while foraging. However, our observations were made during a brief period of the annual cycle. It is possible that foraging behavior changes seasonally, reflecting changes in food resource distribution, so our data are not conclusive evidence for the absence of functionality of the tail spines in foraging. The only real way to understanding the function of the tail spines is through experimental manipulation of their length.

Elongated tail feathers of Rayaditos had a slightly higher coefficient of variation than other morphological characters, as is characteristic of epigamic traits (Møller & Pomiankowski 1993, Fitzpatrick 1997). Rectrix and spine lengths were similar in both sexes, but males were significantly larger, thus the elongated central rectrices are apparently not a partially sex-limited trait that exists in females merely as a genetically correlated response to selection on males (Lande 1980). The longest central rectrices and spines measured actually belonged to females. The elaborate tails of furnariids may be an expression of mutual sexual selection in this family that is characterized by social monogamy and the absence of plumage dimorphism. The long tail of Rayaditos could also function in social contexts affecting both sexes, for example in competition for food or territories (West-Eberhard 1983, Kraaijeveld *et al.* 2004). Only experimental tail manipulations and detailed studies of social behavior and mate choice in this and

TABLE 3. Hourly number of provisioning visits and removal of fecal sacs at 3 and 14–17 days of nestling age, and hourly brooding time and mean brooding session duration at 3 days of age (means ± SE, number of individuals in parentheses), by female and male Thorn-tailed Rayaditos on Chiloé Island, Chile. Significant differences between the sexes after Bonferroni correction are denoted with an asterisk.

	Female	Male	*F*	*P*
Provisioning rates, 3 days (h^{-1})	10.6 ± 1.6 (5)	10.1 ± 1.7 (4)	0.04	0.85
Fecal sacs, 3 days (h^{-1})	1.1 ±03 (5)	1.7 ± 0.4 (4)	1.3	0.29
Brooding time, 3 days (min h^{-1})	13.1 ± 1.4 (5)	17.3 ± 1.6 (4)	3.9	0.09
Mean brooding session (min)	4.7 ± 0.6 (5)	7.1 ± 0.7 (4)	6.0	0.04
Provisioning rates, 14–17 days (h^{-1})	21.2 ± 1.5 (18)	20.6 ± 1.7 (15)	0.1	0.79
Fecal sacs, 14–17 days (h^{-1})	4.2 ± 0.3 (18)	2.1 ± 0.3 (15)	20.0	<0.001*

other furnariids can confirm if tails play a role in signaling in this large, understudied family.

ACKNOWLEDGMENTS

The project was financed through grants to RAV (FONDECYT 1060186, 1020550, and 7020550) and JM (CGL2004-00787/BOS); a grant from Consejo Superior de Investigaciones Científicas–Universidad de Chile (2003-04-09) to JM, SM, RAV, and J. Armesto; and the 2004 Banco Bilbao-Vizcaya-Argentaria (BBVA) Foundation prize in Research on Conservation Biology (to RAV). Fundación Senda Darwin and Mary F. Willson kindly allowed us to work in Senda Darwin Biological Station and Inés Hannig granted permission to work in Fundo "Los Cisnes." This paper forms part of the research activities of the Institute of Ecology and Biodiversity (Grant No. ICM-05-002), and is a contribution to the research program of Senda Darwin Biological Station, Chiloé Island. Andrea Gallardo, Silvina Ippi, Rocío Jaña, Natalia Márquez, Daniela Parra, René Quispe, and Camila Villavicencio helped with fieldwork. Elena Arriero and Sonia García-Fraile advised on DNA analyses. Isabel López helped with spectrophotometric analyses in the lab. The comments of two anonymous reviewers were very helpful. The capture, banding, and handling of birds was authorized by Servicio Agrícola y Ganadero of Chile.

REFERENCES

ANDERSSON, M. 1982. Female choice selects for extreme tail length in a widowbird. *Nature* 361: 628-631.

BOLAND, C.R.J., M.C. DOUBLE & G.B. BAKER. 2004. Assortative mating by tail streamer length in Red-tailed Tropicbirds *Phaethon rubricauda* breeding in the Coral Sea. *Ibis* 146: 687-690.

CUTHILL, I.C., A.T.D. BENNETT, J.C. PARTRIDGE & E.J. MAIER. 1999. Plumage reflectance and the objective assessment of avian sexual dichromatism. *American Naturalist* 153: 183-200.

FITZPATRICK, S. 1997. Patterns of morphometric variation in birds' tails: length, shape and variability. *Biological Journal of the Linnean Society* 62: 145-162.

FRAGA, R. 1980. The breeding of Rufous Horneros (*Furnarius rufus*). *Condor* 82: 58-68.

FRIDOLFSSON, A.K. & H. ELLEGREN. 1999. A simple and universal method for molecular sexing of non-ratite birds. *Journal of Avian Biology* 30: 116-121.

GRAFEN, A. 1990. Biological signals as handicaps. *Journal of Theoretical Biology* 144: 517-546.

IPPI, S. & A. TREJO. 2003. Dinámica y estructura de bandadas mixtas de aves en un bosque de lenga (*Nothofagus pumilio*) del noroeste de la Patagonia argentina. *Ornitología Neotropical* 14: 353-362.

JOHNSON, A.W. & J.D. GOODALL. 1967. *The Birds of Chile and Adjacent Regions of Argentina, Bolivia and Peru.* Platt Establecimientos Gráficos S.A., Buenos Aires.

JOHNSTONE, R.A. 1995. Sexual selection, honest advertisement and the handicap principle: reviewing the evidence. *Biological Reviews* 70: 1-65.

KENDEIGH, S.C. 1952. *Parental Care and its Evolution in Birds.* The University of Illinois Press, Urbana, Illinois.

KRAAIJEVELD, K., J. GREGURKE, C. HALL, J. KOMDEUR & R.A. MULDER. 2004. Mutual ornamentation, sexual selection, and social dominance in the Black Swan. *Behavioral Ecology* 15: 380-389.

LANDE, R. 1977. On comparing coefficients of variation. *Systematic Zoology* 26: 214-217.

LANDE, R. 1980. Sexual dimorphism, sexual selection, and adaptation in polygenic characters. *Evolution* 34: 292-305.

MØLLER, A.P. & A. POMIANKOWSKI. 1993. Why have birds got multiple sexual ornaments? *Behavioral Ecology and Sociobiology* 32: 167-176.

MORENO, J., S. MERINO, R.A. VÁSQUEZ & J.J. ARMESTO. 2005. Breeding biology of the Thorn-tailed Rayadito (Furnariidae) in south-temperate rainforests of Chile. *Condor* 107: 69-77.

NORES, A.I. & M. NORES. 1994. Nest building and nesting behavior of the Brown Cacholote. *Wilson Bulletin* 106: 106-120.

ÖDEEN, A. & O. HÅSTAD. 2003. Complex distribution of avian color vision systems revealed by sequencing the SWS1 Opsin from total DNA. *Molecular Biology and Evolution* 20: 855-861.

REMSEN, J.V. 2003. Family Furnariidae (ovenbirds). Pp. 162-239 *in Handbook of the Birds of the World. Vol. 8.* J. del Hoyo, A. Elliott & D. Christie (eds.). Lynx Edicions, Barcelona.

RICE, W.R. 1989. Analyzing tables of statistical tests. *Evolution* 43: 223-225.

SELANDER, R.K. 1966. Sexual dimorphism and differential niche utilization in birds. *Condor* 68: 113-151.

SKUTCH, A.F. 1967. Life histories of Central American highland birds. *Publications of the Nuttall Ornithological Club*, No. 7.

SKUTCH, A.F. 1996. *Antbirds and Ovenbirds*. University of Texas Press, Austin, Texas.

SVENSSON, L. 1984. *Identification guide to European passerines*. Lars Svensson, Stockholm, Sweden.

THOMAS, A.L.R. 1993. On the areodynamics of birds' tails. *Philosophical Transactions of the Royal Society of London B* 340: 361-380.

VUILLEUMIER, F. 1967. Mixed species flocks in Patagonian forests, with remarks on interspecies flock formation. *Condor* 69: 400-404.

WEST-EBERHARD, M.J. 1983. Sexual selection, social competition, and speciation. *Quarterly Review of Biology* 58: 155-183.

WINKER, K., VOELKER, G.A. & J.T. KLICKA. 1994. A morphometric examination of sexual dimorphism in the *Hylophilus*, *Xenops*, and an *Automolus* from southern Veracruz, Mexico. *Journal of Field Ornithology* 65: 307-323.

Manuscript received 2 May 2006; accepted 5 December 2006. E-mail: jmoreno@mncn.csic.es.

13 HAEMATOZOA IN FOREST BIRDS FROM SOUTHERN CHILE: LATITUDINAL GRADIENTS IN PREVALENCE AND PARASITE LINEAGE RICHNESS

SANTIAGO MERINO[1*], JUAN MORENO[1], RODRIGO A. VÁSQUEZ[2], JAVIER MARTÍNEZ[3], INOCENCIA SÁNCHEZ-MONSÁLVEZ[3], CRISTIÁN F. ESTADES[4], SILVINA IPPI[2], PABLO SABAT[5,6], RICARDO ROZZI[2,7,8,9], AND STEVEN MCGEHEE[9]

[1]Departamento de Ecología Evolutiva, Museo Nacional de Ciencias Naturales, Consejo Superior de Investigaciones Científicas, José Gutiérrez Abascal 2, 28006 Madrid, Spain (Email: santiagom@mncn.csic.es), [2]Instituto de Ecología y Biodiversidad, Departamento de Ciencias Ecológicas, Facultad de Ciencias Universidad de Chile, Santiago, Chile, [3]Departamento de Parasitología, Universidad de Alcalá, Alcalá de Henares, Spain, [4]Departamento de Manejo de Recursos Forestales, Facultad de Ciencias Forestales, Universidad de Chile, Chile, [5]Departamento de Ciencias Ecológicas, Facultad de Ciencias, Universidad de Chile Santiago, Chile, [6]Center for Advanced Studies in Ecology and Biodiversity, Facultad de Ciencias Biológicas, Pontificia Universidad Católica de Chile, Santiago, Chile, [7]Universidad de Magallanes, Punta Arenas, Chile; [8]University of North Texas, Denton, Texas,USA, & [9]Parque Etnobotánico Omora, Puerto Williams, Chile

Austral Ecology 33: 329-340 (2008)

ABSTRACT. Haematozoa in forest birds from southern Chile: latitudinal gradients in prevalence and parasite lineage richness. The existence of latitudinal gradients in species richness and their abundance is known for many free living organisms, but few cases have been reported for parasitic diseases. In addition, asymmetries between the Northern and Southern hemispheres in several characteristics may affect the distribution and diversity of species at all ecological levels. In this respect, we studied the distribution of several genera of blood parasites infecting birds along a latitudinal gradient that included the world's southernmost forests ecosystems. Birds were mist-netted and sampled for blood in localities across Chile ranging from 33°S to 55°S during the years 2003-06. Overall, 26 bird species were sampled and 27 parasite lineages were identified. The latter belonged to three genera: *Plasmodium* (8), *Haemoproteus* (8), and *Leucocytozoon* (11). We found a positive significant relationship between prevalence and latitude for *Leucocytozoon* lineages and a negative relationship for *Haemoproteus*, *Plasmodium,* and mixed infections. However, we did not find a significant relationship between parasite diversity and latitude. We found 18 lineages infecting only one species of host, and 19 lineages appear in only one of the localities of sampled. This pattern implies that some parasite lineages may have evolved in isolation in some species/localities. In addition, specificity at the host-family level was only found for *Haemoproteus* lineages, infecting birds in the family Emberizidae. Individuals of the long distance migrant bird White-crested Elaenia (*Elaenia albiceps*), were found infected by the same parasite lineages in localities separated by 20° of latitude. Infections by these lineages were detected in other sedentary birds including juveniles and nestlings of different bird species. Therefore, long distance migrants are able to distort the presence of latitudinal gradients of diseases due to the potential role of migrants in spreading infections. Geographical gradients in prevalence of avian haematozoa differ between parasite genera and hemispheres, probably in relation to the existence of appropriate

Key words: blood parasites, DNA analysis, host specificity, migratory birds, vector-parasite-host interactions, vectors.

RESUMEN. Hematozoos en aves de bosque del sur de Chile: gradientes latitudinales en la prevalencia y riqueza de linajes de parásitos. La existencia de gradientes latitudinales en la riqueza de especies y sus abundancias es conocida en muchos organismos de vida libre, pero pocos casos han sido reportados para enfermedades parasitarias. Adicionalmente, las asimetrías entre el hemisferio norte y sur en varias características pueden afectar la distribución y diversidad de especies en todos los niveles ecológicos. En este sentido, estudiamos la distribución de varios géneros de parásitos sanguíneos que infectan aves, a lo largo de un gradiente latitudinal que incluye los ecosistemas forestales más australes del mundo. En localidades a lo largo de Chile entre los 33°S hasta los 55°S, durante los años 2003-06, las aves fueron capturadas con redes de neblina y se les tomó una muestra de sangre. En total fueron muestreadas 26 especies de aves, identificándose 27 linajes de parásitos. Estos últimos pertenecían a tres géneros: *Plasmodium* (8), *Haemoproteus* (8) y *Leucocytozoon* (11). Encontramos una relación significativa positiva entre la prevalencia y la latitud para el linaje *Leucocytozoon* y una relación negativa para *Haemoproteus*, *Plasmodium* e infecciones mixtas. Sin embargo, no encontramos una relación significativa entre la diversidad de parásitos y latitud. Encontramos 18 linajes que infectan a una sola especie de huesped y 19 linajes aparecieron en sólo una localidad del muestreo. Este patrón implica que algunos linajes de parásitos pueden evolucionar de forma aislada en algunas especies/localidades. Además, la especificidad a nivel familia del huésped sólo fue encontrada para el linaje *Haemoproteus*, infectando a aves de la familia Emberizidae. Individuos de fío-fío (*Elaenia albiceps*), una especie de ave migratoria de larga distancia, se encontraron infectados por los mismos linajes de parásitos en localidades separadas por 20° de latitud. Infecciones por estos linajes fueron detectadas en otras aves residentes, incluyendo juveniles y polluelos de diferentes especies de aves. Por lo tanto, las aves migratorias de larga distancia son capaces de modificar la presencia de los gradientes latitudinales de enfermedades, debido a su rol potencial en la propagación de infecciones. Gradientes geográficos en la prevalencia de hematozoos aviares difieren entre géneros de parásitos y hemisferios, probablemente en relación con la existencia de las interacciones apropiadas de vector-parásito-huésped.

Palabras clave: análisis de ADN, aves migratorias, especificidad del hospedero, interacciones vector-parásito-hospedero, parásitos sanguíneos, vectores.

INTRODUCTION

Although many studies have shown the existence of latitudinal gradients in species richness of free living organisms (Pianka 1966, Rohde 1992, Stevens 1992, Huston 1999, Chown & Gaston 2000), knowledge about these potential gradients in pathogenic organisms is scarce (see Hillebrand *et al.* 2001, Curtis *et al.* 2002, Nee 2003, Guernier *et al.* 2004, Guégan *et al.* 2005). The scarcity of studies addressing latitudinal variation in pathogenic microorganisms is especially marked for the Southern Hemisphere. For example, Guernier *et al.* (2004) showed a decline in the number of parasitic and infectious diseases affecting humans at higher latitudes in both hemispheres, but far more data and range of latitude were provided for populations in the north. In this respect, the different distribution of landmasses between both hemispheres may cause differences in biodiversity distribution as pointed out by several authors (see Chown *et al.* 2004). One of the main asymmetries between the Northern and Southern Hemispheres is their proportion of oceanic to land surface. Thus, most of the world's emerged land is situated between latitudes 30 and 60° north. This produces a ratio of area covered by water to area covered by land of 1:1 in comparison with a ratio of approximately 16:1 for the same range of latitudes in the southern hemisphere (Chown *et al.* 2004). The result is a marked oceanicity of southern land-masses that has a considerable effect on climatic conditions and consequently the distribution and diversity of species at all ecological levels, and microorganism species are not an exception (Merino & Potti 1996, Bonan 2002, Harvell *et al.* 2002).

South America is the only continental landmass reaching latitudes further south than 40 degrees where coexisting forests, forest birds, and their blood parasites can be found. Moreover, the Chilean forest ecosystems are especially interesting due to their isolation from other forest regions. Great orographic and climatic barriers separate them from the nearest tropical forests by at least 1500 km. To the north lies the Atacama Desert, the driest in the world, to the west and south the Pacific Ocean, and to the east rise the high Andean Mountains. In this way, the Chilean temperate forests become a biogeographical island (Medel & Vásquez 1994, Armesto *et al.* 1998, Rozzi *et al.* 2000). Toward the extreme south, the Chilean forest region becomes an archipelago composed of numerous islands of different size, very close to each other and to the continent. Moreover, with the exception of *Elaenia albiceps*, there are no migratory forest birds that could spread diseases from the tropical areas to the extreme south.

Recently, avian diseases have gained attention due to their potential implication in the spread of zoonoses and diseases of economic importance among different countries (Rappole *et al.* 2000, Chen *et al.* 2006). Avian haematozoa commonly infect birds across the world (Bennett *et al.* 1982b, Bishop & Bennett 1992, Valkūnas 2005) and represent good candidates to study latitudinal distribution of diseases in wild animals. Moreover, some phylogeographical studies have shown that parasite diversification is the result of frequent host-switching, rapid parasite dispersal, and local specificity in new hosts (Fallon *et al.* 2005). Thus, there exists a potential effect of migrant birds working as a disruptive factor of latitudinal gradients, as they may spread parasite lineages along their migratory routes, which later specialise and diversify in other host species.

In this study, we present data on the latitudinal incidence and lineage richness of three blood protozoa genera (*Plasmodium*, *Haemoproteus*, and *Leucocytozoon*) infecting forest birds in Chile between latitudes 33 and 55°S. We specifically study (i) the existence of latitudinal gradients in prevalence and richness of three different genera of blood parasites; (ii) the presence of asymmetries between hemispheres in latitudinal gradients for these blood parasites in birds; (iii) the potential effect of host family specificity in reducing options of parasite lineage spreading; and (iv) the potential effect of migrants as disruptors of these gradients through spreading parasites.

METHODS

Birds were trapped with mist-nets situated at the edge of forest areas in several localities across Chile during 2003, 2004, and 2005 (Table 1). Each bird was individually marked with numbered or coloured bands to avoid re-sampling. For each bird, a drop of blood was obtained from the brachial vein and preserved in a buffer composed of 100 mM TRIS pH = 8.0, 100 mM of ethylenediaminetetraacetic acid (EDTA), 2% sodium dodecyl sulphate (SDS) (see Jarvi *et al.* 2002) for DNA analyses. We used the same protocol in the spring of 2005/2006 except that samples were stored on FTA classic cards (Whatman International Ltd. UK, see Gutiérrez-Corchero *et al.* 2002). For a subsample of birds, another drop of blood was smeared on a slide, air-dried and fixed in 96% ethanol. Blood smears were stained with Giemsa stain (1/10 v/v) for 40 min and scanned in search of blood parasites following the methods described in Merino *et al.* (1997).

Five areas of study were selected for the capture and sampling of birds (Fig. 1, Table 1): (i) Temperate areas: (1) In *Rinconada de Maipú* (nearby Santiago), birds were captured during May–August 2005. This area is covered by dry sclerophyllous shrubland dominated by *Acacia caven* and the climate is Mediterranean (Tapia 2005). (2) *Pantanillos* is comprised of a mosaic of deciduous *Nothofagus glauca* forests and pine (*Pinus radiata*) plantations. The climate is humid with a short dry summer (Márquez 1992). Birds in this area were captured in November 2004. (3) *Ancud*, Senda Darwin Biological Station and Fundo Los Cisnes, Chiloé Island. The area is composed of second-growth forests with a mixture of several evergreen species (Gajardo 1994). The climate is mild and wet. Birds were captured during November and December of 2003 and 2004. (ii) Extreme South areas: (4) *Outskirts of Punta Arenas* on an ecotonal zone between *Nothofagus* deciduous forests and the Patagonian steppe (Gajardo 1994). The climate is cold with low rainfall and snow in the winter. Birds were mist-netted near

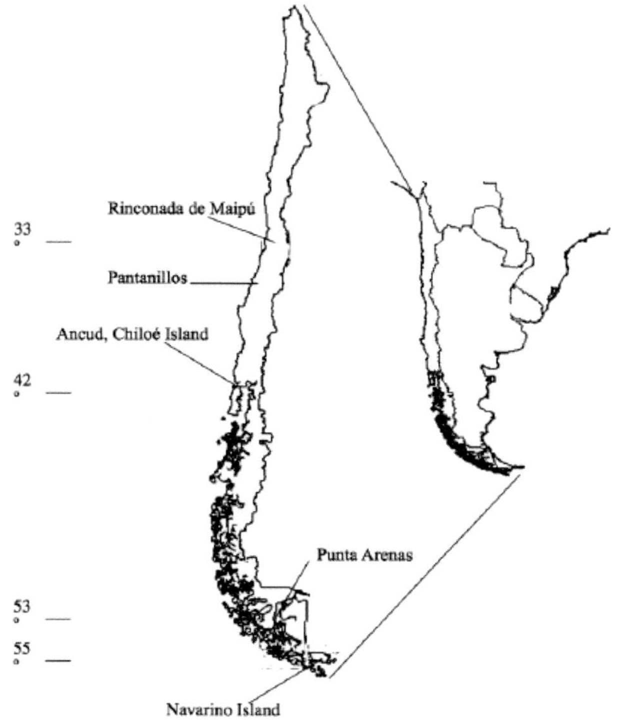

Fig. 1. Map of Chile showing the localities of sampling and latitude.

Punta Arenas during December 2005. (5) Omora Ethnobotanical Park on *Navarino Island* is a mosaic of evergreen broadleaf forests dominated by *Nothofagus betuloides* and *Drimys winteri*, deciduous *Nothofagus* forests of *Nothofagus pumilio* and *Nothofagus antarctica*, and mixed patches of evergreen (*N. betuloides*) and deciduous *Nothofagus* (Pisano 1977, Rozzi *et al.* 2006). Rainfall is distributed more or less uniformly throughout the year and part of the precipitation is in the form of snow. Birds were captured on Navarino Island during December 2005 and January 2006. Bird nomenclature follows Jaramillo *et al.* (2003). In the (3) Chiloé Island area of study, 54 Southern House Wrens (*Troglodytes musculus*) nestlings and 89 Thorn-tailed Rayaditos (*Aphrastura spinicauda*) nestlings from a population breeding in nest boxes (Moreno *et al.* 2005) were sampled. Data from nestlings were treated separately in the analyses as they do not share the same probability of exposure and/or time to develop infections as compared to adults.

Although our areas of study include several localities on islands, the proximity to mainland and size and habitat complexity on them reduce the potential on an island effect on the diversity of parasites.

DNA analyses

Genomic DNA from samples conserved in lysis buffer was obtained using the UltraClean DNA BloodSpin kit (MO BIO Laboratories, Inc., Solana Beach, CA, USA). Genomic DNA on FTA cards was

TABLE 1. Localities of sampling, indicating latitude, longitude, number of birds infected and sampled individuals (%), number of birds infected by each parasite genus and mixed infections (%) and number of lineages identified for each parasite genus; (ChH: *Haemoproteus*, ChP: *Plasmodium*, ChL: *Leucocytozoon*).

Locality	Latitude and Longitude	Birds infected/ sampled	H	P	L	Mixed	Number of host species	Parasite lineages
1. Rinconada de Maipú	33°32´S 70°06´W	13/61 (21.3%)	12 (19.7%)	6 (9.8%)	3 (4.9%)	7 (11.5%)	12	ChH: 5, 6, 7, 8 ChP: 6 ChL: 7, 8, 11
2. Pantanillos	35°25´S 72°08´W	20/79 (25.3%)	14 (17.7%)	8 (10.1%)	6 (7.6%)	7 (8.9%)	10	ChH:1, 6 ChP: 1, 4, 6, 7 ChL: 3, 6, 9
3. Ancud, Isla de Chiloé	41°51´S 73°49´W	22/165 (13.3%)†	9 (5.5%)	15 (9.1%)	10 (6.1%)	9 (5.5%)	15	ChH: 1, 6 ChP: 1, 2, 3, 4, 8 ChL: 1, 4, 5, 9, 10
4. Punta Arenas	53°08´S 70°55´W	3/42 (7.1%)	2 (4.8%)	1 (2.4%)	1 (2.4%)‡	1 (2.4%)	7	ChH: 2, 4 ChP: 5 ChL:?
5. Navarino Island	54°56´S 67°36´W	37/270 (13.7%)	3 (1.1%)	1 (0.4%)	34 (12.6%)	1 (0.4%)	13	ChH: 1, 3, 6 ChP: 1 ChL: 1, 2, 3
Total		95/617 (15.4%)	40 (6.5%)	31 (5.0%)	55 (8.9%)	25 (11.5%)	26	8ChH, 8ChP, 11ChL

† 1/143 (0.7%) for nestlings sampled from nest-boxes (see Methods). ‡ Detected only in blood smears.

extracted to a soluble solution before polymerase chain reaction (PCR) using the following protocol: cored samples were transferred to collection vials with 250 μL of SET buffer (0.15 M NaCl, 0.05 M Tris, 0.001 M EDTA, pH = 8) at 4°C for 6 h. Afterwards, SDS 20% (7 μL) and proteinase K (50 μg) were added to the vials and incubated at 55°C overnight. After incubation, ammonium acetate 4 M (250 μL) was added to the vials at room temperature for 30 min. Subsequently, vials were centrifuged at 13 000 g for 10 min. After removing the pellet, DNA was precipitated with ethanol and re-suspended in sterile water.

Initially, partial amplification of the cytochrome B gene was accomplished by PCR using the non-specific primers 3760F (5'-GAG TGG ATG GTG TTT TAG AT-3') and 4292Rw (5'-TGG AAC AAT ATG TAR AGG AGT-3') as previously used for detection of *Haemoproteus/Plasmodium* (see Beadell *et al.* 2004). Among samples found to be positive for infection by using those primers, we tried to detect double infections by using newly designed specific primers: PF (5'-GGA TTT GTG GTG GAT ATC TTG-3') and 4292Rw for *Plasmodium* and HML (5'-GCT ACT GGT GCT ACA TTT GT-3') and HMR (5'-CCT AAA GGA TTA GAG CTA CC-3') for *Haemoproteus*. These pairs of primers have been found to be specific for the detection of *Plasmodium* or *Haemoproteus* in several bird species (J. Martínez *et al.* unpubl. data). Partial amplification of the cytochrome B gene was accomplished using the newly designed primers LDLd (5'-CAT TCY ACW GGT GCA TCT TT-3') and LDRd (5'-CTG GAT GWG ATA ATG GWG CA-3') for *Leucocytozoon*. The criteria used for selection of these new primer pairs were: (i) a short product (<600 nucleotides), which would be more likely to be recovered in samples with mainly host DNA (Saiki 1990) and (ii) primer sites conserved across the parasites, but different enough from bird DNA to prevent its amplification. PCR reactions consisted of 25-μL reaction volumes containing 20 ng template DNA, 50 mM KCl, 10 mM Tris-HCl, 1.5 $MgCl_2$, 0.2 mM of each dNTP, 1 μM of each primer, and 1.25 U of AmpliTaq Gold (Applied Biosystems, Foster City, CA, USA). The reactions were cycled at the following parameters using a thermal cycler (MasterCycler Personal, Eppendorf): 94°C for 10 min (polymerase activation), 40 cycles at 95°C for 40 s, 58°C for 1 min, 72°C for 1 min, and a final extension at 72°C for 10 min. All the DNA fragments obtained after PCR assays were recovered from agarose gels and subjected to direct sequencing. DNA fragments obtained were sequenced using an ABI 3130 (Applied Biosystems) automated sequencer. For potential cases of under-amplification (*i.e.,* weak signal obtained), we repeated the PCR procedure under the same conditions. All of these samples were considered as negative because the second amplification failed to render a positive band. Thus, all samples rendering a clear PCR product were sequenced and identified to genus. A sample previously identified as infected by both methods of blood smear and molecular sequencing, was included at each PCR as a positive control.

DNA sequences were aligned using the CLUSTALW program (Thompson *et al.* 1994). The BIOEDIT program (Hall 1999) was used to edit the sequences. The MEGA3.1 (Kumar *et al.* 2004) software package was used in phylogram construction/drawing. Computer programs were set at their default parameters in all analyses. Phylogenetic analyses were carried out using the parsimony method and the Kimura model. The phylogenetic tree was estimated by bootstrap analysis with 100 replications. As both methods rendered similar results, we present only the results for the Kimura model. Phylogenetic analysis was carried out using sequences with a length of 405 bp after removing columns containing gaps or missing data. The length of sequences from lineages ChP7 and ChP8 were 315 bp and 259 bp, respectively. Thus, both lineages were excluded in phylogenetic analysis. Several GenBank sequences from parasites recovered from birds of the same families were included in the phylogenetic trees to compare affinities between our samples and those from other localities.

Statistics

We looked for relationships between prevalence and parasite richness as dependent variables, and latitude and the two measures of sampling effort (number of samples at each locality and number of host species sampled at each locality; see Gregory 1990) as independent variables by means of GLM (general linear model). Variables were eliminated step-by-step from the model when $P > 0.05$ and models retained all significant variables. Prevalences were arcsine square root transformed. As some of the variability in prevalence between localities may be due to the host species composition at each locality, we conducted similar analyses but using only the three most frequently sampled bird species (*Aphrastura spinicauda*, *Elaenia albiceps*, *Troglodytes musculus*) from each of four localities (all except locality 1; Rinconada where these species were scant; see Appendix 1 at http://www.ecolsoc.org.au/ What%20we%20do/Publications/Austral%20Ecology/AE.html). In these analyses, variables to control for sampling effort only included number of samples as the same number of hosts species (three) were used for each locality. Latitude was introduced as its absolute value that ranged from 33.32°S to 54.56°S (Table 1). We also made comparisons of prevalence of infections between temperate areas (localities 1, 2, and 3) *versus* extreme south areas (localities 4 and 5) by means of χ^2 tests.

RESULTS

Using DNA analyses, 95 samples from 617 adult birds and one sample from 143 nestlings were found to be infected (Table 1). We detected 40 birds (6.5%) belonging to several species, infected by *Haemoproteus* (excluding nestlings), 31 birds (5.0%) infected by *Plasmodium,* and 55 birds (8.9%) infected by *Leucocytozoon*. Mixed infections (several lineages of blood parasites; Table 1) were carried by 25 birds. Eight lineages of *Plasmodium*, eight of *Haemoproteus,* and 11 of *Leucocytozoon* were identified as different lineages when their sequences differed in at least one bp (see Appendix 2 at http://www.ecolsoc.org.au/What%20we%20do/Publications/Austral%20Ecology/AE.html). Minimum and maximum lineage divergence was 0.3–7.5% (1–31 bp) for *Haemoproteus*, 1.5–7.7% (6–32 bp) for *Plasmodium,* and 0.3–13.9% (1–58 bp) for *Leucocytozoon*. Five lineages of *Plasmodium*, five of *Haemoproteus,* and eight of *Leucocytozoon* appeared as host-specific, as they were found in only one bird species (see Appendix 2 at http://www.ecolsoc.org.au/What%20we%20do/Publications/Austral%20Ecology/AE.html). In addition, 19 lineages were detected in only one of the localities (five of *Plasmodium*, six of *Haemoproteus*, and eight of *Leucocytozoon*) (see Appendix 2 at http://www.ecolsoc.org.au/What%20we%20do/Publications/Austral%20Ecology/AE.html).

For the subsample of blood smears, we detected 17 infections from 198 birds (8.6%), four corresponding to *Haemoproteus* (2.0%), five to *Plasmodium* (2.5%), and eight to *Leucocytozoon* (4.0%). All infections detected in blood smears were also detected by amplification, of parasite DNA except one infection by *Leucocytozoon* in a Patagonian Sierra Finch (*Phrygilus patagonicus*) from Punta Arenas. However, by molecular methods, we detected 20 other infections by the different genera of parasites that were not detected in blood smears.

Overall prevalence of infection (by the three genera of parasites together) was negatively related to latitude, but not significantly, once the number of hosts sampled and the number of samples examined were eliminated from the model (Table 2). When we analysed the prevalence by each parasite genus separately, we found a significant negative relationship between *Haemoproteus* prevalence and latitude (Fig. 2a, Table 2) when we controlled for the number of hosts sampled. The number of samples was not retained in the model for *Haemoproteus* prevalence. A similar result was obtained for the relationship between latitude and *Plasmodium* prevalence (Fig. 2b, Table 2), but the number of samples and the number of hosts sampled were not retained in the model. In addition, we found a significant positive relationship between *Leucocytozoon* prevalence and latitude (Fig. 2c, Table 2) when we controlled for the positive effect of number of hosts sampled and number of samples examined. A significant negative relationship between mixed infections and latitude was also detected (Fig. 2d, Table 2) after the number of hosts sampled and number of samples examined were eliminated from the model. Parasite richness was not significantly related to latitude and was the second variable eliminated from the model, which did not retain significant variables (Table 2). Parasite richness for each parasite genus was not significantly related to any of the variables in the model (Table 2).

A comparison of prevalences between temperate (localities 1, 2, and 3) *versus* extreme south areas (localities 4 and 5) also showed latitudinal differences, with *Haemoproteus*, *Plasmodium*, and mixed infections being more common in temperate than in extreme south areas (Yates χ^2_1 = 23.19, P < 0.0001, Yates χ^2_1 = 23.58, P < 0.0001, and Yates χ^2_1 = 17.15, P = 0.0001, respectively). However, *Leucocytozoon* infections were more common in extreme south areas than in temperate ones (Yates χ^2_1 = 4.20, P = 0.04). Parasite richness was higher in temperate (28 lineages) than in extreme south localities (11 lineages), but not significantly so (Mann–Whitney *U*-test, Z = 1.73, P = 0.08).

For the subsample of three host species being sampled at each of four localities, we obtained a significant and negative relationship between prevalence of *Plasmodium* and latitude ($F_{1,3}$ = 27.75, P = 0.034). In addition, a similar pattern was found for relationships between latitude and prevalence of mixed infections, but did not reach significance ($F_{1,3}$ = 16.77, P = 0.08). Other relationships were not significant (data not shown). The comparison between temperate and extreme south areas for the pool of these three host species also indicated some differences in prevalences: higher prevalence of *Haemoproteus* and *Plasmodium* were found in temperate areas (Yates χ^2_1 = 8.66, P = 0.003 and Yates χ^2_1 = 10.67, P = 0.001, respectively). In addition, higher prevalence of *Leucocytozoon* was found in extreme south localities (Yates χ^2_1 = 7.57, P = 0.006).

Two migratory species were present in our sample: the Chilean Swallow (*Tachycineta meyeni*) with migratory populations in southern Chile and resident populations in central Chile, and the White-crested Elaenia (*Elaenia albiceps*), which is a long distance migrant with wintering grounds in northern South America. Swallows were not captured in localities 1 and 2 where the species is resident, and 26 individuals were found uninfected by the three parasite genera in the other three localities. In contrast White-crested Elaenias were captured in all localities except in the northernmostone, Rinconada de Maipú. Eight different lineages were detected in this species

TABLE 2. Results of GLMs (general linear models) relating prevalence and diversity for each parasite genera and latitude, number of samples and number of host sampled. Variables were eliminated step-by-step from the model when P > 0.05.

	Latitude		Number of samples		Number of hosts	
	F	*P*	*F*	*P*	*F*	*P*
Prevalence						
H	144.78	0.01	0.07	0.83	19.82	0.05
P	22.96	0.02	0.36	0.61	0.67	0.56
L	386.29	0.03	854.45	0.02	242.18	0.04
Mixed	57.06	0.005	1.74	0.32	0.002	0.97
Total	6.57	0.08	83.11	0.07	57.28	0.08
Diversity						
H	0.23	0.66	0.11	0.78	0.0006	0.98
P	0.68	0.47	0.005	0.96	0.34	0.62
L	3.25	0.21	0.81	0.53	18.60	0.02
Total	2.93	0.23	0.02	0.92	5.50	0.10

Statistical values before to be eliminated from each model are presented for-non-significant variables. H, *Haemoproteus*; L, *Leucocytozoon*; Mixed, Mixed infections; P, *Plasmodium*; Total, sum of infections by each parasite genera.

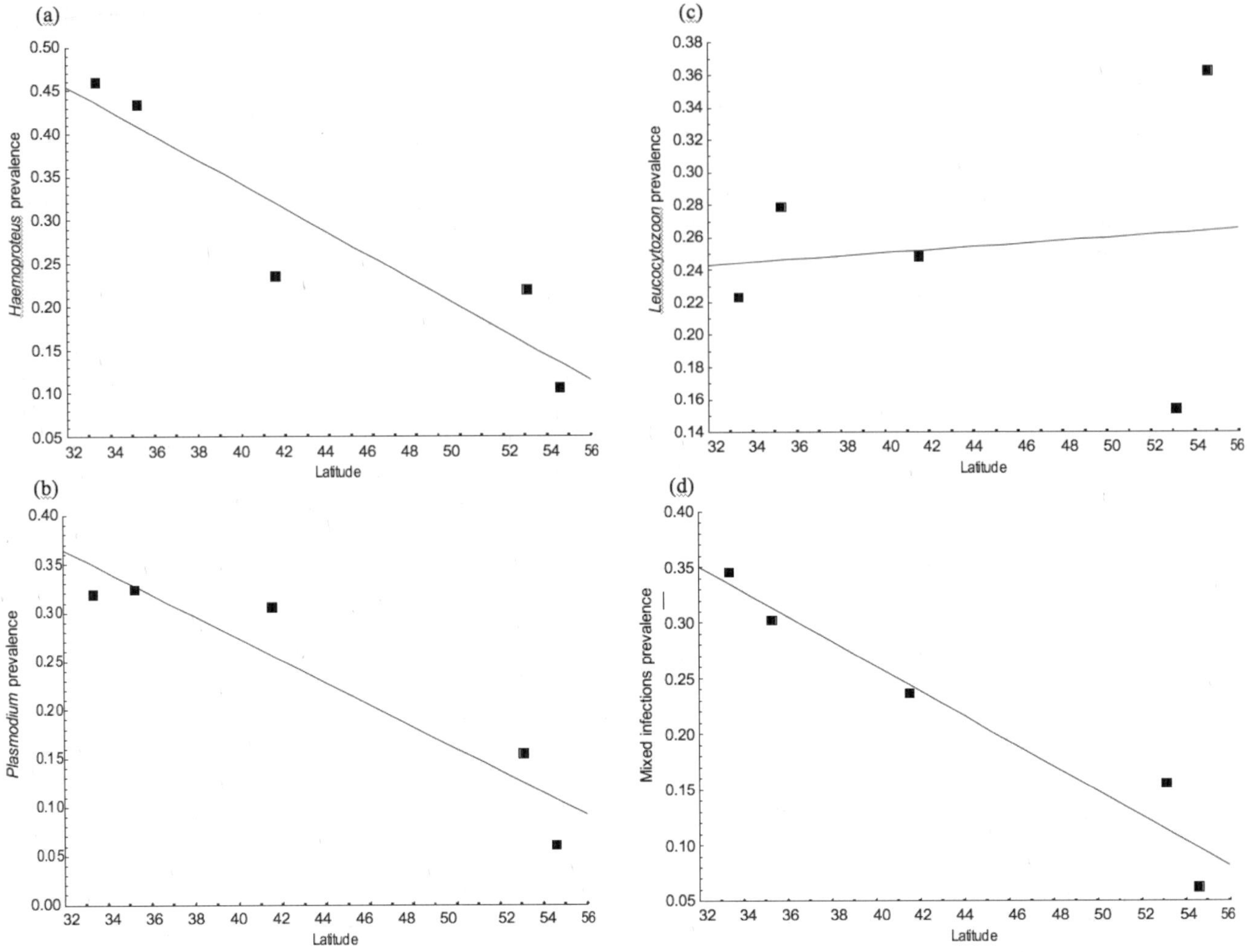

Fig. 2. Relationship between prevalence (arcsine square root transformed) for each parasite genus and latitude for the localities included in the study. (a) *Haemoproteus*, (b) *Plasmodium*, (c) *Leucocytozoon*, (d) Mixed infections. Lines indicate the pattern of the relationship. Raw data are shown.

(three *Haemoproteus*, four *Plasmodium,* and one *Leucocytozoon*) and only two were species-specific (ChH2 and ChP7). Of the other six lineages infecting this species, ChH1 was the most frequently found and it infected Elaenias in localities 2, 3, and 5. This parasite lineage also infected one adult and one nestling Thorn-tailed Rayadito at localities 2 and 3, respectively; two Rufous-crowned Sparrows and one Green-backed Firecrown at locality 2, and one Southern House Wren at locality 3 were also infected (see Appendix 2 at http://www.ecolsoc.org.au/What%20we%20do/Publications/Austral%20Ecology/AE.html).

Phylogenetic trees showed certain specificity at the family level for *Haemoproteus* infections within the avian family Emberizidae, as most of the lineages from birds of this family were grouped with a bootstrap value of 88 (see Fig. 3). No clear host family pattern appears for the *Plasmodium* and *Leucocytozoon* lineages.

DISCUSSION

The latitudinal range covered in this study included almost the entire distribution of the native forests of Chile, bounded to the north by the Atacama Desert and to the south by Pacific and Atlantic oceans (*i.e.,* the Drake Pass). In fact, Navarino Island, one of the largest islands in the Cape Horn Biosphere Reserve, includes the southernmost forests of the world (Rozzi *et al.* 2006).

With respect to our first objective, the existence of latitudinal gradients in prevalence and richness of three different genera of blood parasites, prevalences of *Haemoproteus*, *Plasmodium,* and mixed infections increased at lower latitudes (Fig. 2a,b,d). Similar results were obtained for prevalence of *Plasmodium* when only three host species captured at each of four localities of sampling were analysed. In addition, there were clear differences between temperate (localities 1–3) and extreme south areas (localities 4-5), even when the analyses were restricted to the three common host species at four localities. Higher prevalence of *Plasmodium* and *Haemoproteus* parasites occurred in temperate areas as compared to extreme south areas. These results were in agreement with previous studies for other diseases in humans and wild animals. For example, Nunn *et al.* (2005) concluded either that the greater abundance and diversity of biting arthropods in the tropics or that the climatic effects on vector behaviour and parasite development were responsible for the relationship between latitude and parasite richness increasing towards the equator. Guernier *et al.* (2004) showed a similar latitudinal pattern for several human diseases across the world. The greater abundance and/or diversity of vectors toward the equator may also explain our results for the relationship of prevalences of *Plasmodium*, *Haemoproteus,* and mixed infections and latitude. Contrary to those previous studies, we failed to detect a significant negative relationship between latitude and parasite richness. This may partly be due to the isolation of Chile from the rest of the continent due to geographical and ecological barriers (Andes Mountains, Atacama Desert, Pacific Ocean), thus preventing the arrival of a higher diversity of parasites. Also diversity gradients may be more intense at lower latitudes. Although the use of smaller screening fragments could allow detection of more parasite lineages, we expect that the probability of missed lineages was similar at the different areas of study and therefore should not affect our results.

The relationship found here between the prevalence of *Haemoproteus* and latitude agrees with data from other localities in South America. Durrant *et al.* (2006) found a higher prevalence of *Haemoproteus* and *Plasmodium* in 53 bird species from Guyana when compared to 111 bird species from Uruguay. Overall prevalence and parasite richness in the study by Durrant *et al.* (2006) was higher than in our study and, contrary to our results, they reported a higher prevalence of infection by *Plasmodium* than by *Haemoproteus*. In Uruguay, which is situated at the latitude of our northernmost study site (sampled localities in Uruguay varied from 30.19 to 34.39°S) the prevalence of *Haemoproteus* was clearly lower (3.4%) than in our study area in Rinconada de Maipú (19.7%, Table 1) but the prevalence of *Plasmodium* was higher (14.9% in Uruguay versus 9.8% in Rinconada). These patterns may imply a different vector abundance or activity in Chile and Uruguay for blood-sucking mosquitoes (Diptera: Culicidae) that are vectors of *Plasmodium*, and biting midges (Diptera: Ceratopogonidae) and hippoboscid flies (Diptera: Hippoboscidae) that are vectors of *Haemoproteus* (Valkiūnas 2005). These differences are probably due to differences in habitat and climate conditions between the Atlantic and the Pacific sides of the continent even at the same latitude.

We failed to find similar studies based on DNA analyses for *Leucocytozoon* species in South America, so we can only compare our results with studies based on the analyses of blood smears. White *et al.* (1978) reviewed the literature based on studies of blood smears of the haematozoa from Neotropical birds and reported an overall prevalence of 10.5% for 35555 birds. They found that *Haemoproteus* was the most prevalent parasite genus (7.4%), followed by *Plasmodium* (1.9%), with *Leucocytozoon* appearing in only 0.2% of the birds. These numbers were not very different from those reported by Valkiūnas (2005, see their Table 2, p. 117). White *et al.* (1978) also reported that the Neotropics showed much lower parasite prevalence and a near absence of *Leucocytozoon* in comparison with a similar review of Nearctic avian haematozoan distribution (Greiner *et al.* 1975). In another study conducted in the semiarid northernmost area of Chile, Forrester *et al.* (1977) reported data that also fit well with the patterns of higher *Haemoproteus* prevalence and lower *Leucocytozoon* prevalence at lower latitudes. Our results for the subsample of blood smears showed lower prevalences for *Haemoproteus* and higher prevalences for *Plasmodium* and especially for *Leucocytozoon* than those reported by these authors.

In relation to our second objective, the presence of asymmetries between hemispheres in latitudinal gradients, we found that our data also differed from those reported from North America by Greiner *et al.* (1975), who found 19.5% prevalence for *Haemoproteus*, 17.7% for *Leucocytozoon*, and 3.8% for *Plasmodium*. These authors also failed to find a clear pattern of distribution of prevalence among different regions of North America other than a higher prevalence for the northwestern Montane region and a lower one for the Arctic barrens region. They found a more or less homogeneous distribution of *Haemoproteus* across most regions (except the Arctic region) and concluded that vector potential, and thus transmission of this parasite, was rather uniform throughout the continent. Greiner *et al.* (1975) also concluded that transmission of *Leucocytozoon* was restricted by the presence of suitable breeding streams for the vectors (mainly simulids), and therefore depended on the topography of the region. That is, *Leucocytozoon* prevalence is expected to be higher in montane regions. The picture for *Plasmodium* was not so clear and the relatively short duration of a patent infection was stated as the most important factor for the lower number of infections diagnosed by blood smears (Greiner *et al.* 1975).

The high prevalence of *Leucocytozoon* in our study as compared to that of White *et al.* (1978) may also be related to topography as suggested by Greiner *et al.* (1975) as the highest prevalence for this parasite genus appeared in Navarino, where mountain streams are common in comparison with other study areas. The second locality with high prevalence for this parasite was Pantanillos, which is also a relatively mountainous area. However, data from Forrester *et al.* (2001) did not agree with that explanation as they reported a prevalence of 87% for *Leucocytozoon toddi* in blood smears of 15 Chimango Caracaras (*Milvago chimango*), a bird of prey sampled on Chiloé Island, one of our areas of study not specially characterized by elevated topography. The significant relationship between *Leucocytozoon* prevalence and latitude found in our study may be influenced by the higher prevalence of this parasite genus in Navarino Island. In this respect, it is clear that latitudinal gradients in protozoan infections may be distorted due to distributions of their vectors. However, the abundance of vectors is also high close to the equator and this may reveal an absence of a clear latitudinal gradient for some parasite genera (*e.g., Leucocytozoon*) when a higher range of latitudes than those reported in the present study are included. This appears to be the case for studies carried out in Europe, where low prevalences of blood parasites in central Europe in comparison with prevalences in northern and southern populations of birds support the hypothesis that destruction of natural habitats has led to a decline of vector populations in central Europe (Bennett *et al.* 1982a, Merilä *et al.* 1995; but see Scheuerlein & Ricklefs 2004). In any case, it appears that latitudinal gradients for haemoparasites of birds differ among genera of parasites and also differ between North and South America, with less clear gradients in the Northern Hemisphere. In fact, Ricklefs *et al.* (2005) reported a lower incidence of haematozoa in tropical areas than in temperate areas of North America based on data obtained from blood smears, thus

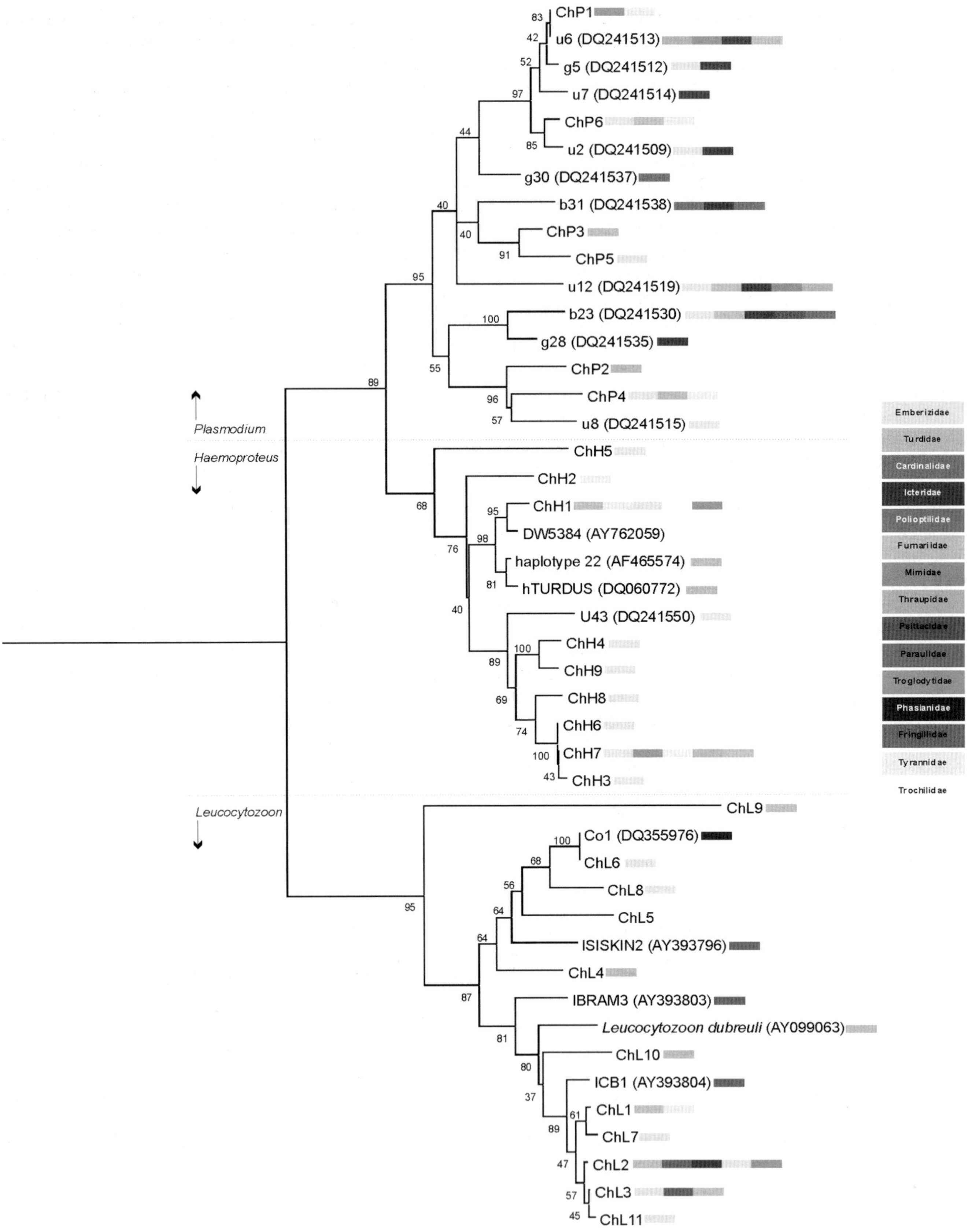

Fig. 3. Phylogenetic tree obtained with the MEGA3 software package using the model of Kimura. Bootstrap values (with 100 replications) are shown at the corresponding nodes. Numbers in brackets after the species name indicate the GenBank accession number of the lineage number for samples from this study.

also pointing to a difference in latitudinal gradients in haematozoa between hemispheres.

In reference to our third objective on host family specificity, *Haemoproteus* and *Leucocytozoon* parasites have been considered as specific at the family level in their avian hosts (Valkiūnas 2005). Several classic experiments using cross-infection between birds of the same and different families have supported this theory (Fallis & Bennett 1960, Atkinson 1986, Valkiūnas 2005). This fact may reduce possibilities of parasite lineage spreading if they represent parasite species adapted to infect one host family.

Our data show a relatively tight grouping for *Haemoproteus* lineages infecting birds in the family Emberizidae with the exception of those birds infected by the lineages ChH1 and ChH5 (Fig. 3). However, the bird species infected by lineage ChH5 have been proposed to be transferred from Emberizidae to the family Thraupidae based on genetic data (Bledsoe 1988, Burns *et al.* 2002, 2003) and this may explain why ChH5 did not appear grouped with parasites infecting birds in the Emberizidae. On the other hand, ChH1 appeared to be a generalist parasite lineage infecting several bird species, and it is possible that it did not originate in the Emberizidae. This possibility requires further research. The high number of lineages specific at the level of host species and even at the level of locality may imply that parasites show high local and species adaptation. Specificity appears to be linked to birds within a region or within a well-interconnected region, for example via migratory corridors, although genetic similarity among potential hosts could allow successful transmissions (see Engelstädter & Hurst 2006).

Finally, on the potential effect of migrant hosts as disruptors of parasitic gradients (fourth objective), we must emphasize that bird migration has been considered as a potential way of spreading these protozoan diseases (Borg 1992) and data based on molecular detection of blood parasites support this assumption (Waldenström *et al.* 2002). Although we failed to confirm infection by any of the three parasite genera here studied in one of the two migrant species sampled, several parasite lineages were found infecting the White-crested Elaenias populations across Southern Chile, and most of these lineages were not specific for this species. Elaenias are long-distance migrants and therefore coexist with many more species than resident birds, or even than short-distance migrants, which increases the chances of becoming infected with a higher diversity of lineages (Waldenström *et al.* 2002). Additionally, this bird species may act as a vehicle to transport some parasite lineages between regions within South America. Although molecular studies of these parasites have shown that at least some parasite lineages are capable of infecting birds from different families (Bensch *et al.* 2000), not all of the bird–parasite interactions detected will be stable through time. The fact that one parasite was found in one host may not reflect long-term coevolution between them, but it might be more indicative of the feeding habits of its invertebrate vector, as mosquitoes and biting midges are generalist feeders (Beadell *et al.* 2004). In this respect, the lineage ChH1 that was found in several individuals of White-crested Elaenias across Chile infected one nestling and one juvenile Thorn-tailed Rayadito and one juvenile Green-backed Firecrown, but it was absent in adult populations of these species; thus indicating a failure to establish themselves in hosts from the Furnariidae and Trochilidae families. Potential specificity in a family or at given localities and the spread of parasite lineages carried by migratory species may ultimately depend on the availability of appropriate vectors. This appeared to be the most important factor shaping the latitudinal gradients in prevalence observed for these parasites.

In conclusion, distribution of blood parasites across a latitudinal gradient in Chile varied among parasite genera, which was probably a function of the availability and abundance of appropriate vectors. These gradients seem to be absent in North America where a more homogeneous distribution of parasites appears to be the rule. In addition, although we found high specificity among the lineages of parasites in the three genera, some of these are capable of infecting avian species of very different genera examined, while some of the other lineages appeared restricted to geographical areas, thus pointing to a coadaptation between lineages and the bird fauna present in each locality. Migratory species may disrupt these patterns of coadaptation, spreading these lineages across southern South America, but successful and stable infections probably depend on the existence of appropriate parasite–vector–host associations.

ACKNOWLEDGEMENTS

This study has been partially financed by the following projects: grants FONDECYT 1020550, 1060186 and 7020550, travel grants from CSIC (grants CSIC-Univ. de Chile 2003 CL0012 and 2004 CL0033) and projects BOS2003-05724 and CGL2006-14129-C02-01/BOS from the Spanish Ministry of Education and Science. This study forms part of the research activities of the Institute of Ecology and Biodiversity (ICMP05-002). We gratefully acknowledge the owners of Rinconada de Maipú Field Station in the Metropolitan region, Pantanillos Field Station in the Maule region, Fundo 'Los Cisnes' and Senda Darwin Biological Station in Chiloé Island, and Hacienda 'Puro Chile' and Omora Ethnobotanical Park in Magallanes for permission to work on their lands. We gratefully acknowledge M.A. Rodríguez-Girones, J. Martínez de la Puente, E. Lobato, S. García-Fraile, M.A. Vukasovic, R. Quispe, and C. Villavicencio for their help during field and lab work. This is a contribution to the research program of Senda Darwin Biological Station, Chiloé Island, and the Omora Ethnobotanical Park in the Cape Horn Biosphere Reserve. Two anonymous referees greatly improved the manuscript with their constructive comments.

REFERENCES

ARMESTO, J.J., R. ROZZI, C. SMITH-RAMÍREZ & M.T.K. ARROYO. 1998. Effective conservation targets in South American temperate forests. *Science* 282: 1271-2.

ATKINSON, C.T. 1986. Host specificity and morphometric variation of *Haemoproteus meleagridis* Levine, 1961 (Protozoa: Haemosporina) in gallinaceous birds. *Canadian Journal of Zoology* 64: 2634-8.

BEADELL, J.S., E. GERING, J. AUSTIN *et al.* 2004. Prevalence and differential host-specificity of two avian blood parasite genera in the Australo-Papuan region. *Molecular Ecology* 13: 3829-44.

BENNETT, G.F., F. THOMMES, J. BLANCOU & M. ARTOIS. 1982a. Blood parasites of some birds from the Lorraine region, France. *Journal of Wildlife Diseases* 18: 81-8.

BENNETT, G.F., M. WHITEWAY & C.B. WOODWORTH-LYNAS. 1982b. *A host-parasite catalogue of the avian haematozoa.* Memorial University of Newfoundland Occasional Papers in Biology 5. Memorial University of Newfoundland, Newfoundland, Canada.

BENSCH, S., M. STJERNMAN, D. HASSELQUIST *et al.* 2000. Host specificity in avian blood parasites: a study of *Plasmodium* and *Haemoproteus* mitochondrial DNA amplified from birds. *Proceedings Royal Society London B* 267: 1583-9.

BISHOP, M. & G.F. BENNETT. 1992. *Host-parasite catalogue of the avian haematozoa Supplement 1.* Memorial University of Newfoundland Occasional Papers in Biology 15. Memorial University of Newfoundland, Newfoundland, Canada.

BLEDSOE, A.H. 1988. Nuclear DNA evolution and phylogeny of the New World nine-primaried oscines. *Auk* 105: 504-15.

BONAN, G. 2002. *Ecological Climatology: Concepts and Applications.* Cambridge University Press, Cambridge.

BORG, K. 1992. Blood parasites and their mode of spread among birds. *Ornis Svecisa* 2: 45-54. (In Swedish with English summary).

BURNS, K.J., S.J. HACKETT & N.K. KLEIN. 2002. Phylogenetic relationships and morphological diversity in Darwin's finches and their relatives. *Evolution* 56: 1240-52.

BURNS, K.J., S. J. HACKETT & N.K. KLEIN. 2003. Phylogenetic relationships of Neotropical honeycreepers and the evolution of feeding morphology. *Journal of Avian Biology* 34: 360-70.

CHEN, H., G.J.D. SMITH, K.S. LI *ET AL.* 2006. Establishment of multiple sublineages of H5N1 influenza virus in Asia: implications for pandemic control. *Proceedings of the National Academy of Sciences USA* 103: 2845-50.

CHOWN, S.L. & K.J. GASTON. 2000. Areas, cradles and museums: the latitudinal gradient in species richness. *Trends in Ecolology and Evolution* 15: 311-15.

CHOWN, S.L., B.J. SINCLAIR, H.P. LEINAAS & K.J. GASTON. 2004. Hemispheric asymmetries in biodiversity. A serious matter for ecology. *PLoS Biology* 2: 1701-7.

CURTIS, T.P., W.T. SLOAN & J.W. SCANNELL. 2002. Estimating prokaryotic diversity and its limits. *Proceedings of the National Academy of Sciences USA* 99: 10494-9.

DURRANT, K.L., J.S. BEADELL, F. ISHTIAQ *et al.* 2006. Avian hematozoa in South America: a comparison of temperate and tropical zones. *Ornithological Monographs* 60: 98-111.

ENGELSTÄDTER, J. & G.D.D. HURST. 2006. The dynamics of parasite incidence across host species. *Evolutionary Ecology* 20: 603-16.

FALLIS, A.M. & G.F. BENNETT. 1960. Description of *Haemoproteus canachites* n. sp. (Sporozoa: Haemoproteidae) and sporogony in *Culicoides* (Diptera: Ceratopogonidae). *Canadian Journal of Zoology* 38: 455-64.

FALLON, S.M., E. BERMINGHAM & R.E. RICKLEFS. 2005. Host specialization and geographic localization of avian malaria parasites: a regional analysis in the Lesser Antilles. *American Naturalist* 165: 466-80.

FORRESTER, D.J., E.C. GREINER & R.W. MCFARLANE. 1977. Blood parasites of some columbiform and passeriform birds from Chile. *Journal of Wildlife Diseases* 13: 94-6.

FORRESTER, D.J., G.W. FOSTER & J.L. MORRISON. 2001. *Leucocytozoon toddi* and *Haemoproteus tinnunculi* (Protozoa: Haemosporina) in the Chimango Caracara (*Milvago chimango*) in Southern Chile. *Memorias Instituto Oswaldo Cruz* 96: 1023-4.

GAJARDO, R. 1994. *La Vegetación Natural de Chile. Clasificación y Distribución Geográfica.* Editorial Universitaria, Santiago.

GREGORY, R.D. 1990. Parasites and host geographic range as illustrated by waterfowl. *Functional Ecology* 4: 65-54.

GREINER, E.C., G.F. BENNETT, E.M. WHITE & R.F. COOMBS. 1975. Distribution of the avian hematozoa of North America. *Canadian Journal of Zoology* 53: 1762-87.

GUÉGAN, J.F., S. MORAND & R. POULIN. 2005. Are there general laws in parasite community ecology? The emergence of spatial parasitology and epidemiology. Pp. 22-42 *in Parasitism and Ecosystems*, F. Thomas, F. Renaud & J.F. Guégan (eds.). Oxford University Press, Oxford.

GUERNIER, V., M.E. HOCHBERG & J.F. GUÉGAN. 2004. Ecology drives the worldwide distribution of human diseases. *PLoS Biology* 2: 740-6.

GUTIÉRREZ-CORCHERO, F., M.V. ARRUGA, L. SANZ, C. GARCÍA, M.A. HERNÁNDEZ & F. CAMPOS. 2002. Using FTA cards to store avian blood samples for genetic studies. Their application in sex determination. *Molecular Ecology Notes* 2: 75-7.

HALL, T. A. 1999. BioEdit: a user-friendly biological sequence alignment editor and analysis program for Windows 95/98/NT. *Nucleic Acids Symposium Series* 41: 95-8.

HARVELL, C.D., C.E. MITCHELL, J.R. WARD *et al.* 2002. Climate warming and disease risk for terrestrial and marine biota. *Science* 296: 2158-62.

HILLEBRAND, H., F. WATERMANN, R. KAREZ & U.G. BERNINGER. 2001. Differences in species richness patterns between unicellular and multicellular organisms. *Oecologia* 126: 114-24.

HUSTON, M.A. 1999. Local processes and regional patterns: appropriate scales for understanding variation in the diversity of plants and animals. *Oikos* 86: 393-401.

JARAMILLO, A., P. BURKE & D. BEADLE. 2003. *Birds of Chile.* Helm Field Guides. Christopher Helm, an imprint of A & C Black Publishers Ltd., London.

JARVI, S.I., J.J. SCHULTZ & C.T. ATKINSON. 2002. PCR diagnostics underestimate the prevalence of avian malaria (*Plasmodium relictum*) in experimentally-infected passerines. *Journal of Parasitology* 88: 153-8.

KUMAR, S., K. TAMURA & M. NEI. 2004. MEGA3: Integrated software for molecular evolutionary genetics analysis and sequence alignment. *Briefing in Bioinformatics* 5: 150-63.

MÁRQUEZ, O. 1992. *Bases para la Formulación de un Plan de Desarrollo en los Predios Experimentales Las Brisas y Pantanillos, VII Región.* Memoria Ing. Forestal. Universidad de Chile, Facultad de Ciencias Agrarias y Forestales, Santiago.

MEDEL, R.G. & R.A. VÁSQUEZ. 1994. Comparative analysis of harvester ant assemblages of Argentinian and Chilean arid zones. *Journal of Arid Environments* 26: 363-71.

MERILÄ, J., M. BJÖRKLUND & G.F. BENNETT. 1995. Geographic and individual variation in haematozoan infections in the greenfinch, *Carduelis chloris*. *Canadian Journal of Zoology* 73: 1798-804.

MERINO, S. & J. POTTI. 1996. Weather dependent effects of ectoparasites on their bird host. *Ecography* 19: 107-13.

MERINO, S., J. POTTI & J.A. FARGALLO. 1997. Blood parasites of some passerine birds from central Spain. *Journal of Wildlife Diseases* 33: 638-41.

MORENO, J., S. MERINO, R.A. VÁSQUEZ & J.J. ARMESTO. 2005. Breeding biology of the Thorn-tailed Rayadito (Furnariidae) in south-temperate rainforests of Chile. *Condor* 107: 69-77.

NEE, S. 2003. Unveiling prokaryotic diversity. *Trends in Ecology and Evolution* 18: 62-3.

NUNN, C.L., S.M. ALTIZER, W. SECHREST & A.A. CUNNINGHAM. 2005. Latitudinal gradients of parasite species richness in primates. *Diversity and Distributions* 11: 249-56.

PIANKA, E.R. 1966. Latitudinal gradients in species diversity: a review of concepts. *American Naturalist* 100: 33-46.

PISANO, E. 1977. Fitogeografía de Fuego-Patagonia Chilena. I. Comunidades vegetales entre las latitudes 52° y 56°S. *Anales del Instituto de la Patagonia* (Chile) 8: 121-250.

RAPPOLE, J.H., S.R. DERRICKSON & Z. HUBÁLEK. 2000. Migratory birds and spread of West Nile Virus in the Western Hemisphere. *Emerging Infectious Diseases* 6: 319-28.

RICKLEFS, R.E., S.M. FALLON, S. LATTA, B. SWANSON & E. BERMINGHAM. 2005. Migrants and their parasites: a bridge between two worlds. Pp. 210-21 *in Birds of Two Worlds. The Ecology and Evolution of Migration*, R. Greenberg & P.P. Marra (eds.). John Hopkins University Press, Baltimore.

ROHDE, K. 1992. Latitudinal gradients in species diversity: the search for the primary cause. *Oikos* 65: 514-27.

ROZZI, R., J. SILANDER, J.J. ARMESTO, P. FEINSINGER & F. MASSARDO. 2000. Three levels of integrating ecology with the conservation of South American temperate forests: the initiative of the Institute of Ecological Research Chiloé, Chile. *Biodiversity & Conservation* 9: 1199-217.

ROZZI, R., F. MASSARDO, C.B. ANDERSON, K. HEIDINGER & J.A. SILANDER, JR. 2006. Ten principles for biocultural conservation at the southern tip of the Americas: the approach of the Omora Ethnobotanical park. *Ecology & Society* 11: 43. [Cited 29 Nov 2007.] Available from URL: http://www.ecologyandsociety.org/vol11/iss1/art43/

SAIKI, R.K. 1990. Amplification of genomic DNA. Pp. 13-20 *in PCR Protocols*, M.A. Innis, D.H. Gelfand, J.J. Sninsky & T.J. White (eds.). Academic Press, San Diego.

SCHEUERLEIN, A. & R.E. RICKLEFS. 2004. Prevalence of blood parasites in European passeriform birds. *Proceedings of the Royal Society of London B* 271: 1363-70.

STEVENS, G.C. 1992. The elevation gradients in altitudinal range: an extension of Rapoport's latitudinal rule to altitude. *American Naturalist* 140: 893-911.

TAPIA, D. 2005. Comunidades vegetales de la quebrada de La Plata, Región Metropolitana (Chile). *Chloris Chilensis Año* 8: 2. [Cited 29 Nov 2007.] Available from URL: http://www.chlorischile.cl

THOMPSON, J.D., D.G. HIGGINS & T.J. GIBSON. 1994. CLUSTALW: improving the sensitivity of progressive multiple sequence alignment through sequence weighting, positions-specific gap penalties and weight matrix choice. *Nucleic Acids Research* 22: 4673-80.

VALKIŪNAS, G. 2005. *Avian Malaria Parasites and Other Haemosporidia*. CRC Press, Boca Raton.

WALDENSTRÖM, J., S. BENSCH, S. KIBOI, D. HASSELQUIST & U. OTTOSSON. 2002. Cross-species infection of blood parasites between resident and migratory songbirds in Africa. *Molecular Ecology* 11: 1545-54.

WHITE, E.M., E.C. GREINER, G.F. BENNETT & C.M. HERMAN. 1978. Distribution of the haematozoa of Neotropical birds. *Revista de Biología Tropical* 26: 43-102.

Accepted for publication July 2007.

PART III

SEASONAL DYNAMICS OF THE SUB-ANTARCTIC BIRD COMMUNITIES IN DIFFERENT HABITAT TYPES OF THE CAPE HORN BIOSPHERE RESERVE

14 INTRODUCTION TO THE SEASONAL DYNAMICS OF THE SUB-ANTARCTIC BIRD COMMUNITIES IN DIFFERENT HABITAT TYPES OF THE CAPE HORN BIOSPHERE RESERVE

Christopher S. Elphick, Jaime E. Jiménez, Ronnie Reyes, and Ricardo Rozzi

The four papers that make up this section all provide descriptive information about the bird assemblages of the Cape Horn region. Attaining a solid understanding of the species composition and basic natural history of this area's main habitats is a crucial first step towards developing more detailed questions in both basic and applied ecology.

For much fundamental ecological research to proceed, researchers need to know the interactions that individual species are likely to have with one another so that studies can be planned appropriately. This requires that we know which species are present in a given place at any given time, in what numbers they occur, and what they are likely to be doing. For birds – with their propensity to move seasonally between habitats, altitudes, latitudes, and sometimes all three – this information is especially important. Northern Hemisphere researchers tend to take it for granted that this information is already available for the majority of the species that they are likely to study. Indeed, hundreds of pages have been filled with this basic natural history information in book projects such as the Birds of the Western Palearctic (Cramp *et al.* 1977-1994) and the Birds of North America (Poole 2011) series. But, in many tropical and southern temperate areas, even the basics are poorly known.

With the many changes that are affecting habitats worldwide, the need for this basic ecological knowledge is becoming more acute. Obtaining a baseline understanding of how species live their lives and interact with one another is fundamental to devising studies for how emerging conditions will change natural ecosystems. This is true whether one is interested in the introduction of non-native species, which both homogenizes global species assemblages and creates entirely new ones; or, the consequences of habitat loss and change due to development, extraction, and other human activities; or, the effects of climate shifts, which are likely to be especially severe at higher latitude sites, such as the Cape Horn Biosphere Reserve.

Two of the papers in this section focus on forest ecosystems and two on wetlands. Anderson and Rozzi (2000) documents the *Nothofagus* forest avifauna at sites on Isla Navarino and Península Antonio Varas, describing differences between the two sites and between forest interior, forest edge, and open patch habitats within sites. Focusing on the Isla Navarino site, found in the southernmost expanse of forest on Earth, Ippi *et al.* (2009) describe phenological changes in bird occurrence and abundance throughout the annual cycle. The general results of these studies are not surprising: fewer species were found at the more southernmost sites, species richness peaked in summer and declined as birds migrated north for the winter, and species composition differed between forest interior and edge habitats. But the details of how individual species contributed to these changes clarify our understanding of range limits, migration timing, and habitat use. The two wetland papers focus on the temporal and spatial factors that influence species occurrence (Ibarra *et al.* 2009) and on the importance of peat bogs (Ibarra *et al.* 2010). Wetland-associated species make up a large proportion of the Cape Horn avifauna and the total species richness of this habitat is more than double that found in the forest studies. Many of these species are coastal birds and, although Isla Navarino is not a large island, distance from the coast had a notable effect on species occurrence and abundance in wetlands. Relatively small changes in elevation also affected wetland bird assemblages, perhaps also because of the loss of coastal species at higher sites. Thirdly, wetlands that contain *Sphagnum*-dominated peat bogs appear to have bird assemblages that were distinctive compared to wetlands without bogs.

Although these studies have provided information that we previously lacked, there is much that we still do not know. To better understand the ecological factors influencing species assemblages, much more comprehensive sampling is needed. Ideally, one would like to see standardized sampling across the region, for example in the form of an "atlas" project (Donald & Fuller 1998). Implementing such large-scale projects can be difficult; however, this often relies on experienced volunteers to help gather data. The rapid development of on-line systems for compiling distributional data perhaps provides a cost-effective alternative. For instance, the Cornell Laboratory of Ornithology's eBird project <http://ebird.org/content/ebird/>, which allows birdwatchers to enter lists of species seen during their birding activities and links them to specific geographic locations in a single database, has great potential for providing fine-scale range descriptions. The recent expansion of this system to allow sightings from anywhere in the world will prove especially valuable in places that lack detailed geographic information about species occurrence. Between 2000 and 2010, for example, the number of checklists entered from the Magallanes region increased ten-fold from 18 to 196 per year (Fig. 1). As awareness of this resource spreads, we can anticipate much more growth in its use. Exploration of the data s*et al.*o could be used to guide directed surveys to places of particular interest, such as those where unusual species have been reported or where sampling has been especially limited. These sighting data, linked with habitat information gathered from remote sensing sources, would allow the development of predictive distribution maps for most of the region's species.

Disentangling the importance of different potential mechanisms that underlie the variation in assemblage composition among sites is another important next step. Identifying these causal relationships will allow better predictions to be made about how assemblages will change as future conditions shift. As data accumulate, additional analyses can also be used to guide conservation decisions. For example, nestedness analysis (Ulrich *et al.* 2009) would allow one to determine whether variation among wetlands is caused simply by the loss of coastal specialists as one moves away from the coast, or whether interior and higher elevation sites have distinct assemblages of their own. If the former is true, and assemblages are highly nested, then protection should focus primarily on the most species-rich sites, which will generally be those close to the coast. But, if species-poor sites offer something distinct to the region's avifauna, then this should be accounted for, complicating conservation decisions.

Over the coming decades we can expect to see substantial changes to the Cape Horn Archipelago. Climate change is at the forefront of most people's minds, and is likely to be a major factor shaping the region's biota. Without major interventions, non-native species, such as beavers (*Castor canadensis*) and minks (*Neovison vison*) are likely to continue to spread. Land use changes, such as grazing and peat extraction, may also play an increasing role. Even relatively benign activities, such as increased ecotourism, may prove harmful if their effects are not closely monitored and regulated. Consequently, taking the current studies and building on them to develop sufficient understanding to design effective management strategies is essential if we are to prevent biodiversity loss and the erosion of ecosystem services.

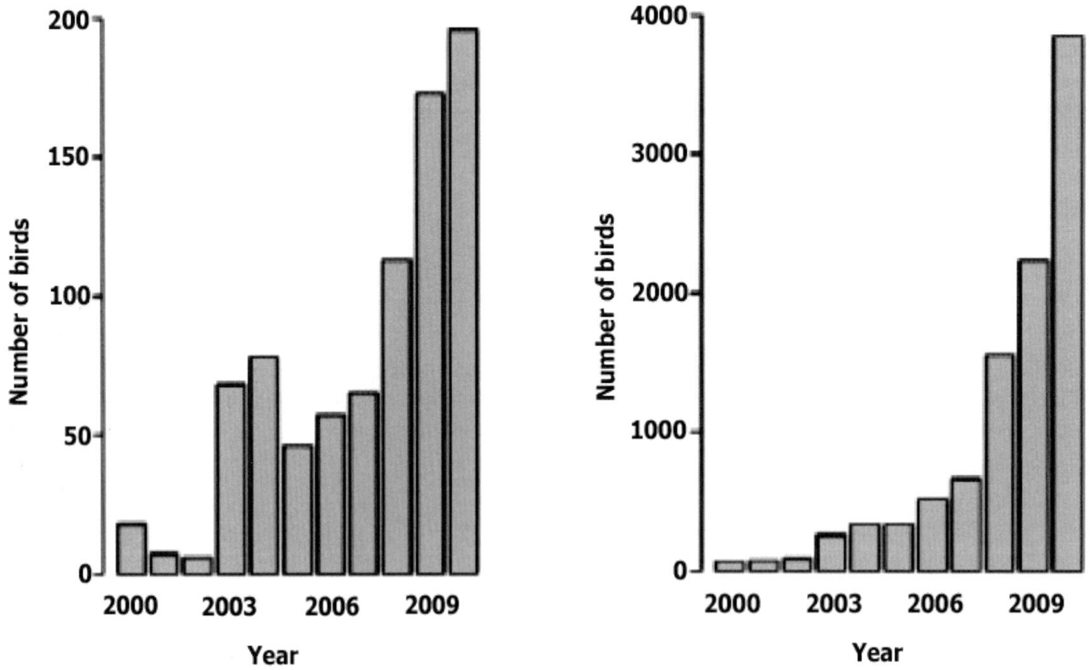

Fig. 1. Annual number of bird lists submitted to the eBird website of the Cornell Laboratory of Ornithology (http://ebird.org/content/ebird/) for (a) Magellanes and (b) Chile, showing the rapid growth in this program and its potential to compile information on bird distribution.

REFERENCES

ANDERSON, C. & R. ROZZI. 2000. Bird assemblages in the southernmost forests of the world: methodological variations for determining species composition. *Anales del Instituto de la Patagonia, Serie Ciencias Naturales* 28: 89-100.

CRAMP, S., K.E.L. SIMMONS & C.M. PERRINS. 1977-1994. *The Birds of the Western Palearctic.* Oxford University Press, Oxford, United Kingdom.

DONALD, P.F. & R.J. FULLER. 1998. Ornithological atlas data: a review of uses and limitations. *Bird Study* 45: 129-145.

IBARRA, J.T., R. ROZZI, H. GILABERT, C.B. ANDERSON, S. MCGEHEE & C. BONACIC. 2009. Dinámica estacional y patrones de distribución de la avifauna asociada a humedales subantárticos en la Reserva de Biosfera Cabo de Hornos (54-55ºS), Chile. *Ornitología Neotropical* 20: 321-337.

IBARRA, J.T., C.B. ANDERSON, T.A. ALTAMIRANO, R. ROZZI & C. BONACIC. 2010. Diversidad y singularidad de la avifauna en turberas esfagnosas del Cabo de Hornos. *Ciencia e Investigación Agraria* 37: 29-43.

IPPI, S., C.B. ANDERSON, R. ROZZI & C.S. ELPHICK. 2009.Variación anual de la abundancia y la composición del ensamble de aves de bosque en la isla Navarino, Reserva de la Biósfera Cabo de Hornos, Chile. *Ornitología Neotropical* 20: 231-245.

POOLE, A. (ED.). 2011. *The Birds of North America Online.* Cornell Laboratory of Ornithology, Ithaca, New York, U.S.A. Available in: http://bna.birds.cornell.edu/bna/

SULLIVAN, B.L., C.L. WOOD, M.J. ILIFF, R.E. BONNEY, D. FINK & S. KELLING. 2009. eBird: a citizen-based bird observation network in the biological sciences. *Biological Conservation* 142: 2282-2292.

ULRICH, W., M. ALMEIDA-NETO & N.J. GOTELLI. 2009. A consumer's guide to nestedness analysis. *Oikos* 118: 3-17.

15 BIRD ASSEMBLAGES IN THE SOUTHERNMOST FORESTS OF THE WORLD: METHODOLOGICAL VARIATIONS FOR DETERMINING SPECIES COMPOSITION

Christopher Anderson[1] and Ricardo Rozzi[2]

[1]Institute of Ecosystem Studies, Box AB, Millbrook, NY 12545, USA (chrisinchile@yahoo.com)
[2]Department of Ecology and Evolutionary Biology, University of Connecticut, 75 North Eagleville Road, Storrs, Connecticut 06269-3043, USA (rozzir@ecostudies.org); Parque Etnobotánico *Omora,* Puerto Williams, Magallanes, Chile

Anales del Instituto de la Patagonia (Chile) 28: 89-100 (2000)

ABSTRACT. Bird assemblages in the southernmost forests of the world: methodological variations for determining species composition. During the austral summer of 2000, we studied the avifauna in coastal forests of Magallanes, Chile at two study sites: Antonio Varas Peninsula (52ºS) and Navarino Island (55ºS). For each of these sites we investigated the species composition of the woodland bird assemblage with mist-nets, point-count stations along transects, and oportunistic observations. Then, we compared the results obtained by each method, as well as with species lists obtained from bibliographic sources. Our species lists corresponded to 95% of bibliographic accounts. At Navarino we detected the presence of two species which were previously unreported: *Colorhamphus parvirostris* and *Sephanoides galeritus*. Transect and mist-net methodologies accounted for just part of the bird species assemblage, but tracked important variations in abundance among species.

Compared to avian assemblages recorded in Chilean forests at lower latitudes, the Magellan area contains fewer species. This confirms previous statements about the decrease in bird species with increasing latitude along these forests. There are also some major changes in relative abundance of species. For example, the hummingbird *S. galeritus* is the most abundant species in forests at Chiloé, but in Magallanes it was never greater than 5% of all birds recorded (*e.g., Elaenia albiceps, Turdus falcklandii, Aphrastura spinicauda,* and *Phrygilus patagonicus*) remain very similar as compared to Chiloé. We also compared bird assemblages of three types of habitats: forest interior, forest border, and open areas. These last two habitat types host greater abundance and species richness, including birds that are characteristic of forest interior habitats. This result needs to be carefully interpreted at the landscape level, because areas opened by human activity in Magallanes are still embedded in a region that conserves large expanses of primary forest.

Key words: bird censuses, Chile, avifauna, Magellanic forests, mist-netting, Navarino Island, *Nothofagus*, transects.

RESUMEN. Ensambles de aves en los bosques más australes del mundo: variaciones metodológicas para la determinación de la composición de especies. Durante el verano del año 2000 se investigó la avifauna de bosques costeros de Magallanes, Chile, en dos sitios de estudio: la península Antonio Varas (52ºS) y la isla Navarino (55ºS). En cada sitio, se investigó la composición de especies de las comunidades de aves por medio de redes de neblina, estaciones de observación/escucha a lo largo transectas y listados de todas las especies observadas -incluyendo aquellas detectadas fuera de transectas. Luego, se compararon los resultados obtenidos con cada metodología, como también con listas de especies elaboradas a partir de revisión bibliográfica. Nuestras listas de especies incluyeron el 95% de las especies documentadas y para Navarino se detectó la presencia de dos especies no registradas previamente: *Colorhamphus parvirostris* y *Sephanoides galeritus.* Los métodos de captura con redes de neblina y censos en transectas detectaron sólo parte del total de especies de aves, pero detectaron las importantes variaciones de abundancia entre ellas.

Las comunidades de aves de Magallanes presentaron menos especies que las de bosques chilenos en latitudes más bajas, confirmando propuestas anteriores acerca de la disminución de especies de aves con el aumento de latitud en estos bosques. También hubo variaciones en la abundancia relativa de las especies. Por ejemplo, el picaflor es la especie más abundante en bosques de Chiloé, mientras que en Magallanes nunca superó el 5%. Otras especies (*e.g., Elaenia albiceps, Turdus falcklandii, Aphrastura spinicauda* y *Phrygilus patagonicus*), sin embargo, presentaron abundancias relativas similares. Se compararon también las comunidades de aves en tres tipos de hábitats: interior de bosque, margen de bosque y terrenos abiertos. Los dos últimos tipos de hábitat tuvieron mayor número de especies y abundancia de aves, e incluyeron especies características del interior de bosque. Este resultado debe ser cuidadosamente interpretado a la escala de paisaje, puesto que las áreas abiertas por actividad humana se encuentran inmersas en una región donde todavía quedan vastas extensiones de bosque primario.

Palabras clave: avifauna, bosques magallánicos, censos de aves, Chile, Isla Navarino, *Nothofagus*, redes de neblina, transectos.

INTRODUCTION

Species richness diminishes with latitude for a variety of groups of organisms (Brown & Gibson 1983, Gaston & Spicer 1998), such as butterflies (Sutton & Collin 1991), plants (Gaston *et al.* 1995), amphibians (Williams *et al.* 1997), mammals (Rosenzweig 1992), and birds (Cook 1969). Schlatter *et al.* (1997) proposed for the avian assemblages in the *Nothofagus* forests of southern Chile that bird species do not vary significantly, but there is a tendency to decrease in species number. This suggests that the species inhabiting the southernmost forests correspond to a smaller subset of representatives found toward the north, and bird species of the Magallanes region may constitute a fraction of the whole assemblage of the bird species inhabiting the temperate forests of South America. In addition some major variations in relative abundance could occur along the latitudinal gradient. For example, Charles Darwin (1839) was surprised by the abundance of *Scytalopus magellanicus* in the forests of Tierra del Fuego, while on Chiloé he was startled by the abundance of the hummingbird *Sephanoides galeritus* (see Rozzi *et al.* 1996a).

Other than latitude, habitat is a factor that affects avian assemblage parameters such as species richness (Remsen & Parker 1983, Cody 1985). Although frequently thought of as *terra incognita*, Magellanic forests do not escape human influences that today affect the landscape at a global scale. Until 150 years ago, *Nothofagus* forests covered most of the southern tip of South America (Pisano 1977). However, with European colonization, the landscape began to change dramatically. In 1877, when cutting and burning removed large tracts of forest, vast expanses of land were converted to pasture for sheep estancias (Martinic 1973, 1985, McEwan *et al.* 1997, Rozzi *et al.* 2000). Since its peak, though, the extent of ranching decreased during the last 30 years. As the acreage under hoof and tooth decreases, secondary forests have become a major landscape feature, presenting structurally shorter habitats and higher patchiness than the original primary forests (Silander *et al.* in preparation). Hence, to investigate fully the forest avifauna of Magallanes, we deemed it necessary to incorporate the human dimension by comparing habitats subject to different degrees of human influence.

The woodland birds of the temperate forests of the southern tip of Chile bear the distinction of inhabiting the most austral forests in the world. Although species composition seems not to change greatly from site to site, at the regional level they host a relatively high degree of endemism (around 30%; Vuilleumier 1985, Rozzi *et al.* 1996a). In spite of these distinctive attributes of the Magellan avifauna has been left out from major comparisons between South and North American temperate forest avifauna (see Jaksic & Feinsinger 1991, Willson 1991). Notwithstanding the extensive ornithological surveys performed in the forests of the Magellan Region of Chile (see Barros 1971, 1976, Sielfeld 1977, Venegas 1981, 1991), the above reviews made no mention of the most southern *Nothofagus* forests (*i.e., N. betuloides*, *N. pumilio*, and *N. antarctica*). They considered only the birds of the Valdivian rainforests. With the aim to contribute to future comparisons, we investigated the austral avifauna with similar methodologies to those employed for the study of bird species composition in the Valdivian rainforests (see García 1982, Erazo 1984, Sabag 1993, Willson *et al.* 1994, Rozzi *et al.* 1996a,b).

Within this context, we compare three factors that could determine variations in bird species composition and richness in the austral forests of Magallanes: (1) comparison of bird assemblages at two latitudes, considering site sampling and bibliographic information about forest birds within each general area, because the regional pool of species can be important for species composition at local sites (Ricklefs & Schulter 1993), (2) comparison of bird assemblages in primary forests and in anthropogenic habitats, as some forest specialist birds, such as the Magellanic Woodpecker (*Campephilus magellanicus*) are sensitive to human impact (Willson *et al.* 1994, McBride 2000) and (3) comparison of bird assemblages described by diverse methods, which can account for part of the variations in the species composition at different locations (Gibbons *et al.* 1997).

For the latitudinal comparisons, we studied the woodland birds at two sites: Peninsula Antonio Varas (52°S) and Navarino Island (55°S). For the comparison of human impact, at each site we censused birds in primary forests (interior and border) and anthropogenic open lands. For the methodological comparisons, we studied the bird species composition through four methods: (i) captures with mist-nets (including capture and recapture); (ii) point-count stations along transects (recording all birds seen or heard); (iii) species list prepared with all forest birds observed during the field work period (including those species not detected with mist-nets or at point-count stations); and (iv) bibliographic review. We expect that the addition of these baseline data to previous ornithological surveys will assist future comparisons, as well as long-term monitoring of forest bird populations, especially in the wake of current global change.

Study Area

During the austral summer of 2000, we performed sampling with mist-nets and point-count stations along transects in two areas of the Magellan region in southern Chile: Peninsula Antonio Varas (52°S), and Navarino Island (55°S) (Fig. 1). In both areas, studies were conducted at sites close to the coastline that are currently, or were formerly, dominated by evergreen forests of *Nothofagus betuloides*. The censusing locations included a range of anthropic influence, derived mostly from cutting and burning for pastureland. The land's history of use generated a mosaic of habitats, which include: (a) anthropogenic park formations characterized by large trees of Magellanic Coigüe (*Nothofagus betuloides*) and Lenga (*Nothofagus pumilio*), growing among burned trunks, logs, snags, regenerating saplings, and abundant shrubs of *Berberis buxifolia*, and *Ribes magellanicum*; (b) open areas covered by low shrubby Ericacea (*Gaultheria mucronata* and *Empetrum rubrum*); (c) recently burned sites dominated by *Chiliotrichum diffusum*; (d) naturally or anthropically perturbed areas dominated by thickets of the Protecaeae tree *Embothrium coccineum* (Notro or Ciruelillo), which has red tubular flowers rich in nectar; and (e) areas of different sizes covered by continuous old-growth forests dominated by *N. betuloides* and *N. pumilio*, with an understory of several species that produce fleshy fruit, such as the tree species of *Drimys winteri*, and *Maytenus magellanica*; the shrubby species of *Berberis buxifolia*, *B. ilicifolia*, *Empetrum rubrum*, *Fuchsia magellanica*, *Gaultheria mucronata*, *Maytenus*

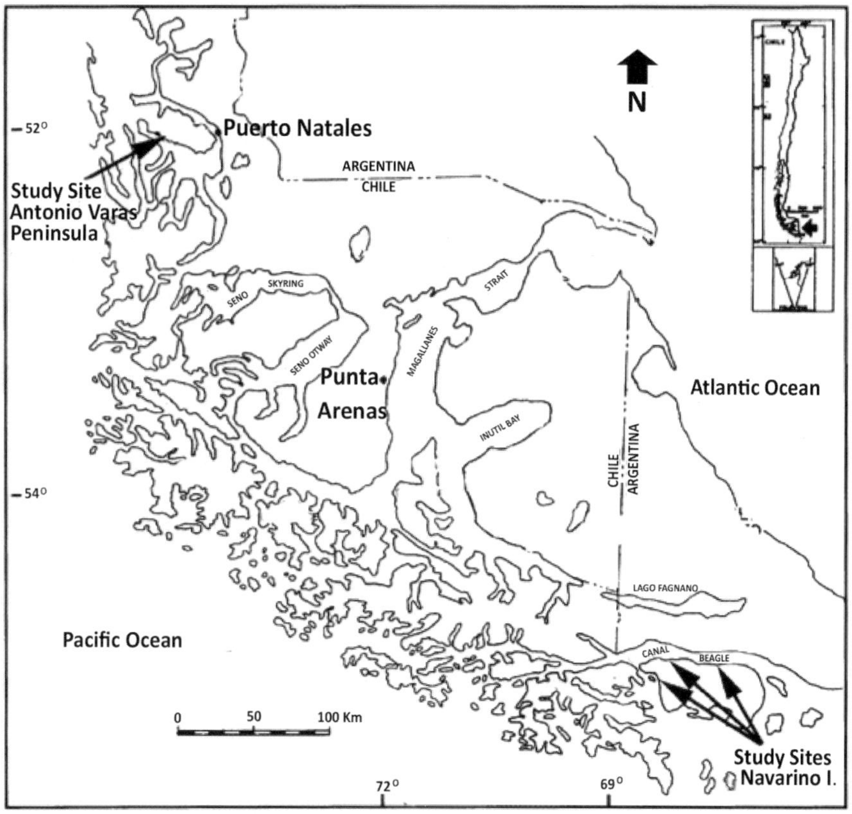

Fig. 1. Map of the southern part of the Magellan Region, indicating the two study areas.

disticha, and *Ribes magellanicum*; and the woody-herbs *Gunnera magellanica*, *Luzuriaga marginata*, and *Rubus geoides*. For the analyses we defined three types of habitats: (I) forest border, (2) forest interior, and (3) open habitat (which includes the habitat types a-d described above).

Peninsula Antonio Varas, is located near the town of Puerto Natales, across the Golfo Almirante Montt. The study sites border the coastline in the western part of the peninsula in the area of Bahia Talcahuano (51°44'S; 72°53'W), which is found south of the Southern Patagonian Icefields, and constitutes part of fiord region of the western archipelago of southern Chilean Patagonia. This area receives between 1000 and 1500 mm of precipitation annually and has a mean annual temperature of 7°C (McCulloch *et al.* 1997). Navarino Island is situated further south between the Cape Horn archipelago and Tierra del Fuego. The principal study sites on Navarino Island were near the coast at Caleta Robalo (54°57'S, 67°39'W), Mejillones (54°55'S, 67°54'W), Puerto Inutil (54°59'S, 68°13'W) and Wulaia (55°03'S, 68°09'W). Navarino Island receives about 650 mm of precipitation annually and has a mean annual temperature of 6°C. Monthly mean temperatures range from 2°C in June/July to 9°C in December/January/February (di Castri & Hajek 1976).

Although both Peninsula Antonio Varas and Navarino Island are found within the region dominated by the evergreen forest type of *Nothofagus betuloides* (see Pisano 1977), plant species abundance and diversity is higher in the area of Peninsula Antonio Varas. For example, *Fuchsia magellanica*, a key species for plant-bird interactions, was found only at this northern location (Rozzi, in preparation). Other than influence from livestock ranching and wood-cutting, Navarino Island has been subject to a recently introduced invasive mammal species, which drastically modifies the hydrology and forest structure: the beaver (*Castor canadensis*). This "ecosystem engineer" was brought in the 1950's to Tierra del Fuego to begin a fur industry in Argentina, but soon individuals escaped. It was first detected on Navarino in 1962, where it has since come to occupy every available watershed (Sielfeld & Venegas 1980).

MATERIALS AND METHODS

Methodological approach

We considered three field methods to investigate the bird species in forest sites: (1) bird mist-netting, (2) point-count stations along transects, and (3) list of observed species. In addition, we accumulated species lists for each site based on bibliographic sources. Each of these methods presents benefits and disadvantages. (i) Mist-netting allows collection of diverse, valuable information on individual birds (such as morphological measurements, diet analysis, and pollination sampling) and the recording of secretive and non-territorial species. However, this method is very selective. For example, captures rates are higher for birds that inhabit the lower 2-3 m of the habitat (Karr 1981). Also, factors such as weather and net tension influence netting efficiency (see Karr 1981, Silkey *et al.* 1999). (ii) Point-count stations along transects allow one to detect a wider range of bird species, but data collected during transects are only as reliable as the observer. It is possible to over count particularly loud or mobile species (Karr 1981). These two first methods are, however, systematic methods of censusing populations, allowing comparisons and estimations of density, diversity, richness, and composition. They also permit comparisons based on habitat preferences. (iii) A less systematic method consisted of recording any forest bird species observed while working in the field. Lists of observed species are beneficial when trying to document the whole assemblage of species (including rare or infrequent species) or determine range limits for a given species. (iv) Bibliographic accounts allowed us to use the accumulated information, collected over many years (Humphrey *et al.* 1970, Barros 1971, 1976, Sielfeld 1977, Venegas 1981, Venegas & Sielfeld 1998, Couve 1991, 1998). However, bibliographic data may not encompass all parts of the bird assemblage or include references from comparable locations or habitats.

On the one hand, the increasing degree of subjectivity from the first to the third method is a negative aspect in this sequence. Although, the range of species detected also grows from the first to the third, representing the positive aspect of the last method. So, comparisons of all four of these methods allow a better description of the avifauna of Magallanes. Further, a comparison of methodologies is germane to future work and determining appropriate sampling regimes.

Mist-netting

In Peninsula Antonio Varas, and Navarino Island we completed a total of 279.21 hours of mist-netting, distributed in 30 sampling days, between January 6 and March 23, 2000. Each sampling day, 2 mist-nets were installed: one 6 m long by 2.6 m high and the other of 12 x 2.6 m. Both mist-nets had a grid of squares with a diagonal length of 30 mm. The nets were checked every 15 to 30 min. To standardize the sampling effort for each site, we considered the size of the net and the time of display. The products of the net-area (m^2) and the number of hours of net display (based on the time of installation and removal of each net) were calculated for each net and sampling day (see Appendix 1). At each netting site, habitat was described. The habitat was classified as forest border, forest interior, or open site.

Captured birds were banded to initiate long term study of these populations. Specimens were banded with aluminum rings of 2.78 mm in diameter bearing sequential numbers followed by RROO (*i.e.,* Ricardo Rozzi 2000). In the case of the hummingbird, individuals were banded with small, blue-plastic rings that were sequentially numbered.

Point-count stations along transects

In both study sites, we recorded all seen or heard individual birds during 5 minutes at point count stations, along transects. A total of 10 transects with 93 stations were completed. Stations were separated by at least 200 m. Each observation included the number of individuals of the species, microhabitat, activity, and time of day.

RESULTS

Bird Species in Magellanic Forests

Table 1 shows all of the birds recorded through our four methods of analysis. From literature citings, we considered 22 species as possible inhabitants of the forests of Antonio Varas Peninsula (AV) and 20 on Navarino Island (NI) (Humphrey *et al.* 1970, Barros 1971, 1976, Sielfeld 1977, Venegas 1981, 1991, Couve & Vidal 1999, Venegas & Schlatter 1999). *Colaptes pitius*, *Picoides lignarius*, *Sicalis lebruni* and *Sylviorthorhynchus desmursii* were seen in the northern Magallanes and the Antonio Varas area (AV), but not at Navarino Island (NI). Likewise, *Colorhamphus parvirostris* and *Pygochelidon cyanoleuca* were both observed in NI, but not AV. Species previously reported in various bibliographic lists provide a good idea of the total species that inhabit the Magellan forests.

Of the 22 forest bird species, only 6 species were not passerines, and they included a kingfisher (Alcedinidae), a parrot (Psittacidae), a hummingbird (Trochilidae), and 3 woodpeckers (Picidae) (Table 1). The 18 species of Passeriformes were taxonomically diverse including 18 genera, and 13 families. Among the families, Furnaridae and Tyranniidae were the most represented, each with 4 species.

In general, our own species list corresponds with previous work. At both AV and NI 95% of the species previously reported were found during our study, as well. However, a few differences occurred. At AV we found *C. parvirostris*, which was previously undescribed for the northwestern archipelago area of Magallanes. But in this area, we did not find *P. lignarius*, which is listed by Couve (1998). At NI, both the bibliography and our observations recorded fewer species than at AV. Absent from our NI list were *P. lignarius*, *C. pitius*, *S. desmursii*, and *S. lebruni*. Only one species was described previously, but not found during our stay on NI; *P. cyanoleuca*. However, on Navarino Island we detected the presence of two previously unreported species; *Colorhamphus parvirostris*, and *Sephanoides galeritus*. Furthermore, we established a new southern limit for *S. galeritus* at Bahia Douglas on the southwestern corner of the island. Point-count stations at AV recorded 555 individuals from 18 species, during 5.25 hours of observations along transects. At NI, 223 individuals that belonged to 16 bird species were detected during 2.42 hours of observations. As with the previous two methods, AV had greater species richness than NI. Considering the proportion of species detected by point-count stations compared to the species recorded in the bibliography, in AV, 81.8% were surveyed along transects, and in NI, 75% of the possible species were detected with this method (Table 1). It is necessary to mention that there were problems regarding the distinction between *Phrygilus patagonicus* and *P. gayi*. Because the former species is dominant in forest habitats, we decided to count all *Phrygilus* as *P. patagonicus*.

With mist-nets, a total of 193 individuals were captured in AV, and they belonged to 12 species. In NI, the captures included 83 individuals that belonged to 9 species (Table 1). Again, like with the previous methods, AV exhibits greater species richness than NI. However, netting proved to be the least effective method for describing the whole assemblage of forest avifauna. At both AV and NI, 45% of possible species actually were captured in the nets.

TABLE 1. Bird species reported by previous studies (bibliography) and recorded during the summer 2000 by different field methods (list of all observed birds, point-count stations along transects, and mist-nets) for the coastal forests of the Peninsula Antonio Varas and Navarino Island.

Species			Area of Antonio Varas (52°S)				Area of Navarino Island (55°S)			
Common name	**Scientific name**	**Family**	**Bibliography***	**List**	**Transect**	**Net**	**Bibliography****	**List**	**Transect**	**Net**
Martín pescador	*Ceryle torquata* (Linné)	Alcedinidae	a	X	X		1,2,5,7	X	X	
Cachaña	*Enicognathus ferrugineus* (Müller)	Psittacidae	a, b	X	X	X	1,2,5,7	X	X	
Picaflor chico	*Sephanoides galeritus* (Molina)	Trochilidae	b	X	X	X	1,6,7	X		X
Carpintero negro	*Campephilus magellanicus* (King)	Picidae	a, b	X	X		1,2,5,7	X	X	
Pitío	*Colaptes pitius* (Molina)	Picidae	a, b	X						
Carpintero chico	*Picoides lignarius* (Molina)	Picidae	b							
Tordo	*Curaeus curaeus* (Molina)	Icteridae	a, b, c	X	X		1,2,4,5	X	X	
Chirihue	*Sicalis lebruni* (Oustalet)	Emberizidae	a	X	X					
Chincol	*Zonotrichia capensis* (Müller)	Emberizidae	a	X		X	1,2,5,7	X		X
Cometocino	*Phrygilus patagonicus* (Lowe)	Fringillidae	a		X	X	1,2,4,5,7	X	X	X
Jilguero	*Carduelis barbata* (Molina)	Fringillidae	a	X	X	X	1,2,4,5	X	X	X
Rayadito	*Aphrastura spinicauda* (Gmelin)	Furnariidae	a, b, c	X	X	X	1,2,4,5,7	X	X	X
Churrete	*Cinclodes patagonicus* (Gmelin)	Furnariidae	a, b, c	X	X	X	1,3,4,6	X		
Comesebo grande	*Pygarrhichas albogularis* (King)	Furnariidae	a, b	X	X	X	1,2,3,4,6,7	X	X	X
Colilarga	*Sylviorthorhynchus desmursii* (Des Murs)	Furnariidae	a	X	X					
Golondrina	*Tachycineta meyeni* (Cabanis)	Hirundinidae	a, b	X		X	1,2,4,5,7	X		
Golondrina	*Pygochelidon cyanoleuca* (Vieillot)	Hirundinidae					2,6			
Zorzal	*Turdus falcklandii* (Quoy & Gaimard)	Turdidae	a, b, c	X	X	X	1,2,4,5,7	X	X	
Churrín	*Scytalopus magellanicus* (Gmelin)	Rhinochryptidae	a, b	X	X		1,2,3,4,6,7	X	X	
Chercán	*Troglodytes musculus* (Vieillot)	Troglodytidae	a, b	X	X	X	1,2,4,5,7	X	X	X
Cachudito	*Anairetes parulus* (Kittlitz)	Tyrannidae	a, b	X	X		1,2,3,6,7	X	X	
Viudita	*Coloramphus parvirostris* (Darwin)	Tyrannidae		X	X		1,2,7	X	X	X
Fío-fío	*Elaenia albiceps* (Lafresnaye & d' Orbigny)	Tyrannidae	a, b	X	X	X	1,2,3,4,5,6,7	X	X	X
Diucón	*Xolmis pyrope* (Kittlitz)	Tyrannidae	a, b	X			1,2,4,5,7	X		
Total number of species		22	22	22	18	12	20	19	16	9

*a=Venegas 1994, b=Couve & Vidal 1999, c=Texera 1972; **1=Humphrey *et al.* 1970, 2=Venegas 1994, 3=Venegas 1991, 4=Barros 1971, 5=Barros 1976, 6=Sielfeld 1977, 7=Venegas & Schlatter 1999.

DISCUSSION

The combination of species lists, point count-stations along transects, and mist-netting, allowed us to record 95% of the species previously reported for the areas of Antonio Varas Peninsula and Navarino Island (Table 1). We also added the observation of *Sephanoides galeritus* on Navarino Island, sighted as far south as Bahía Douglas, and of *Colorhamphus parvirostris* in AV, where it was not previously reported.

Consistently, all methods show that AV hosts 10% more bird species than NI. Two species seem to be restricted to the forest region north of the Magellan Strait: *Colaptes pitius* and *Picoides lignarius*. McBride (2000) discusses in detail the important ecological role of *Campephilus magellanicus*, as the single woodpecker in the southernmost forests. The relative high abundance of *C. magellanicus* at NI highlights the importance that the austral forests can have for the conservation of one of its vulnerable bird species.

A comparison with bird assemblages farther north, such as on Chiloé Island, reveals a higher contrast in the number of species. On Chiloé, 30 bird species were found to inhabit primary rainforest (Rozzi *et al.* 1996a). Chiloé is at 42°S latitude, which is 10 degrees of latitude north of AV. In the 3 degrees of latitudinal difference between AV and NI the number of species decreased by two or 10%. In 10 degrees latitude the reduction was of ten species or 33%. These data support the statement about the latitudinal decline in species number along the temperate forests of Chile made by Schlatter *et al.* (1997).

A major family that experiences a drastic diminution in species from Chiloé to Magallanes is Rhinocryptidae. It has four abundant species in Chiloé, but only one, *S. magellanicus*, at AV and NI. Other conspicuous bird species at Chiloé that are absent in southern Magellanic forests are *Columba araucana* and *Enicognathus leptorhynchus*. Additionally, some major differences occur in the relative abundance of bird species between Chiloé and Magallanes. For example, the hummingbird is among the most abundant species on Chiloé, reaching up to 22.64% of all individuals (Rozzi *et al.* 1996a). Magellanic forests host far fewer hummingbirds. The highest relative abundance found for *S. galeritus* was 5.56%, using mist-nets at Peninsula Antonio Varas. The relative abundance of other common species was, however, strikingly similar between Chilote and Magellanic forests. In both regions, *P. patagonicus*, *A. spinicauda*, *T. falcklandii*, and *E. albiceps* are dominant species, accounting for more than 50% of all forest birds.

With regards to the habitat's influence on avian fauna, we found that species richness and abundance consistently were lowest in the interior forests. Both of our methodologies supported this trend. We would expect mist-netting to be biased in interior habitats. As stated previously, Karr (1981) reported that nets mainly capture

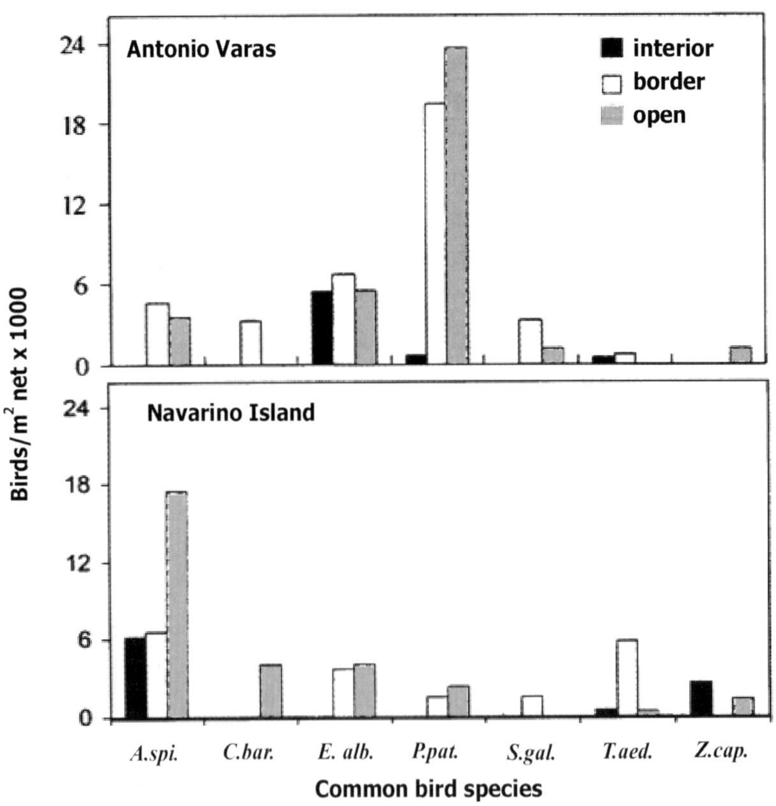

Fig. 2. Number of captured birds per hour and square meter of displayed mist-net, for the most frequently caught bird species at forest interior, edge, and open lands in Peninsula Antonio Varas and Navarino Island. *A.spi* = *Aphrastura spnicauda*, *C.bar.* = *Carduelis barbata*, *E.alb.* = *Elaenia albiceps*, *P.pat.* = *Phrygillus patagonicus*, *S.gal.* = *Sephanoides galeritus*, *T.aed.* = *Troglodytes aedon*, *Z.cap.* = *Zonotrichia capensis*.

birds in the lower 2-3 m of strata. Because interior forests provide a greater vertical structure, we would expect our mist-netting data to be bias in that respect. However, the same trend is observed with transects. The fact that the two methods corroborate one another gives the data added confidence. Perhaps, open habitats and edge ecotones provide more diverse resource opportunities. For example, the abundance of fruit and seed producing plants in these early successional areas is often times greater (Armesto *et al.* 1987, Armesto & Rozzi 1989, Rozzi *et al.* 1996b, Willson *et al.* 1996).

How does this relate to the question about human impact?

Prior to colonization the west coast of Magallanes was continuously covered by *Nothofagus* forests (Pisano 1977). Open sites for the most part are, therefore, of anthropogenic origin. Our censuses detected higher abundance and richness of bird species in open sites. However, it must be taken into account that these open patches are still placed within a landscape with vast forest expanses. This landscape level perspective is critical for understanding ecological processes, as well as to discuss conservation questions regarding the forest ecosystem of southern Chile (Rozzi *et al.* 1994, Armesto *et al.* 1998).

The comparison of census methodologies provides some valuable insights. At first, it seems that mist-netting, as well as transects, are of little value for determining species composition of forest bird assemblages, because they leave out a considerable number of species. However, when analyzing relative abundance of species the usefulness of these two methods is underscored. In terms of number of individuals, more than 80% are concentrated in just 5 (app.) species. This dominance in terms of individuals seems to be a common phenomenon in Chilean forests (Rozzi *et al.* 1996b, Cofré 1999) and emphasizes the importance of quantitative censuses to describe bird assemblages in these ecosystems. Because species richness is decreased in southern Chile and a few species are very common, quantitative censusing methods (like mist-netting and point-count stations) are especially appropriate. On the other hand, species lists continue to be important for detecting uncommon species, especially those not captured with mist-nets.

ACKNOWLEDGMENTS

We would like to thank the Francis L. Phillips Scholarship for providing funding for this research. We are appreciative also to the Universidad de Magallanes and the Instituto de la Patagonia for logistic support and for the kindness of faculty members, especially Orlando Dollenz, Sylvia Oyarzún, and Claudio Venegas. An anonymous reviewer and Francisca Massardo provided valuable comments and assistance with the preparation of the manuscript and figures. We, of course, are grateful to the many people who assisted in the fieldwork; most notably Ezio Firmani, don Santiago Frez, don José Provoste, and David and Matías Li. A special thanks is due to Governor Eduardo Barros and Lorena Peñaranda from the Gobernación Provincia Antártica Chilena and to Maurice van de Maele from the Martín Gusinde Museum. This research is part of ongoing investigations associated with the Parque Etnobotánico *Omara*, Puerto Williams, Chile.

REFERENCES

ARMESTO, J.J., R. ROZZI, P. MIRANDA & C. SABAG. 1987. Plant/frugivore interaction in South American temperate forests. *Revista Chilena de Historia Natural* 60: 321-336.

ARMESTO, J.J. & R. ROZZI. 1989. Seed dispersal in the rain forests of Chiloé: a survey of syndromes in comparison with other temperate and tropical forests. *Journal of Biogeography* 16: 219-226.

ARMESTO, J.J., R. ROZZI, C. SMITH-RAMÍREZ & M.T.K. ARROYO. 1998. Effective conservation targets in South American temperate forests. *Science* 282: 1271-1272.

BARROS, V.A. 1971. Aves observadas en las islas Picton, Nueva, Lennox y Navarino oriental. *Anales del Instituto de la Patagonia* (Chile) 2: 166-180.

BARROS, V.A. 1976. Nuevas aves observadas en las islas Picton, Nueva, Lennox y Navarino oriental. *Anales del Instituto de la Patagonia* (Chile) 7: 190-193.

BROWN, J. & A. GIBSON. 1983. *Biogeography.* The C.V. Mosby Co. Missouri, USA.

CODY, M.L. (ED.). 1985. *Habitat Selection in Birds.* Academic Press, Inc., New York.

COFRÉ, H. 1999. Patrones de rareza de las aves del bosque templado de Chile: implicancias para su conservación. *Boletín Chileno de Ornitología* 6: 8-16.

COOK, R. 1969. Variation in species density of North American birds. *Systematic Zoology* 18: 63-84.

COUVE, E. & C. VIDAL. 1999. *Dónde Observar Aves en al Parque Nacional Torres del Paine, Guía de Identificación.* Fantástico Sur, Punta Arenas, Chile. 238 pp.

DARWIN, C. 1839. *The Voyage of the Beagle.* Reprint, London: Everyman's Library, 1975.

DI CASTRI, F. & E.R. HAJEK. 1976. *Bioclimatología de Chile.* Vicerrectoría Académica, Universidad Católica de Chile, Santiago.

ERAZO, L.A. 1984. Análisis de censos de avifauna realizados en un rodal boscoso de olivillo, Valdivia, Chile. X Región. *Revista Geográfica de Valparaíso* 15: 49-71.

GARCÍA, J.A. 1982. *Comunidad Avifaunística del Delta del Río Gol-Gol, una Necesidad de Conservación.* Tesis de Ingeniería Forestal, Universidad Austral de Chile. Valdivia, Chile.

GASTON, K., P. WILLIAMS, P. EGGLETON & C. HUMPHREIS. 1995. Large scale patterns of biodiversity: spatial variation in family richness. *Proceedings of the Royal Society of London,* Series B 260: 149-164.

GASTON, K. & J. SPICER. 1998. *Biodiversity.* Blackwell Science Ltd. London, UK.

GIBBONS, D., D. HILL & W. SUTHERLAND. 1997. Birds. Pp. 227-259 *in Ecological Census Techniques*, W. Sutherland (ed.). Cambridge University Press. Cambridge, UK.

HUMPHREY, P.S., D. BRIDGE, P.W. REYNOLDS & R.T. PETERSON. 1970. *Birds of Isla Grande (Tierra del Fuego).* Preliminary Smithsonian Manual, Museum of Natural History, University of Kansas, Lawrence, Kansas, USA. 411 pp.

JAKSIC, F.M. & P. FEINSINGER. 1991. Bird assemblages in temperate forests of North and South America: a comparison of diversity, dynamics, guild structure, and resource use. *Revista Chilena de Historia Natural* 64: 491-510.

KARR, J.R. 1981. Surveying birds with mist-nets. *Studies in Avian Biology* 6: 62-67.

MARTINIC, M. 1973. *Crónica de las Tierras del Sur del Canal Beagle.* Editorial Francisco de Aguirre, S.A. Buenos Aires-Santiago de Chile. 236 pp.

MARTINIC, M. 1985. *Ultima Esperanza en el Tiempo.* Ediciones de la Universidad de Magallanes. Punta Arenas, Chile. 276 pp.

MCBRIDE, P. 2000. *Magellanic Woodpecker (*Campephilus magellanicus*) Habitat Selection in Deciduos* Nothofagus *Forests of Tierra del Fuego.* M.Sc. Dissertation, University of Washington. Seattle, Washington, USA.

MCCULLOCH, R.D., C.M. CLAPPERTON, J. RABASSA & A.P. CURRANT. 1997. The glacial and post-glacial environmental history of Fuego-Patagonia. Pp. 12-31 *in Patagonia. Natural History, Prehistory and Ethnography*, C. McEwan, L.A. Borrero & A. Prieto, (eds.). Princeton Univ. Press, Princeton, New Jersey.

MCEWAN, C., L.A. BORRERO & A. PRIETO (EDS.). 1997. *Patagonia. Natural History, Prehistory and Ethnography.* Princeton Univ. Press, Princeton, New Jersey.

PISANO, E. 1977. Fitogeografía de Fuego-Patagonia chilena. I. Comunidades vegetales entre las latitudes 52ºS y 56ºS. *Anales del Instituto de la Patagonia* (Chile) 8: 121-250.

REMSEN, J.V. & T.A. PARKER, III. 1983. Contributions of river-created habitats to bird species richness in Amazonia. *Biotropica* 15: 223-231.

RICKLEFS, R. & D. SCHULTER (EDS.). 1993. *Species Diversity in Ecological Communities. Historical and Geographical Perspectives.* The University of Chicago Press. Chicago, USA.

ROSENZWEIG, M. 1992. Species diversity gradients: we know more and less than we thought. *Journal of Mammalogy* 73: 715-730.

ROZZI, R., J.J. ARMESTO & J. FIGUEROA. 1994. Biodiversidad y conservación de los bosques nativos de Chile: una aproximación jerárquica. *Bosque* 15: 55-64.

ROZZI, R., J.J. ARMESTO, A. CORREA, J.C. TORRES-MURA & M. SALLABERRY. 1996a. Avifauna de bosques primarios templados en islas deshabitadas del archipiélago de Chiloé, Chile. *Revista Chilena de Historia Natural* 69: 125-139.

ROZZI, R., D. MARTÍNEZ, M.F. WILLSON & C. SABAG. 1996b. Avifauna de los Bosques Templados de Sudamérica. Pp. 135-152 *in Ecología de los Bosques Nativos de Chile*, J.J. Armesto, C. Villagrán & M.T. Kalin (eds.). Editorial Universitaria, Santiago, Chile.

ROZZI, R., J. SILANDER, J.J. ARMESTO, P. FEINSINGER & F. MASSARDO. 2000. Three levels of integrating ecology with the conservation of South American temperate forests: The initiative of the Institute of Ecological Research Chiloé, Chile. *Biodiversity and Conservation* 19: 71-90.

SABAG, C. 1993. *El Rol de las Aves en la Dispersión de Semillas en el Bosque Templado Secundario de Chiloé (42ºS).* Tesis Magíster. Facultad de Ciencias. Universidad de Chile. 79 pp.

SCHLATTER, R., A. SIMEONE & C. VENEGAS. 1997. Avian assemblage and dynamics in southern Chilean forests: a comparative approach between 37° and 54ºS lat. *Noticiero de Biología de Chile* 5: 121.

SIELFELD, W. 1977. Reconocimiento macrofaunistico terrestre en el área de Seno Ponsonby (isla Hoste). *Anales del Instituto de la Patagonia* (Chile) 8: 275-297.

SIELFELD, W. & C. VENEGAS. 1980. Poblamiento e impacto ambiental de *Castor canadensis* Kuhl en Navarino Island, Chile. *Anales del Instituto de la Patagonia* (Chile) 11: 247-257.

SILKEY, M., N. NUR & G.R. GEUPEL. 1999. The use of mist-net capture rates to monitor annual variation in abundance: A validation study. *Condor* 100: 288-298.

SUTTON, C. & N. COLLINS. 1991. Insects and tropical forest conservation. Pp. 405-424 *in The Conservation of Insects and their Habitats*, N. Collins & J. Thomas (eds.). Academic Press. London, UK.

TEXERA, W.A. 1972. Distribución y diversidad de mamíferos y aves en la Provincia de Magallanes. I. Análisis preliminar de la diversidad ecológica y variación taxonómica. II. Algunas notas ecológicas sobre los canales patagónicos. *Anales del Instituto de la Patagonia* (Chile) 3: 171-305.

VENEGAS, C. 1981. Aves de las islas Wollaston y Bayly, archipiélago del Cabo de Hornos. *Anales del Instituto de la Patagonia* (Chile) 12: 213-219.

VENEGAS, C. 1991. Ensambles avifaunísticos estivales del archipiélago Cabo de Hornos. *Anales del Instituto de la Patagonia, Serie Ciencias Naturales* (Chile) 20: 69-81.

VENEGAS, C. 1994. *Aves de Magallanes.* Ediciones de la Universidad de Magallanes, Punta Arenas, Chile. 158 pp.

VENEGAS, C. & R. SCHLATTER. 1999. Efecto de la intervención silvícola en bosques de *Nothofagus pumilio* sobre ensambles avifaunísticos estivales en Tierra del Fuego (Chile). *Anales del Instituto de la Patagonia Serie Ciencias Naturales* (Chile) 27: 41-50.

VUILLEUMIER, F. 1985. Forest birds of Patagonia: ecological geography, speciation, endemism, and faunal history. *Neotropical Ornithology, Ornithological Monographs* 36: 255-304.

WILLIAMS, P.H., K.J. GASTON & C.J. HUMPHRIES. 1997. Mapping biodiversity value worldwide: combining higher-taxon richness from different groups. *Proceedings of the Royal Society, Biological Sciences* 264: 141-148.

WILLSON, M.F. 1991. Dispersal of seeds by frugivorous animals in temperate forests. *Revista Chilena de Historia Natural* 64: 537-554.

WILLSON, M.F., T.L. DE SANTO, C. SABAG & J.J. ARMESTO. 1994. Avian communities of fragmented south-temperate rainforests in Chile. *Conservation Biology* 8: 508-520.

WILLSON, M.F., C. SMITH-RAMÍREZ, C. SABAG & J.F. HERNÁNDEZ. 1996. Mutualismos entre plantas y animales en bosques templados de Chile. Pp. 251-264 *in Ecología de los Bosques Nativos de Chile*, J.J. Armesto, C. Villagrán & M.T. Kalin (eds.). Editorial Universitaria, Santiago, Chile.

16 SEASONAL DYNAMICS AND DISTRIBUTION PATTERNS OF BIRDS ASSOCIATED WITH SUB-ANTARCTIC WETLANDS IN THE CAPE HORN BIOSPHERE RESERVE (54-55°S), CHILE

JOSÉ TOMÁS IBARRA[1,2], RICARDO ROZZI[2,3], HORACIO GILABERT[4], CHRISTOPHER B. ANDERSON[2], STEVEN M. MCGEHEE[2,5], AND CRISTIÁN BONACIC[1]

[1]Laboratorio Fauna Australis, Programa de Recursos Naturales, Facultad de Agronomía e Ingeniería Forestal, Pontificia Univ. Católica de Chile, Casilla 306-22, Santiago, Chile. *E-mail:* jtibarra@uc.cl,
[2]Parque Etnobotánico Omora (Sede Puerto Williams Universidad de Magallanes, Instituto Milenio de Ecología y Biodiversidad, Fundación Omora), Puerto Williams, Chile,
[3]Department of Philosophy & Religion Studies, Univ. of North Texas, Denton, Texas, USA,
[4]Programa de Recursos Naturales, Facultad de Agronomía e Ingeniería Forestal, Pontificia Univ. Católica de Chile, [5]Department of Biology, Univ. of Victoria, Victoria, B.C., Canada

Ornitología Neotropical 20: 321-337 (2009) (Originally published in Spanish)

ABSTRACT. Seasonal dynamics and distribution patterns of birds associated with sub-Antarctic wetlands in the Cape Horn Biosphere Reserve (54-55°S), Chile. Little is known about bird ecology in sub-Antarctic ecosystems of the Cape Horn Biosphere Reserve (CHBR), Chile. We examined the role of altitude, seashore distance, lake size, season, and vegetation cover in affecting bird diversity and abundance on Navarino Island (54-55°S, 67°W) in the CHBR. We determined avian community composition, species richness, and relative abundance for wetlands and their adjacent habitats on a seasonal basis. A total of 56 species belonging to 10 orders and 25 families were recorded, and significant seasonal changes for bird species ($P = 0.001$) and abundance ($P < 0.01$) were observed, with 33.9% residents, 48.2% winter migrants, and 17.9% with unknown migratory status. The distance of wetlands from the coastline was the main predictor variable for annual species richness ($P < 0.01$) and total avian abundance ($P < 0.05$). Wetlands far from the seashore and at higher altitudes had lower diversity and abundance, although the change in elevation was relatively slight (<300 m). These data show the seasonal patterns and dynamics, as well as the major underlying mechanisms, for birds of wetlands and adjacent habitats in the CHBR.

Key words: Avifauna, Cape Horn, Navarino Island, seasonal variations, sub-Antarctic, wetlands.

RESUMEN. Dinámica estacional y patrones de distribución de la avifauna asociada a humedales subantárticos en la Reserva de Biosfera Cabo de Hornos (54-55°S), Chile. El rol de las aves en la ecología de los ecosistemas subantárticos de la Reserva de Biosfera Cabo de Hornos (RBCH), Chile, ha sido poco estudiado. Se estudió el efecto de la altitud, distancia al borde costero, tamaño del espejo de agua, estación del año y cobertura vegetacional sobre la diversidad y abundancia de aves de humedales y de sus hábitats adyacentes en isla Navarino (54-55°S, 67°W) en la CHBR. Se cuantificó la composición específica, riqueza de especies y abundancias relativas de aves de humedales y de sus hábitats adyacentes, de manera estacional. Se registró un total de 56 especies pertenecientes a 10 órdenes y 25 familias, con una marcada estacionalidad en riqueza ($P = 0{,}001$) y número total de individuos ($P < 0{,}01$). De ellas, 33,9% fueron residentes, 48,2% llegan en primavera-verano y para un 17,9% se desconoce su estacionalidad. La distancia a la costa marina es la variable que mejor explica tanto la riqueza ($P < 0{,}01$) como el número total de individuos ($P < 0{,}05$). Humedales a mayor distancia de la costa y altitud, presentaron menor diversidad de aves, a pesar de que los cambios en altitud fueron bajos (< 300 m). Este trabajo muestra los patrones de estacionalidad y dinámica, así como los mecanismos subyacentes, de la avifauna de humedales y sus hábitats adyacentes en la CHBR.

Palabras clave: avifauna, Cabo de Hornos, humedales, isla Navarino, subantártico, variaciones estacionales.

INTRODUCTION

Wetlands are crucial for the life cycles of multiple plants and animals, composing a habitat of a great diversity of organisms, especially of aquatic and terrestrial migratory birds (Hauenstein *et al.* 2002). In Chile, the avifauna of continental aquatic environments comprises 29% of all birds registered in the country (Martínez & González 2004). In the case of the Cape Horn Archipelago, located in one of the 24 most pristine ecoregions on the planet (Mittermeier *et al.* 2002, Rozzi *et al.* 2006), birds make up the group of terrestrial vertebrates with the highest species richness and abundance (Anderson *et al.* 2006). Previous studies of the assemblage of birds in the extreme American south have mainly focused on its description through detailed sampling, like what was done on Hoste Island (*e.g.*, Oustalet 1891, Siefeld 1977, Venegas & Siefeld 1998), and Diego Ramírez Island (*e.g.*, Schlatter & Riveros 1987). These studies only present some of the birds of freshwater and marine-coastal habitats in sub-Antarctic environments. In other areas of Patagonia, studies describe birds of forests, steppes and Magellanic tundra (including *Sphagnum* peat bogs) (*e.g.*, Venegas 1976, Guzmán *et al.* 1985, Vuilleumier 1998, Blanco *et al.* 2004, Schlatter 2004). Garay *et al.* (1991) present the first systematic study of wetland birds of lakes and lagoons and their seasonality in Torres del Paine National Park (51°S).

On Navarino Island (54°55'S, 67°36'W), Barros (1971, 1976) provides the first listing of avifauna observed in the eastern zone of the island where some wetland species are mentioned. Following these descriptions, the level of understanding the forest birds of Navarino Island has increased significantly in the last decades (*e.g.*, Anderson & Rozzi 2000, Anderson *et al.* 2002, Rozzi & Massardo 2002, Ippi & Rozzi 2004). Despite this, the information generated by the Omora Ethnobotanical Park long-term ornithological project (see Anderson *et al.* 2002) has not dealt with the conformation of wetland and terrestrial birds assemblages that use adjacent habitats, nor their seasonal dynamics, which is little known because systematic investigations of this information do not exist to date. On the other hand, various authors have studied environmental factors that could explain the diversity patterns of wetland avifauna, like the association between water surface area and avifauna diversity (*e.g.*, Sillén & Solbreck 1977, Garay *et al.* 1991, Elmberg *et al.* 1994). However, the mentioned authors only studied this relation in some species, and only the study of Garay *et al.* (1991) took place in Chilean Patagonia. At the same time, studies of avifaunal diversity along altitudinal gradients have taken place in other parts of the world (*e.g.*, Navarro 1992, Rahbek 1997, Hawkins 1999, Kessler *et al.* 2001), but comparative studies do not exist in the extreme American south.

Given the importance of avifauna in the representation of vertebrates in the Cape Horn Biosphere Reserve (CHBR), it is a priority to know their diversity and abundance to establish conservation guidelines, monitoring, and management (Heywood 1995). Similarly, it is necessary to understand their attributes, such as distribution ranges, habitat use and seasonality, many of which are unknown, even for such a conspicuous group like birds. In this context, the present study hopes to confront composition patterns, structure, and dynamics of the group of birds of wetlands and adjacent habitats in the sub-Antarctic archipelago, permitting the current or future recognition of variations in their composition and/or structure due to natural or anthropogenic events.

Specifically, this study quantifies species composition, richness and abundance of birds of wetlands and adjacent habitats. In order to explain the patterns found by the study, we test various possible mechanisms that underlie them, including a) seasonality, b) seashore distance, c) water surface area, and d) vegetation cover of wetlands. Additionally, we classify birds inhabiting sub-Antarctic environments on Navarino Island based on habitat requirements, and discuss their usage patterns of the aquatic component of wetlands and its seasonality.

MATERIALS AND METHODS

From a total of 49 wetlands visited on Navarino Island (Fig. 1), we randomly selected 10 of them. These sites spread from the coastline to the interior of the mountainous zones of the study area; through which the distance to the coast increases upon ascent through waterways towards the mountainous Dientes de Navarino, from sea level to an altitude of 379 m (Table 1). The wetlands were separated by a minimal distance of 1.5 km and a maximum of 7.5 km, with a distance of 31.5 km between the two furthest points. Hills, dense vegetation, and abrupt relief separated the bodies of water, and the only wetlands 1.5 km apart (the smallest distance registered between wetlands) were separated by the city of Puerto Williams. The studied wetlands correspond to lakes (wetlands 2, 8 and 9), lagoons (wetlands 3, 4, 5, 6, 7 and 10), and a coastal wetland (wetland 1) (Table 1, see Ibarra 2007 for more detail).

We conducted point counts of birds along perimeter runs and at fixed points in each one of the 10 bodies of water between 06:00 and 11:30 h, using binoculars (7-20 x 21, 10 x 30 and 10 x 42). Sampling took place in the months of January (summer), April (fall), August (winter), and October (spring) of 2006. We used point counts with a fixed radius of 25 m (Bibby *et al.* 1992, Elmberg *et al.* 1994, Rozzi *et al.* 1996, González *et al.* 2004). Points at the edge of the water's surface included all birds within 100 m inward of the coast and within a 25 m outward radius. Counts commenced upon arriving at each station (Ralph *et al.* 1995, Jiménez 2000) and at each one we registered the birds seen or heard from the center of the point for a period of 5 minutes. This period of time is convenient since a longer recording time increases the chance of double counting (Bibby *et al.* 1992). Additionally, we listed all species spotted outside the listening points and perimeter runs throughout the census at each wetland to create a list of bird species of the study area.

We chose a total of 12 points per wetland, organized systematically with a minimum distance of 100 m between each one (Bibby *et al.* 1992, Rozzi *et al.* 1995). The layout of the points resulted from lineal transects of 200 m starting from the edge of each body of water, laid out according to the cardinal directions. At each transect, the first point was located at the water's edge (northern, southern, eastern, and western points). From that, we established two more points at 100 and 200 meters from the first point, further from the water.

The composition, species richness, and relative abundances were determined with respect to all registered individuals in the wetlands during each season of the year. We calculated the Shannon-Weaver or H' diversity index, using natural logarithm

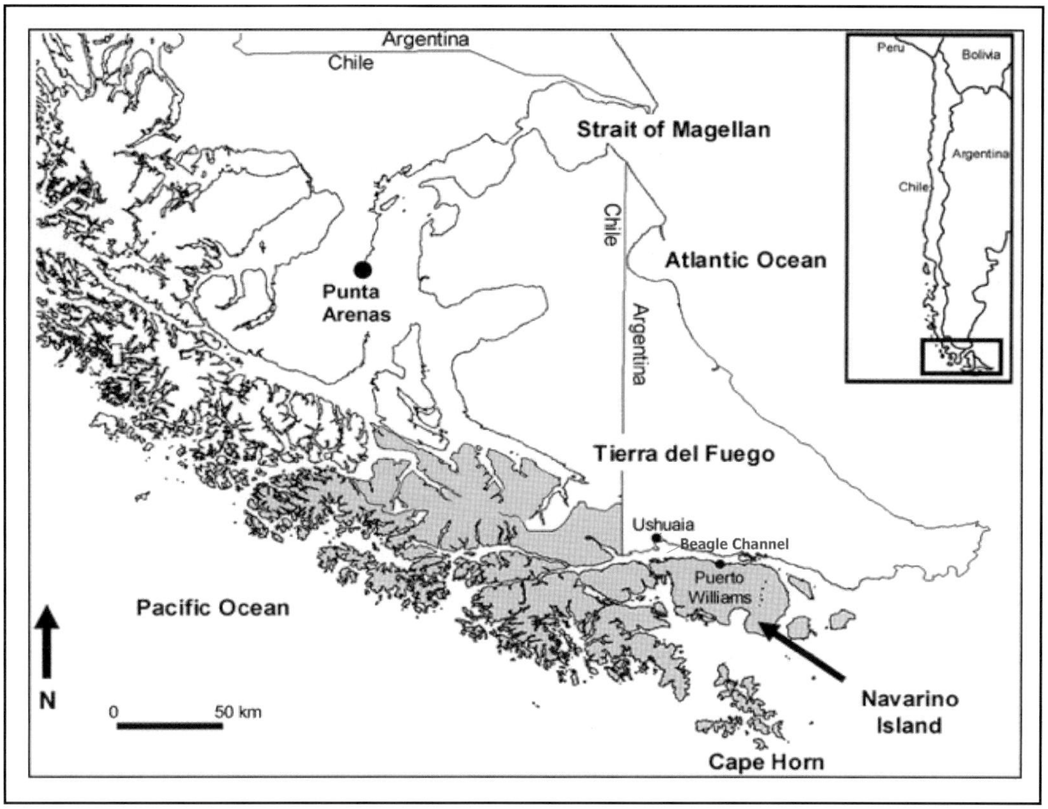

Fig. 1. Map of southern South America, with the Cape Horn Biosphere Reserve in grey. Study sites are located on Navarino Island (54-55°S, 67°W), Chile.

TABLE 1. General characteristics of the 10 seasonally studied wetlands in relation to the assemblage of birds on Navarino Island.

Wetland number Coordinates	UTM[1] (masl)	Altitude sea shore (m)	Distance to the	Water surface (ha)[2] winter (%)	Frozen in
1	3911991 594707	4	26.1	7.38	100
2	3912036 586920	10	146.1	23.84	100
3	3911164 587149	26	455.7	2.44	100
4	3910637 588622	31	821.4	3.04	100
5	3912079 581337	12	368.9	1.48	100
6	3916006 566397	41	401.5	0.83	100
7	3912282 563368	220	3525.9	2.93	100
8	3912708 564743	214	2826.2	12.04	100
9	3912265 573934	131	2154.4	20.90	85
10	3906193 583626	379	5398.8	6.13	100

[1]Coordinates in DATUM WGS 84 and correspond to the northern point of each wetland. [2]Discrimination between lakes and lagoons according to Garay *et al.* (1991). Water surface area ≥ 10ha corresponds to a lake.

and base 10, since both are utilized in similar studies (Zar 1984, Schemnitz 1980). We also calculated Simpson's Inverse Index or C_{inv}. These two indices, H' and C_{inv}, do not exactly respond in the same way to changes in species richness or evenness. C_{inv} is more sensitive to changes in evenness, whereas H' is more sensitive to changes in species richness (Feisinger 2003). For this reason, the calculation and comparison of both indices allow a better understanding of avifaunal community structure.

On the other hand, we determined habitat use based on microhabitat requirements through direct observation, complemented by published literature (Rottmann 1995, Araya & Millie 1996, González *et al.* 2004, Martínez & González 2004). This way, we essentially classified the birds in two large groups: (1) aquatic and (2) terrestrial. The first group completely or partially requires wet environments (associated with bodies of water) in their life cycles (Rottmann 1995, González *et al.* 2004). The second includes birds that have a terrestrial life cycle, but use wetlands and adjacent environments opportunistically (Figueroa *et al.* 2000). Since at 0 m from the water's edge we considered a larger counting area than at 100 and 200 m (birds on the water's surface up to 100 m from the edge were counted), we did not make comparisons between point counts in terms of richness or abundance of individuals. However, we determined if birds classified as aquatic, terrestrial or aquatic-terrestrial varied between counting points at 0, 100 and 200 m, in terms of species composition.

Species richness and the total number of individuals were compared through repeated measures analysis of variance of sampling over time with SIMPLE contrast for species number and the total number of individuals per season of the year (Scheiner & Gurevitch 1993). We tested the normality assumptions, sphericity, and fit tests using Pillai's Trace and Wilks' Lambda. Records were complemented with information on the seasonality of birds in the southern region (Garay *et al.* 1991, 8-year unpublished database of Omora Park's banding program). A species was considered a resident when individuals were recorded permanently in the study area during the coldest months (Newton 1998).

Through multiple regression analysis we determined whether the annual richness and total number of individuals have a significant correlation with wetland altitude (msnm), the distance of the body of water from the coast (m), water surface area (ha), vegetation cover (%), and the types of vegetation present (tall woody plants, small woody plants, herbaceous, and peat), and the general median cover of adjacent habitats for each of the 10 wetlands (Bennett *et al.* 2004, Mason & Macdonald 2005). We used Pearson's correlation analysis to analyze the distance to the marine coast and the diversity indices of Shannon-Weaver (H') and Simpson's Inverse (C_{inv}).

Statistical analysis was conducted using the program SPSS version 10.0 (SPSS 1999).

RESULTS

Taxonomic diversity and habitat use classification. We registered a total of 56 species that used wetlands and/or their surrounding habitats on Navarino Island in 2006 (Appendix 1). This represents approximately 12% of all birds registered in Chile (Martínez & González 2004) and 36% of all birds registered in the Cape Horn Archipelago (Couve & Vidal 2000).

The birds registered pertained to 10 orders and 25 families. The order Passeriformes was the most represented with 23 species, the families Tyrannidae, Furnariidae and Emberizidae being the most numerous of this group (Appendix 1). The order Anseriformes was the second most numerous with 10 species. Another important order was Charadriiformes with 9 species pertaining to 5 families. At the family level, Anatidae was the most numerous of all families identified in wetland habitats with 10 species, representing 34.5% of the anatids registered at the national level (Vilina & Cofré 2006). Among the anatids registered, *Tachyeres pteneres*, *Tachyeres patachonicus*, *Lophonetta specularioides*, *Anas sibilatrix*, *Anas flavirostris*, *Anas georgica*, and *Anas platalea* were only observed on the water (station at 0 m) (Appendix 1).

Using this species classification in function with habitat use, 13 of the 56 bird species (23.2%) were only registered on the water's surface and can be considered strictly aquatic. These birds include all species of the families Podicipedidae, Rallidae, Phalacrocoracidae and the majority of Anatidae. Among the latter, the geese *Chloephaga picta* and *C. poliocephala* frequently used habitats bordering the water's surface (100 and 200 m). Among the 13 species of strictly aquatic birds, three species (*Chloephaga hybrida*, *T. pteneres* and *Phalacrocorax atriceps*) were associated with coastal marine habitats (Appendix 1). A 26.8% (15 species) of the 56 species registered used terrestrial and aquatic habitats interchangeably. Among these birds, one species, *Haematopus leucopodus*, was present near coastal marine habitats (Appendix 1). The majority of the families of birds registered present species that were spotted at 0, 100, and 200 m from the water's surface, displaying a more generalist lifestyle in terms of habitat use. Based on these records and the natural history of each species, we determined that 50% (28 species) of the species registered have primarily terrestrial habits (Appendix 1). Therefore, the assemblage of birds associated with sub-Antarctic wetlands in Cape Horn include strictly aquatic species, as well as birds that use water bodys opportunistically.

Seasonality of avifauna and relationships between habitat and diversity. Avifaunal diversity presented a distinct seasonality en terms of species richness ($F_{3\text{-}7}$ = 19.164, P = 0.001, Pillai's Trace = 0.891, Wilks Lambda = 0.109) and the total number of individuals ($F_{3\text{-}7}$ = 13.837, P = 0.003, Pillai's Trace= 0.856, Wilks' Lambda = 0.144) (Table 2, Fig. 2). Species richness in the summer (13.8 ± 5.5) and in the spring (13.4 ± 5.0) is higher than in the fall (8.9 ± 4.5, F = 22.77, P = 0.001), and reaches a minimum in winter (5.9 ± 3.4, F = 48.9, P < 0.001). Winter species richness presents only 42.7% of the richness recorded in the summer. For bird abundances, the average value for the total number of counted individuals was highest in summer (80.1 ± 14.5) and spring (65.3 ± 12.1), and decreased markedly in the fall (48.2 ± 14.1, F = 7.1, P = 0.026), reaching a minimum in winter (25.9 ± 5.0, F = 21.8, P = 0.001). In winter we registered 32.3% of the average number of counted individuals in the summer (Fig. 2, Appendix 1).

Of all species recorded in the wetlands, 19 (33.9%) species were residents and 27 (48.2%) arrived during the summer season. For 10 species (17.9%) it was not possible to determine their seasonality due to lack of data or mentions in related literature (Appendix 1).

The terrestrial migrant *Tachycineta meyeni* was the most abundant during summer (12.7%), whereas the terrestrial resident *Aphrastura spinicauda* took first place in the fall (19.7%) and winter (35.9%), and the terrestrial migrant *Zonotrichia capensis* in spring (17.2%). For primarily aquatic species, the most abundant in each season were: *C. picta* in the summer (6.4%) and spring (10.4%), *A. sibilatrix* in the fall (10.6%), and *T. patachonicus* in winter (4.2%).

The sites with the highest number of species and diversity index values were wetlands close to the coast (1 and 2), and those with the lowest values were located furthest into the interior (9 and 10) (Table 2). The environmental variables that resulted significant in the explanation of this pattern were: (a) the wetland's distance from the coast negatively explained annual species richness ($r^2_{adj} = 0.544$, N = 10, $P = 0.009$) and the total number of individuals ($r^2_{adj} = 0.349$, N = 10, $P = 0.042$) (Fig. 3), and (b) water surface area, which determined the increase in species richness and total number of individuals as surface area increased (Table 3).

We observed a negative correlation between distance from the coast with respect to both diversity indices (Table 4). The C_{inv} index presented the largest decrease moving toward the island's interior in relation to the other index, which did not have a clear tendency (Fig. 4).

DISCUSSION

Avifaunal composition and seasonality. The assemblage of birds in wetlands and adjacent habitats present a clear seasonality on Navarino Island, CHBR. In particular, aquatic species demonstrate this pattern strongly, while terrestrial species tend to be more residents. This difference between the two groups of birds using wetland habitats could be explained by the freezing of the aquatic component of wetlands between the months of May and September. The seasonality of aquatic birds is a pattern shared with other parts of Patagonia, for example Torres del Paine (Garay *et al.* 1991). In this case, aquatic birds begin migration in the month of March, and in May-June have disappeared from the lakes and lagoons, when the wetlands there have frozen over.

Our sampling did not allow us to establish the exact month that the species undertake their migrations, but in the month of August (winter) we registered individuals of the migratory species *C. picta* (8-year database of Omora Park's banding program). Its presence at the end of winter could mark the beginning of its reproductive stage, the moment when paired adults of this species arrive in the southern zone (returning from migration) with the objective of searching for a territory to reproduce (Summers & McAdam 1993). Even more, *C. picta* was the most abundant in wetland habitats, in accordance with data from Garay *et al.* (1991) in the sector of Torres del Paine. This species, along with *C. poliocephala*, were the only species of the family Anatidae that we observed at 100 and 200 m from the water. This is explained in that both are dependent on wetlands, but also make use of pastures and peat bogs (Rottmann 1995), and in the case of *C. poliocephala*, makes use of forested habitats (for reproduction, resting, and foraging) in sub-Antarctic ecosystems (Schlatter *et al.* 2002).

We also registered individuals of *Colorhamphus parvirostris* in the winter (August), indicating a contradiction to preliminary results regarding this species in the CHBR that classified it as migratory (McGehee *et al.* 2004). *C. parvirostris* would be a resident species (confirmed by the Omora Ethnobotanical Park's long-term banding program, unpublished data).

The clear seasonality of Cape Horn's avifauna has significant implications on the food availability of predators, including the American mink (*Mustela vison*). It has been demonstrated that for this recently arrived exotic species in the CHBR, birds contribute a fundamental component of their summer and spring diet, and depend on introduced rodents as a food source in the fall and winter (Ibarra *et al.* 2009). It is also interesting to consider that birds made up a fundamental element of human populations that had inhabited the archipelago since at least 7000 years ago (Legoupil 1993). In archeological deposits of Navarino Island, Legoupil (2003) identified a significant portion of bony remains of aquatic birds in ancient shell middens. These precedents suggest a long history of interaction between wetland avifauna and this particular human population of Cape Horn.

Underlying mechanisms of avifaunal diversity patterns. Species richness and total number of individuals in this environment are explained largely by the distance between the coast and the wetlands. For example, wetlands closer to the coast were the most diverse because marine, freshwater, and terrestrial species were present. In the case of the coastal wetland, we observed species characteristic of marine-coastal environments like *C. hybrida*, *H. leucopodus*, *Leucophaeus scoresbii*, *P. atriceps*, and *T. pteneres*.

Distance from the coast had a larger effect on the Simpson Inverse diversity index (C_{inv}) than on the Shannon-Weaver diversity index (H'), because both indices respond differently to species

TABLE 2. Species richness, total abundances and diversity indices for 10 wetlands and their adjacent habitats on Navarino Island in 2006.

Wetland	1	2	3	4	5	6	7	8	9	10
Number of species (S)	31	36	20	23	21	20	13	19	11	9
Number of individuals (N)	416	475	192	244	174	182	102	214	112	84
H' (log_{10})	1.23	1.3	1.07	1.16	1.11	1.11	0.99	1.09	0.82	0.84
H' (log_e)	2.82	2.98	2.46	2.68	2.56	2.50	2.29	2.50	1.88	1.95
C_{inv}	12.47	12.36	7.74	11.07	9.67	8.65	8.42	9.29	4.25	6.26

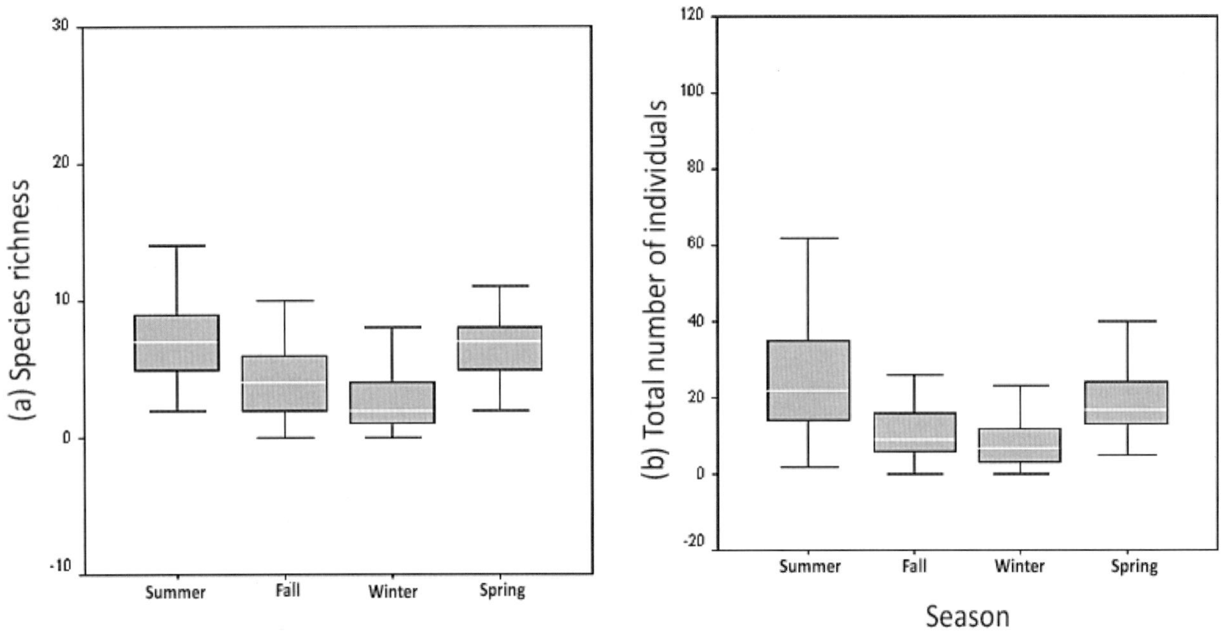

Fig. 2. Box-plots of (a) Species richness and (b) Total number of individuals, seasonally recorded at 10 wetlands and their adjacent habitats on Navarino Island in 2006. Seasons of the year were significantly different (ANOVA test of repeated measures over time) for species richness ($F_{3\text{-}7}$ 19.164, $P = 0.001$) and for the total number of individuals ($F_{33\text{-}7}$ 13.837, $P = 0.003$). Bars indicate confidence intervals (95%).

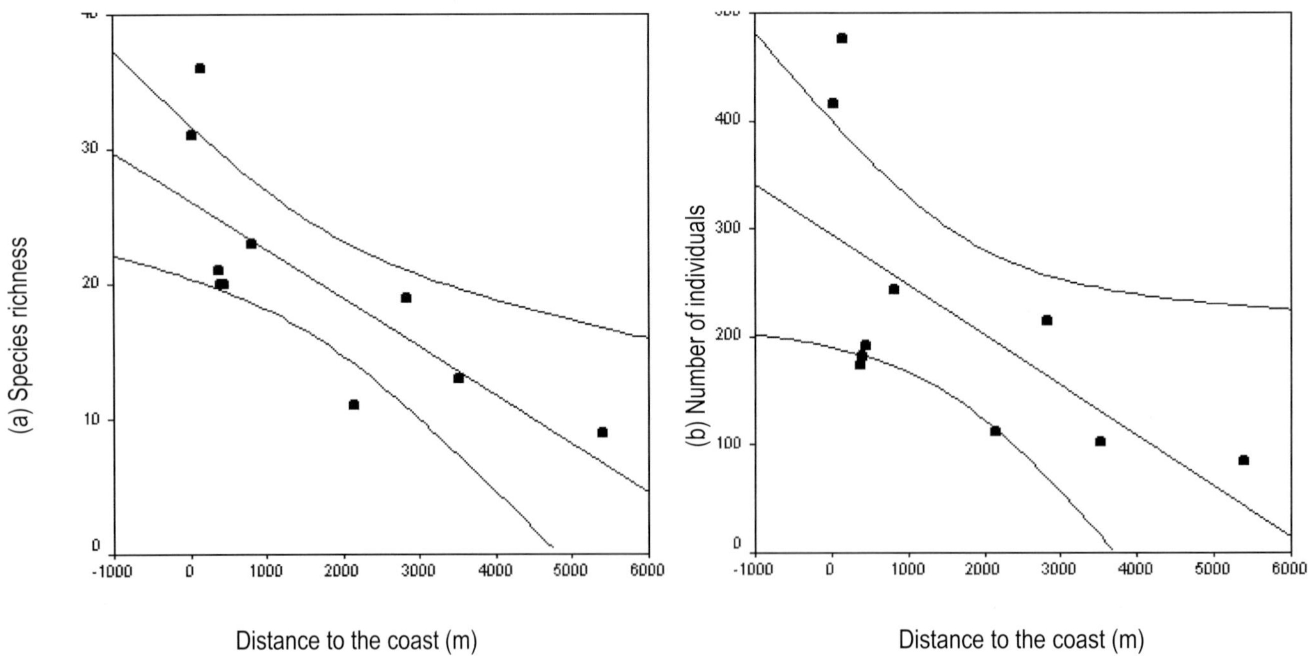

Fig. 3. Variations in (a) Species richness (multiple regression analysis, $r^2_{adj} = 0.544$, N = 10, $P = 0.009$) and (b) Total number of individuals (multiple regression analysis, $r^2_{adj} = 0.349$, N = 10, $P = 0.042$), at different distances from the coast (m), for 10 wetlands and their adjacent habitats on Navarino Island. Trend line and confidence intervals (95%) are shown.

TABLE 3. Variables that explain species richness and the total number of individuals of birds in wetlands and adjacent habitats of Navarino Island. The values r^2_{adj} and the significant *P* are shown (multiple regression analysis, $P < 0.05$).

Species richness				Number of individuals			
Explanatory variable		r^2_{adj}	*P*	Explanatory variable		r^2_{adj}	*P*
Season	Summer	n.s.	n.s.	Season	Summer	n.s.	n.s.
	Fall	0.5392	<0.0001		Fall	0.3951	0.0003
	Winter	0.4227	<0.0001		Winter	0.2094	0.0005
	Spring	n.s.	n.s.		Spring	n.s.	n.s.
Water surface area (ha)		0.5763	0.0019	Water surface area (ha)		0.449	0.0403

TABLE 4. Correlation values between distance from the coast and the diversity indices of Shannon-Weaver and Simpson (Pearson correlation analysis, $P < 0.05$).

Variable		H'_{log10}	H'_{loge}	C_{inv}
Distance to the coast	r^2	-0.75	-0.74	-0.72
	P	0.01	0.01	0.02

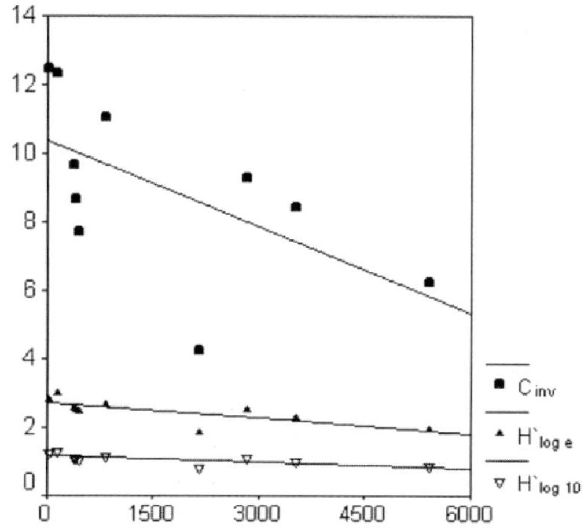

Fig. 4. Response of the diversity indices of birds for 10 wetlands and adjacent habitats, at different distances from the coast (m) on Navarino Island (Pearson correlation analysis, $P < 0.05$). Note: C_{inv} most appropriately reflects the relationship between distance from the coast (m) and decrease in bird diversity indices.

richness or evenness. C_{inv} is more sensitive to changes in evenness or equity, whereas H' is more sensitive to changes in species richness (Feinsinger 2003), where the maximum possible diversity for a given number of species occurs if all are present in equal numbers, corresponding to equity (Schemnitz 1980). This observed pattern suggests that in wetlands and their adjacent habitats on Navarino Island, distance to the coast has a greater effect on evenness of the number of species than solely on species richness. For some species we were able to observe that they primarily use wet habitats close to the marine coast, but at the same time disperse in fewer numbers toward the interior of the island. To exemplify this pattern it is interesting to consider the case of *C. picta*. Individuals of this species arrive in great numbers to Navarino Island during the southern reproductive period, concentrating in coastal zones and wetlands near the Beagle Channel, but some individuals reproduce in interior wetlands of the island during the summer. This same pattern was observed for other anatids in the area like *T. patachonicus* and *L. specularioides*. One frequent species during the reproductive period, although we observed it in few numbers during systematic counts, was *Charadrius modestus*. This species resides in both Chile and Argentina, dispersing to Uruguay and Brazil, and migrates to reproduce in southern Patagonia, reaching the wetlands of Cape Horn (Couve & Vidal 2000). This species arrived during the spring to reproduce in sphagnum wetlands of the island's interior.

On Navarino Island we observed the highest diversity of birds in wetlands near the sea, and the lowest in wetlands of the interior, establishing an avifaunal diversity gradient along an altitudinal gradient of wetlands. It is worth reminding that interior wetlands ascend through basins going further away from the coast, and such distance from the sea positively correlates with altitude. This determines that species diversity decreases as distance from the coast and altitude increase. This could be due to habitat conditions (lower temperature, longer annual freezing time of water, less food availability, less aquatic vegetation and/or more scrubby vegetation) acting as attributes that may determine adverse conditions for nesting, refuge and foraging for some species. Permanence in areas neighboring high-altitude wetlands could be of shorter duration in comparison with wetlands close to sea level. This altitudinal pattern between species diversity and elevation over sea level is a common aspect in the determination of biotic community structures (Kessler *et al.* 2001) and has been described as an analogous pattern to the latitudinal gradient of species richness (Stevens 1989, 1992).

This latitudinal pattern also makes sense in the interpretation of our results, upon comparing the species richness of birds (18 species) found by Garay *et al.* (1991) in Torres del Paine (51°S), which is lower on Navarino Island (54-55°S) with only 11 strictly aquatic species. Aquatic species commonly found in northern Magallanes and that do not reach the planet's southernmost forested basins in the CHBR correspond to *Anas specularis*, *Anas versicolor*, *Coscoroba coscoroba*, *Cygnus melanocoryphus* and *Oxyura jamaicensis* (Anatidae); *Phoenicopterus chilensis* (Phoenicopteridae), and *Podiceps occipitalis* (Podicipedidae). All of these species are present in lakes and lagoons of Tierra del Fuego (Venegas 1986), immediately north of the study site. In the case of *P. occipitalis*, there exists one record by Barros (1971) on Picton Island, situated to the east of Navarino Island and *C. melanocoryphus* has been recorded on some occasions, including on the Antarctic Peninsula (Orgeira & Fogliatto 1991, Quintana *et al.* 2000).

In conclusion, studies of birds in wetland environments and their adjacent habitats in Patagonia are scarce. However, this study demonstrates how diverse and abundant these ecosystems are, including their surrounding habitats, in the CHBR. It also shows that the use of these environments by birds is clearly seasonal, since many species (especially the aquatic species) are migratory and the CHBR may be their southern distribution limit.

This study investigated some aspects of the composition, structure, and dynamics of the avifaunal assemblage of wetlands and their adjacent habitats in the CHBR. The development of follow-up studies, and long-term monitoring, and investigations that include other parts of the archipelago rich in wetlands that consider population characteristics, migration patterns, local distributions, and threats that could affect the birds and wetlands of the extreme American south are needed.

ACKNOWLEDGMENTS

This study was funded by the Comisión Nacional del Medioambiente (CONAMA), project 12-001-06 and by the Wildlife Trust Alliance. We thank the Instituto Milenio de Ecología y Biodiversidad (IEB) (ICM P05-001) and the Puerto Williams branch of Universidad de Magallanes for all its logistical and financial support of our team. This investigation was made possible by those who accompanied us on the field: Tomás Altamirano, Antonia Barreau, Germán González, and Ezio Firmani. A special thanks to Santiago Imberti, Harvey Tastuyasu, and André Weller for their comments on our manuscript. This project is part of the long-term conservation and research programs of Omora Park UMAG-IEB in the CHBR.

REFERENCES

ANDERSON, C.B. & R. ROZZI. 2000. Bird assemblages in the southernmost forests in the world: Methodological variations for determining species composition. *Anales del Instituto de la Patagonia* 28: 89-100.

ANDERSON, C.B., R. ROZZI, C. ELPHICK & S.M. MCGEHEE. 2002. El programa Omora de anillamiento de aves en los bosques subantárticos: la estandarización del tamaño de anillos apropiados para las aves de la región de Magallanes. *Boletín Chileno de Ornitología* 9: 2-11.

ANDERSON, C.B., R. ROZZI, J.C. TORRES-MURA, S.M. MCGEHEE, M.F. SHERRIFFS, E. SCHÜTTLER & A.D. ROSEMOND. 2006. Exotic vertebrate fauna in the remote and pristine sub-Antarctic Cape Horn Archipelago region of Chile. *Biodiversity and Conservation* 10: 3295-3313.

ARAYA, B., M. BERNAL, R. SCHLATTER & M. SALABERRY. 1995. *Lista Patrón de las Aves de Chile*. Editorial Universitaria, Santiago, Chile.

ARAYA, B. & G. MILLIE. 1996. *Guía de Campo de las Aves de Chile*. 4ª ed. Editorial Universitaria, Santiago, Chile.

BARROS, A. 1971. Aves observadas en las islas Picton, Nueva, Lennox y Navarino oriental. *Anales del Instituto de la Patagonia* 2: 166-180.

BARROS, A. 1976. Nuevas aves observadas en las islas Picton, Nueva, Lennox y Navarino oriental. *Anales del Instituto de la Patagonia* 7: 189-193.

BENNETT, A.F., S.A. HINSLEY, P.E. BELLANY, R.D. SWETNAM & R. MACNALLY. 2004. Do regional gradients in land-use influence richness, composition and turnover of bird assemblages in small woods? *Biological Conservation* 119: 191-206.

BIBBY, C.J., N.D. BURGESS & D.A. HILL. 1992. *Bird Census Techniques*. Academic Press, London, UK.

BLANCO, D.E., V.M. DE LA BALZE & G. PUGNALI. 2004. Avifauna de las turberas del centro de la Isla de Tierra del Fuego, Argentina. Pp. 113-118 *in Los Turbales de la Patagonia: Bases para su Inventario y la Conservación de la Biodiversidad*, D.E. Blanco & D.M. de la Balze (eds.). Wetlands International, Publicación 19, Buenos Aires, Argentina.

COUVE, E. & C. VIDAL. 2000. *Aves del Canal Beagle y Cabo de Hornos*. Editorial Fantástico Sur Birding LTDA, Punta Arenas, Chile.

ELMBERG, J., P. NUMMI, H. POYSA & K. SJOBERG. 1994. Relationships between species, lake size and resource diversity in assemblages of breeding waterfowl. *Journal of Biogeography* 21: 75-84.

FEINSINGER, P. 2003. *Diseño de Estudios de Campo para la Conservación de la Biodiversidad*. The Nature Conservancy. FAN-Wildlife Conservation Society, Santa Cruz de la Sierra, Bolivia.

FIGUEROA, R., C. BRAVO, E. CORALES, R. LÓPEZ & S. ALVARADO. 2000. Avifauna del Santuario de la Naturaleza Los Huemules del Niblinto, región del Bío-Bío. *Boletín Chileno Ornitología* 7: 2-12.

GARAY, G., W.E. JOHNSON & W.L. FRANKLIN. 1991. Relative abundance of aquatic birds and their use of wetlands in the Patagonia of southern Chile. *Revista Chilena de Historia Natural* 64: 127-137.

GONZÁLEZ, D., C. BENAVENTE & R. FIGUEROA. 2004. Avifauna de la laguna Santa Elena, región del Bío-Bío. *Boletín Chileno de Ornitología* 10: 13-18.

GUZMÁN, L., A. ATALA & C. VENEGAS. 1985. Composición específica y estructura de la comunidad de aves de verano en el complejo de la Tundra Magallánica. *Anales del Instituto de la Patagonia* 16: 75-86.

GROSSMAN, G.D., D.M. NICKERSON & M.C. FREEMAN. 1991. Principal component analysis of assemblage structure data: utility of tests based on Eigenvalues. *Ecology* 72: 341-347.

HAUENSTEIN, E., M. GONZÁLEZ, F. PEÑA-CORTÉS & A. MUÑOZ-PEDREROS. 2002. Clasificación y caracterización de la flora y vegetación de los humedales de la costa de Toltén (IX Región, Chile). *Gayana Botánica* 59: 87-100.

HAWKINS, A.F.A. 1999. Altitudinal and latitudinal distribution of east Malagasy forest bird communities. *Journal of Biogeography* 26: 447-458.

HEYWOOD, C.A. 1995. *Global Biodiversity and Assessment*. Cambridge Univ. Press, New York, New York.

HOOVER, J.P. 2006. Water depth influences nest predation for a wetland-dependent bird in fragmented bottomland forests. *Biological Conservation* 127: 37-45.

IBARRA, J.T. 2007. *El Visón en* Upushwaia*: Variaciones Estacionales de la Avifauna en Ambientes de Humedal y de su Depredación por la Especie Exótica Invasora* Mustela vison *(Schreber) en Isla Navarino, Reserva de Biosfera Cabo de Hornos, Chile*. M.Sc. Tesis Recursos Naturales, Facultad de Agronomía e Ingeniería Forestal, Pontificia Univ. Católica de Chile, Santiago, Chile.

IBARRA, J.T., L. FASOLA, D.W. MACDONALD, R. ROZZI & C. BONACIC. 2009. Invasive American mink *Mustela vison* in wetlands of the Cape Horn Biosphere Reserve, southern Chile: what are they eating? *Oryx* 43: 87-90.

IPPI, S. & R. ROZZI. 2004. Actividad diurna y nocturna del concón (*Strix rufipes*) en los bosques del Cabo de Hornos. *Boletín Chileno de Ornitología* 10: 9-12.

JIMÉNEZ, J.E. 2000. Effect of sample size, plot size, and counting time on estimates of avian diversity and abundance in a Chilean rainforest. *Journal of Field Ornithology* 71: 66-87.

KESSLER, M., S.K. HERZOG, J. FJELDSA & K. BACH. 2001. Species richness and endemism of plant and bird communities along two gradients of elevation, humidity and land use in the Bolivian Andes. *Diversity and Distributions* 7: 61-77.

KOENEN, M.T. & S. GALE. 2000. Effects of fire on birds in paramo habitat of northern Ecuador. *Ornitología Neotropical* 11: 155-163.

LEGOUPIL, D. 1993. El archipiélago del Cabo de Hornos y la costa sur de la isla Navarino: poblamiento y modelos económicos. *Anales del Instituto de la Patagonia* 22: 101-121.

LEGOUPIL, D. 2003. Cazadores-recolectores de Ponsonby (Patagonia Austral) y su paleoambiente desde VI al III milenio A.C. *Magallania* 31, Volumen especial.

MCGEHEE, S., R. ROZZI, C.B. ANDERSON, S. IPPI, R.A. VÁSQUEZ & S. WOODLAND. 2004. Late-summer presence of the Patagonian tyrant, *Colorhamphus parvirostris* (Darwin) in Navarino Island, Cape Horn county, Chile. *Anales del Instituto de la Patagonia* 32: 25-33.

MAGURRAN, A. 1988. *Ecological Diversity and its Measurement*. Princeton Univ. Press, Princeton, New Jersey.

MARTÍNEZ, D. & G. GONZÁLEZ. 2004. *Las Aves de Chile: Nueva Guía de Campo*. Ediciones del Naturalista, Santiago, Chile.

MASON, C.F. & S.M. MACDONALD. 2005. The bird assemblage of coastal borrowdykes in relation to habitat features. *Folia Zoologica* 54: 293-300.

MITTERMEIER, R.A., C.G. MITTERMEIER, P. ROBLES-GIL, J.D. PILGRM, W.R. KONSTANT, G.A.B. DA FONSECA & T.M. BROOKS (EDS.). 2002. *Wilderness: Earth's Last Wild Places*. CEMEX-Conservation International, Mexico.

NAVARRO, A.G. 1992. Altitudinal distribution of birds in the Sierra Madre del Sur, Guerrero, Mexico. *Condor* 94: 29-39.

NEWTON, I. 1998. *Population Limitation in Birds*. 1st ed. Academic Press, San Diego, California.

OLROG, C.C. 1950. Mamíferos y aves del archipiélago de Cabo de Hornos. *Acta Zoológica Lilloana* 9: 505-532.

ORGEIRA, J.L. & O.N. FOGLIATTO. 1991. Presence of the Black-necked Swan *Cygnus melanocoryphus* in Antarctica. *Marine Ornithology* 19: 140-143.

OUSTALET, E. 1891. Oiseaux mission scientifique du Cap Horn 1882-1883. *Zoologie* 6: 1-341.

QUINTANA, R.D., V. CIRELLI & J.L. ORGEIRA. 2000. Abundance and spatial distribution of bird populations at Cierva point, Antarctic península. *Marine Ornitholology* 28: 21-27.

RAHBEK, C. 1997. The relationship among area, elevation, and regional species richness in Neotropical birds. *American Naturalist* 149: 875-902.

RALPH, C.J., S. DROEGE & J. R. SAUER. 1995. Managing and monitoring birds using point counts: Standards and applications. Pp. 161-168 *in Monitoring Bird Populations by Point Counts*, C.J. Ralph, J.R. Sauer & S. Droege (eds.). Pacific Southwest Research Station, Forest Service, U.S. Department of Agriculture, Albany, California.

REYNOLDS, P.W. 1935. Notes on the birds of Cape Horn. *Ibis 5, Ser.* 13: 65-101.

ROTTMANN, J. 1995. *Guía de Identificación de Aves de Ambientes Acuáticos*. Unión de Ornitólogos de Chile, Santiago, Chile.

ROZZI, R., D. MARTÍNEZ, M.F. WILLSON & C. SABAG. 1995. Avifauna de los bosques templados de Sudamérica. Pp. 135-152 *in Ecología de los Bosques Nativos de Chile*, J.J. Armesto, C. Villagrán & M.T.K. Arroyo (eds.). Editorial Universitaria, Santiago, Chile.

ROZZI, R., J. ARMESTO, A. CORREA, J.C. TORRES-MURA & M. SALLABERRY. 1996. Avifauna de bosques primarios templados en islas deshabitadas del archipiélago de Chiloé, Chile. *Revista Chilena de Historia Natural* 69: 125-139.

ROZZI, R. & F. MASSARDO. 2002. *Antecedentes de Biodiversidad y Sitios Prioritarios en la Comuna Cabo de Hornos: Recopilación de Información Sobre los Ecosistemas Subantárticos en Apoyo a la Estrategia y Plan de Acción para la Biodiversidad*. Informe Técnico. Comisión Nacional del Medio Ambiente XII Región, Punta Arenas, Chile.

ROZZI, R., F. MASSARDO, A. BERGHÖFER, C.B. ANDERSON, A. MANSILLA, M. MANSILLA, J. PLANA, U. BERGHÖFER, P. ARAYA & E. BARROS. 2006. *Reserva de Biosfera Cabo de Hornos: Documento de Base para la Incorporación del Territorio Insular del Cabo de Hornos a la Red Mundial de Reservas de Biosfera*. Programa MaB-Unesco. Ediciones de la Universidad de Magallanes, Punta Arenas, Chile.

SCHEINER, S.M. & J. GUREVITCH. 1993. *Design and analysis of ecological experiments*. 2a ed. Oxford Univ. Press, Oxford, UK.

SCHEMNITZ, S.D. 1980. *Wildlife Management Techniques Manual.* 4th ed. The Wildlife Society, Washington, D.C.

SCHLATTER, R.P. & G.M. RIVEROS. 1987. Historia natural del archipiélago Diego Ramírez, Chile. *Serie Científica INACH* (Chile) 47: 87-112.

SCHLATTER, R.P., P. VERGARA & M. BRIONES. 2002. El Canquén (*Chloephaga poliocephala*) en bosques de Tierra del Fuego: distribución y depredadores. *Anales del Instituto de la Patagonia* 30: 61-66.

SCHLATTER, R.P. 2004. Fauna de turberas de la XII Región y Tierra del Fuego. Pp. 107-112 *in Los Turbales de la Patagonia: Bases para su Inventario y la Conservación de la Biodiversidad*, D.E. Blanco & D.M. de la Balze (eds.). Wetlands International, Publicación 19, Buenos Aires, Argentina.

SIELFELD, W.H. 1977. Reconocimiento macrofaunístico terrestre en el área de Seno Ponsonby (isla Hoste). *Anales del Instituto de la Patagonia* 8: 275-296.

SILLÉN, B. & C. SOLBRECK. 1977. Effects of area and habitat diversity on bird species richness in lakes. *Ornis Scandinavica* 8: 185-192.

SPSS Inc. 1999. *SPSS Base 10.0 for Windows User's Guide*. SPSS Inc., Chicago, Illinois.

STEVENS, G.C. 1989. The latitudinal gradient in geographical range: how so many species coexist in the tropics. *American Naturalist* 133: 240-256.

STEVENS, G.C. 1992. The elevational gradient in altitudinal range: an extention of Rapoport`s latitudinal rule to altitude. *American Naturalist* 140: 893-911.

SUMMERS, R. & J. MCADAM. 1993. *The Upland Goose*. Bluntisham Books, Huntingdon, UK.

VENEGAS, C. 1976. Observaciones ornitológicas en la Tundra Magallánica I. Recuento descriptivo del área y de las observaciones aviales entre los paralelos 51°31'S y 52°09'S. *Anales del Instituto de la Patagonia* 7: 171-184.

VENEGAS, C. 1981. Aves de las islas Wollaston y Bayly, Archipiélago del Cabo de Hornos. *Anales del Instituto de la Patagonia* 12: 213-219.

VENEGAS, C. 1986. *Aves de Patagonia y Tierra del Fuego Chileno-Argentina*. Ediciones de la Univ. de Magallanes, Punta Arenas, Chile.

VENEGAS, C. 1991. Ensambles avifaunísticos estivales del archipiélago Cabo de Hornos. *Anales del Instituto de la Patagonia* 20: 69-82.

VENEGAS, C & W. SIELFELD. 1998. *Catálogo de los Vertebrados de la Región de Magallanes y Antártica Chilena*. Ediciones de la Univ. de Magallanes, Punta Arenas, Chile.

VILINA, Y. & H. COFRÉ. 2006. Aves acuáticas continentales. Pp. 270-277 *in Biodiversidad de Chile: Patrimonio y Desafíos*, CONAMA (eds.). Ocho Libros Editores, Santiago, Chile.

VUILLEUMIER, F. 1998. Avian biodiversity in forest and steppe communities of Chilean Fuego-Patagonia. *Anales del Instituto de la Patagonia* 26: 41-57.

WRIGHT, J.P., C.G. JONES & A.S. FLECKER. 2002. An ecosystem engineer, the beaver, increases species richness at the landscape scale. *Oecologia* 132: 96-101.

ZAR, J.H. 1984. *Biostatistical Analysis*. 2nd ed. Prentice Hall, Englewood Cliffs, New Jersey.

Accepted May 20th 2009.

APPENDIX 1. Composition, total number and relative abundances (%) of species during 4 seasons of the year in 2006 in wetlands of Navarino Island (+ indicates its presence outside systematic sampling). [1]A: annual, S: summer, U: unknown; [2]A: aquatic, T: terrestrial, *exclusively marine.

FAMILY	SPECIES	Nº SUMMER	Nº FALL	Nº WINTER	Nº SPRING	SEASONALITY[1]	HABITAT[2]
Anatidae	*Anas flavirostris*	27 (0.034)	8 (0.017)	+	26 (0.040)	S	A
	Anas georgica	25 (0.031)	28 (0.058)	0	14 (0.021)	S	A
	Anas platalea	15 (0.019)	11 (0.023)	0	0	S	A
	Anas sibilatrix	40 (0.050)	51 (0.106)	0	13 (0.020)	S	A
	Chloephaga hybrida	0	9 (0.019)	4 (0.015)	+	A	A*
	Chloephaga picta	51 (0.064)	43 (0.089)	9 (0.035)	68 (0.104)	S	A/T
	Chloephaga poliochepala	+	2 (0.004)	0	6 (0.009)	S	A/T
	Lophonetta specularioides	6 (0.007)	7 (0.015)	10 (0.039)	19 (0.029)	A	A
	Tachyeres patachonicus	0	12 (0.025)	11 (0.042)	14 (0.021)	A	A
	Tachyeres pteneres	0	5 (0.010)	4 (0.015)	0	A	A*
Podicipedidae	*Podiceps major*	3 (0.004)	0	0	3 (0.005)	S	A
	Rollandia rolland	+	1 (0.002)	0	1 (0.002)	S	A
Phalacrocoracidae	*Phalacrocorax atriceps*	2 (0.002)	2 (0.004)	1 (0.004)	1 (0.002)	A	A*
	Phalacrocorax brasilianus	+	3 (0.006)	0	1 (0.002)	A	A
Ardeidae	*Nycticorax nycticorax*	+	2 (0.004)	0	+	U	A/T
Threskiornithidae	*Theristicus melanopis*	5 (0.006)	2 (0.004)	0	15 (0.023)	S	T
Accipitridae	*Accipiter bicolor*	+	0	0	0	U	T
	Geranoaetus melanoleucus	0	0	0	+	U	T
Falconidae	*Caracara plancus*	9 (0.011)	19 (0.039)	17 (0.066)	8 (0.012)	A	T
	Falco sparverius	7 (0.009)	+	0	1 (0.002)	U	T
	Milvago chimango	8 (0.010)	21 (0.044)	19 (0.073)	41 (0.063)	A	T
Rallidae	*Fulica armillata*	3 (0.004)	1 (0.002)	0	0	S	A
Haematopodidae	*Haematopus leucopodus*	0	0	0	+	S	A/T*
Charadriidae	*Charadrius modestus*	0	0	0	4 (0.006)	S	A/T
	Vanellus chilensis	10 (0.012)	0	0	3 (0.005)	S	A/T
Scolopacidae	*Gallinago paraguaiae*	+	+	0	6 (0.009)	S	A/T
	Tringa melanoleuca	+	0	0	0	S	A/T
Stercorariidae	*Stercorarius chilensis*	1 (0.001)	0	0	+	U	A/T
Laridae	*Larus dominicanus*	9 (0.011)	21 (0.044)	7 (0.027)	31 (0.047)	A	A/T

APPENDIX 1. Continued.

FAMILIY	SPECIES	Nº SUMMER	Nº FALL	Nº WINTER	Nº SPRING	SEASONALITY[1]	HABITAT[2]
Laridae	*Leucophaeus scoresbii*	+	0	0	2 (0.003)	U	A/T*
	Sterna hirundinacea	6 (0.007)	0	0	+	S	A/T
Psittacidae	*Enicognathus ferrugineus*	8 (0.010)	6 (0.012)	0	+	S	T
Picidae	*Campephilus magellanicus*	2 (0.002)	1 (0.002)	1 (0.004)	5 (0.008)	A	T
Furnariidae	*Aphrastura spinicauda*	95 (0.119)	95 (0.197)	93 (0.359)	57 (0.087)	A	T
	Cinclodes fuscus	1 (0.001)	1 (0.002)	0	+	U	A/T
	Cinclodes patagonicus	3 (0.004)	0	1 (0.004)	1 (0.002)	U	A/T
	Pygarrhichas albogularis	6 (0.007)	1 (0.002)	4 (0.015)	2 (0.003)	A	T
Rhinocryptidae	*Scytalopus magellanicus*	1 (0.001)	1 (0.002)	0	+	A	T
Tyrannidae	*Anairetes parulus*	1 (0.001)	3 (0.006)	11 (0.042)	2 (0.003)	A	T
	Colorhamphus parvirostris	0	+	2 (0.008)	0	S	T
	Elaenia albiceps	45 (0.056)	0	0	5 (0.008)	S	T
	Lessonia rufa	+	0	0	11 (0.017)	S	T
	Muscisaxicola macloviana	2 (0.002)	0	0	30 (0.046)	S	T
	Xolmis pyrope	31 (0.039)	22 (0.046)	3 (0.012)	10 (0.015)	A	T
Hirundinidae	*Pygochelidon cyanoleuca*	3 (0.004)	0	0	0	S	T
	Tachycineta meyeni	102 (0.127)	0	0	+	S	T
Troglodytidae	*Cistothorus platensis*	1 (0.001)	2 (0.004)	0	4 (0.006)	S	A/T
	Troglodytes musculus	59 (0.074)	4 (0.008)	0	43 (0.066)	S	T
Turdidae	*Turdus falcklandii*	48 (0.060)	21 (0.044)	33 (0.127)	29 (0.044)	A	T
Motacilidae	*Anthus correndera*	0	0	0	+	U	T
Emberizidae	*Melanodera xanthogramma*	+	0	0	0	U	T
	Phrygilus patagonicus	50 (0.062)	44 (0.091)	25 (0.097)	41 (0.063)	A	T
	Zonotrichia capensis	61 (0.076)	0	0	112 (0.172)	S	T
Icteridae	*Curaeus curaeus*	+	0	0	10 (0.015)	A	T
	Sturnella loyca	+	0	0	1 (0.002)	S	T
Fringillidae	*Carduelis barbata*	55 (0.069)	33 (0.068)	4 (0.015)	13 (0.020)	A	T
Species number		36	32	19	37		
Individual number		801	482	259	653		

17 ANNUAL VARIATION OF ABUNDANCE AND COMPOSITION IN FOREST BIRD ASSEMBLAGES ON NAVARINO ISLAND, CAPE HORN BIOSPHERE RESERVE, CHILE

SILVINA IPPI[1,5], CHRISTOPHER B. ANDERSON[1,2], RICARDO ROZZI[1,2,3], AND CHRIS S. ELPHICK[1,4]

[1]Parque Etnobotánico Omora, Universidad de Magallanes, Puerto Williams, Chile. *E-mail*: silvippi@yahoo.com
[2]Instituto de Ecología y Biodiversidad, Casilla 653, Santiago, Chile
[3]Department of Philosophy and Religion Studies, University of North Texas, Denton, TX 76201, Texas, USA
[4]Department of Ecology and Evolutionary Biology and Center for Conservation and Biodiversity, 75 North Eagleville Rd. U-3043, University of Connecticut, Storrs, CT 06269-3042, USA
[5]Current address: Departamento de Ecología Terrestre, Facultad de Ciencias, Universidad de Chile, Las Palmeras 3425, Santiago, Chile

Ornitología Neotropical 20: 231-245 (2009)

ABSTRACT. Annual variation of abundance and composition in forest bird assemblages on Navarino Island, Cape Horn Biosphere Reserve, Chile. The structure and dynamics of avian communities in the temperate forests of southern South America have been generally studied during the breeding season (November-March), and reports about seasonal variations and migratory behavior of species are almost lacking. This study examined intra-annual variations on a monthly basis in bird species composition and abundance in the world's southernmost forested ecosystems, found in the Cape Horn Biosphere Reserve, Chile (55°S). Combining data obtained through point-count surveys, mist-netting, and checklist methods, we recorded a total of 34 bird species belonging to 20 families. Eighteen of these species were Passeriformes, and the most abundant year-round resident species were the Thorn-tailed Rayadito (*Aphrastura spinicauda*) and Patagonian Sierra-Finch (*Phrygilus patagonicus*). The most abundant seasonal migrants species were White-crested Elaenia (*Elaenia albiceps*) and Southern House Wren (*Troglodytes musculus*). Point-count and mist-netting methods showed a reduction in species richness and abundance during winter, with one third of the bird assemblage absent. This proportion was lower than that documented for more northern *Nothofagus* forests. Additionally, our data provided no evidence that intra-annual patterns of avian community structure were significantly affected by a recently implemented tourism trail through the Omora Park. We believe that understanding seasonal patterns of richness, abundance, and migratory status of Chilean temperate forests birds will help to better understand the world's southernmost forest ecosystem, implement effective conservation, facilitate sustainable ecotourism, and lead to new ecological and evolutionary research.

Key words: Cape Horn, Chile, forest birds, point-count surveys, sub-Antarctic forests.

RESUMEN. Variación intra anual de la abundancia y la composición del ensamble de aves de bosque en la isla Navarino, Reserva de la Biósfera Cabo de Hornos, Chile. La estructura y dinámica de las comunidades de aves en los bosques templados del sur de Sudamérica han sido estudiadas generalmente sólo durante la época reproductiva (noviembre-marzo); por lo tanto, los estudios sobre variaciones estacionales y de comportamiento migratorio son casi inexistentes. Este trabajo se enfoca en la variación intra anual de la abundancia y la composición del ensamble de aves forestales en los bosques más australes del mundo, en el Parque Etnobotánico Omora dentro de la Reserva de la Biósfera del Cabo de Hornos, Chile (55°S). Combinando los datos obtenidos a través de censos en estaciones de escucha, capturas con redes de niebla y listas de especies, registramos un total de 34 especies de aves pertenecientes a 20 familias. Dieciocho de éstas fueron Passeriformes y, entre las especies residentes, las más abundantes fueron el rayadito (*Aphrastura spinicauda*) y el cometocino patagónico (*Phrygilus patagonicus*). Entre las aves migratorias, las especies más abundantes fueron el fío-fío (*Elaenia albiceps*) y el chercán (*Troglodytes musculus*). Tanto la riqueza de especies como la abundancia disminuyen durante el invierno; sólo un tercio del ensamble de aves de los bosques subantárticos se ausenta durante el invierno. Esta proporción es menor que la documentada para los bosques de *Nothofagus* de menor latitud. Nuestros datos no mostraron que los patrones intra-anuales de la estructura de la comunidad de aves se vean afectados significativamente por el sendero turístico del Parque Etnobotánico Omora. El monitoreo de patrones estacionales de la riqueza, abundancia y migraciones de las aves del bosque templado chileno es necesario para comprender más integralmente los ecosistemas boscosos más australes e implementar estrategias de conservación efectivas y prácticas sustentables de ecoturismo.

Palabras clave: aves de bosque, bosque subantártico, Cabo de Hornos, Chile, censos en puntos de conteo.

INTRODUCTION

The temperate forests of southwestern South America extend from approximately 30º-56ºS along Chile and Argentina. The most Austral portion of this biome, known as the Magellanic subantarctic forest (48º-56ºS), is dominated by broadleaf trees from the genus *Nothofagus* (Armesto *et al.* 1996). Compared to the Northern Hemisphere, the avian communities that inhabit the southern temperate forests are characterized by a higher degree of endemism and relatively fewer migratory species (Rozzi *et al.* 1996a), a combination that arises from the region's biogeographic isolation and the influence of glacial events that reached their maxima 15,000 years ago (Armesto *et al.* 1996, Vuilleumier 1998). Following the general rule of decreasing species diversity with increasing latitude (Alekseev 1982, Gaston *et al.* 1995), bird species richness in the Magellanic subantarctic forests is less than at lower latitudes farther north, such as Chiloé Island (42°S) (Rozzi *et al.* 1996b, Anderson & Rozzi 2000).

Bird community structure, however, is dynamic within and among years (Holmes *et al.* 1986, Holmes 1988, Rozzi *et al.* 1996a), and may vary among habitat types (Maia-Gouvea *et al.* 2005), successional stages (Stiles 1980), or with the degree of anthropogenic disturbance (Newmark 2006). To detect and describe the fluctuations and patterns of avian diversity, long-term and year-round studies are needed (Holmes 1988). Nonetheless, most ornithological field studies are conducted over short time periods, usually in the breeding season (Karr & Freemark 1983). This pattern applies in the southern South America temperate forest biome as well, and even basic information is often lacking about seasonal patterns of bird species occurrence, abundance, and diversity.

For the breeding season, short-term studies have described bird assemblages for South American temperate forests at latitudes extending from the Cape Horn Biosphere Reserve (55–56°S) to Fray Jorge National Park (30°S) (*e.g.*, Sielfeld 1977, Venegas 1981, 1991, Jaksic & Feinsinger 1991, Willson *et al.* 1994, Rozzi *et al.* 1996b, Cofré 1999, Estades & Temple 1999, Anderson & Rozzi 2000, Anderson *et al.* 2002, Cornelius *et al.* 2000, Figueroa Rojas *et al.* 2000, Reid *et al.* 2002, Becerra Serial & Grigera 2005, Díaz 2005, Díaz *et al.* 2005). While most of these studies have been descriptive accounts of the breeding season avifauna, some point-count surveys have shown a reduction in abundance and species diversity in winter (García 1982, Sabag 1993, Cornelius 2000, Becerra Serial & Grigera 2005). On Chiloé Island (41ºS), both abundance and diversity showed a marked increase when fruits and flowers were available for the arrival of migratory birds in spring (Smith-Ramírez & Armesto 1994, Rozzi *et al.* 1996a). In the remnant forest patches at the northern end of the Chilean temperate forest's range (30ºS), which were isolated from the main expanse of temperate forest following incremental aridization during the Quaternary period (Núñez-Avila & Armesto 2006), birds exhibit much less pronounced seasonal patterns (Reid *et al.* 2002).

To investigate intra-annual variability in the structure of bird assemblages in the southernmost forests, we initiated a year-round study at the Omora Ethnobotanical Park (55ºS) on Navarino Island in the Cape Horn Biosphere Reserve. These are the world's southernmost forests, and the archipelago is considered one of the most pristine wilderness areas remaining on the planet (Mittermeier *et al.* 2002). Nonetheless, portions of Navarino Island and adjacent areas in Tierra del Fuego have been subjected to anthropogenic impacts since the late 1800s, including disturbance associated with the introduction of cattle, horses, feral dogs, cats, and the invasive North American beaver (*Castor canadensis*; Anderson *et al.* 2006a). Recently, the American mink (*Mustela vison*), an invasive exotic predator, has also extended its range from Tierra del Fuego to several smaller islands south of the Beagle Channel, including Navarino and Hoste islands, where bird assemblages evolved without significant terrestrial mammalian predators (Rozzi & Sherriffs 2003, Anderson *et al.* 2006a). As a result, the effects of mink are expected to significantly affect the reproductive success of native ground-nesting songbirds and waterbirds. Tourism is also increasing in the Cape Horn Biosphere Reserve, often without sufficient planning, which has the potential to cause negative effects on birds as seen elsewhere (Cornelius *et al.* 2001, Reid *et al.* 2002, Laiolo 2004). We believe that a better understanding of the seasonal avian community patterns will aid authorities in the management of these forests, as birds can be useful indicators of habitat disturbance (Newmark 2006). In addition, such baseline information will prove useful in the planning of long-term ecological and evolutionary research on subantarctic ecosystems. Therefore, in this study, we set out to characterize the monthly variation in forest bird assemblages over the course of a year in the Cape Horn Biosphere Reserve, Chile. We examined temporal abundance patterns, described variations in the use of different micro-habitats within the forest, and tested for an impact of tourism by comparing the bird assemblages found near and far from the interpretive trails in the Omora Ethnobotanical Park.

METHODS

Study area. This research was conducted as part of the long-term forest bird mist-netting program conducted in the Omora Ethnobotanical Park (54º57'S, 67º39'W). The data presented here were taken from November 2002 to December 2003. The Omora Park is located 3 km west from the town of Puerto Williams on the north coast of Navarino Island (Fig. 1). Precipitation in the area reaches approximately 450 mm per year, and the mean annual temperature is 6ºC, ranging from 2ºC in winter (July) to 10ºC in summer (January) (Rozzi *et al.* 2006b). Within the study area, habitat types include secondary and primary *Nothofagus* forests and anthropogenic shrubland, composed of *Berberis buxifolia*, *Chiliotrichum diffusum*, *Ribes magellanicum*, and *Embothrium coccineum*. *Nothofagus* forests can be subdivided further into evergreen stands (*N. betuloides* and *Drimys winteri*) and deciduous forests (*N. pumilio* and *N. antarctica*) (for more details see Rozzi *et al.* 2006a).

Characterizing the forest bird community. Bird assemblages were described using point-counts, mist-netting, and checklists of all birds seen and heard while working in the field. These three methods were complementary and can help to more fully describe the species composition of forest birds than could any one technique alone (Anderson & Rozzi 2000). The observer had previous experience identifying forest birds; nevertheless, pre-sample point-counts were done to standardize the protocol and

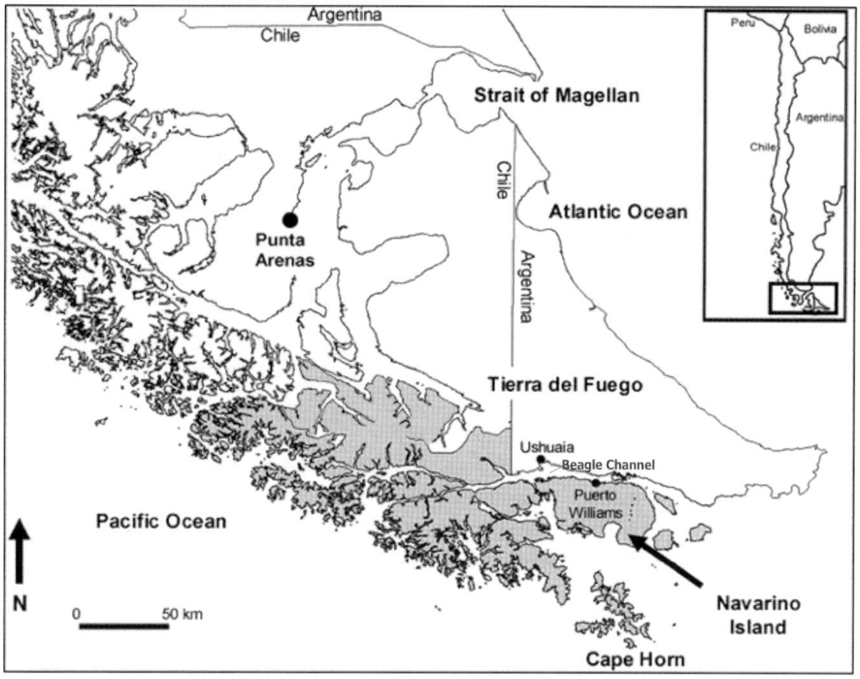

Fig. 1. Map showing the study area in the Omora Ethnobotanical Park on the north coast of Navarino Island in the Cape Horn Biosphere Reserve (shaded area), Chile (55°S).

to calculate the real time consumed in moving from one point to another.

Point-counts were conducted at 20 stations, all at least 150 m apart, and grouped into two transects. Ten stations were located along the visitors' trail that runs through the Omora Park ("near stations"), while a second transect of ten stations was located a minimum of 150 m from the trail in similar habitats ("far stations"). The appropriate radius of point-counts and general methods for birds surveys have been the subject of numerous studies (*e.g.*, Emlen 1971, Reynolds *et al.* 1980, Ralph *et al.* 1993, 1995, Jiménez 2000, Simons *et al.* 2007). In choosing our method, we considered characteristics of the habitat of Magellanic subantarctic forests to select the parameters of the survey protocol. In contrast with more northern temperate forests in Chile, such as the rainforests of Chiloé (Jiménez 2000), these forests are relatively open and unstratified with a very sparse understory and a higher proportion of deciduous trees. Therefore, Silvina Ippi conducted all counts, and recorded all birds within a radius of 50 m for 6 min at each station (Reynolds *et al.* 1980, Ralph *et al.* 1993), with counts starting upon the observer's arrival at the station (Ralph *et al.* 1995, Jiménez 2000). Surveys were conducted during the morning in the second half of each month from December 2002 to November 2003. We could not conduct surveys at six stations in April due to a severe snowstorm. Given the climate of the area, counts were sometimes conducted during light rain, with clouds and slight wind, as well as during good weather.

Three to five mist-nets were deployed monthly (6–12 m x 2.6; 3 x 3 cm mesh) at fixed locations in two habitats (44 days in forests and 46 days in shrublands/forest edge) for a total of three days per site each month from May 2003 to November 2003. During the remaining months, we sampled birds for a minimum of seven days (January = 7, February = 12, March = 13, and April = 11), except in December 2003 when there were only four capturing days. From January 2003 to September 2003, the habitat immediately around each net was classified as either interior forest, shrubland, forest edge, or open, and this subclassification of habitat was used only to evaluate habitat use by the four most captured species: Thorn-tailed Rayadito (*Aphrastura spinicauda*), Patagonian Sierra-Finch (*Phrygilus patagonicus*), White-crested Elaenia (*Elaenia albiceps*), and Black-chinned Siskin (*Carduelis barbata*). Each day, mist-nets were opened for a minimum of 5 h and checked every 15 to 30 min. Each captured bird was banded with a uniquely numbered ring (see Anderson & Rozzi 2000, Anderson *et al.* 2002), and the concerning habitat was recorded. Additionally, within the sampled habitats we recorded all birds seen or heard during visits to the Omora Ethnobotanical Park from December 2002 to December 2003. This method was also used to register the first and last sightings of migratory species.

Data analysis. The average number of individuals detected at points along the two transects were calculated for each month. Methods to correct for detection errors were not used here because their assumptions were not matched due to the study conditions (*e.g.*, distance methods; see also Elphick 2008) or logistical constraints (*e.g.*, multiple observer methods), respectively. Mean species richness for each transect was determined as the average number of species detected in each month (with points as replicates) and averaged throughout the year to compare the near and far transects. Relative abundance was estimated as the percentage of individuals of each species relative to the total number of individuals. To standardize mist-netting sample effort, we calculated the product of the net area (m^2) and the time the nets were open (h), to give the number of birds per m^2 and h for each day (Anderson & Rozzi 2000).

Point-count data (number of individuals and number of species per transect and month) were analyzed with a repeated-measures analysis of variance (ANOVA), using "near" and "far" as the between factor and months as the within subject factor. A post-hoc Tukey test for multiple comparisons was conducted on the significant results of the ANOVA (Quinn & Keough 2002). While normality was not obtained (Shapiro-Wilks test, $P < 0.05$), both homoscedasticity (Levene test) and sphericity (Mauchley sphericity test) assumptions were met. ANOVA is robust to non-normality, especially when there is a balanced design (Sokal & Rohlf 1995), so we used the parametric test. Because some stations could not be visited in April due to heavy snow, this month was eliminated from all analysis of point-count data. Mist-netting data were analyzed using the non-parametric Kruskal-Wallis and Mann-Whitney *U*-tests to compare variation in the number of species and the number of individuals among months and between forest and shrubland habitats, respectively (Siegel & Castellan 1988), as these data failed a Shapiro-Wilk test for normality and Levene test for homocedasticity ($P < 0.05$). The STATISTICA 6.0 software package was used for all analyses and significance was considered with a $P < 0.05$.

RESULTS

Subantarctic forest avian assemblage. Combining the data obtained from all methods, we recorded a total of 34 bird species from 20 families (Table 1). Eighteen species were Passeriformes, and eight were birds of prey (Falconiformes and Strigiformes). Twenty-four (71%) of these species nest and/or forage principally in the forest interior (Table 1). The most speciose families were Furnariidae (five species), Tyrannidae (four), Falconidae, Anatidae, and Strigidae (three each) (Table 1). Most families (65%) were represented in this community by only one species.

Point-count transects and the effects of tourism trail and season on community structure. During the point-count surveys we recorded 1596 birds belonging to 24 different species and 16 families (Table 1). At stations near the trail, we made 743 observations of 22 species, while 853 observations of 20 species were made at the stations farther from the trail. The mean monthly richness was 9.5 (± 2.8 SD) and 9.0 (± 1.7 SD) for the near and far transects, respectively. When data from both transects were combined, the lowest species richness was observed in March (9 species), and the highest in October and December (15 species each). Species richness varied significantly among months ($F_{10, 180} = 10.90$, $P < 0.00001$), but not between transects ($F_{1, 18} = 0.77$, $P = 0.39$) (Fig. 2), with richness peaking in midsummer (November-January; Fig. 2a). Abundance also varied significantly among months ($F_{10, 180} = 3.211$, $P = 0.0008$) with most birds in summer (October–February; Fig. 2b). Abundance at point-count stations near the tourism trail did not differ significantly from those on the transect farther away ($F_{1, 18} = 1.972$, $P = 0.177$; Fig. 2b). Moreover, there was no significant correlation between the month and trail proximity variables for either species richness ($F_{10, 180} = 0.874$, $P = 0.559$) or abundance ($F_{10, 180} = 1.260$, $P = 0.257$).

Mist-net data. During the year-round mist-netting program, we captured 639 individuals belonging to 19 species and 13 families (Table 1). In addition, there were 123 recaptures (16% of all captures). Species richness ($H = 25.0$, $P = 0.0093$) and avian abundance ($H = 29.7$, $P = 0.0018$) varied significantly among months; both were lowest during winter (June-August) and highest during the Austral summer (November-February) (Fig. 3).

We found no differences in species richness ($U = 854$, $P = 0.2021$) when comparing forest and shrubland habitats, but abundance was significantly greater in shrublands ($H = 749$, $P = 0.0337$). Mist-netting also showed that shrubland and forest edge plots had greater abundance than forest interior or open sites for the four most common captured species (Thorn-tailed Rayadito, Patagonian Sierra-Finch, White-crested Elaenia, Black-chinned Siskin) (Fig. 4).

Migratory status of subantarctic forest birds. Based on the pattern of presence and absence of each species throughout the year, we identified eight forest species (33%) as probable migrants (Table 2). Of these, Black-faced Ibis (*Theristicus melanopis*), Rufous-collared Sparrow (*Zonotrichia capensis*), Blue-and-white Swallow (*Pygochelidon cyanoleuca*), Chilean Swallow (*Tachycineta meyeni*), and White-crested Elaenia are categorized as migratory in southern Chile by Jaramillo (2003). Green-backed Firecrown (*Sephanoides sephaniodes*) and Southern House Wren (*Troglodytes musculus*) were previously classified as resident species in Navarino Island (Jaramillo 2003). However, we detected Southern House Wren only in spring and summer, and the hummingbird was recorded only in late summer. Patagonian Tyrant (*Coloramphus parvirostris*) was an uncommon species during our study, making it impossible to properly determine its migratory status. Three other species not generally considered forest species were Upland Goose (*Chloephaga picta*), Ashy-headed Goose (*Chloephaga poliocephala*), and Southern Lapwing (*Vanellus chilensis*) were found sometimes to use forested habitat, but can be regarded as migrants.

On the other hand, Fire-eyed Diucon (*Xolmis pyrope*) was present in our study area throughout the year, despite having been cited as an Austral summer migrant in the south of Chile (Jaramillo 2003) and Tierra del Fuego (54°S) (Schlatter 2004). Overall, the most abundant species were year-round residents and included Thorn-tailed Rayadito, Patagonian Sierra-Finch, and Black-chinned Siskin, which together accounted for 60% of the average total counts in the subantarctic forest avian community across the year (Table 2).

DISCUSSION

Seasonal patterns in the subantarctic bird community structure. It is well-known that species richness diminishes with latitude for numerous taxa of woody plants, invertebrates, and vertebrates (Alekseev 1982, Rosenzweig 1992, Gaston *et al.* 1995, Williams *et al.* 1997), even though the non-vascular flora in Chile provides a significant exception (Rozzi *et al.* 2008). Regarding birds in the temperate forests of South America, the predominant species are common throughout the latitudinal range of *Nothofagus* forests (33-56°S), but there is still a tendency for a decreasing number of species with increasing latitude (Anderson & Rozzi 2000). Our results also indicate that in high latitude, subantarctic forests, a higher percentage of the bird community are year-round residents

TABLE 1. Bird species recorded through mist-netting, point-count surveys, and checklists in the Omora Ethnobotanical Park from November 2002 to December 2003. Migratory status was taken from Jaramillo (2003) and complemented with data from the present study. Abbreviations: M = migrant; R = resident; [A] = scarce information, possibly a portion of the population migrate; [B] = information from bibliography.

Species	Family	Mist-net	Census	Checklist	Migratory status
Ashy-headed Goose	Anatidae		X	X	M
Upland Goose	Anatidae		X	X	M
Black-crowned Night-Heron	Ardeidae			X	R
Black-faced Ibis	Threskiornithidae		X	X	M
Chilean Hawk	Accipitridae			X	R[B]
American Kestrel	Falconidae			X	R[B]
Chimango Caracara	Falconidae	X	X	X	R
Southern Caracara	Falconidae		X	X	R
Southern Lapwing	Charadriidae		X		M
Austral Parakeet	Psittacidae		X	X	R
Barn Owl	Tytonidae			X	R[B]
Magellanic Horned Owl	Strigidae			X	R[B]
Austral Pygmy-Owl	Strigidae	X	X	X	R
Rufous-legged Owl	Strigidae			X	R[B]
Green-backed Firecrown	Trochilidae	X		X	M
Magellanic Woodpecker	Picidae	X	X	X	R
Thorn-tailed Rayadito	Furnariidae	X	X	X	R
Bar-winged Cinclodes	Furnariidae	X		X	M[B]
Grey-flanked Cinclodes	Furnariidae			X	M[B]
Dark-bellied Cinclodes	Furnariidae		X	X	R
White-throated Treerunner	Furnariidae	X	X	X	R
Magellanic Tapaculo	Rhinochryptidae	X	X	X	R
Tufted Tit-Tyrant	Tyrannidae	X	X	X	R
Patagonian Tyrant	Tyrannidae	X	X	X	R[A]
White-crested Elaenia	Tyrannidae	X	X	X	M
Fire-eyed Diucon	Tyrannidae	X	X	X	R
Blue-and-white Swallow	Hirundinidae			X	M
Chilean Swallow	Hirundinidae	X	X	X	M
Southern House Wren	Troglodytidae	X	X	X	M
Austral Thrush	Turdidae	X	X	X	R
Rufous-collared Sparrow	Emberizidae	X	X	X	M
Austral Blackbird	Icteridae	X	X	X	R
Black-chinned Siskin	Fringillidae	X	X	X	R
Patagonian Sierra-Finch	Fringillidae	X	X	X	R
	Total	**19**	**24**	**34**	

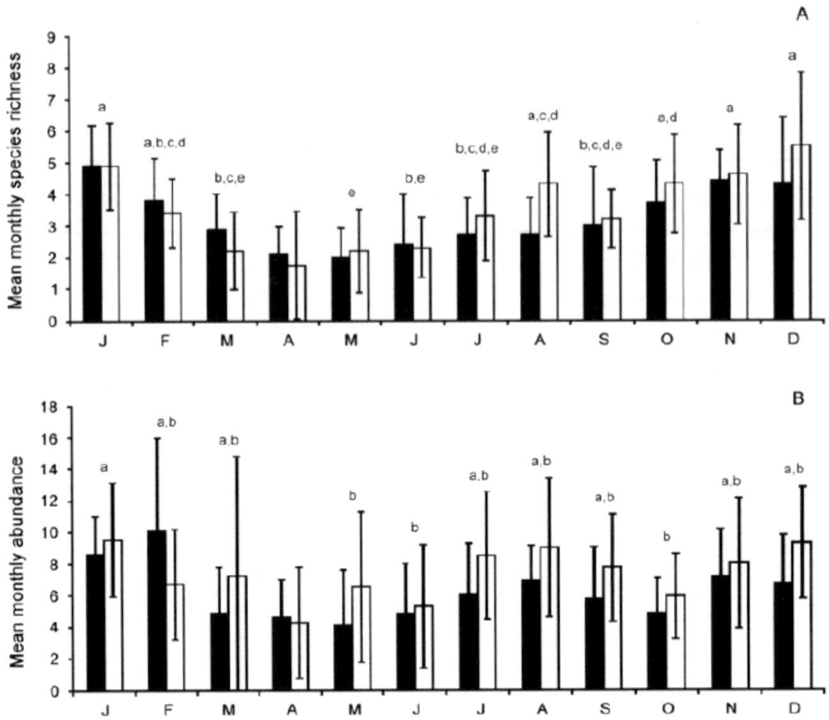

Fig. 2. Mean (± SD) monthly species richness (A) and abundance (B) of forest birds counted at stations "near" (0–5 m; black columns) and "far" (> 150 m; open columns) from the visitor's trail in the Omora Park. Point-counts were conducted from December 2002 to November 2003. In April, only four "far" stations could be visited due to inclement weather. Small letters above columns indicate significant differences (post-hoc Tukey test) in richness or abundance, respectively, among months.

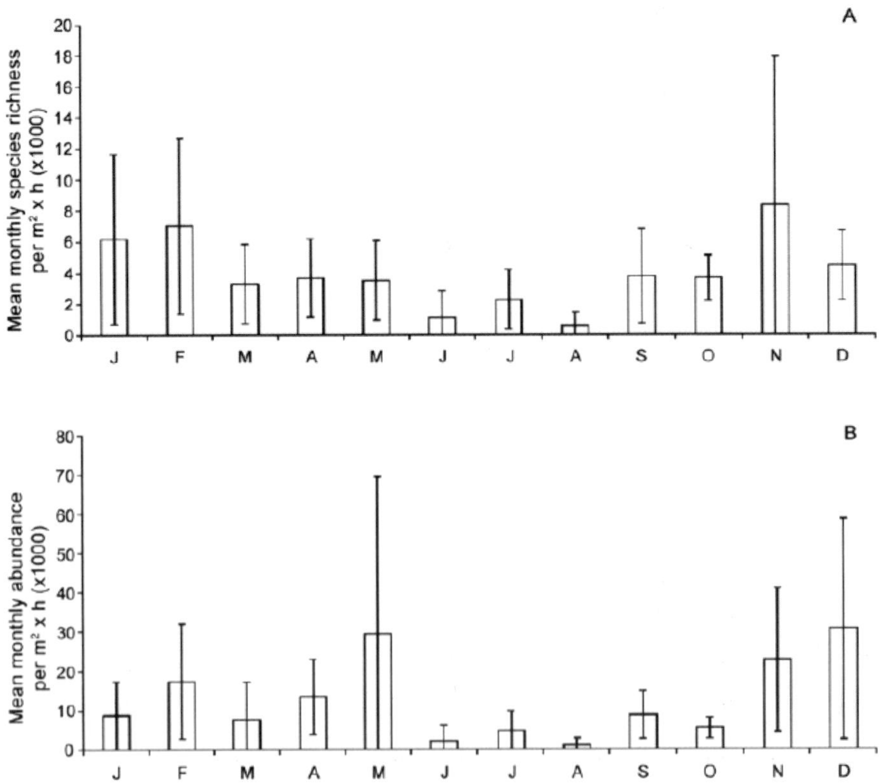

Fig. 3. Mean (±SD) monthly species richness (A) and abundance (B) of forest birds based on captures with mist-nets.

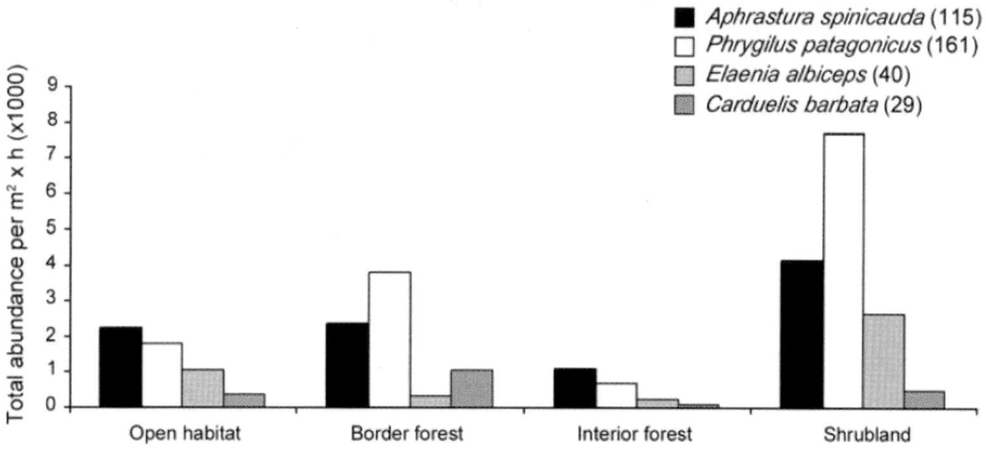

Fig. 4. Abundance of the four most captured species in four different microhabitats based on data from January 2003 to September 2003.

than in similar forests farther north. In the Cape Horn Biosphere Reserve, migratory species comprised one third of the forest bird assemblage, whereas about half of the species in forests farther north in Chile were found to be migrants (Jaksic & Feinsinger 1991). Therefore, our data suggested that not just the total number of migrants, but their proportion of the total assemblage also decreases with latitude, perhaps as a consequence of such barriers as marine channels and mountains (Henningsson & Alerstam 2005) or a more benign than expected maritime, subantarctic climate.

It was expected that during the winter avian abundance and richness would be lower than in summer, due to migration. However, there was only a moderate monthly variation throughout the year, differing from lower latitude *Nothofagus pumilio* forests (41°S) where 63% of the forest bird species were absent in winter (Becerra Serial & Grigera 2005). At the northern end of the Chilean temperate forests, in a relict forest stand at Fray Jorge National Park (30°S), many species that were absent from our study area in winter were residents (*e.g.*, Green-backed Firecrown, Southern House Wren, Rufous-collared Sparrow, Chilean Swallow) (Cornelius *et al.* 2000). Because results of the present research include only one year and only one place (*i.e.*, Omora Park), it is possible that migration of some species involves only a portion of the entire population or movements among different habitats inside the island.

While the four most abundant birds (Thorn-tailed Rayadito, Patagonian Sierra-Finch, Black-chinned Siskin, Austral Thrush *Turdus falcklandii*) in our study were resident species, two migratory species became seasonally important in summer (Southern House Wren, White-crested Elaenia). In other temperate forest sites at lower latitudes in Chile, short-term studies conducted during the summer months have found that Green-backed Firecrown and Thorn-tailed Rayadito were the most abundant species on Chiloé Island (42ºS) (Sabag 1993, Rozzi *et al.* 1996a,b), in the northernmost Chilean temperate forest at Fray Jorge (Cornelius *et al.* 2000), and at Santa Inés (32°S) (Reid *et al.* 2002). However, because recent studies have detected high levels of overestimation for bird populations with point-counts of 50 m of radius (Simons *et al.* 2007), caution is appropriate when our results are compared with those of other studies.

Compared to earlier works, our year-round study allowed us to detect a greater number of forest bird species for the Cape Horn Biosphere Reserve. While previous reports accounted for a total of 19 (Anderson & Rozzi 2000) to 20 "forest" species (Venegas 1991, 1994), excluding raptors, we surveyed a total of 34 species, including eight raptors. In our study, the additional bird species using forest ecosystems were registered only through the point-count surveys and included: Ashy-headed Goose, Upland Goose, Black-faced Ibis, Southern Lapwing, Chimango Caracara (*Milvago chimango*), Southern Caracara (*Caracara plancus*), Austral Pygmy-Owl (*Glaucidium nanum*), Dark-bellied Cinclodes (*Cinclodes patagonicus*), and Blue-and-white Swallow. Aquatic and semi-aquatic species, namely geese, lapwings, and cinclodes, are uncommon in the forest, but may use such habitats when they are close to freshwater or coastal ecosystems. This greater species richness for forest sites in our study was a direct consequence of the year-long survey that provided a better detection of species without conspicuous behavioral or song characteristics, such as owls, or of those that use the forest only occasionally or specialize in rare microhabitats (*e.g.*, freshwater edges).

Factors affecting forest bird assemblages in the Cape Horn Biosphere Reserve. The habitat with the highest avian abundance was shrubland, which differs from findings on Chiloé Island where more individuals were recorded at forest edges (Rozzi *et al.* 1996b). On Navarino Island, however, shrubland ecosystems are rather discrete habitat patches, surrounded by forests, whereas on Chiloé shrublands tend to be larger because of the more intensive history of agricultural and livestock activities on this island. As a result, the landscape of Chiloé is much more fragmented and shrublands perhaps less suitable for birds due to the longer distances between forest fragments. In Navarino, both resident (Thorn-tailed Rayadito and Patagonian Sierra-Finch) and migratory (White-crested Elaenia and Southern House Wren) species are more abundant in shrublands than in forests. However, while the Southern House Wren is also common in forests on Navarino Island, on Chiloé Island this species prefers open habitat and is not found in the forest interior (Rozzi *et al.* 1996b, Díaz *et al.* 2005, SI pers. obs.). At the

TABLE 2. Relative abundance (%) of forest bird species surveyed with point-counts in each of the seasons of the year between December 2002 and November 2003 in the Omora Ethnobotanical Park, Chile (55°S), with species listed from most to least abundant. N = number of individuals recorded.

Species	Year	Percentage (%)			
		Summer (Dec21-Mar20)	Fall (Mar21-Jun20)	Winter (Jun21-Sep20)	Spring (Sep21-Dec20)
Thorn-tailed Rayadito	28.5	19.6	30.7	40.5	26.0
Patagonian Sierra-Finch	16.9	13.4	19.7	17.5	18.6
Black-chinned Siskin	14.6	17.1	23.8	15.6	3.6
Austral Thrush	8.6	5.7	12.1	8.1	10.5
Southern House Wren	7.3	12.6	0.0	0.0	13.3
White-crested Elaenia	6.8	14.1	0.0	0.0	9.4
Rufous-collared Sparrow	3.0	1.4	0.0	3.0	7.4
Austral Parakeet	2.4	7.1	0.0	0.5	0.3
Chimango Caracara	2.3	1.4	4.5	2.7	1.5
White-throated Treerunner	1.8	1.4	2.1	3.2	0.5
Magellanic Woodpecker	1.6	0.8	2.4	2.2	1.5
Tufted Tit-Tyrant	1.4	0.2	0.7	3.5	1.3
Chilean Swallow	0.9	2.4	0.0	0.0	0.5
Black-faced Ibis	0.8	0.6	0.0	0.0	2.3
Fire-eyed Diucon	0.7	0.4	1.4	1.2	0.0
Magellanic Tapaculo	0.4	0.6	0.3	0.2	0.3
Austral Blackbird	0.4	0.0	0.7	0.0	1.0
Southern Caracara	0.3	0.0	0.0	1.0	0.3
Patagonian Tyrant	0.1	0.0	0.7	0.0	0.0
Cinclodes spp.	0.1	0.0	0.0	0.2	0.3
Dark-bellied Cinclodes	0.1	0.0	0.0	0.2	0.0
Upland Goose	0.1	0.0	0.0	0.0	0,5
Ashy-headed Goose	0.1	0.2	0.0	0.0	0.0
Austral Pygmy-Owl	0.1	0.0	0.0	0.2	0.0
Southern Lapwing	0.1	0.0	0.0	0.0	0.3
Unidentified species	0.8	1.2	1.0	0.0	0.8
Total N	**1596**	**509**	**290**	**405**	**392**

same time, White-throated Treerunner (*Pygarrhichas albogularis*) is the only passerine species that is an obligate inhabitant of the forest interior, being totally absent from shrubland and open habitats.

Regarding tourism, we did not find a significant impact of the tourist trail in the Omora Park in 2003. Species richness and abundance was equivalent near and far from the trail, throughout the year. However, given the limited scope of our analysis with only one trail studied, additional monitoring of bird assemblages will be required in order to evaluate the long-term impact of increasing tourism and plans for its anticipated growth in the archipelago.

ACKNOWLEDGMENTS

Field work for the 2003 portion of the Omora mist-netting program was financed by the BIOKONCHIL project (FKZ 01 LM 0208 BMBF). Captures were authorized by the Chilean Agriculture and Livestock Service's Department of Renewable Natural Resource Protection (Resolución N° 986, DEPROREN, Servicio Agrícola y Ganadero). S. McGehee and numerous volunteers provided valuable assistance in the field. M. Willson, R. Figueroa R., and one anonymous reviewer provided useful comments on an early draft of the manuscript. This paper is a contribution to the research activities of the Omora Ethnobotanical Park University of Magallanes (www.omora.org & www.umag.cl/ williams), a member of the Chilean Network of Long-Term Socio-Ecological Study Sites, coordinated by the Institute of Ecology and Biodiversity and financed by ICM P05-002 and PFB-23. A. Rivera-Rei assisted with statistical analyses. SI acknowledges support from a CONICYT scholarship.

REFERENCES

ALEKSEEV, V.V. 1982. Model of the number of species in an ecosystem in dependence on geographical latitude. *Ecological Modelling* 17: 107-112.

ANDERSON, C.B. & R. ROZZI. 2000. Bird assemblages in the southernmost forests in the world: Methodological variations for determining species composition. *Anales del Instituto de la Patagonia, Serie Ciencias Naturales* 28: 89-100.

ANDERSON, C.B., R. ROZZI, C. ELPHICK & S. MCGEHEE. 2002. El programa Omora de anillamiento de aves en los bosques subantárticos: la estandarización del tamaño de anillos apropiados para las aves de la región de Magallanes. *Boletín Chileno de Ornitología* 9: 2-11.

ANDERSON, C.B., C.R. GRIFFITH, A.D. ROSEMOND, R. ROZZI & O. DOLLENZ. 2006a. The effects of invasive North American beavers on riparian vegetation communities in Cape Horn, Chile: Do exotic beavers engineer differently in sub-Antarctic ecosystems? *Biological Conservation* 128: 467-474.

ARMESTO, J.J., P.L. LOBOS & M.K. ARROYO. 1996. Los bosques templados del sur de Chile y Argentina: Una isla biogeográfica. Pp. 23-28 *in Ecología de los bosques de Chile*, J.J. Armesto, C. Villagrán & M.K. Arroyo (eds.). Editorial Universitaria, Santiago, Chile.

BECERRA SERIAL, R.M. & D. GRIGERA. 2005. Dinámica estacional del ensamble de aves de un bosque norpatagónico de lenga (*Nothofagus pumilio*) y su relación con la disponibilidad de sustratos de alimentación. *Hornero* 20: 131-139.

COFRÉ, H.L. 1999. Patrones de rareza de las aves del bosque templado de Chile: implicancias para su conservación. *Boletín Chileno de Ornitología* 6: 8-16.

CORNELIUS, C., H. COFRÉ & P.A. MARQUET. 2000. Effects of habitat fragmentation on bird species in a relict temperate forest in semiarid Chile. *Conservation Biology* 14: 534-543.

CORNELIUS, C., S.A. NAVARRETE & P.A. MARQUET. 2001. Effects of human activity on the structure of coastal marine bird assemblages in Central Chile. *Conservation Biology* 15: 1396-1404.

DÍAZ, I.A. 2005. Historia natural, diversidad y conservación de las aves en bosques de la Cordillera de la Costa de la región de Los Lagos, Chile. Pp. 456-467 *in Historia, Biodiversidad y Ecología de los Bosques Costeros de Chile*, C. Smith-Ramírez, J.J. Armesto & C. Valdovinos (eds.). Editorial Universitaria, Santiago, Chile.

DÍAZ, I.A., J.J. ARMESTO, S. REID, K.E. SIEVING & M.F. WILLSON. 2005. Linking forest structure and composition: Avian diversity in successional forests of Chiloé Island, Chile. *Biological Conservation* 123: 91-101.

ELPHICK, C.S. 2008. How you count counts: the importance of methods research in applied ecology. *Journal of Applied Ecology* 45: 1313-1320.

EMLEN, J.T. 1971. Population densities of birds derived from transect counts. *Auk* 88: 323-342.

ESTADES, C.F. & S.A. TEMPLE. 1999. Deciduous forest bird communities in a fragmented landscape dominated by exotic pine plantations. *Ecological Applications* 9: 573-585.

FIGUEROA ROJAS, R.A., C. BRAVO VIVANCO, E.S. CORALES STAPPUNG, R. LÓPEZ RÜBCKE & S. ALVARADO ORELLANA. 2000. Avifauna del santuario de la naturaleza Los Huemules del Niblinto, región del Bío-Bío, Chile. *Boletín Chileno Ornitología* 7: 2-12.

GARCÍA, J.A. 1982. *Comunidad Avifaunística del Delta del Río Gol-Gol, una Necesidad de Conservación.* Undergraduate thesis, Universidad Austral de Chile, Valdivia, Chile.

GASTON, K., P. WILLIAMS, P. EGGLETON & C. HUMPHRIES. 1995. Large scale patterns of biodiversity: spatial variation in family richness. *Proceedings of the Royal Society of London B* 260: 149-164.

HENNINGSSON, S.S. & T. ALERSTAM. 2005. Barriers and distances as determinants for the evolution of bird migration links: the Arctic shorebird system. *Proceedings of the Royal Society of London B* 272: 2251-2258.

HOLMES, R.T. 1988. Community structure, population fluctuations, and resource dynamics of birds in temperate deciduous forests. Pp. 1318-1327 *in Acta XIX Congressus Internationalis Ornithologici. Natl. Mus. Nat. Sci.*, H. Ouellet (ed.). Ottawa, Ontario, 1986.

HOLMES, R.T., T.W. SHERRY & F.W. STURGES. 1986. Bird community dynamics in a temperate deciduous forest: long-term trends at Hubbard Brook. *Ecological Monographs* 56: 201-220.

KARR, J.R. & K.E. FREEMARK. 1983. Habitat selection and environmental gradients: dynamics in the "stable" tropics. *Ecology* 64: 1481-1494.

JARAMILLO, A. 2003. *Birds of Chile.* Princeton Univ. Press, Princeton, New Jersey.

JAKSIC, F.M. & P. FEINSINGER. 1991. Bird assemblages in temperate forests of North and South America: a comparison of diversity, dynamics, guild structure, and resource use. *Revista Chilena de Historia Natural* 64: 491-510.

JIMÉNEZ, J.E. 2000. Effect of sample size, plot size, and counting time on estimates of avian diversity and abundance in a Chilean rainforest. *Journal of Field Ornithology* 71: 66-87.

LAIOLO, P. 2004. Diversity and structure of the bird community overwintering in the Himalayan subalpine zone: is conservation compatible with tourism? *Biological Conservation* 115: 251-262.

MAIA-GOUVEA, E.R., E. GOUVEA & A. PIRATELLI. 2005. Understory bird community in a surrounded area of Itatiaia National Park, Rio de Janeiro, Brazil. *Revista Brasileira de Zoologia* 22: 859-866.

MITTERMEIER R., C. MITTERMEIER, P. ROBLES-GIL, J. PILGRIM, G. FONSECA, J. BROOKS & J. KONSTANT. 2002. *Wilderness: Earth's Last Wild Places.* Cemex and Conservation International, Washington, D.C.

NEWMARK, W.D. 2006. A 16-year study of forest disturbance and understory bird community structure and composition in Tanzania. *Conservation Biology* 20: 122-134.

NÚÑEZ-AVILA, M. & J.J. ARMESTO. 2006. Relict island of the temperate rainforest tree *Aextoxicon punctatum* (Aextoxicaceae) in semi-arid Chile: genetic diversity and biogeographic history. *Australian Journal of Botany* 54: 733-743.

QUINN, G.P. & M.J. KEOUGH. 2002. *Experimental Design and Data Analysis for Biologists*. Cambridge Univ. Press, Cambridge, Massachusetts.

RALPH, C.J., G.R. GEUPEL, P. PYLE, T.E. MARTIN & D.F. DE SANTE. 1993. *Handbook of Field Methods for Monitoring Landbirds*. G. Tech. Rep. PSW-GTR-144. Pacific Southwest Research Station, Forest Service, U.S. Department of Agriculture, Albany, California.

RALPH, C.J., S. DROEGE & J.R. SAUER. 1995. Managing and monitoring birds using point-counts: Standards and applications. Pp. 161-168 *in Monitoring Bird Populations by Point Counts*, C.J. Ralph, J.R. Sauer & S. Droege (eds.). Gen. Tech. Rep. PSW-TR-149. Pacific Southwest Research Station, Forest Service, U.S. Department of Agriculture, Albany, California.

REID, S., C. CORNELIUS, O. BARBOSA, C. MEYNARD, C. SILVA-GARCÍA & P.A. MARQUET. 2002. Conservation of temperate forest birds in Chile: implications from the study of an isolated forest relict. *Biodiversity & Conservation* 11: 1975-1990.

REYNOLDS, R.T., J.M. SCOTT & R.A. NUSSBAUM. 1980. A variable circular-plot method for estimating bird numbers. *Condor* 82: 309-313.

ROSENZWEIG, M. 1992. Species diversity gradients: we know more and less than we thought. *Journal of Mammalogy* 73: 715-730.

ROZZI, R. & M.F. SHERRIFFS. 2003. El visón (*Mustela vison* Schreber, Carnivora: Mustelidae), un nuevo mamífero exótico para la isla Navarino. *Anales del Instituto de la Patagonia, Serie Ciencias Naturales* 31: 97-104.

ROZZI, R., D. MARTÍNEZ, M.F. WILLSON & C. SABAG. 1996a. Avifauna de los bosques templados de Sudamérica. Pp. 135-151 *in Ecología de los Bosques de Chile*, J.J. Armesto, C. Villagrán & M.K. Arroyo (eds.). Editorial Universitaria, Santiago, Chile.

ROZZI, R., J.J. ARMESTO, A. CORREA, J.C. TORRES-MURA & M. SALLABERRY. 1996b. Avifauna de bosques primarios templados en islas deshabitadas del archipiélago de Chiloé, Chile. *Revista Chilena de Historia Natural* 69: 125-139.

ROZZI, R., F. MASSARDO, C.B ANDERSON, K. HEINDINGER & J.A. SILANDER, JR. 2006a. Ten principles for biocultural conservation at the southern tip of the Americas: the approach of the Omora Ethnobotanical Park. *Ecology & Society* 11: 43. Downloaded on 10 May 2008 from http://www.ecologyandsociety.org/vol11/iss1/art43/

ROZZI, R., F. MASSARDO, C.B. ANDERSON, A. BERGHOEFER, A. MANSILLA, M. MANSILLA, J. PLANA, U. BERGHOEFER, E. BARROS & P. ARAYA. 2006b. *The Cape Horn Biosphere Reserve*. Ediciones de la Universidad de Magallanes, Punta Arenas, Chile.

ROZZI, R., J.J. ARMESTO, B. GOFFINET, W. BUCK, F. MASSARDO, J. SILANDER, M.T.K. ARROYO, S. RUSSELL, C.B. ANDERSON, L.A. CAVIERES & J.B. CALLICOT. 2008. Changing lenses to assess biodiversity: patterns of species richness in sub-Antarctic plants and implications for global conservation. *Frontiers in Ecology and the Environment* 6: 131-137.

SABAG, C.E. 1993. *El Rol de las Aves en la Dispersión de Semillas en un Bosque Templado Secundario de Chiloé (42°S)*. M.Sc. thesis, Univ. de Chile, Santiago, Chile.

SCHLATTER, R.P. 2004. Fauna de turberas de la XII región y Tierra del Fuego. Pp. 107-112 *in Los Turbales de la Patagonia*, D.E. Blanco & V.M. de la Balze (eds.). Bases para su inventario y la conservación de la biodiversidad. Wetland International, Buenos Aires, Argentina.

SIEGEL, C. & N.J. CASTELLAN, JR. 1988. *Nonparametric statistics for the behavioral sciences*. McGraw-Hill International Editions, New York, New York.

SIELFELD, W.K. 1977. Reconocimiento macrofaunístico terrestre en el área de Seno Ponsonby (Isla Hoste). *Anales del Instituto de la Patagonia, Serie Ciencias Naturales* 8: 275-296.

SIMONS, T.R., M.W. ALLDREDGE, K.H. POLLOCK & J.M. WETTROTH. 2007. Experimental analysis of the auditory detection process on avian point counts. *Auk* 124: 986-999.

SMITH-RAMÍREZ, C. & J.J. ARMESTO. 1994. Flowering and fruiting patterns in the temperate rainforest of Chiloé, Chile ecologies and climatic contraints. *Journal of Ecology* 82: 353-365.

SOKAL, R.R. & F.J. ROHLF. 1995. *Biometry. The Principles and Practice of Statistics in Biological Research*. W.H. Freeman and Company, New York, New York.

STILES, E.W. 1980. Bird community structure in older forests in Washington. *Condor* 82: 20-30.

VENEGAS, C. 1981. Aves de las islas Wollaston y Bayly, archipiélago del Cabo de Hornos. *Anales del Instituto de la Patagonia, Serie Ciencias Naturales* 12: 213-219.

VENEGAS, C. 1991. Ensambles avifaunísticos estivales del Archipiélago Cabo de Hornos. *Anales del Instituto de la Patagonia, Serie Ciencias Naturales* 20: 69-82.

VENEGAS, C. 1994. *Aves de Magallanes*. Ediciones de la Universidad de Magallanes, Punta Arenas, Chile.

VUILLEUMIER, F. 1998. Avian biodiversity in forest and steppe communities of Chilean Fuego-Patagonia. *Anales del Instituto de la Patagonia, Serie Ciencias Naturales* 26: 41-57.

WILLIAMS, P.H., K.J. GASTON & C.J. HUMPHRIES. 1997. Mapping biodiversity value worldwide: combining higher-taxon richness from different groups. *Proceedings of the Royal Society of London B* 264: 141-148.

WILLSON, M.F., T.L. DE SANTO, C. SABAG & J.J. ARMESTO. 1994. Avian communities of fragmented south-temperate rainforests in Chile. *Conservation Biology* 8: 508-520.

Accepted 14 April 2009.

18

DIVERSITY AND SINGULARITY OF THE AVIFAUNA IN THE AUSTRAL PEAT BOGS OF THE CAPE HORN BIOSPHERE RESERVE, CHILE

José Tomás Ibarra[1,2*], Christopher B. Anderson[2], Tomás A. Altamirano[1], Ricardo Rozzi[2,3], and Cristián Bonacic[1]

[1]Laboratorio Fauna Australis, Programa de Recursos Naturales, Facultad de Agronomía e Ingeniería Forestal, Pontificia Universidad Católica de Chile, Casilla 306-22, Santiago, Chile
[2]Parque Etnobotánico Omora (Instituto de Ecología y Biodiversidad, Universidad de Magallanes), Reserva de Biosfera Cabo de Hornos, Puerto Williams, Chile
[3]Department of Philosophy and Religion Studies, University of North Texas, Denton, Texas, USA
*Corresponding author: jtibarra@uc.cl

Ciencia e Investigación Agraria 37: 29-43 (2010)

ABSTRACT. Diversity and singularity of the avifauna in the austral peat bogs of the Cape Horn Biosphere Reserve, Chile. *Sphagnum*-dominated peat bogs that are strongly embedded within the southern temperate forest matrix are increasingly being used for agriculture. Nevertheless, little is known about their biodiversity. Moreover, the remote areas of southern Chile where peat bogs are found, such as the Cape Horn Biosphere Reserve (CHBR, 54-55°S), where birds are the most diverse and best represented group of vertebrates, have not been well-investigated. With the aim to broaden this knowledge in the CHBR, we studied the diversity of the avian assemblage in peat bogs on Navarino Island. We compared the composition of avian species between wetlands with and without peat bogs to test if *Sphagnum* bogs represented a singular habitat for birds in this area. Furthermore, the 37 bird species recorded in these habitats were classified according to guild structure. The community similarity values showed that peat bogs hosted a bird composition that was different from that present in wetlands without *Sphagnum*, suggesting that peat bogs are a singular type of habitat for birds in the CHBR. The most frequently feeding groups recorded in these wetlands were insectivores (48.7%), followed by omnivores (23.1%). Our results showed that, in contrast to previous studies of birds in peat bogs, these environments constituted a distinct wetland habitat for feeding, reproduction, and sheltering for some species in the CHBR. Thus, plans for the conservation and rational use of peat ecosystems should consider the high value of these habitats for biodiversity on a landscape scale, especially for birds of the southernmost extreme of the Americas.

Key words: Assemblage, Cape Horn Biosphere Reserve, Navarino Island, Patagonia, *Sphagnum* spp., wetlands.

RESUMEN. Diversidad y singularidad de la avifauna en turberas esfagnosas australes de la Reserva de Biosfera Cabo de Hornos, Chile. Las turberas esfagnosas que se encuentran fuertemente embebidas en la matriz de bosques templados australes, se utilizan de forma creciente en agricultura. Sin embargo, el conocimiento de su biodiversidad es escaso. Más aún, zonas remotas del sur de Chile donde se encuentran las turberas, como la Reserva de Biosfera Cabo de Hornos (RBCH, 54-55°S), donde las aves constituyen el grupo de vertebrados más diverso y representativo, aún han sido escasamente estudiadas. Con el objetivo de ampliar este conocimiento en la RBCH, se estudió la diversidad del ensamble de aves en turberas esfagnosas de la isla Navarino. Comparamos la composición de especies de aves entre humedales con y sin turberas para evaluar si las turberas esfagnosas representan un hábitat singular para las aves en esta área. Se clasificó a las 37 especies de aves registradas en estos hábitats de acuerdo a la estructura de gremios. Los valores de similitud comunitario mostraron que las turberas albergaban una composición de aves que era diferente de aquella presente en humedales sin *Sphagnum*, sugiriendo que este es un tipo de hábitat singular para las aves en la RBCH. Los grupos tróficos más frecuentemente registrados fueron los insectívoros (48,7%), seguidos por los omnívoros (23,1%). Nuestros resultados muestran que, a diferencia de lo señalado en trabajos previos sobre aves de turberas, estos ambientes constituyen un hábitat de humedal singular para la alimentación, reproducción y refugio de algunas especies en la RBCH. De esta manera, los planes para la conservación y uso racional de los ecosistemas de turberas deben considerar el alto valor de estos hábitats para la biodiversidad a escala de paisaje, especialmente para las aves del extremo sur americano.

Palabras clave: ensamble, humedales, isla Navarino, Patagonia, Reserva de Biosfera Cabo de Hornos, *Sphagnum* spp.

Not registered in *Sphagnum* bogs but in other wetland types:

Anatidae	*Anas platalea*
	Chloephaga hybrida
	Tachyeres pteneres
Podicipedidae	*Podiceps major*
	Rollandia rolland
Phalacrocoracidae	*Phalacrocorax atriceps*
	Phalacrocorax brasilianus
Accipitridae	*Accipiter bicolor*
	Geranoaetus melanoleucus
Rallidae	*Fulica armillata*
Haematopodidae	*Haematopus leucopodus*
Scolopacidae	*Tringa melanoleuca*
Stercorariidae	*Stercorarius chilensis*
Laridae	*Leucophaeus scoresbii*
	Sterna hirundinacea
Hirundinidae	*Pygochelidon cyanoleuca*
Motacillidae	*Anthus correndera*
Emberizidae	*Melanodera xanthogramma*
Icteridae	*Sturnella loyca*

+, indicates presence, but out of the systematic sampling.

In relation to the categories of activity, birds were observed flying in 40% of the cases. *Anas sibilatrix*, *Caracara plancus*, *Carduelis barbata*, *Enicognathus ferrugineus*, and *L. dominicanus* were the species that presented the highest proportion of birds performing this activity, reaching 74% of the records. The second most represented category, in relative terms, was the sound record (23%). *Colorhamphus parvirostris*, *Elaenia albiceps*, *Pygarrhichas albogularis*, and *Scytalopus magellanicus* were detected in high proportions by means of their vocalization (>74%). Resting (15%) was the third most represented category: *A. georgica*, *Curaeus curaeus*, and *T. patachonicus* showed this activity in more than 63% of the cases. Two species, *Aphrastura spinicauda* and *T. patachonicus*, often showed an alarm behavior (>32%). *Campephilus magellanicus* (38%), *Cinclodes patagonicus* (33%), *L. specularioides* (50%), and *V. chilensis* (50%) were the four species with the highest dedication to feeding. Only one species, *Chloephaga poliocephala* (33%), was observed fighting. Finally, only *Cistothorus platensis* (20%) presented courting behavior during the sampling (Table 4).

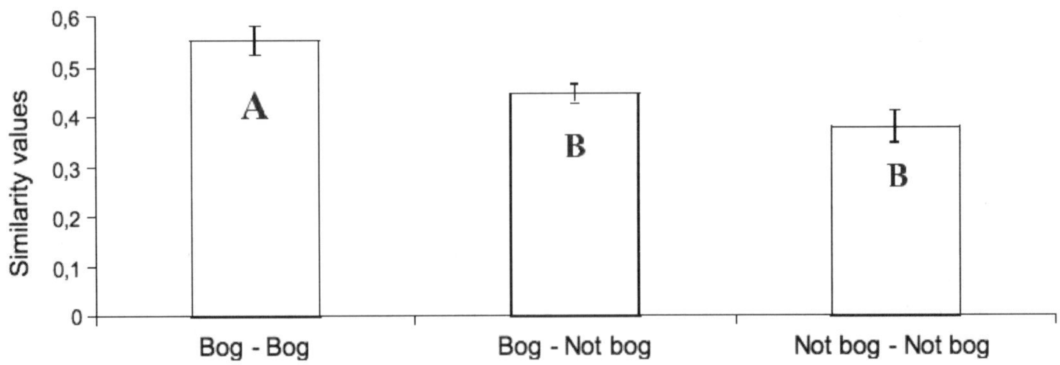

Fig. 2. Similarity values (Jaccard Index) for the bird community compositions in wetlands with and without *Sphagnum*-dominated peat bogs. Bars with distinct letters were significantly different according to the post-hoc analysis of Tukey ($F_{2,42}$=9.028; p = 0.0005). Small bars are ± standard deviation.

TABLE 3. Habitat, feeding habits and microhabitat for birds in peat bogs of the Cape Horn Biosphere Reserve.

Species	Habitat[1]	Feeding habits[2]	Micro-habitat
Anas flavirostris	A	O	Open waters
Anas georgica	A	O	Open waters
Anas sibilatrix	A	H	Open waters
Chloephaga picta	A/T	H	Shores/Open waters
Chloephaga poliocephala	A/T	H	Shores/Open waters
Lophonetta specularioides	A	O	Open waters
Tachyeres patachonicus	A	O	Open waters
Nycticorax nycticorax	A/T	C	Shores/Coasts
Theristicus melanopis	T	I	
Caracara plancus	T	C	
Falco sparverius	T	C-I	
Milvago chimango	T	C-I	
Charadrius modestus	A/T	O	Shores/Coasts
Vanellus chilensis	A/T	O	Shores
Gallinago paraguaiae	A/T	I	Rushes
Larus dominicanus	A/T	O	Shores/Coasts/Open waters
Enicognathus ferrugineus	T	H	
Campephilus magellanicus	T	I	
Aphrastura spinicauda	T	I	
Cinclodes fuscus	A/T	I	Shores
Cinclodes patagonicus	A/T	I	Shores/Coasts
Pygarrhichas albogularis	T	I	
Scytalopus magellanicus	T	I	
Anairetes parulus	T	I	
Colorhamphus parvirostris	T	I	
Elaenia albiceps	T	I	
Lessonia rufa	T	I	
Muscisaxicola macloviana	T	I	
Xolmis pyrope	T	I	
Tachycineta meyeni	T	I	
Cistothorus platensis	A/T	I	Rushes
Troglodytes musculus	T	I	
Turdus falcklandii	T	O	
Phrygilus patagonicus	T	H	
Zonotrichia capensis	T	H	
Curaeus curaeus	T	O	
Carduelis barbata	T	H	

[1] A: aquatic, T: terrestrial.

[2] C: carnivores, H: herbivores (including granivores), I: insectivores and other invertebrates, O: omnivores.

TABLA 4. Relative frequency of activity categories for each bird species recorded in peat bogs of the Cape Horn Biosphere Reserve.

Species	n	Fighting	Feeding	Flying	Alarming	Courting	Resting or swimming	Vocalizing record	Total
Anairetes parulus	14		0.18	0.36				0.45	1
Anas flavirostris	22			0.47			0.53		1
Anas georgica	31			0.36			0.64		1
Anas sibilatrix	2			1.00					1
Aphrastura spinicauda	197		0.10	0.17	0.47			0.26	1
Campephilus magellanicus	8		0.38	0.50				0.13	1
Caracara plancus	17			0.75			0.19	0.06	1
Carduelis barbata	58		0.04	0.76			0.08	0.12	1
Charadrius modestus	4			0.50			0.50		1
Chloephaga picta	62		0.05	0.55			0.40		1
Chloephaga poliochepala	2	0.33		0.33			0.33		1
*Cinclodes fuscus**	0								0
Cinclodes patagonicus	4		0.33	0.50			0.17		1
Cistothorus platensis	3		0.20	0.40		0.20	0,.20		1
Colorhamphus parvirostris	2							1.00	1
Curaeus curaeus	3						1.00		1
Elaenia albiceps	30			0.09				0.91	1
Enicognathus ferrugineus	3			1.00					1
Falco sparverius	3			0.67			0.33		1
Gallinago paraguaiae	2			0.50			0.50		1
Larus dominicanus	4			1.00					1
*Lessonia rufa**	0								0
Lophonetta specularioides	5		0.50				0.50		1
Milvago chimango	34			0.71			0.14	0.14	1
*Muscisaxicola macloviana**	0								0
*Nycticorax nycticorax**	0								0
Phrygilus patagonicus	68		0.07	0.66	0.03			0.24	1
Pygarrhichas albogularis	7		0.25					0.75	1
Scytalopus magellanicus	2							1.00	1
Tachycineta meyeni	42		0.19	0.71			0.05	0.05	1
Tachyeres patachonicus	4				0.33		0.67		1
Theristicus melanopis	4			0.67			0.33		1
Troglodytes musculus	48		0.08	0.15	0.21		0.03	0.54	1
Turdus falcklandii	88		0.02	0.34	0.27		0.23	0.14	1
Vanellus chilensis	10		0.50	0.25	0.25				1
Xolmis pyrope	38		0.06	0.44			0.25	0.25	1
Zonotrichia capensis	73		0.21	0.34			0.25	0.21	1
Total number of birds	894								

*Species recorded out of systematic counts.

DISCUSSION

This work provides new information on the natural history, diversity and community structure of the birds that inhabit *Sphagnum*-dominated peat bogs in the CHBR. Our results show that, unlike previous works on the avifauna present in peat bogs in other zones in Patagonia (Sielfeld 1977, Venegas 1976, 1981, 1991, Guzmán *et al.* 1986, Venegas & Sielfeld 1998, Blanco *et al.* 2004, and Schlatter 2004), these environments shelter a significant portion of the avian diversity found on Navarino Island in the Cape Horn Biosphere Reserve. Our records of species richness (N = 37) in peat bogs is much higher than those obtained in studies conducted farther north in the Magallanes Region (*e.g.*, Venegas 1976, N = 3 species; Guzmán *et al.* 1986, N = 6 species; Blanco *et al.* 2004, N = 6 species, and Schlatter 2004, N = 25 species). In general, the previously mentioned studies consisted of specific samplings of a few days during one season (except Schlatter 2004), which could explain the lower numbers of species recorded, and in turn highlight the value of long-term research programs (Anderson *et al.* 2008). Furthermore, ecotone zones were considered in our study (*e.g.*, bog-forest, bog-shrub, and bog-grassland), and freshwater birds were included; both of these aspects could have influenced the high number of species observed in this study.

The increasing exploitation and consecutive isolation of peat lands have been reported as main factors affecting bird diversity at the landscape scale (Calmé & Desrochers 2000). In contrast, in the CHBR the mosaic of different habitat types represents a very rare example of the non-fragmented areas in the world (Rozzi *et al.* 2006). This characteristic of the study area, in addition to our results showing a higher diversity of birds in peat bogs within the CHBR than in other nearby areas, suggests that the importance of peat bog ecosystems for a particular area is probably contingent on the matrix of habitats found in the broader landscape. Consequently, in the CHBR, where peat bogs are more abundant and connected with other heterogenic and undisturbed habitats, this environment could occupy a more prominent role in determining the landscape-level species assemblages of avifauna. Therefore, landscape measures are necessary, but probably not sufficient, to describe the patterns of bird diversity found in the CHBR.

Based on our findings, we put forward that *Sphagnum*-dominated peat bogs represent a unique wetland habitat for a significant portion of the avian community in the CHBR. This position is supported by the fact that the assemblages found in these peat bogs are highly divergent from those of other wetland types (Similarity index, Fig. 2). At the same time, the species composition within peat bogs tends to be very uniform (*i.e.*, displaying a high degree of similarity among these peat bogs). This result could be interpreted at least in two different ways: (1) when the rest of the wetlands are grouped as "of other type" their intrinsic variability is not being considered, as the different types of associated habitats (shrubs, rushes, forest, grassland, etc.) should be considered; this information would in turn define different types of assemblages that do not necessarily need to be similar; or (2) that peat bogs actually tend to be more similar to each other than do "other types" of wetlands in terms of their specific composition. Consequently, and regardless the first alternative, our study indicates that peat bogs represent a distinct habitat within the sub-Antarctic landscape matrix whose biodiversity is insufficiently studied and has previously been unappreciated. However, the underlying mechanisms explaining the biodiversity present in peat bogs, and the organization and processes involved, still need to be further studied (Arroyo *et al.* 2005).

The bird assemblage that used peat bogs and their ecotones included mainly insectivores (*e.g.*, *G. paraguaiae*) but also species from a complete suite of other feeding guilds, including omnivores (*e.g.*, *C. modestus*), herbivores (*e.g.*, *C. picta*), granivores (*e.g.*, *Phrygilus patagonicus*), and carnivores (*e.g.*, *Nycticorax nycticorax*). Community composition is known to be strongly related to food availability (Cody 1981), a factor that might be relevant for birds in the *Sphagnum* wetlands as well (Guzmán *et al.* 1986). This finding also agrees with what has been reported by Schiavini (2000), who associates the presence of birds in peat bogs mainly to food searching.

In general, in peat bog systems small birds (insectivores) are predominant; these species must feed frequently and, therefore, must have a food source that allows them to feed at regular intervals (Griffiths *et al.* 1982, Guzmán *et al.* 1986). The presence of a high proportion of small birds in peat bogs may indicate that these environments have a reliable food source, at least for the predominant insectivorous species. These species form the dominant feeding guild in sub-Antarctic wetlands during the coldest months, when wetlands are completely frozen (Ibarra *et al.* 2009b). In the case of the herbivorous species (excluding the granivorous species such as *P. patagonicus* or *C. barbata*) and omnivores species, fundamentally the Anatidae, food is frequently obtained from the open waters or shores where there might be a higher production of food resources, in contrast to the fact that Magellanic peat bogs have been considered to be a system of low productivity where plant species present scarce palatability (Pisano 1977).

In the case of reproduction, Matteazzi (1997) and Schiavini (2000) pointed out that the peat bog-forest ecotone is a nesting site for non-passeriform species such as *C. picta* and *G. paraguaiae*. *C. picta* usually nests in inland peat bog wetlands in the CHBR (Ibarra *et al* 2009b), while *G. paraguaiae* was commonly recorded in peat bogs during the spring and summer seasons, in twilight, performing nuptial flights with a characteristic buzz that is made with the tail. This reproductive behavior is characteristic for this Scolopacidae (Rottmann 1995). The results for the activity categories also allowed for the identification of *C. platensis* while performing courting activity and *C. poliocephala* while fighting for a territory during the breeding season. Likewise, the records of *C. magellanicus* were obtained in a transition bog-forest, even though this species is always associated with the presence of old-growth trees. This Picidae is usually found in the old-growth forests of Cape Horn (Arango *et al.* 2007), which constitute the main matrix of *Sphagnum*-dominated peat bogs in the Magallanes (Arroyo *et al.* 2005).

In light of the fact that, currently, *Sphagnum*-dominated peat bogs are under a growing pressure for extraction in southern South America, conservation-oriented policies and policies for sustainable use should consider their high value for feeding, reproduction and sheltering for an important diversity of birds. According to Arroyo *et al.* (2005), peat bogs and forests are intrinsically linked at the ecosystem level in austral South America; therefore, the destiny for peat bogs will determine the future of Southern temperate forests, and vice versa.

ACKNOWLEDGMENTS

This work was financed by the project FPA 12-001-06 of CONAMA, granted to the Omora Foundation, and by Wildlife Trust Alliance. The Institute of Ecology and Biodiversity (IEB) also provided financial contributions through grant ICM P05-002, and the Puerto Williams Branch-UMAG provided logistic support to the team. Special thanks to A. Barreau, E. Firmani, G. González and S. McGehee for their support in the field. The comments by B. Latorre helped to improve preliminary versions of the manuscript. J.T. Ibarra and T.A. Altamirano are CONICYT research grant holders. This study is part of the long-term programs of conservation and research of the Omora Ethnobotanical Park, the scientific research center of the Cape Horn Biosphere Reserve and the founding member of Chile's long-term socio-ecological research network (www.ieb-chile.cl/ltser).

VENEGAS, C. & W. SIELFELD. 1998. *Catálogo de los Vertebrados de la Región de Magallanes y Antártica Chilena*. Ediciones de la Universidad de Magallanes, Punta Arenas, Chile. 122 pp.

ZAR, J.H. 1984. *Biostatistical Analysis*. 2nd edition. Prentice Hall. NJ, USA. 71 pp.

PART IV

THE MINK (*NEOVISON VISON*), THE IMPACT OF A CARNIVORE ON THE BIRDS OF THE CAPE HORN BIOSPHERE RESERVE

19 INTRODUCTION TO THE AMERICAN MINK (*NEOVISON VISON*): IMPACT OF AN INVASIVE CARNIVORE ON BIRDS IN ONE OF THE MOST "PRISTINE" ZONES OF THE PLANET

José Tomás Ibarra and Christopher B. Anderson

The increasing spread of organisms by humans across natural barriers that once seemed insurmountable, such as oceans, mountain ranges, deserts, rivers, and inhospitable climate zones, has become a global phenomenon with both ecological and socio-cultural causes and consequences at local scale. Even places relatively free of other human impacts have not escaped the introduction, naturalization, and spread of alien species. For example, from the long-term research implemented in the Cape Horn archipelago with the establishment of Omora Park, since 2000, we have recorded that this remote southern part of the American continent has also seen the arrival and establishment of many alien species (Anderson *et al.* 2006). In turn, the introduction of this new flora and fauna would be a reflection of an "exotic mentality" that prioritizes and often values more non-native species, primarily of European or American origin (Rozzi *et al.* 2003). Thus, biotic homogenization would be in part, a consequence of cultural homogenization.

The new findings of alien species in the southernmost part of the American continent spark reflection about the role, both academic and social, that the sites of long-term research such as the Omora Park can play (Anderson *et al.* 2010). If we evaluate the claim that the sub-Antarctic forest is one of the most pristine ecoregions of the planet, we find that this "recognition" can be dated to a publication in the prestigious journal *Proceedings of the National Academy of Sciences* in 2003. A group of researchers from Conservation International, based in New York, stated that this ecoregion, along with 23 other areas, was one of the last "wilderness areas" of the planet. This identification was based on three criteria that can be measured with remote sensing and demographic information, which are: i) a high percentage of intact vegetation cover, ii) low human density, and iii) a large area of the ecoregion (Mittermeier *et al.* 2003). The authors also highlighted the high percentage of protected areas in the region (> 50% of its total area).

However, this study did not use knowledge based on "the local," such as the history of colonization and human occupation from about 7,000 years ago, or the historical and current extractive uses of these forests and their associated environments, such as fishing in the marine area. Nor did it mention that protected areas in the south of America lack management plans, rangers, transport and infrastructure to access, monitor, and protect these parks that are among the largest and most remote of Latin America. Thus, there is clearly a need to complement the global analyses with long-term research in situ. With the latter approach, Omora Park researchers were able to determine, for example, that islands in parts of the archipelago are full of introduced species. Even today, the number of species of exotic mammals is greater than the number of natives (Anderson *et al.* 2006). This "discovery" has stimulated not only research, but also support decision-making in the implementation of management plans and control of damaging wildlife, that are all pioneers in Chile, in partnership with other academic and government institutions.

The critical analysis of how we know and, therefore, how we prioritize some areas for the conservation of the world, does not intend to underestimate the idea that the sub-Antarctic ecoregion is one of the world's most important reservoirs by being relatively little affected by the modern industrial society. What we want to emphasize, on the contrary, is the importance of context and complement the global analysis with local information: both scientific knowledge, as well as historically and locally traditional.

Today, the integration of the local and the global is essential as processes of biotic and cultural homogenization operate at multiple scales. That's why within their lines of socio-ecological research, Omora Park has prioritized long-term study of introduced species as a phenomenon that links spatial and temporal scales with its social and ecological dimensions.

Within the assembly of introduced mammals, the North American mink (*Neovison vison*) is of particular interest, as it is the latest exotic mammal that has arrived at the Cape Horn Biosphere Reserve (CHBR) (Rozzi & Sherriffs 2003). Thus, this new biological invasion offers the opportunity to understand the response mechanisms of native biodiversity, especially birds, to the arrival of a new predator. The avifauna of this island system has at least two attributes with serious implications for the interaction Mink-birds: 1. Birds are the most species rich and abundant group of terrestrial vertebrates in the area, as amphibians and reptiles are not represented, and native mammals are only ten species (Anderson *et al.* 2006); 2. They have evolved almost free of carnivorous terrestrial vertebrates (with the exception of the Culpeo fox, *Lycalopex culpaeus lycoides*, in some areas of Hoste Island and Tierra del Fuego), which would have generated that several species nest at or near ground level, being then even more vulnerable to predation by this mustelid (Maley *et al.* 2011).

discuss the implications of its expansion on the conservation of austral biodiversity.

Background of the American mink in Magallanes and the austral zone of Argentina

American minks were introduced to Chile for the first in Punta Arenas time during the years 1934 and 1936. Mr. Edmundo Pisano B. imported them from the United States of America and maintained a breeding center until 1950, when he closed the breeding facility due to low profitability (Sandoval 1994). Part of the stock was killed to sell the skins and the rest was sold to a new breeding facility that functioned until 1970. Between 1968 and 1970, Mr. Hernán Pisano and other partners attempted to start a new breeding center in Punta Arenas, but it failed (Jaksic *et al.* 2002). Sandoval (1994) offers a historic account of the introduction of the American mink that occurred north of Magallanes in the region of Aysén and later in Los Lagos Region. Recently, in the Aysén Region, the American mink has become a plague and 223 individuals have been captured only in the areas south of Coyhaique during fourth months (SAG & CIR La Senda Ltda. 1999).

Parallel to its introduction in Chile, the American mink was imported by commercial breeders to various provinces of southern Argentina since the 1930s (Lizarralde 2000). In 1959 there were more than 60 mink breeding centers in southern Argentina, which ultimately led to the establishment of wild populations due to escapes. Lizarralde (2000) specifies that during the 1940s and 1950s in Tierra del Fuego various breeding centers that were established and closed released individuals into the Irigoyen River. Until the end of the 1980s, its presence was only detected on the Mitre Peninsula, but since 1990 sightings of the American mink have become more common in several areas of the large island of Tierra del Fuego (Massoia & Chébez 1993, Lizarralde 2000).

Background of the American mink in Cape Horn County

The first record of *M. vison* in Cape Horn County corresponds to a specimen captured 2 km northwest of the city of Puerto Williams by Jorge Quelín (see Appendix 1), who on the 21st of November 2001 brought the specimen to the governor's office of the Chilean Antarctic Province where the individual was determined by Ricardo Rozzi. Mr. Quelín brought the specimen saying it was a "strange otter," which suggests that the American mink corresponds to an unknown species for the people of this area or the animal went unspotted until that date. For this reason we conducted interviews with Mr. Quelín and 10 people older than 40 of the Yagán indigenous community, born in diverse sectors of Cape Horn County. All remember the arrival of other exotic species, especially the muskrat (*Ondatra zibethica*) and the North American Beaver (*Castor canadensis*). Even one of them participated in the muskrat fur trade. However, none of them remembered having seen an American mink on Navarino Island or other islands prior to the capture in 2001. Even more, there were no recorded bite marks on dogs like those that had been seen during previous years, nor American mink skeletons. Only Germán González Calderón expressed doubts regarding a "strange otter" specimen observed in port Eugenia and in a lagoon near the airport of Puerto Williams in 1993. On the other hand, two recent zoological expeditions (but prior to 1990) in Cape Horn do not mention the American mink; in contrast, they highlight the presence of *Castor canadensis* and *Ondatra zibethica* (Siefeld 1977, Siefeld & Venegas 1980). Therefore, *M. vison* would have arrived on Navarino Island in the 1990s, or at the beginning of the current decade: it is the exotic mammal species that has arrived most recently to Cape Horn County.

After the detection of the American mink on Navarino Island, the community was requested to report any sighting or capture of the mammal in the County to the scientists of Omora Ethnobotanical Park and/or to the Agriculture and Livestock Bureau office. Locations of sightings or captures were often visited to verify the presence of the Mink. With this method information was gathered concerning 30 individuals captured or observed in diverse sectors and environmental types on Navarino Island and, probably, also on Hoste Island (Appendix 1). From these 30 individuals, 3 specimens were donated to the zoological collection of the Patagonian Institute, the governor's office of the Antarctic Province, and to the collection of the Omora Ethnobotanical Park.

Capture and sighting data of the American mink (Appendix 1) suggest a rapid population growth on Navarino Island (Fig. 1). Even more, they would have expanded their distribution range (Fig. 2). Mauricio Zárraga points out a probable sighting in Orange Bay south of Hoste Island. According to the data collected, the American mink uses a large variety of habitats.The coastal (35.5%) and riparian (32.3%) environments are the most frequent. In sections of riparian forests of Evergreen Beech (*Nothofagus betuloides*), long dens of the Mink were found (>3m), but were superficial and visible as bumps in the soil. The American mink was also found in forests (16.1%) and interior lagoons (9.7%) (Fig. 3). Likewise, 25.8% of the American minks detected were associated with rural houses in Puerto Williams.

Regarding information collected on the American mink's diet, one specimen was captured while hunting a Magellanic Penguin (*Spheniscus magellanicus*) on the coast of Guerrico (54°55'S, 67°54'W) in the north of Navarino Island. Two individuals captured in houses located in the mouth of the Lum River (54°56'S, 68°03'W) had been attracted by poultry. One specimen captured in Mejillones Bay (54°55'S, 68°01'W) in December of 2002 had destroyed the nest of at least two pairs of Upland Goose (*Chloephaga picta*). Finally, near the dens of *M. vison* in the forests of *N. betuloides* in the Omora Ethnobotanical Park (54°57'S, 67°39'W), bones and other fish remains pertaining to unknown species were found.

Figure 2 shows the detection sites of the American mink on Navarino Island, including the record in Orange Bay, Hoste Island. The map also indicates the locations where SAG has installed the first traps to capture the mink. These traps consist of PVC pipes 1 m in length with a 16 cm diameter, installed at a 50° angle with bait at the end. This Geographic Information System map, like the Mink searching system, corresponds to a Mink control program that SAG has initiated in collaboration with the provincial government, the NGO Omora and the University of Magallanes, which requires immediate expansion to control this new exotic invasive species.

Exotic and Native Mammal Species in Cape Horn County

Currently, the number of exotic mammal species present in Cape Horn County surpasses the number of native mammals. Table 1 presents

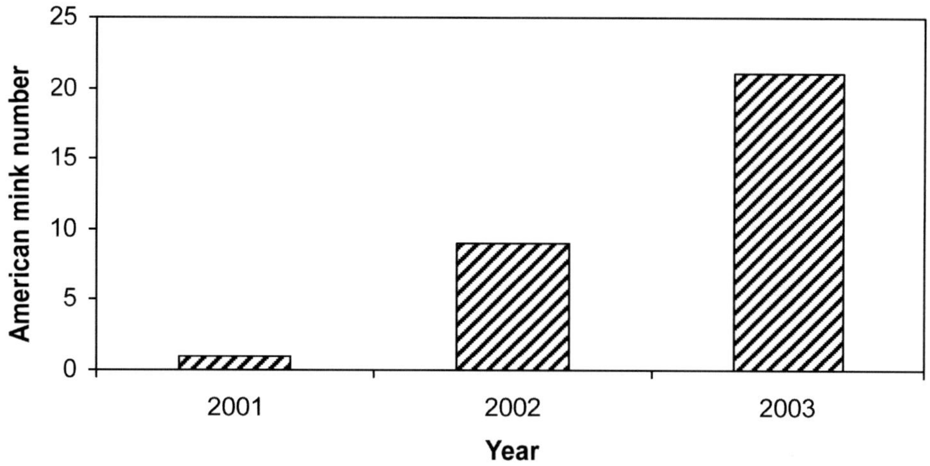

Fig. 1. The number of American minks (*Mustela vison*) captured or observed on Navarino Island and other sites in Cape Horn County between November 2001 and May 2003. The method and intensity of recording is identical during all periods considered.

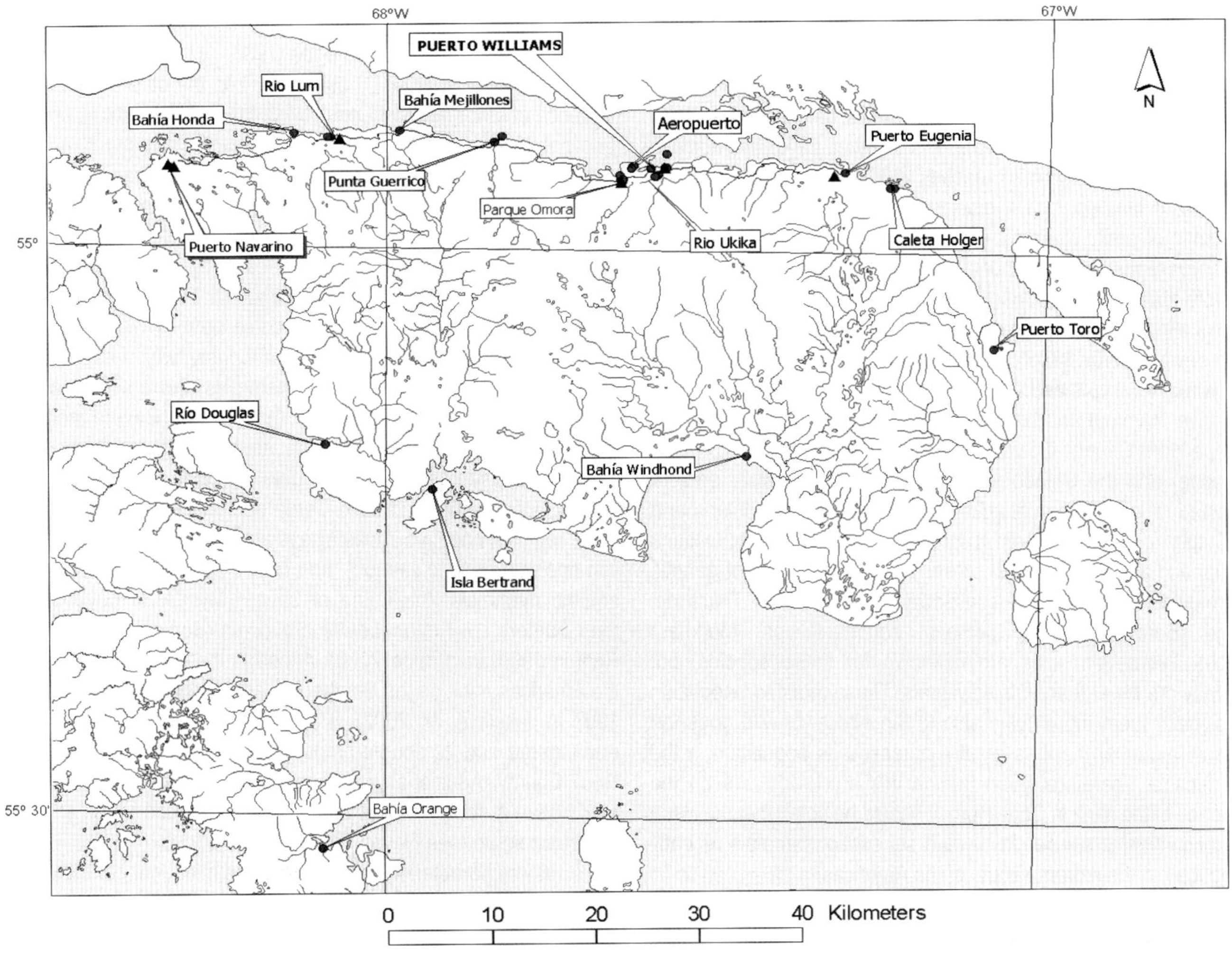

Fig. 2. Map indicating the locations of sightings or captures of the American mink (*Mustela vison*) (black circles) on Navarino Island, including the record in Orange Bay, Hoste Island. We also indicate the locations where SAG has installed the first traps to capture the American mink (black triangles).

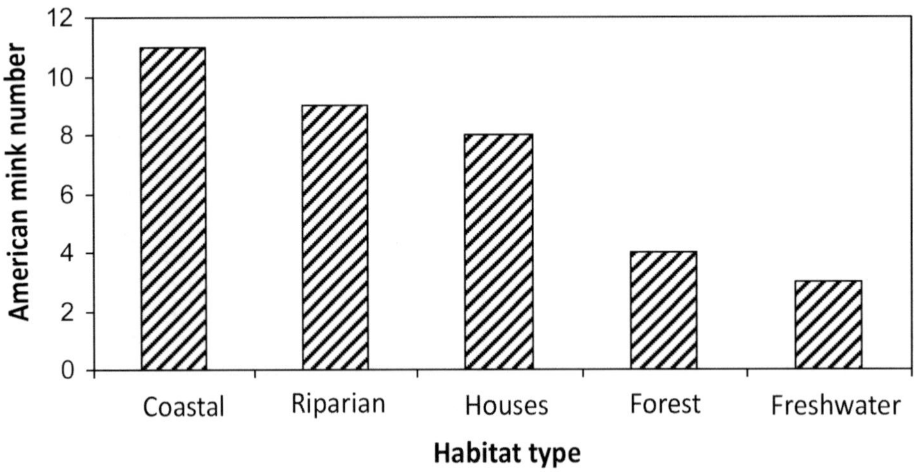

Fig. 3. The number of American minks (*Mustela vison*) captured or observed in diverse habitat types on Navarino Island, between November 2001 and May 2003. The sampling effort is similar in all habitat types.

the species of native and exotic terrestrial mammals recorded in the Omora Ethnobotanical Park's database and bibliographical sources. In Cape Horn County, 7 species of native terrestrial mammals exist pertaining to 7 genera, 4 families, and 4 orders (Artiodactyla, Carnivora, Chiroptera, Rodentia). On the other hand, the number of exotic terrestrial mammals reaches at least 9 species, including 9 genera, 8 families, and 4 orders. Therefore, in Cape Horn the taxonomic diversity of exotic mammals is greater than the native diversity in terms of the number of species, genera, and families.

Note that Table 1 includes 7 species of exotic mammals whose presence in Cape Horn has not been previously published. Among these newly recorded exotic species, there are two other species of carnivores: feral cats and dogs, which could have a severe impact on native mammal and bird species. Regarding native mammals in Cape Horn County, the two largest species, the guanaco (*Lama guanicoe*) and the Culpeo fox (*Pseudalopex culpaeus*), present restricted ranges. The guanacos only inhabit Navarino Island and the Culpeo fox is only present on Hoste Island. Both the populations of guanaco and culpeo could present a certain degree of genetic differentiation compared to conspecific populations in Tierra del Fuego (Benito González personal communication, Markham 1971b). Population sizes are unknown for these species, but currently are lower than 500 individuals. These reduced population sizes make them more vulnerable to the effects of wild dogs, an invasive mammal that has greatly increased its population in the last decade. Therefore, even though this article focuses on the presence of the mink in Cape Horn, it should be noted that the issue of exotic invasive species involves a set exotic mammals recently introduced to the extreme south of the Americas.

Implications of the Presence of the American mink in Cape Horn on the Conservation of Austral Biodiversity

Among the invasive exotic species found in Cape Horn, the beaver is the most well-known because its abundance and impact on nearly every forested watershed on Navarino Island has been studied by scientists (Sielfeld & Venegas 1980, Lizarralde & Venegas 2001, Briones *et al.* 2001). Less known, on the other hand, is the presence of other exotic mammals like the American mink, and the report of its presence should alert the scientific community and the authorities responsible for environmental protection in Chile.

The American mink possesses many characteristic attributes of invasive exotic species: a high reproductive rate, a generalist in habitat and/or diet, a high dispersion capacity, and an association with human beings (Lodge 1993, Sandoval 1994). The American mink is a voracious predator that could have a highly negative impact on the avifauna in Cape Horn County, which has evolved in absence of native carnivorous mammals and presents a high number of species that nest on the ground. Severe negative impacts of the American mink on aquatic avifauna have been reported in Aysén and southern Argentina (Lizarralde 2000, reports by SAG-Coyhaique, and Julio Cerda personal communication). In these regions, negative impacts have also been reported on poultry, fish (including exotic salmon), and on the population of two native aquatic mammals: the nutria or *coypu* (*Myocastor coypus*) and the Southern river otter (*Lontra provocax*) (Bonino 1995, SAG-XI Region 1999, Lizarralde 2000). Although the degree of impact of the American mink on the Southern river otter is debatable (Medina 1997 in Jaksic *et al.* 2002), it should be noted that this otter is endangered and its southernmost populations are found in Cape Horn County. Additionally, in the near future the American mink could have a strong impact on the native avifauna in the extreme American south (see Navas 1991).

Another characteristic of the American mink that contributes to its invasive capacity corresponds to its ability to swim across large bodies of water. In Aysen it has colonized Magdalena Island, and recently has arrived to the Guaitecas Islands, crossing distances greater than 20 km (SAG-XI Region, Julio Cerda personal communication). This swimming capacity could potentially explain its arrival on Navarino Island from Tierra del Fuego. The distances

TABLE 1. Native and exotic terrestrial mammal species present in the Cape Horn County.

	Order	Family	Species	Common Name	Location	Habitat	References
NATIVE	Artiodactyla	Camelidae	*Lama guanicoe*	Guanaco	Navarino Island	Open habitat	0,5
	Carnivora	Canidae	*Pseudalopex culpaeus*	Culpeo fox	Hoste Island	Coastal and forest	0,4,8
	Chiroptera	Vesper-tilionidae	*Histiotus montanus*	Small Long-eared bat	Navarino Island (Omora Park, Wulaia, Bertrand, Pto. Toro) and Grevy Island	Forest	0,1,2
			Myotis chiloensis	Southern Mouse-eared bat	Navarino Island (Douglas River and Omora Park) and Wollaston	Forest	0,1,2
	Rodentia	Cricetidae	*Euneomys chinchilloides*	Silky Mouse	Hermite Island, Bahía Orange	Open habitat	3
			*Oligoryzomys magellanicus**	Long-tailed Magellanic rat	Navarino Island (Omora Park), Hoste Island and Cape Horn	Forest	0,3,4
			Abrothrix xanthorhinus	Yellow-nosed mouse	Navarino Island (Omora Park), Hoste Island	All	0,4
	Total	4	7				
EXOTIC	Lagomorpha	Leporidae	*Oryctolagus cuniculus*	European rabbit	Navarino Island	Open habitat	0
	Carnivora	Canidae	*Canis lupus familiaris*	Feral dog	Navarino Island	Open habitat	0
		Mustelidae	*Mustela vison*	American mink	Navarino Island	All	0
		Felidae	*Felis silvestris* &	Feral cat	Navarino Island	All	0
	Artiodactyla	Suidae	*Sus scrofa*	Feral pig	Navarino Island	All	0
	Rodentia	Castoridae	*Castor canadensis*	American beaver	Navarino, Gable and Hoste Islands	Freshwater	0,5,9
		Cricetidae	*Ondatra zibethica*	Musk rat	Navarino and Hoste Islands	Freshwater	0,4,6,7
		Muridae	*Rattus norvegicus*	Norway rat	Navarino and Hoste Islands	All	0
			Mus musculus	House mouse	Navarino Island (P. Williams)	All	0
	Total	8	9				

* Confirmation of the identity of this species being analyzed at the Museo de Historia Natural de Chile, since it could be *Oligoryzomys longicaudatus* (Long-tailed mouse). & Also called *Felis domesticus*
0 = Omora database BIOKONCHIL, 1 = Olrog 1950, 2 = Peña & García 1972, 3 = Thomas 1890, 4 = Sielfeld 1977, 5 = Massoia & Chébez 1993, 6 = Pine *et al.* 1978, 7 = Markham 1971a, 8 = Markham 1971b, 9 = Sielfeld & Venegas 1980.

that separate both islands along the Beagle Canal are less than 5 km at numerous points, with less than 2 km of separation at Gable Island (Table 2). The point of Tierra del Fuego closest to Navarino Island corresponds to the southernmost part of Gable Island, only 1.4 km from Chilean territory. In turn, Gable Island is only 0.4 km from Tierra del Fuego. This would be the shortest means of arrival. The shortest distance between Navarino Island and Hoste Island is found in Murray Channel, with 0.4 km separating the coasts. The shortest distance between Navarino Island and Picton Island is 4.6 km located between Puerto Toro (Navarino Island) and Caleta Piedras Cove (Picton Island). The two closest river mouths between Tierra del Fuego and Navarino Island correspond to the Róbalo River (Navarino) and the Almanza River (T. del Fuego), a distance of 6.2 km. These short distances could explain why the American mink has colonized Navarino Island, probably via Tierra del Fuego. At the same time, the short distances between islands warn of the

TABLE 2. Minimum distances between Navarino Island sites and Tierra del Fuego sites and other islands, at narrow points.

Navarino Island	Tierra del Fuego or another island	Distance (km)
Punta Oriental (Navarino)	Punta Southwest, Isla Hoste	0.4
Gable Island (closest point)	Tierra del Fuego, closest point	0.4
Gable Island (closest point)	Tierra Mayor	1.0
Closest Chilean point	Gable Island	1.4
Caleta Mitchell	Hoste Island (Caleta Hahn)	1.6
Punta Trucco	Gable Island	1.8
Caleta Mejillones	Punta San Juan	4.6
Puerto Toro	Picton Island (Caleta Piedras)	4.6
Punta Guerrico	Closest point to Tierra del Fuego	4.8
Punta Guerrico	Punta Remolino	5.6
Punta Gusano	Puerto Almanza	5.6
Río Róbalo mouth	Río Almanza mouth	6.2
Punta Ganado (Picton)	Closest Argentinean point	7.0
Puerto Williams	Almanza	7.6
Puerto Navarino	Punta Ushuaia (closest point)	8.5
Puerto Navarino	Ushuaia	13.2

immediate danger of the spread of this exotic mammal to Hoste Island and even the Archipelago of the Wollaston Islands, which belong to the Cape Horn National Park whose mission is to protect the biota at the southern end of the American continent.

Lastly, the American mink possesses a high reproductive capacity. They mate at the end of the winter or at the beginning of spring and produce an offspring of 2 to 10 youth, with an average litter of 5 to 6 individuals (Fur Commission 2002). Juveniles begin to hunt at 5 months and reach reproductive maturity at 10 months of age. These characteristics permit a rapid population growth, where both adults and juveniles represent aggressive, generalist predators. In North America, their native territory, the mink population is controlled principally by wolves, coyotes and bobcats (Fur Commission 2002). Conversely, the predators of the mink are absent at the southern end of the Americas. This absence of predators in Cape Horn County could facilitate its population expansion, as it has been the case for the demographic explosion of the American beaver (Lizarralde 2000).

The American mink already has a long history as an invasive exotic species around the world. It was introduced in many parts of Europe as a source of fur, which largely resembles the case in southern South America. Escapes from farms and intentional release in Russia in the 1930s and 40s led to its naturalization in many European countries, where it has caused a population decrease of the European Mink, a related native species that is now considered the European carnivore with the greatest risk of extinction (Fur Commission 2002).

Currently large amounts of money are spent to control invasive species (Primack *et al.* 2001). The reduction of the rate of invasive species is a priority of conservation efforts, requiring effective and cheap control options. However, once invasive species are introduced, the initiation of control and eradication programs is crucial. The sooner such programs are started upon the arrival of invasive species, the greater the probability of success and the lower the cost. The detection of the American mink on Navarino Island is an alarm for austral biodiversity conservation, due to two main reasons: i) this voracious exotic carnivore has had a high impact on the avifauna in the regions of Aysen and Los Lagos in Chile (Sandoval 1994); ii) the avifauna of Navarino Island has evolved in absence of native terrestrial carnivorous mammals (Rozzi 2001). The presence of the American mink on Navarino Island has motivated the rapid alert of the national and regional SAG, and a campaign for the detection and elimination of the mink in collaboration with the local population. However, it is necessary to reiterate that both research and control programs of this voracious invasive species need to be reinforced. For successful control, we must take advantage of its early detection.

Before concluding we want to point out that the life of every being and every biological species has a value. The problem of the American mink, and of exotic species in general, does not lie in its exotic nature *per se*. The problem arises when exotic species generate a process of ecological invasion that eliminates local biota and accelerates the anthropogenic biotic homogenization of the planet, even at the most remote areas of the word, such as Cape Horn (Rozzi *et al.* 2000).

ACKNOWLEDGMENTS

To Mr. Agustín Iriarte, Mr. Carlos Rowland, Mr. Nicolás Soto, and Mr. Cristián Osorio of the Agriculture and Livestock Bureau, with whom it has been possible to alert the region and nation of the presence of the American mink on Navarino Island. To Mr. Eduardo Barros, Governor of the Chilean Antarctic Province, who began this alert program, as part of his government program marking that values environmental sustainability in the extreme south of Chile. To Mr. Jorge Quelín, provincial government official, for his continued cooperation on field work. To Mr. Germán González, Mr. Mauricio Zárraga, and other members of the Yahgan Indigenous Community that collaborated in the capture and detection of individuals in Cape Horn. To Francisca Massardo, Christopher Anderson, Oliver Vogel, and Steven McGehee, scientists of the NGO Omora that helped while conducting field work. María Rosa Gallardo (CERE-University of Magallanes) who prepared the GIS map of Figure 2. The authors appreciate the valuable commentaries on the manuscript by Jaime Jiménez and an anonymous reviewer. Margaret Sherriffs appreciates the support of the Gordon Fellowship-Yale University and Ricardo Rozzi appreciates the BIOKONCHIL project (German Ministry of Science and Education, BMBF, FKZ 01 LM 0208) and the Investigation Program (247) of the University of Magallanes "Spatial Support Systems for Decision Making." This is a contribution of the research program of the Omora Ethnobotanical Park, Chilean Antarctic Province.

REFERENCES

BONINO, N.A. 1995. Introduced mammals in Patagonia, southern Argentina: consequences, problems, and management considerations. Pp. 406-409 *in Proceedings of the First International Wildlife Management Congress*, J.A. Bissonette & P.R. Krausman (eds.). The Wildlife Society, Maryland, USA.

BRIONES, M., R. SCHLATTER, A. WOLODARSKY & C. VENEGAS. 2001. Clasificación ambiental para habitats de *Castor canadensis* (Kuhl 1820, Rodentia), de acuerdo a características de cuencas en un sector de Tierra del Fuego. *Anales Instituto de la Patagonia* 29: 75-93.

CREE, A., C.H. DAUGHERTY & J.M. HAY. 1995. Reproduction of a rare New Zealand reptile, the tuatara *(Sphenodon punctatus)*, on rat-free and rat-inhabited islands. *Conservation Biology* 9: 373-384.

DRAKE, J.A. *et al.* (EDS.). 1989. *Biological Invasions: A Global Perspective*, SCOPE Report No. 37. John Wiley, Nueva York.

ELTON, C.S. 1958. *The Ecology of Invasions*. John Wiley, Nueva York. Fur Commission 2002. Mink Biology.

HEDGEPETH, J.W. 1993. Foreign invaders. *Science* 261: 34-35.

FUR COMMISSION. 2002. www.furcommission.com

JAFFE, M. 1994. *And no Bird Sing*. Simon and Schuster, Nueva York.

JAKSIC, F., A. IRIARTE, J. JIMÉNEZ & D. MARTÍNEZ. 2002. Invaders without frontiers: cross-border invasions of exotic mammals. *Biological Invasions* 4: 153-173.

LIZARRALDE, M. 1993. Current status of the introduced beaver (*Castor canadensis*) population in Tierra del Fuego, Argentina. *Ambio* 22: 351-358.

LIZARRALDE, M. 2000. Mamíferos exóticos en Tierra del Fuego. *Ciencia Hoy* 10 (56): 21-28.

LIZARRALDE, M., G. DEFERRARI, S. ALVAREZ & J. ESCOBAR. 1996. Effects of beaver (*Castor canadensis*) on the nutrient dynamics of the Southern Beech forests of Tierra del Fuego (Argentina). *Ecología Austral* 6: 101-105.

LIZARRALDE, M. & C. VENEGAS. 2001. El castor: un ingeniero exótico en las tierras más australes del planeta. Pp. 231-233 *in Fundamentos de Conservación Biológica: Perspectivas Latinoamericanas*, R. Primack, R. Rozzi, P. Feinsinger, R. Dirzo & F. Massardo (eds.). Fondo de Cultura Económica, México.

LODGE, D.M. 1993. Species invasions and deletions: Community effects and responses to climate and habitat changes. Pp. 367-387 *in Biotic Interactions and Global Change*, P.M. Kareiva, J.G. Kingsolver & R.B. Huey (eds.). Sinauer Associates, Sunderland, Massachusetts, U.S.

MARKHAM, B.J. 1971a. *Catálogo de los Anfibios, Reptiles, Aves y Mamíferos de la Provincia de Magallanes (Chile)*. Publicación del Instituto de la Patagonia. Punta Arenas.

MARKHAM, B.J. 1971b. Presencia de *Dusicyon culpaeus* en Isla Hoste. *Anales del Instituto de la Patagonia* 8: 275-297.

MASSOIA, E. & J.C. CHEBEZ. 1993. *Mamíferos Silvestres del Archipiélago Fueguino*. L.O.L.A. Buenos Aires, Argentina.

MITTERMEIER, R.A. *Et al.* 2002. *Wilderness: Earth's Last Wild Places*. CEMEX-Conservation International, Washington DC.

NAVAS, J.R. 1991. Los vertebrados exóticos introducidos en Argentina. *Revista del Museo Argentino de Ciencias Naturales, Zoología* 14(2): 7-37.

OLROG, C. 1950. Notas sobre mamíferos y aves del archipiélago Cabo de Hornos. *Acta Zoológica Lilloana* (Tucumán) 9: 505-532.

PEÑA, L. & G. GARCÍA. 1972. Presencia de *Histiotus montanus magellanicus* Philp. y de *Myotis ch. chiloensis* Waterhouse (Chiroptera) al sur del Estrecho de Magallanes. *Anales del Museo de Historia Natural de Valparaíso* (5), Chile.

PINE, R., P. ANGLE & D. BRIDGE. 1978. Mammals from the sea, mainland and islands at the southern tip of South America. *Mammalia* 42: 105-114.

PRIMACK, R., R. ROZZI, P. FEINSINGER, R. DIRZO & F. MASSARDO. 2001. *Fundamentos de Conservación Biológica: Perspectivas Latinoamericanas*. Fondo de Cultura Económica, México.

ROZZI, R., J. SILANDER, J.J. ARMESTO, P. FEINSINGER & F. MASSARDO. 2000. Three levels of integrating ecology with the conservation of South American temperate forests: The initiative of the Institute of Ecological Research Chiloé, Chile. *Biodiversity and Conservation* 9: 1199-1217.

ROZZI, R. 2001. *Biological and Cultural Conservation in the Archipelago Forest Ecosystems of Southern Chile. Ph.D. Dissertation*. Department of Ecology and Evolutionary Biology, University of Connecticut, USA.

ROZZI, R., F. MASSARDO, J. SILANDER, JR., C. ANDERSON & A. MARIN. 2003a. Conservación biocultural y ética ambiental en el extremo austral de América: oportunidades y dificultades para el bienestar ecosocial. Pp. 51-85 *in Biodiversidad y Globalización*, E. Figueroa & J. Simonetti (eds.). Editorial Universitaria, Santiago, Chile.

ROZZI, R., F. MASSARDO, J. SILANDER, JR., C. ANDERSON, O. DOLLENZ & A. MARIN. 2003b. El Parque Etnobotánico Omora: una alianza público-privada para la conservación biocultural en el confín del mundo. *Ambiente y Desarrollo* 19 (1): 43-55.

SAG-XI REGIÓN .1999. *El Visón (*Mustela vison*)*. Cartilla de Servicio Agrícola y Ganadero, Coyhaique, Chile.

SAG & CIR LA SENDA LTDA. 1999. *Informe Final Proyecto Investigación para el Control del Visón en la XI Región*. Coyhaique, XI Región, Chile.

SANDOVAL, R. 1994. *Estudio Ecológico del Visón Asilvestrado (*Mustela vison*, Schreber) en la XI Región*. Tesis de grado en Medicina Veterinaria, Universidad Austral de Chile, Valdivia, Chile.

SAVIDGE, J.A. 1987. Extinction of an island forest avifauna by an introduced snake. *Ecology* 68: 660-668.

SIELFELD, W. 1977. Reconocimiento macrofaunistico terrestre en el área de Seno Ponsonby (Isla Hoste). *Anales del Instituto de la Patagonia* 8: 275-297.

SIELFELD, W. & C. VENEGAS 1980. Poblamiento e impacto ambiental de *Castor canadensis* en Isla Navarino, Chile. *Anales del Instituto de la Patagonia* 1: 247-257.

THOMAS, O. 1890. Pp. 23-32 *in Misión Scientifique du Cap Horn*, 1882-1883. Volumen 6, Zoología, Mamíferos. Gauthier-Villars et fils, Imprijeurs-Libraires. Paris, Francia.

VITOUSEK, R.M., C.M. D'ANTONIO, L.L. LOOPE & R. WESTERBROOKS. 1996. Biological invasions as global environmental change. *American Scientist* 84: 468-478.

Appendix 1. Details of the 31 American minks (*Mustela vison*) caught or seen on Navarino Island and other islands of the Cape Horn county, between November 2001 and May 2003.

Date	Location	Latitude/Longitude	Habitat	Color	Collector	Captured/Observed
21-Nov-01	Puerto Williams Airport	54°56'S,67°38'W	Lagoon	Light gray	Jorge Quelín-Ricardo Rozzi	C
06-Jan-02	Omora Park	54°56'S,67°39'W	Coigüe forest	Dark brown	Oliver Vogel-Francisca Massardo	O[&]
06-Jan-02	Omora Park	54°57'S,67°39'W	Coigüe forest	Dark brown	Oliver Vogel-Francisca Massardo	O[&]
06-Jan-02	Omora Park	54°57'S,67°39'W	Coigüe forest	Dark brown	Oliver Vogel-Francisca Massardo	O[&]
16-Jan-02	Ukika River Mouth	54°56'S,67°35'W	Houses	Dark brown	Ezio M. Firmani-Ricardo Rozzi	O
13-Feb-02	Ukika River Mouth	54°56'S,67°35'W	Houses	Dark brown	Jorge Quelín-Ricardo Rozzi	C
13-Feb-02	Puerto Williams	54°56'S,67°35'W	Turbal-Houses	Dark brown	Ricardo Rozzi	O
13-Aug-02	Bertrand Island	55°13'S,67°56'W	Coastal shrub	Dark brown	Mauricio Zárraga	O
06-Nov-02	Omora Park	54°56'S,67°39'W	Coigüe forest-River	Dark brown	Alfonso Flores-Leslie Flores	O
01-Dec-02	Mussels	54°54'S,68°01'W	River	Dark brown	Germán González-Ricardo Rozzi	C
30-Jan-03	Lum River Mouth	54°54'S,68°53'W	Houses-River	Dark brown	Maria Unión-Francisca Massardo	C[&]
30-Jan-03	Lum River Mouth	54°54'S,68°55'W	Houses-River	Dark brown	Maria Unión-Francisca Massardo	C[&]
03-Feb-03	Douglas River	55°10'S,68°55'W	River	Light gray	Carlos Barría	C
09-Feb-03	Orange Bay, Hoste Island	55°32'S,68°55'W	Coast	Dark brown	Mauricio Zárraga	O
13-Jan-03	Puerto Toro	55°50'S,67°48'W	Coastal shrub	Dark brown	Carlos Barría	C
13-Feb-03	Punta Guerrico	54°54'S,67°50'W	Coastal shrub	Dark brown	Patricio Quelín-Ricardo Rozzi	C
25-Feb-03	Punta Guerrico	54°54'S,67°50'W	Coastal shrub	Black	Patricio Quelín-Ricardo Rozzi	C
07-Mar-03	Ukika River Mouth	54°56'S,67°35'W	Houses	Dark brown	Germán González-Ricardo Rozzi	C*
22-Mar-03	Winhond Bay	55°11'S,67°27'W	River-Turbal	Dark brown	Mauricio Navarro	O
30-Mar-03	Santa Rosa	54°54'S,68°86'W	Coastal shrub-Houses	Dark brown	Francisco Filgueira	C
05-Apr-03	Vuelta del Perro Peninsula	54°56'S,67°19'W	Coastal shrub	Black	Germán González-Ricardo Rozzi	C
15-Apr-03	Caleta Holger	54°57'S,67°14'W	Coigüe forest	Dark brown	Humberto Maricoy	C
17-Apr-03	Ukika River	54°56'S,67°36'W	River	Dark brown	Mario Chiquay	C
25-Apr-03	Punta Guerrico	54°55'S,67°51'W	Coastal shrub	Dark brown	Agusto Quelín-Ricardo Rozzi	C
26-Apr-03	Caleta Holger	54°57'S,67°15'W	Coast	Dark brown	Humberto Maricoy	O
29-Apr-03	Ukika River	54°56'S,67°36'W	River	Light gray	Roberto González	O
29-Apr-03	Ukika River	54°22'S,67°36'W	River	Dark brown	Jose Cárcamo	O
05-May-03	Puerto Williams (Micalvi)	54°56'S,67°36'W	Houses-Puerto	Dark brown	Jorge Quelín-Ricardo Rozzi	O
15-May-03	Puerto Williams Airport	54°56'S,67°38'W	Lagoon	Dark brown	Jorge Quelín-Ricardo Rozzi	O
25-May-03	Puerto Williams Airport	54°56'S,67°38'W	Lagoon	Light brown	Jorge Quelín	C
29-May-03	Omora Park	54°57'S,67°39'W	River	Dark brown	Jorge Quelín-Cristián Osorio	O

* Skin deposited in the Zoological Colection of the Instituto de la Patagonia

[&] Adults and juveniles

INTRODUCTION

Invasive exotic species and their impacts are currently regarded as one of the major causes of anthropogenic global change (Sala *et al.* 2000) and biodiversity loss (Vitousek *et al.* 1997). In particular, invasive predators can have severe impacts on native prey populations, especially on remote islands due to prey naivety or a lack of their natural predators or competitors that would have otherwise limited their success (Elton 1958, Macdonald & Thom 2001). In the case of evolutionary isolation of native species, introductions of carnivore species to island ecosystems can even lead to local extinctions, as it has been documented for several bird species (Courchamp *et al.* 2003).

The American mink (*Mustela vison* Schreber, 1777) is a successful alien predator in most European countries, where it has established feral populations following its introduction from North America for the purpose of fur farming (reviews in Macdonald & Harrington 2003, Bonesi & Palazon 2007). In South America the mink was introduced to Chile and Argentina in the 1930s (Jaksic *et al.* 2002), where populations established in southern Chile (Sandoval 1994, Rozzi & Sherriffs 2003, Anderson *et al.* 2006a) and Argentina (Pagnoni *et al.* 1986). Mink are medium-sized mustelids with a body weight of about 1 kg. They are semi-aquatic mammals found associated with marine shore habitats, river banks, lake shores, freshwater and saltwater marshes. As generalist predators their diet includes prey from both aquatic and terrestrial sources in variable proportions and strongly reflects local and seasonal availability of prey (Dunstone 1993).

Being a highly adaptable and opportunistic predator various studies in Europe have shown that mink can be detrimental to native species (Macdonald & Harrington 2003). The most apparent impact of the mink is a reduction in the range or population size of native prey as has been well documented for seabird colonies on islands (Clode & Macdonald 2002, Nordström *et al.* 2004), ground-nesting inland water birds (Ferreras & Macdonald 1999), intertidal marine communities (Delibes *et al.* 2004), rodents (Jefferies 2003), and amphibians (Ahola *et al.* 2006). Negative impacts due to competition are subject to discussion for the European mink (Maran *et al.* 1998, Sidorovich *et al.* 2001). In South America mink have caused reductions in waterbird populations (Lizarralde & Escobar 2000). They are also considered to be responsible for the decline of the River otter *Lontra provocax* (Thomas, 1908) (Previtali *et al.* 1998), although Medina (1997) found little support on competition for space and food.

The case of Navarino Island represents a recent invasion of the mink, where it was first recorded in 2001 (Rozzi & Sherriffs 2003). Navarino Island is located south of Tierra del Fuego, from which it is separated by the Beagle Channel. At numerous points this channel is less than 5 km wide. Mink that escaped or were released from mink farms on Tierra del Fuego might have swum across the Beagle Channel, reaching Navarino Island, and other islands of the Cape Horn Biosphere Reserve (Rozzi & Sherriffs 2003). Here, they represent a new guild because Navarino Island lacks native terrestrial mammalian predators. Thus among the mechanisms by which introduced carnivores affect the local biota (competition, disease, interbreeding, predation, Macdonald & Thom 2001) predation is the most expected mechanism for mink on Navarino, with possible indirect effects on trophic webs.

On Navarino Island the most diverse and abundant group of vertebrates are birds (Rozzi *et al.* 2006a). Many of them are ground-nesting, and expected to be especially vulnerable to predation from introduced mink as behavioral adaptations to terrestrial predators might lack (Anderson *et al.* 2006a, Soto & Cabello 2007). Although the Cape Horn region is part of one of the most pristine areas in the world (Mittermeier *et al.* 2003), this archipelago has not only been invaded by mink, but also by feral domestic animals, rodents and two more North American wild fur mammals: the American beaver (*Castor canadensis* Kuhl, 1820) and the muskrat (*Ondatra zibethicus* Linnaeus, 1766). In total, the assemblage of exotic terrestrial mammals on the island outnumbers their native counterparts (Anderson *et al.* 2006a). Some of these exotic species of rodents might represent prey for the mink, thereby generating possible predator-prey interactions among exotic species.

The prime purpose of this study is to quantify the composition of the diet of the southernmost population of American mink in its initial stage of colonization, considering seasonal and habitat variations, in order to provide an initial baseline diagnosis about its potential impact on native and exotic fauna of Navarino Island, and the Cape Horn Biosphere Reserve. Our hypotheses are as follows: First, given that birds are the most abundant group of terrestrial vertebrates on Navarino, birds could represent the main prey group in the diet of the mink. Second, given that in Cape Horn several bird species and/or populations are winter migrants, we expect birds to be the main prey especially during the breeding season and in marine coastal habitats where bird populations are particularly abundant and diverse. Third, given that the exotic muskrat is an important prey for mink in its original distribution range; this rodent species could also constitute an important prey for mink populations in the area of Cape Horn, thus facilitating the arrival of the newcomer mink (*i.e.*, invasional meltdown hypothesis by Simberloff & Von Holle 1999). We will try to estimate the absolute number of birds, mammals, and fish that mink remove monthly and discuss these results in the context of conservation and management with a special focus on the potential impact that the recently introduced mink might have on native bird, especially ground-nesting, and mammal populations.

MATERIAL AND METHODS

Study area

The study was carried out on Navarino Island (2,528 km^2), located at the extreme southern tip of South America (Fig. 1). The island forms part of the Cape Horn Biosphere Reserve (54°-56°S) and belongs to the Magellanic Sub-Antarctic Evergreen Rainforest ecoregion, recently identified as one of the 24 most pristine wilderness areas of the world (Mittermeier *et al.* 2003). The main habitats include (i) evergreen and deciduous forests dominated by the genus *Nothofagus*, (ii) peatlands, moorlands, and bogs, (iii) alpine communities dominated by cushion plants and lichens, (iv) streams and lakes, and (v) thickets or scrublands in naturally or anthropogenically disturbed areas (Pisano 1977, Rozzi *et al.* 2006b). The climate is oceanic, with a low annual thermic fluctuation (< 5°C), a mean annual temperature of 6°C, and an annual precipitation of 467.3 mm (Pisano 1977). During the winter, streams and lakes are icebound. The human population is concentrated in the town of Puerto Williams, the capital city of the Chilean Antarctic Province, on the northern coast of Navarino. Access to the town, rural settlements and navy stations relies mostly on access from the sea, except for a dirt road that connects the entire northern coast of Navarino Island.

Potential vertebrate prey on Navarino Island

With 154 species, birds represent the most abundant and diverse group of terrestrial vertebrates on the island (Couve & Vidal 2003, Rozzi *et al.* 2006a). Native mammals include only five species: one Artiodactyla, two Chiroptera, and two Rodentia. At present, exotic mammals include eleven species, therefore outnumbering native species. They include four species of rodents: *Castor canadensis*, *Mus musculus* (Linnaeus, 1758), *Ondatra zibethicus*, and *Rattus norvegicus* (Berkenhout, 1769) (Anderson *et al.* 2006a). *Ondatra zibethicus* is an important native prey to mink in North America (Dunstone 1993). As far as freshwater fish are concerned, Navarino Island only hosts one common native species (*Galaxias maculatus* Jenyns, 1842), two extremely rare native species (*Applochiton* spp.) and two exotic species of trout, *Salvelinus fontinalis* (Mitchill, 1814), and *Oncorhynchus mykiss* (Walbaum, 1792) (Moorman 2007). However, the marine fish fauna in the Beagle Channel is rich and includes more than 50 species (López *et al.* 1996). There are no amphibians or reptiles present on Navarino Island (Anderson *et al.* 2006a).

Dietary analysis

Diet was analyzed by examining 512 feces (= scats) collected from 36 sites on Navarino Island between April 2005 and March 2007. The sites included different semi-aquatic habitats: 13 marine coastal sites, 14 sites along lake shores, and nine sites along riparian shores. The majority of the sites (28 sites) were located at the northern margin of Navarino Island due to the limited accessibility of the southern areas. To ensure that collected scats represented different individuals we sampled
sites that were separated by at least three km (the average linear territory size described for the American mink, Dunstone 1993). We searched for scats at all sites over the summer. Due to climatic conditions and accessibility, we only conducted scat searches over the spring, autumn, and winter seasons at coastal sites. All scats were frozen for long-term storage.

Collected scats were thawn and soaked in water overnight prior to sieving (0.3 mm). The washed scats were then dried at 50°C (24 h) and stored in paper bags. We sorted the undigested prey remains into six categories (mammals, birds, fish, insects, crustaceans, mollusks) using a binocular microscope. Seeds and plant material were excluded from the analysis as we suspected that this was a result of secondary prey, accidental intake or adhesion after defecation. We estimated the percentage volume of each prey category per scat to the nearest 10% and weighed it to 0.01 g.

For the identification of mammals and birds we used the reference collections of the Instituto de la Patagonia, Universidad de Magallanes, complemented by our own additional collections, and local keys (Reise 1973, Chehébar & Martín 1989, Reyes 1992, Rau & Martínez 2004). Mammals were identified to the species level by examining hair samples; birds were identified to the order level, the taxonomic level a microscopic examination of feathers permits (Day 1966). To address the question of which age category birds are preferably consumed by mink we classified birds that had been identified taxonomically into adults and chicks. We assigned the sample to chicks if at least three of the following features applied: long slender barbs at short rhachis, presence of papillae, truncated

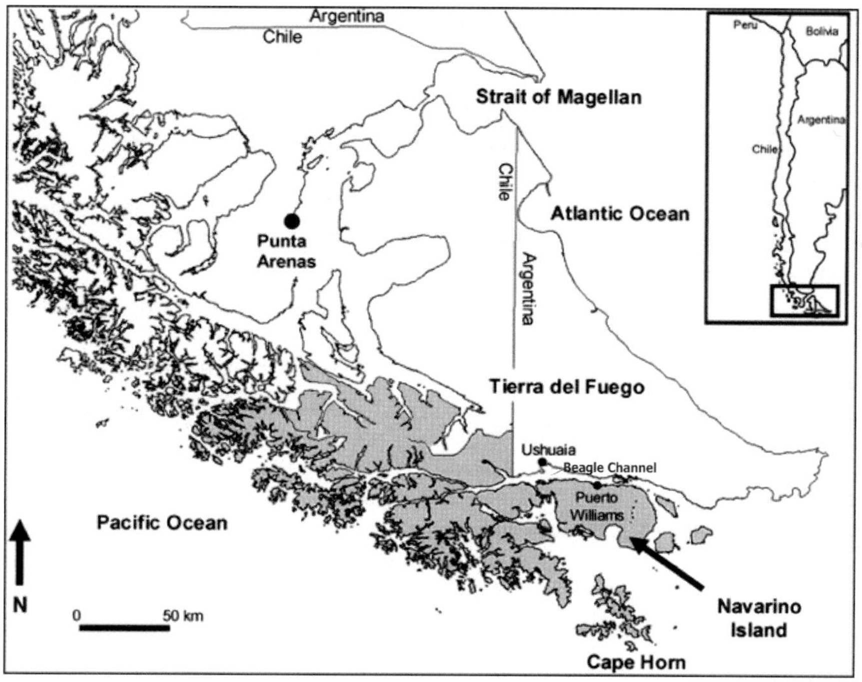

Fig. 1. Map of southern South America with the study area Navarino Island south of the Argentinean portion of Tierra del Fuego (separated by the Beagle Channel).

feather shape, lack of coloration, lack of pennaceous barbs, and a small amount of bones in the sample (Ewart 1921, Busching 2005). Insects were identified by a local entomologist to the species level, and where this was impossible to the next higher taxonomic level (order, subclass).

To quantify the relative contribution of prey groups to mink's diet we applied three indices; the first two are commonly used (Jedrzejewska & Jedrzejewski 1998):

(1) Percentage of relative frequency of occurrence of each food item (RFO), calculated as the number of occurrences of a prey category divided by the number of occurrences of all prey categories. With this index, small prey items tend to be overrated in terms of importance.

(2) Percentage biomass of a given prey item (BIO) estimated by multiplying its dry mass by its empirically determined coefficients of digestibility. This coefficient is measured as the ratio of fresh mass of a given prey to the dry mass of its remains in scats (Jedrzejewska *et al.* 2001). We applied the following correction factors: mammals (17.3), birds (17.2), eggs (687.5), fish (30.8), crustaceans and mollusks (14.8), provided by Brzezinski & Marzec (2003). For insects we used the value (5.0) provided by Lockie (1961). Calculating the biomass reflects the real intake of prey as it takes into account different sizes and digestibility of prey (Brzezinski & Marzec 2003).

(3) Percentage occurrence of the dominant item (POD) with the item constituting the largest volume class considered as the dominant one (Hammershøj *et al.* 2004). This method can compensate for the disadvantage of secondary prey intake.

Statistics

We conducted Spearman's rank correlations between the corresponding prey groups of different indices (RFO, BIO, POD) in order to evaluate whether those indices were comparable with each other. We assessed differences in the diet composition between habitats and seasons using chi-squared tests with Yates' continuity and Bonferroni corrections, Fisher's exact tests when expected values were less than five, and two-sample tests for equality of proportions, all two-sided, running the R version 2.7.1. (R Development Core Team 2008).

In order to measure the degree of specialization of mink in different habitats and seasons we calculated its food-niche breadth for the six defined food categories using Hurlbert's standardization of Levins' B index (Levins 1968, Hurlbert 1978, Krebs 1999):

$$B_s = \frac{(B-1)}{(n-1)}$$

with B_s = standardized niche breadth, B = Levins' index, n = number of food categories,

$$B = \frac{1}{\sum p_i^2}$$

where p_i is the proportion of items in the diet that are of food category i. The proportion was calculated for each of the three indices used in this study (RFO, BIO, POD). The range of Levins' B index is 1 to n. Therefore it depends on the number of food categories. The standardized niche breadth B_s is independent of n and has a range from 0 to 1, with 1 indicating the broadest niche.

We estimated the intake of prey groups by the mink in order to evaluate its impact on prey populations where abundances are known. To assess how many birds, mammals, and fish a single mink might consume during a warm month, we followed the approach developed by Jedrzejewska & Jedrzejewski (1998) in Bartoszewicz & Zalewski (2003). Accordingly, the number of a given prey group eaten per day per mink (N_{pd}) was calculated as:

$$N_{pd} = \frac{DFC \times B_p}{Wt_p}$$

where *DFC* is the average daily food consumption of mink (app. 190 g, 153 ± 48 g for females, 231 ± 72 g for males, estimated from Dunstone 1993), B_p is the fraction of given prey biomass in the mink diet, and Wt_p is the wet mass of a given prey: 75 g for small birds, 35 g for small rodents, 1100 g for the muskrat, and 113 g for fish (Dunstone 1993, Bartoszewicz & Zalewski 2003). We used an overall index of 75 g for birds as our results indicated that bigger birds (*e.g.*, Anseriformes) were almost exclusively consumed as chicks.

RESULTS

The analysis of scat content showed that the mink diet consisted mainly of mammals and birds (Fig. 2). Combined, both taxonomic groups accounted for more than half of the mink diet using the relative frequency of occurrences index (59.3 % RFO), for 68.0 % when estimated by the biomass index (BIO), and for 80.1 % with the dominant item method (POD). In the overall diet, mammals were more important than birds using RFO (χ^2= 7.79, df = 1, P = 0.005), but not significantly different form birds in biomass intake (χ^2 = 0.73, P = 0.394). Fish were the third most important component in the mink diet accounting for 14.1-23.7 %. Overall, we found that terrestrial food (72.5-83.2 % mammals, birds, insects) provided a much more important source of food for mink than did aquatic prey (16.8-27.5 % fish, crustaceans, mollusks) (for all indices χ^2-values ≥ 375.35, all df = 1, all P-values < 0.0001).

Variations in mink diet among habitats

All three indices correlated significantly with each other (Spearman rank correlations, all N = 6 prey groups: RFO with BIO, r_s = 0.94, P = 0.017; RFO with POD, r_s = 1.0, P = 0.003; BIO with POD, r_s = 0.94, P = 0.017). For this reason, we focused on the most conservative index, the dominant item method (POD), to evaluate differences in prey composition between scats collected in different habitat types. Mink diet in the summer varied significantly among different habitat types (Fig. 3). The dominant item index differed between riparian and marine coastal habitats (Fisher's exact test, df = 5, P < 0.0001), and riparian and lake habitats (df = 5, P < 0.0001). Differences between marine coastline and lakes were also significant (df = 5, P = 0.031), but the contrast in values was less pronounced. In lake habitats, the proportion of birds (58.0 %) exceeded the proportion of mammals (34.1 %) significantly (χ^2 = 9.15, df = 1, *P* = 0.002). In contrast, in riparian habitats bird prey was much less frequent than mammal ones (22.4 versus 55.1 %, χ^2 = 9.67, *P* = 0.002). Indeed, in riparian habitats the consumption of birds by the mink was significantly lower than in marine (χ^2 = 6.62, *P* = 0.01) and lake habitats (χ^2 = 10.67, *P* = 0.001). Regarding the other taxonomic groups, fish was only found to be frequent prey in marine coastal habitats (17.4 %). Thus, the consumption of fish by the mink was significantly higher in marine habitats compared to lake (χ^2 = 5.78, *P* = 0.016), and riparian habitats (χ^2 = 6.19, *P* = 0.013). Regarding other prey (insects, crustaceans, mollusks), insects were found to be the dominant item in riparian habitats occurring in 20.4 % of the analyzed scats, while crustaceans were the most important prey in marine coastal habitat accounting for 5.2 %.

Variations in mink diet among seasons

We assessed seasonal differences in the diet of the mink in marine coastal habitats focusing on the POD index as done for differences between habitats. Diet composition between spring-summer and autumn-winter, respectively, was insignificant (Fisher's exact test, df = 5, *P* = 0.119, and *P* = 0.233, respectively) so that data were pooled for the warm season (spring-summer) and the cool season (autumn-winter). However, the diet of the mink in warm and cool seasons varied significantly (df = 5, *P* < 0.0001) (Fig. 4). Scats collected during the warm season at marine coastal sites were approximately equally dominated by mammals and birds (37.6 versus 40.7%, χ^2 = 0.4, df = 1, *P* = 0.528). In contrast, during the cool season, scats collected in marine habitats were dominated by mammal items (59.8% POD), while birds, which were dominant items in only 16.2% of analyzed scats, were significantly less consumed (χ^2 = 45.33, df = 1, *P* < 0.0001). Fish represented the dominant item in 16.7-19.7% of scats in warm and cool seasons, and no significant temporal variation could be determined. Furthermore, no significant seasonal fluctuations for the other prey groups could be found, which made up less than 5 % of the dominant items in warm and cool seasons.

Trophic niche breadth

The standardized trophic niche breadths during the summer season were highest for mink inhabiting marine coastal habitats (Table 1). Lake habitats exhibited low values of trophic niche breadth. Mink diet in lake habitats concentrated on birds (58% POD); fish and other prey groups were almost absent (Fig. 3). In marine habitats, mink diet showed the highest variation during summer, and the lowest during winter when mink relied essentially on mammals (63.6% POD), and entire prey groups (insects, crustaceans, mollusks) lacked in its diet. For spring and autumn the trophic niche breadths were intermediate.

Taxonomic identification

We identified prey groups from 193 scats collected during the warm season (spring-summer) in marine coastal habitat. We focused on this sample as food niche breadth was broadest in marine habitats, and in order to coincide with the breeding season of birds. Birds were the most diverse group of prey on Navarino Island. Seven of the twelve orders of birds breeding on the island were present in the scats analyzed (Table 2). Passeriformes were the most abundant order in the diet, with indices from 10.2-14.8% (RFO, BIO, POD). However, Pelecaniformes, Anseriformes, and eggs also played an important role in terms of biomass intake. Together these four groups accounted for 76.2% of the bird biomass consumed by

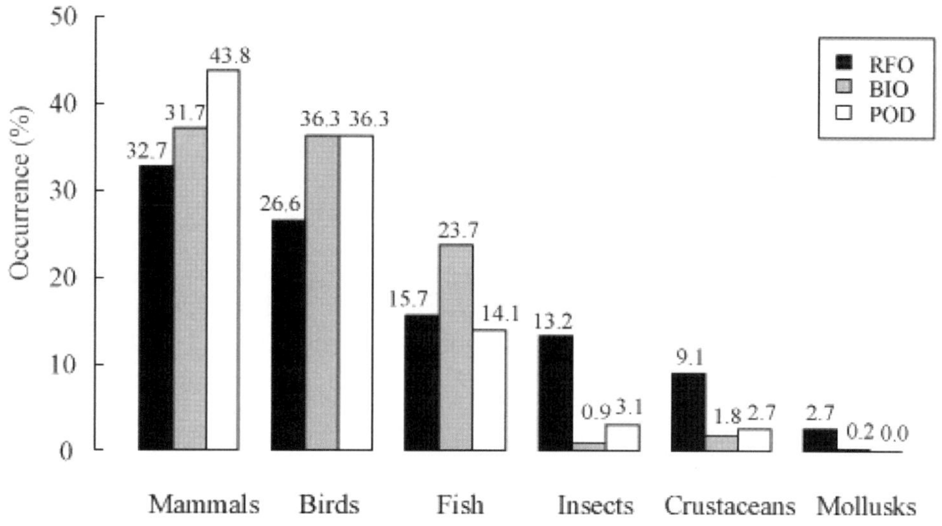

Fig. 2. Overall occurrence of prey categories in scats of American mink on Navarino Island. Percentages are based on data from all scats (n = 512) collected from three types of semi-aquatic habitats (marine, riparian, lake) over the four seasons during a three-year period (2005-2007). RFO = relative frequency of occurrences of each prey category, BIO = percentage of biomass consumed, POD = percentage of occurrence of each prey category as the dominant item.

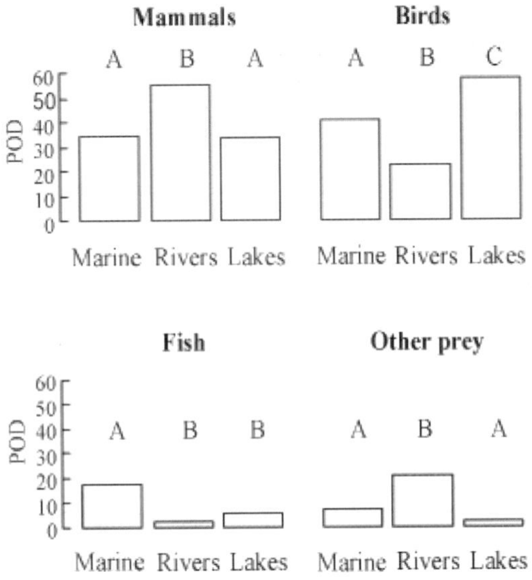

Fig. 3. Differences in the consumption of mammals, birds, fish, and other prey (insects, crustaceans, mollusks) by mink over the summer in different habitat types (n = 292 scats). For each graph different letters above the bars indicate significant differences ($P < 0.05$ with 2-sample tests for equality of proportions) in the percentage of occurrence of the dominant item (POD) in a prey group among habitat types.

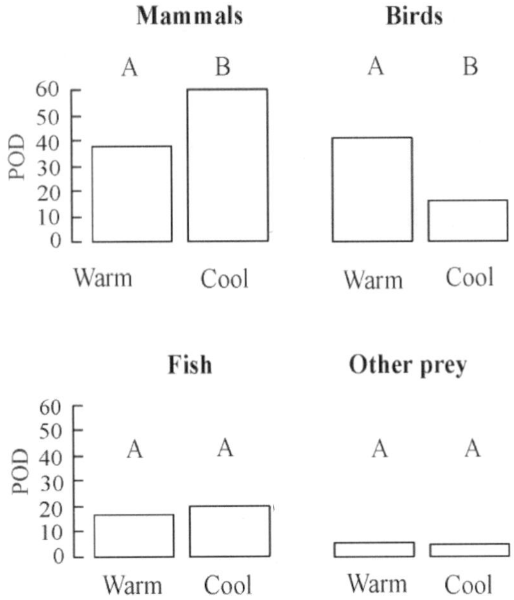

Fig. 4. Differences in the consumption of mammals, birds, fish, and other prey (insects, crustaceans, mollusks) by mink during warm (spring-summer) and cool (autumn-winter) seasons in marine coastal habitat (n = 375 scats). For eachgraph different letters above the bars indicate significant differences ($P < 0.05$ with 2-sample tests for equality of proportions) in the percentage of occurrence of the dominant item (POD) in a prey group among seasons.

mink. Passeriformes were mainly caught as adults, whereas all other bird orders were caught as chicks, with the difference being highly significant ($\chi^2 = 37.9$, df = 1, $P < 0.0001$).

As far as mammals are concerned, the two dominant species were the native rodent *Abrothrix xanthorhinus* (Waterhouse, 1837) and the exotic muskrat (*Ondatra zibethicus*), with 8.9-14.8%, and 8.7-15.2% (RFO, BIO, POD), respectively (Table 2). Together, *A. xanthorhinus* and *O. zibethicus* represented 78.2% of the mammal biomass consumed by the mink on Navarino Island. The relative importance of *Oligoryzomys longicaudatus* (Bennett, 1832), the second native rodent species on the island, was modest. Its relative frequency of occurrence, biomass and dominance as a prey item were less than a third of the values determined for *A. xanthorhinus* and *O. zibethicus*. The presence of the house mouse (*Mus musculus*) in the diet was negligible. It was found only in three samples collected in the vicinity of houses. Noticeably, the two other exotic rodent species, namely the beaver (*Castor canadensis*) and the Norway rat (*Rattus norvegicus*), were not found in any of the scats collected in marine coastal habitats over the spring and summer. The occurrence of mink hair in some of the scats was ascribed to grooming. In summary, the proportion of exotic mammals in terms of biomass was 49.9%, a proportion that exceeds the value of 43.2% estimated for native mammals (the other 6.9% correspond to non identified mammal prey). The contribution of invertebrates to the mink's biomass intake (2.9%) was negligible (Table 2). In spite of this however, mink seem to actively search for insects because all of the insect taxa identified, except for the coleoptera *Microplophorus magellanicus* (Blanchard, 1851), were flightless, and therewith easier to catch. Hence, insects were probably consumed directly by mink (although we cannot exclude consumption by other active hunters).

Monthly food consumption of mink

How hungry are mink? On average the daily food consumption of a mink is approximately 190 g (estimated from Dunstone 1993). Following our biomass fractions (BIO) of prey groups over the spring and summer we can estimate an extrapolated monthly (30 days) consumption, which does not take into account varying energy demands of the animal. Accordingly, a single mink would consume monthly on average: nine Passeriformes, four Anseriformes, six Pelecaniformes, 13 other birds (total birds: 32), 17 *Abrothrix xanthorhinus*, five *Oligoryzomys longicaudatus*, one muskrat, three other small rodents (total mammals: 26), and 13 fish. The small proportion of remaining prey groups (2.9% BIO) was neglected in this estimation. Based on trapping data of mink on Navarino Island, mink relative densities ranged from 0.79-1.32 individuals km^{-1} along marine coastal habitat (Anderson *et al.* 2006a). Applying these densities to a 10 km stretch of coastline eight to thirteen mink living there could roughly consume about 248-415 small birds, 205-343 mammals, and 105-175 fish in one month during the warm season.

TABLE 1. Levins' standardized niche breadth for American mink on Navarino Island, estimated on the basis of scats collected in marine coastal, riparian, and lake habitats over the summer (N = 292), and for scats collected from marine habitats over different seasons of the year (N = 375).

Index	Habitat (summer)			Season (marine coast)			
	Marine	Rivers	Lakes	Spring	Summer	Autumn	Winter
RFO	0.78	0.45	0.47	0.64	0.78	0.52	0.35
BIO	0.43	0.18	0.17	0.39	0.43	0.31	0.40
POD	0.42	0.30	0.24	0.35	0.42	0.28	0.21

TABLE 2. Diet composition of American mink over the warm season (spring-summer) in marine coastal habitat on Navarino Island (N = 193 scats, 2005-2007); RFO = relative frequency of occurrences, BIO = percentage of biomass, POD = percentage of occurrence of dominant item.

Prey	RFO	BIO	POD
Abrothrix xanthorhinus	8.9	10.1	14.8
Mus musculus	0.6	0.1	0.3
Mustela vison	2.4	1.3	3.8
Oligoryzomys longicandatus	2.6	3.2	4.8
Ondatra zibethicus	8.7	13.9	15.2
Unidentified mammals	6.1	2.1	2.8
Total mammals	29.2	30.8	41.7
Anseriformes	3.9	5.1	6.9
Ciconiformes	1.3	1.7	2.4
Coraciiformes	0.4	0.3	0.7
Gruiformes	0.2	0.4	0.3
Passeriformes	10.2	11.3	14.8
Pelecaniformes	3.7	7.5	6.2
Podicipediformes	0.9	1.0	1.7
Eggs	3.9	6.6	-
Unidentified birds	6.3	6.2	6.6
Total birds	30.7	40.0	39.7
Fishes	15.5	26.3	14.8
Arachnida, Acari	0.4	-	-
Coleoptera	0.9	-	-
Hemiptera, Heteroptera	0.2	-	-
Aegorhinus vitulus	0.2	-	-
Microplophorus magellanicus	1.5	0.1	0.3
Pycnosiphorus femoralis	1.7	0.1	0.3
Sinopla perpunctatus	0.2	-	-
Unidentified insects	5.7	-	0.3
Total insects	10.7	0.3	1.0
Crustaceans	10.9	2.4	2.8
Mollusks	3.0	0.2	-
Food niche breadth	0.68	0.42	0.36
Total N items or biomass (g)	541	2,566.6	290

DISCUSSION

Mammals as a reliable source, birds welcome when available

Diet composition showed that mink relied principally on mammalian and avian prey with significant variations over habitats and seasons. In marine coastal habitat and in particular at lakes birds were the principal prey group. Over the warm season mink preyed on both vertebrate groups almost equally, but relied mainly on mammals over the cool season. These results suggest that mammals represent a stable base in the diet compensating for periods with lower bird availability.

Compared to other studies on mink diet worldwide, the proportion of birds in scats from marine coastal habitat on Navarino Island over the summer was exceptionally high. Further north in Argentinean Patagonia coastal mink predominantly prey on crustaceans and insects (Previtali *et al.* 1998). In Eurasia fish and crustaceans seem to be the most important prey for mink in marine habitat (Jedrzejewska *et al.* 2001, Delibes *et al.* 2004), although mammals can also play an important role (Dunstone & Birks 1987). In riparian habitats, fish, mammals, and amphibians are reported as principal prey groups in the diet of introduced mink (Medina 1997, Ferreras & Macdonald 1999, Jedrzejewska *et al.* 2001, Hammershøj *et al.* 2004). Noticeably, few studies report a high consumption of birds, which only occurred in productive waterbird breeding habitats and areas with large wildfowl populations in inland wetlands, especially during the warm season (Ward *et al.* 1986, Arnold & Fritzell 1987, Bartoszewicz & Zalewski 2003). Therefore, the opportunistic diet habits of the mink (Dunstone 1993) seem to reflect the abundance and diversity of prey available in each region.

The exceptionally high abundance of birds in the diet of mink at the southern end of the Americas might be explained by at least three factors: (i) in the insular austral environment amphibians and reptiles as possible prey groups are absent (ii) aquatic prey might play a minor role due to the climate conditions in combination with the greater level of energy expenditure used to catch it (Stephenson *et al.* 1988), and (iii) during the warm season, the density of birds returning from migration is high, the mobility of birds during incubation, brood rearing or moulting is reduced (Bartoszewicz & Zalewski 2003), and many ground-nesting birds on Navarino supply vulnerable offspring.

Impact on native prey populations

In order to assess the direct ecological impact of mink on prey populations we must identify "sensitive species", *i.e.* species with low abundances, low hunting effort for mink, and high energy content (Dunstone 1993). Our identification of bird remains in mink scats is reliable to the level of orders. Consequently, we attempted to identify potentially sensitive species among the main orders we found in the diet of mink on Navarino Island. Passeriformes were the most abundant bird order in the diet, and furthermore affected as being caught as adults, *i.e.,* as individuals contributing to the reproduction of the population. Although we don't know whether these passerines were ground or tree-nesting, we suggest that potentially sensitive species might be those that breed in open nests at low height, and present low abundances in the Cape Horn Archipelago region. Based on avian censuses for Navarino Island (Anderson & Rozzi 2000, Anderson *et al.* 2002, McGehee *et al.* 2004) these criteria target three species: the ground nesting Magellanic Tapaculo (*Scytalopus magellanicus* Gmelin, 1789), the Patagonian Tyrant (*Colorhamphus parvirostris* Darwin, 1839), and in particular the Fire-eyed Diucon (*Xolmis pyrope* Kittlitz, 1830) (Ippi personal communication). In inland wetlands bird species that are locally rare, such as the Fuegian Snipe (*Gallinago stricklandii* Gray, 1845) (Couve & Vidal 2003) might be threatened because in comparison to marine coastal habitats, inland wetlands on Navarino Island seem to harbor less potential prey for mink (lack of crustaceans, marine fish, etc.).

With respect to large ground-nesting birds (*e.g.*, Anseriformes, Pelecaniformes) our results confirm that mink predominantly caught their chicks. Among Anseriformes, two species endemic to the subantarctic archipelago region of southwestern South America (41-56°S, including the Falklands) have small populations that might be particularly vulnerable to mink predation of their offsprings: the Flightless Steamer Duck (*Tachyeres pteneres* Forster, 1844) and the Kelp Goose (*Chloephaga hybrida* Molina, 1782). Both species are strictly coastal, and they inhabit the area all year round (Couve & Vidal 2003). Their population densities along the coasts of the Beagle Channel (0.74, and 0.46 birds km^{-1} for *T. pteneres* and *C. hybrida*, respectively, estimated by Raya & Schiavini 2002) are very low in the light of the numbers of birds consumed by mink in our numeric example (25-42 $month^{-1}$ km^{-1}).

Regarding native mammals, mink hunted native rodents in proportion to their availability. In the spring-summer diet we found three times more *Abrothrix xanthorhinus* than *Oligoryzomys longicaudatus*. This proportion resembles the relative abundances obtained for these two species through censuses with Sherman traps on the study sites on Navarino Island (Gañan *et al.* unpublished results). *A. xanthorhinus* was trapped six times more frequently than *O. longicaudatus*. For *A. xanthorhinus* Gañan *et al.* estimated mean relative densities of 8-39 individuals ha^{-1} (per 100 trap nights, depending on the habitat type: forest, pasture, reed, shrubland), and 0-6 individuals ha^{-1} for *O. longicaudatus*, respectively, during spring to autumn (N = 3,800 trap nights). Hence, based on our estimations for the consumption of native rodents from our biomass fractions in mink diet (17-29 $month^{-1}$ km^{-1}), it seems that *Mustela vison* does not represent a threat to these mammal species, although these rough estimates have to be treated with caution. The indirect impact of mink on changes in the prey offer for autochthonous bird predators requires further analysis.

Exotic mammals as prey for mink

Our results from marine coastal scats over the warm season show a considerable proportion of muskrat among mammal biomass consumed by mink on Navarino Island. Interestingly, this result coincides with studies from North America, where muskrats are the largest and probably most important mammalian prey for mink (Dunstone 1993). Over the winter months in particular young muskrat are vulnerable to mink predation (Errington 1954). Thus, populations of these two North American invasive mammals have reestablished their predator-prey interactions at the southern end of the continent. We lack systematic abundance data of the muskrat, but SAG's Control Program of Exotic Species in Magallanes (Soto & Cabello 2007), reports that 250 Muskrats were captured in Tierra del Fuego and the Cape Horn Region by a single trapper working in lake and riparian habitats over 2005/2006. We therefore assume that muskrat is a reliable and highly energetic source for mink. This result adds one more example of facilitatory interactions between an invasive species already present, the muskrat (arrival in the fifties, Jaksic *et al.* 2002), aiding a new species to establish (Simberloff & Von Holle 1999), although not in a mutual way.

As far as the other species of North American introduced rodents are concerned, we did not find any beaver remains in the diet of mink in marine habitats. This might be due to the fact that beavers are mostly associated with riparian habitats (Anderson *et al.* 2006b). Studies on mink scats (N = 235) collected in inland wetlands on Navarino Island have shown that only 2% contained beaver remains (Ibarra 2007). Again, these results for Navarino Island coincide with those found in North America (Dunstone 1993) and Europe (*e.g.,* Brezezinski & Zurowski 1992) where beavers are not reported as prey species. A probable explanation is that beavers are too large as prey for mink (their body weight ranges from 18-23 kg, Aleksiuk 1968), and that young beavers are sufficiently guarded by their parents (Brezezinski & Zurowski 1992). However, mink might profit from the coexistence with beavers, since their engineering activities provide denning facilities for the mink (Zurowski & Kammler 1987).

Concluding remarks

Birds constituted an important prey group for the introduced mink on Navarino Island, especially in marine coastal habitat and lakes during the breeding season. Studies in other insular ecosystems have shown that the introduction of mink can lead to severe reductions of bird populations (Ferreras & Macdonald 1999, Nordström & Korpimäki 2003). We identified two Anseriformes (*Tachyeres pteneres* and *Chloephaga hybrida*) endemic to southern South America, which might be seriously threatened by the presence of mink on Navarino Island. For these reasons we advise governmental agencies (Iriarte *et al.* 2005) to start controlling mink populations immediately. In order to increase trapping success, control should take place mainly during periods of high activity of mink (spring and autumn, Moore *et al.* 2003), and with particular intensity at nesting habitats of sensitive bird species during their breeding periods (under the consideration of the trapper's influence).

For an integrated management of exotic species on Navarino Island and the Cape Horn Biosphere Reserve it is important to undertake further research on the direct and indirect interactions established here among the various introduced mammal species (Silva & Saavedra 2008). The muskrat might protect native species from being preyed upon by the opportunistic predator, but on the other hand it represents a reliable food source for the mink. Finally, we recommend that long-term monitoring of bird populations should accompany the mink management strategies that are currently being implemented in this remote region of the world. The critical need for conserving the avifauna and ecosystem integrity of the Cape Horn Biosphere Reserve is especially relevant for ecotourism, which currently represents the main option for achieving both economic and environmental sustainability at the southern end of the Americas (Rozzi *et al.* 2004).

ACKNOWLEDGMENTS

We thank José Llaipén for his great support in field work and laboratory analysis. Steven McGehee and Melisa Gañan assisted with the field work. We are grateful to Eduardo Faúndez who identified the insects. Kurt Jax and Jorge Gibbons provided helpful advice regarding scientific questions. We thank Carlos Soto and Alex Muñoz for hosting us in their laboratories. Jana Zschille provided us with experience and bibliographic citations. We express our gratitude to two anonymous reviewers whose comments greatly improved the paper. The manuscript also benefited due to comments from Klaus Henle, Kurt Jax, Steven McGehee, Michael Gerisch, Simone Lampa, and Bianca Bauch. This work was co-sponsored by the German Academic Exchange Service (DAAD), by the Chilean Millennium Institute of Ecology and Biodiversity Contract (ICM, PO2-051-F), the Universidad de Magallanes, and the Omora Foundation.

REFERENCES

AHOLA, M., M. NORDSTRÖM, P.B. BANKS, N. LAANETU & E. KORPIMÄKI. 2006. Alien mink predation induces prolonged declines in archipelago amphibians. *Proceedings of the Royal Society B* 273: 1261-1265.

ALEKSIUK, M. 1968. Scent mound communication, territoriality, and population regulation in the beaver. *Journal of Mammalogy* 49: 759-762.

ANDERSON, C.B. & R. ROZZI. 2000. Bird assemblages in the southernmost forests in the world: methodological variations for determining species composition. *Anales del Instituto de la Patagonia* 28: 89-100.

ANDERSON, C.B., R. ROZZI, C. ELPHICK & S. MCGEHEE. 2002. El programa Omora de anillamiento de aves en los bosques subantárticos: la estandardización del tamaño de anillos apropiados para las aves de la región de Magallanes. *Boletín Chileno de Ornitología* 9: 2-11.

ANDERSON, C.B., R. ROZZI, J.C. TORRES-MURA, S.M. MCGEHEE, M.F. SHERRIFFS, E. SCHÜTTLER & A.D. ROSEMOND. 2006a. Exotic vertebrate fauna in the remote and pristine sub-Antarctic Cape Horn Archipelago, Chile. *Biodiversity and Conservation* 15: 3295-3313.

ANDERSON, C.B., C.R. GRIFFITH, A.D. ROSEMOND, R. ROZZI & O. DOLLENZ. 2006b. The effects of invasive North American beavers on riparian plant communities in Cape Horn, Chile. Do exotic beavers engineer differently in sub-Antarctic ecosystems? *Biological Conservation* 128: 467-474.

ARNOLD, T.W. & E.K. FRITZELL. 1987. Food habits of prairie mink during the waterfowl breeding season. *Canadian Journal of Zoology* 68: 2322-2324.

BARTOSZEWICZ, M. & A. ZALEWSKI. 2003. American mink, *Mustela vison* diet and predation on waterfowl in the Slonsk Reserve, western Poland. *Folia Zoologica* 52: 225-238.

BONESI, L. & S. PALAZON. 2007. The American mink in Europe: Status, impacts, and control. *Biological Conservation* 134: 470-483.

BREZEZINSKI, M. & W. ZUROWSKI. 1992. Spring diet of the American mink *Mustela vison* in the Mazurian and Brodnica Lakelands, northern Poland. *Acta Theriologica* 37: 193-198.

BRZEZINSKI, M. & M. MARZEC. 2003. Correction factors used for estimating prey biomass in the diet of American mink *Mustela vison*. *Acta Theriologica* 48: 247-254.

BUSCHING, W.D. 2005. *Einführung in die Gefieder- und Rupfungskunde*. AULA-Verlag GmbH, Wiebelsheim. 408 pp.

CHEHÉBAR, C. & S. MARTÍN. 1989. Guía para el reconocimiento microscópico de los pelos de los mamíferos de la Patagonia. *Doñana Acta Vertebrata* 16: 247-291.

CLODE, D. & D.W. MACDONALD. 2002. Invasive predators and the conservation of island birds: the case of American mink *Mustela vison* and terns *Sterna* spp. in the Western Isles, Scotland. *Bird Study* 49: 118-123.

COURCHAMP, F., J.L. CHAPUIS & M. PASCAL. 2003. Mammal invaders on islands: impact, control and control impact. *Biological Reviews* 78: 347-383.

COUVE, E. & C. VIDAL. 2003. *Birds of the Beagle Channel and Cape Horn*. Ediciones Fantástico Sur, Punta Arenas, Chile. 265 pp.

DAY, M.G. 1966. Identification of hair and feather remains in the gut and faeces of stoats and weasels. *Journal of Zoology* 148: 201-217.

DELIBES, M., M. CLAVERO, J. PRENDA, M. DEL CARMEN-BLÁSQUEZ & P. FERRERAS. 2004. Potential impact of an exotic mammal on rocky intertidal communities of northwestern Spain. *Biological Invasions* 6: 213-219.

DUNSTONE, N. & J.D.S. BIRKS. 1987. The feeding ecology of mink (*Mustela vison*) in coastal habitat. *Journal of Zoology* 212: 69-83.

DUNSTONE, N. 1993. *The Mink*. T & AD Poyser Ltd., London, United Kingdom. 232 pp.

ELTON, C.S. 1958. *The Ecology of Invasions by Animals and Plants*. Methuen and Company, London, United Kingdom. 181 pp.

ERRINGTON, P.L. 1954. An analysis of mink predation upon muskrats in north-central United States. Iowa State College. *Agricultural Experimental Station Research Bulletin* 320: 798-924.

EWART, C.J. 1921. The nestling feathers of the mallard, with observations on the composition, origin, and history of feathers. *Proceedings of the Zoological Society of London* 2: 609-642.

FERRERAS, P. & D.W. MACDONALD. 1999. The impact of American mink *Mustela vison* on water birds in the upper Thames. *Journal of Applied Ecology* 36: 701-708.

HAMMERSHØJ, M., E.A. THOMSEN & A.B. MADSEN. 2004. Diet of free-ranging American mink and European polecat in Denmark. *Acta Theriologica* 49: 337-347.

HURLBERT, S.H. 1978. The measurement of niche overlap and some relatives. *Ecology* 59: 67-77.

IBARRA, J.T. 2007. *El Visón en Upushwaia: Variaciones Estacionales de la Avifauna en Ambientes de Humedal y de su Depredación por la Especie Exótica Invasora* Mustela vison *(Schreber) en Isla Navarino, Reserva de Biosfera Cabo de Hornos, Chile.* Master thesis, Facultad de Agronomía e Ingeniería Forestal, Pontificia Universidad Católica de Chile, Santiago, Chile. 117 pp.

IRIARTE, J.A., G.A. LOBOS & F.M. JAKSIC. 2005. Invasive vertebrate species in Chile and their control and monitoring by governmental agencies. *Revista Chilena de Historia Natural* 78: 143-154.

JAKSIC, F.M., J.A. IRIARTE, J.E. JIMÉNEZ & D.R. MARTÍNEZ. 2002. Invaders without frontiers: cross-border invasions of exotic mammals. *Biological Invasions* 4: 157-173.

JEDRZEJEWSKA, B. & W. JEDRZEJEWSKI. 1998. *Predation in vertebrate communities.* The Bialowieza Primeval Forest as a case study. Springer, Berlin, Germany. 450 pp.

JEDRZEJEWSKA, B., V.E. SIDOROVICH, M.M. PIKULIK & W. JEDRZEJEWSKI. 2001. Feeding habits of the otter and the American mink in Bialowieza Primeval Forest (Poland) compared to other Eurasian population. *Ecography* 24: 165-180.

JEFFERIES, D.J. 2003. *The Water Vole and Mink Survey of Britain 1996-1998 with a History of the Long Term Changes in the Status of Both Species and their Causes.* The Vincent Wildlife Trust, Ledbury.

KREBS, C.J. 1999. *Ecological Methodology.* Benjamin/Cummings, Menlo Park, California, USA. 620 pp.

LEVINS, R. 1968. *Evolution in Changing Environments: some Theoretical Explorations.* Princeton University Press, Princeton, New Jersey, USA. 132 pp.

LIZARRALDE, M.S. & J.M. ESCOBAR. 2000. Mamíferos exóticos en la Tierra del Fuego. *Ciencia Hoy* (Chile) 10: 52-63.

LOCKIE, J.D. 1961. The food of the Pine Marten *Martes martes* in West Ross Shire, Scotland. *Proceedings of the Zoological Society of London* 136: 187-195.

LÓPEZ, H.L., M.L. GARCÍA & N.A. SAN ROMÁN. 1996. *Lista Comentada de la Ictiofauna del Canal Beagle, Tierra del Fuego, Argentina.* Centro Austral de Investigaciones Científicas (CADIC), Ushuaia, Argentina. 84 pp.

MACDONALD, D.W. & M.D. THOM. 2001. Alien carnivores: unwelcome experiments in ecological theory. Pp. 93-122 *in Carnivore conservation*, J.L. Gittleman, S.M. Funk, D.W. Macdonald & R.K. Wayne, (eds.). Cambridge University Press, Cambridge, United Kingdom.

MACDONALD, D.W. & L.A. HARRINGTON. 2003. The American mink: the triumph and tragedy of adaptation out of context. *New Zealand Journal of Zoology* 30: 421-441.

MARAN, T., H. KRUUK, D.W. MACDONALD & M. POLMA. 1998. Diet of two species of mink in Estonia: displacement of *Mustela lutreola* by *M. vison*. *Journal of Zoology* 245: 218-222.

MCGEHEE, S., R. ROZZI, S. IPPI, C.B. ANDERSON, R. VÁSQUEZ & S. WOODLAND. 2004. Late-summer presence of the Patagonian Tyrant, *Colorhamphus parvirostris* (Darwin) in Navarino Island, Cape Horn county. *Anales del Instituto de la Patagonia* (Chile) 32: 25-33.

MEDINA, G. 1997. A comparison of the diet and distribution of southern River Otter (*Lutra provocax*) and Mink (*Mustela vison*) in southern Chile. *Journal of Zoology London* 242: 291-297.

MITTERMEIER, R.A., C.G. MITTERMEIER, T.M. BROOKS, J.D. PILGRIM, W.R. KONSTANT, G.A.B. DA FONSECA & C. KORMOS. 2003. Wilderness and biodiversity conservation. *Proceedings of the National Academy of Sciences of the United States of America* 18: 10309-10313.

MOORE, N.P., S.S. ROY & A. HELYAR. 2003. Mink (*Mustela vison*) eradication to protect ground nesting birds in the Western Isles, Scotland, United Kingdom. *New Zealand Journal of Zoology* 30: 443-452.

MOORMAN, M.C. 2007. *The Conservation Implications of Introduced Trout and Beaver on Native Fish in the Cape Horn Biosphere Reserve, Chile.* Master thesis, North Carolina State University, Raleigh, North Carolina, USA. 57 pp.

NORDSTRÖM, M. & E. KORPIMÄKI. 2003. Effects of island isolation and feral mink removal on bird communities on small islands in the Baltic Sea. *Journal of Animal Ecology* 73: 424-433.

NORDSTRÖM, M., J. LAINE, M. AHOLA & E. KORPIMÄKI. 2004. Reduced nest defence intensity and improved breeding success in terns as responses to removal of non-native American mink. *Behavioral Ecology and Sociobiology* 55: 454-460.

PAGNONI, G.O., J.L. GARRIDO & M.R. MARÍN. 1986. *Impacto Económico y Ambiental del Visón,* Mustela vison *(Schreber, 1877) en el Norte de la Patagonia.* Centro Nacional Patagónico (CENPAT)-Consejo Nacional de Investigaciones Científicas y Técnicas (CONICET), Dirección de Fauna Silvestre, Provincia de Chubut, Argentina. 20 pp.

PISANO, E. 1977. Fitogeografía de fuego-patagonia chilena. I. Comunidades vegetales entre las latitudes 52° y 56° S. *Anales del Instituto de la Patagonia* (Chile) 8: 121-250.

PREVITALI, A., M.H. CASSINI & D.W. MACDONALD. 1998. Habitat use and diet of the American mink (*Mustela vison*) in Argentinian Patagonia. *Journal of Zoology London* 246: 482-486.

R DEVELOPMENT CORE TEAM. 2008. *R: a language and environment for statistical computing.* R Foundation for Statistical Computing, Vienna, Austria. ISBN 3-900051-07-0, URL http://www.Rproject. org/. Cited 14 Jun 2008.

RAU, J. & D.R. MARTÍNEZ. 2004. Identificación de los órdenes de aves chilenas a través de la microestructura de sus plumas. Pp. 229-234 *in Aves Rapaces de Chile*, A. Muñoz, J. Rau & J. Valenzuela, (eds.). CEA Ediciones, Valdiva, Chile.

RAYA-REY, A.R. & A. SCHIAVINI. 2002. Distribution and density of kelp geese and flightless steamer ducks along the Beagle Channel, Tierra del Fuego, Argentina. *Waterbirds* 25: 225-229.

REISE, D. 1973. Clave para la identificación de los cráneos de marsupiales y roedores chilenos. *Gayana* (Chile) 27: 1-20.

REYES, C.F. 1992. *Clave para la Identificación de los Órdenes de Aves Chilenas: Microestructura de los Nodos de las Bárbulas.* Thesis, Instituto Profesional de Osorno, Osorno, Chile. 81 pp.

ROZZI, R. & M. SHERRIFFS. 2003. El visón (*Mustela vison* Schreber, Carnivora: Mustelidae), un nuevo mamífero exótico para la isla Navarino. *Anales del Instituto de la Patagonia* (Chile) 31: 97-104.

ROZZI, R., F. MASSARDO & C. ANDERSON (EDS.). 2004. *The Cape Horn Biosphere Reserve: A Proposal for Conservation and Tourism to Achieve Sustainable Development at the Southern End of the Americas.* Bilingual English-Spanish Edition. Ediciones Universidad de Magallanes, Punta Arenas, Chile. 264 pp.

ROZZI, R., F. MASSARDO, A. BERGHÖFER, C.B. ANDERSON, A. MANSILLA, M. MANSILLA, J. PLANA, U. BERGHÖFER, P. ARAYA & E. BARROS. 2006a. *Cape Horn Biosphere Reserve: Nomination Document for the Incorporation of the Cape Horn Archipelago Territory into the World Biosphere Reserve Network.* MaB Program-UNESCO. Ediciones de la Universidad de Magallanes, Punta Arenas, Chile.

ROZZI, R., F. MASSARDO, C.B. ANDERSON, K. HEIDINGER & J. SILANDER. 2006b. Ten principles for biocultural conservation at the southern tip of the Americas: the approach of the Omora Ethnobotanical Park. *Ecology and Society* 11: 43 [online] www.ecologyandsociety.org/vol11/iss1/art43/.

SALA, O.E., F.S. CHAPIN-III, J.J. ARMESTO, E. BERLOW, J. BLOOMFIELD, R. DIRZO, E. HUBER-SANWALD, L.F. HUENNEKE, R.B. JACKSON, A. KINZIG, R. LEEMANS, D.M. LODGE, H.A. MOONEY, M. OESTERHELD, N.L. POFF, M.T. SYKES, B.H. WALKER, M. WALKER & D.H. WALL. 2000. Global biodiversity scenarios for the year 2100. *Science* 287: 1770-1774.

SANDOVAL, R.J. 1994. *Estudio Ecológico del Visón Asilvestrado (*Mustela vison*, Schreber) en la XI Región.* Thesis, Universidad Austral de Chile, Valdivia, Chile. 74 pp.

SIDOROVICH, V.E., D.W. MACDONALD, M.M. PIKULIK & H. KRUUK. 2001. Individual feeding specialization in the European Mink, *Mustela lutreola* and the American mink, *M. vison* in north-eastern Belarus. *Folia Zoologica* 50: 27-42.

SILVA, C.A. & B. SAAVEDRA. 2008. Knowing for controlling: ecological effects of invasive vertebrates in Tierra del Fuego. *Revista Chilena de Historia Natural* 81: 123-136.

SIMBERLOFF, D. & B. VON HOLLE. 1999. Positive interactions of nonindigenous species: invasional meltdown? *Biological Invasions* 1: 21-32.

SOTO, N. & J. CABELLO. 2007. *Informe Final: Programa Control de Fauna Dañina en la XIIa Región 2004-2007.* SAG-FONDEMA, Servicio Agrícola y Ganadero, Magallanes y Antártica Chilena, Punta Arenas. 59 pp. + appendix.

STEPHENSON, R., P.J. BUTLER, N. DUNSTONE & A.J. WOAKES. 1988. Heart rate and gas exchange in freely diving American mink (*Mustela vison*). *Journal of Experimental Biology* 134: 435-442.

VITOUSEK, P.M., C.M. D'ANTONIO, L.L. LOOPE, M. REJMÁNEK & R. WESTBROOKS. 1997. Introduced species: a significant component of human-caused global change. *New Zealand Journal of Ecology* 21: 1-16.

WARD, D.P., C.M. SMAL & J.S. FAIRLEY. 1986. The food of Mink *Mustela vison* in the Irish midlands. *Proceedings of the Royal Irish Academy Section B* 86: 169-182.

ZUROWSKI, W. & J. KAMMLER. 1987. American mink (*Mustela vison* Schreber, 1777) in beaver's sites. *Przeglad Zoologiczny* 31: 513-521.

Associate Editor: Luis Ebensperger
Received July 23, 2008; accepted September 23, 2008.

The Cape Horn Archipelago has been identified as one of the 24 most pristine ecoregions of the world (Mittermeier *et al.* 2003), and was recently declared a Biosphere Reserve (Rozzi *et al.* 2006). Although the archipelago is isolated and located in the sub-Antarctic region it has not been spared the blight of biological invasions (Vitousek *et al.* 1996). One of these was by the American mink *Mustela vison*, imported to fur farms in southern Chile and Argentina during 1930–1950 (Jaksic *et al.* 2002). Accidental escapes and intentional releases led to the establishment of feral populations in many areas of both southern Argentina (Chehébar 1985) and Chile (Medina 1997, Jaksic *et al.* 2002). The ca. 2,528 km^2 Navarino Island, part of the Cape Horn Biosphere Reserve, is located south of Tierra del Fuego. Although mink were found in the wild in Tierra del Fuego as early as the 1960s, the first mink on Navarino Island was only recorded in 2001 (Rozzi & Sherriffs 2003). As birds constitute the most diverse vertebrate group in the Reserve, with several ground-nesting species (Rozzi & Sherriffs 2003), there is concern over which species are part of the diet of the alien American mink. Here we quantify the diet of mink and its seasonal variation to examine the potential impact of this alien species on ground-nesting birds in the wetlands of Navarino Island.

Navarino Island, which contains one settlement, Puerto Williams, with 2,262 inhabitants (Fig. 1), consists of thousands of water bodies (lakes, ponds, streams and bogs) and forest areas. The ecoregion corresponds to the Sub-Antarctic Magellanic Forest, dominated by the genus *Nothofagus* within a mosaic of *Sphagnum* spp. bogs. Most of the aquatic bird species inhabiting the wetlands are non-residents and arrive in summer to breed (J.T. Ibarra, R. Rozzi, H. Gilabert, C.B. Anderson, S.M. McGehee & C. Bonacic, unpubl. data). There are 56 species of birds in 25 families in the wetlands and adjacent habitats of Navarino Island. Potential mammalian prey of mink on the island include both native (Yellow-nosed Grass mouse *Abrothrix xanthorhinus* and Long-tailed Pygmy Rice rat *Oligoryzomys longicaudatus*) and exotic rodents (beaver *Castor canadensis*, mouse *Mus musculus*, muskrat *Ondatra zibethica* and Norway rat *Rattus norvegicus*). Aquatic prey include native fish (*Aplochiton taeniatus*, *Aplochiton zebra*, *Galaxias maculatus,* and *Galaxias platei*) and introduced trout species (*Onchorynchus mykiss* and *Salvelinus fontinalis*), which are known to be eaten by mink in other regions of Chile and Argentina (Medina, 1997).

A total of 414 mink scats were collected over January–November 2006 along the shoreline of 27 ponds. Faeces were sorted into seven categories according to their contents: (1) bird, including feathers, bone fragments, and egg shells; (2) mammal, including hair, teeth, and bone fragments; (3) fish, including vertebrae, scales, and otoliths; (4) insects, including appendages and exoskeleton fragments; (5) crustaceans, incuding appendages and exoskeletal fragments; (6) molluscs, including valve parts; (7) plant matter, including seeds and leaves. Mammals and birds were identified to species, if possible, using taxonomic keys and references. In the case of mammals, teeth and the medullar and cuticle scale patterns of hairs were compared to patterns described in guides (Reise 1973, Chehébar & Martín 1989, Pearson 1995). For the few bird species, feather colour patterns were used for identification. We computed frequency of occurrence expressed

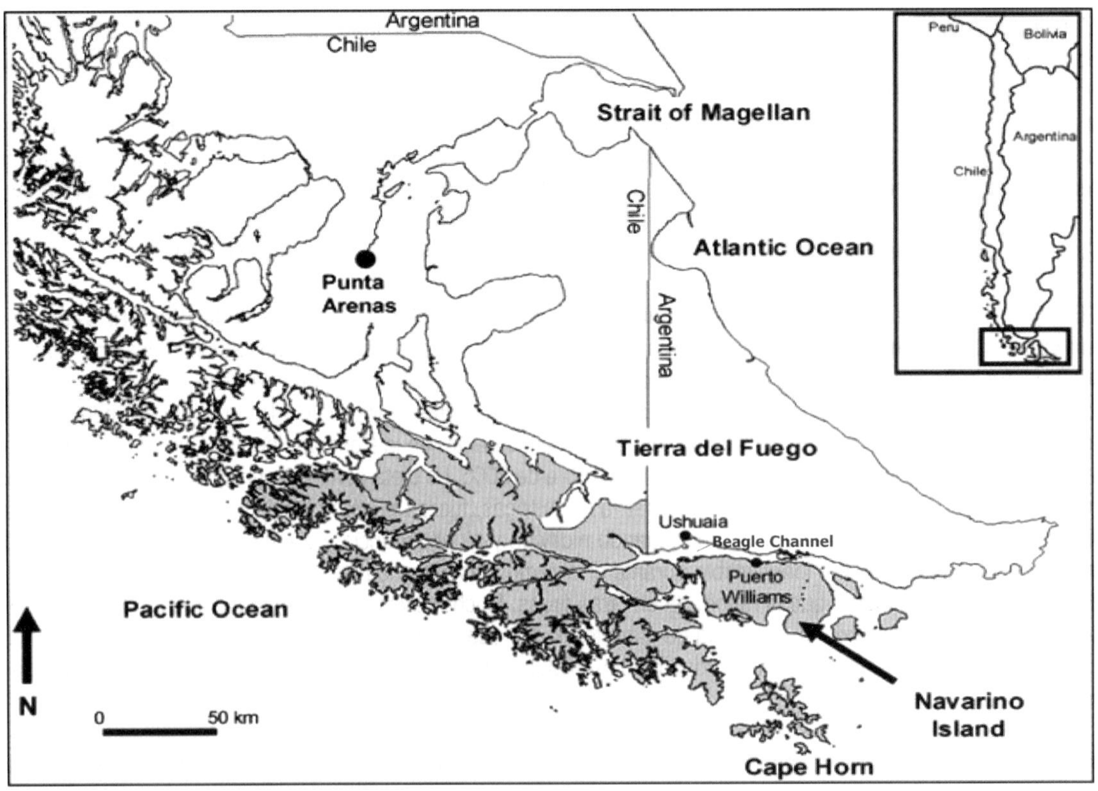

Fig. 1. Southern South America, showing the Cape Horn Biosphere Reserve (shaded in grey) and Navarino Island, Chile.

as a percentage (number of scats with a prey category divided by the total number of scats, by season), and percentage of bulk of a prey category (proportion of volume of a scat with a prey category multiplied by dry weight of the scat, divided by total dry weight of scats, by season).

The diet of mink in Navarino Island's wetlands consisted mainly of mammal, birds, insects and plant matter. Whereas the occurrence of mammals and birds in scats was similar during summer and winter, the representation of mammals in the scats in autumn and spring was almost double that of birds (Fig. 2). Of 272 scats with mammal remains, 235 comprised fragments amenable to further identification, including native, exotic, and unidentified mammals (Table 1). Exotic mammals (identified in 86 scats) were most abundant in the scats during autumn and winter, whereas native mammal species dominated during the summer (Table 1). Hair identification showed that muskrat (N = 78) was the most important exotic mammal prey followed by beaver (N = 4) and murid rodents (N = 4). Bird consumption was higher during the summer when 58.5% of scats contained bird remains (Fig. 2). Bird remains were identified in 202 scats, but only 11.4% contained remains that could be identified to species level. Remains of three species were identified: the ground-nesting Upland Goose *Chloephaga picta*, tree-cavity nesting Thorn-tailed Rayadito *Aphrastura spinicauda*, and Southern House Wren *Troglodytes musculus*. Egg shells were found in only 0.03% of the scats collected during spring and summer (N = 9) but mink eat the contents of eggs without necessarily eating the shell (Ferreras & Macdonald 1999). Plant matter was more frequent in winter scats and seemed to compensate for the decrease in mammal consumption recorded for this season (Fig. 2). For details of the importance of other prey items see Fig. 2.

American mink have only recently invaded Navarino Island and this is the first report of their diet in the early years of their establishment, with a significant representation of native wetland and terrestrial birds. The consumption of birds was highest during the warmer months. Of mammalian prey, introduced species were mainly consumed during autumn and winter. This demonstrates that introduced prey species are available for mink when wild rodents and birds are probably less abundant. The three bird species identified in scats have their southernmost breeding distribution in the Cape Horn Biosphere Reserve.

The consumption of birds by mink in the Reserve appeared to be higher than reported in other parts of Patagonia (Medina 1997, Previtali *et al.* 1998) and elsewhere (Erlinge 1969, Melquist *et al.* 1981, Ferreras & Macdonald, 1999), but comparable to that on Scottish islands in the UK (Clode & Macdonald 1995). Unsurprisingly, birds mostly featured in the diet of mink during the nesting season.

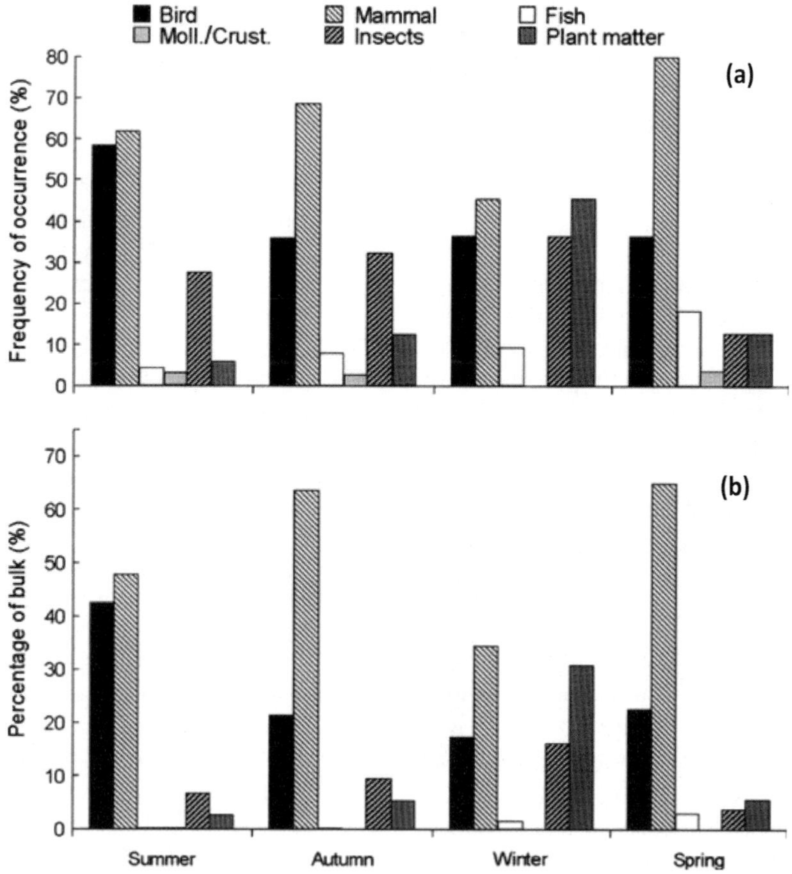

Fig. 2. (a) Percentage frequency of occurrence and (b) percentage of bulk of bird, mammal, fish, mollusc/crustacean and plant remains (see text for further details) in American mink scats collected on Navarino Island (Fig. 1), by season (see Table 1), in 2006.

TABLE 1. Mammalian remains found in mink scats, by season, with % of scats containing exotic, native, and unidentified mammal species, on Navarino Island (Fig. 1).

Season	No. of scats	No. with mammalian remains	% with exotic mammals	% with native mammals*	% with unidentified mammals
Summer (January-February)	236	146	21.2	51.4	27.4
Autumn (April-May)	112	77	46.8	26.0	27.3
Winter (July-August)	11	5	60.0	0.0	40.0
Spring (October-November)	55	44	36.4	38.6	25.0

*Representation of native mammals differed significantly from that of exotic mammals between seasons (χ^2 = 18.97, df = 2; P < 0.0001; analysis excluding winter data). Differences were driven by summer peak in consumption of native mammals (significant differences disappeared when summer was excluded, χ^2 = 2.13, df = 1, P = 0.144).

Because there are no terrestrial mammalian predators native to Navarino Island (Anderson *et al.* 2006), many bird species have evolved ground-nesting strategies (Moore *et al.* 2003). Under these circumstances invasive mammalian predators are particularly damaging (Amarasekare 1993, Macdonald *et al.* 2007). The high proportion of birds in the diet of mink on Navarino Island is only comparable to that recorded for islands sharing the same characteristics (Clode & Macdonald 1995) and where negative effects of mink on native birds were corroborated. Our findings indicate unambiguously that mink on Navarino Island are feeding on native vertebrates. We emphasize the need for further research to reveal whether predation by invasive American mink is affecting the populations of native mammals and birds in the Cape Horn Biosphere Reserve. We are designing studies that will allow us to answer this question.

ACKNOWLEDGEMENTS

This study was funded by the Chilean Environmental National Commission (CONAMA; Project 12-001-06), the Institute of Ecology and Biodiversity (ICM P05-002), University of Magallanes and Wildlife Trust Alliance. We received support from the UK Darwin Initiative (Refs 15006 and 13016). We sincerely thank Antonia Barreau, Tomás Altamirano, Steven McGehee, and Germán González for help in the field and laboratory. This article is part of the ongoing research and conservation programmes conducted at the Omora Ethnobotanical Park (http://www.omora.org) on the Cape Horn Biosphere Reserve.

REFERENCES

AMARASEKARE, P. 1993. Potential impact of mammalian nest predators on endemic forest birds of western Mauna Kea, Hawaii. *Conservation Biology* 7: 316-324.

ANDERSON, C.B., R. ROZZI, J.C. TORRES-MURA, S.M. MCGEHEE, M.F. SHERRIFFS, E. SCHÜTTLER & A.D. ROSEMOND. 2006. Exotic vertebrate fauna in the remote and pristine sub-Antarctic Cape Horn Archipelago, Chile. *Biodiversity and Conservation* 15: 3295-3313.

CHEHÉBAR, C. 1985. A survey of the southern river otter *Lutra provocax* Thomas in Nahuel Huapi National Park, Argentina. *Biological Conservation* 32: 299-307.

CHEHÉBAR, C. & S. MARTÍN. 1989. Guía para el reconocimiento microscópico de los pelos de los mamíferos de Patagonia. *Doñana, Acta Vertebrata* 16: 247-291.

CLODE, D. & D.W. MACDONALD. 1995. Evidence for food competition between mink (*Mustela vison*) and otter (*Lutra lutra*) on Scottish islands. *Journal of Zoology* 237: 435-444.

ERLINGE, S. 1969. Food habits of the otter (*Lutra lutra* L.) and the mink (*Mustela vison* Schreber) in Sweden. *Oikos* 23: 327-335.

FERRERAS, P. & D.W. MACDONALD. 1999. The impact of American mink *Mustela vison* on water birds in the upper Thames. *Journal of Applied Ecology* 36: 701-708.

JAKSIC, F., A. IRIARTE, J. JIMÉNEZ & D. MARTÍNEZ. 2002. Invaders without frontiers: cross-border invasions of exotic mammals. *Biological Invasions* 4: 153-173.

MACDONALD, D.W., C.M. KING & R. STRACHAN. 2007. Introduced species and the line between biodiversity conservation and naturalistic eugenics. Pp. 186-205 *in Key Topics in Conservation Biology*, D.W. Macdonald & K. Services, (eds.). Blackwell Publishing, Oxford, United Kingdom.

MEDINA, G. 1997. A comparison of the diet and distribution of southern river otter (*Lutra provocax*) and mink (*Mustela vison*) in southern Chile. *Journal of Zoology* 242: 291-297.

MELQUIST, W.E., J.S. WHITMAN & M.G. HORNOCKER 1981. Resource partitioning and co-existence of sympatric mink and river otter populations. Pp. 187-221 *in Proceedings of the Worldwide Furbearers Conference 1*, J.A. Chapman & P. Pursley (eds.). R.R. Donnely and Sons Co., Frostburg, USA.

MITTERMEIER, R.A., C.G. MITTERMEIER, T.M. BROOKS, J.D. PILGRIM, W.R. KONSTANT, G.A.B. DA FONSECA & C. KORMOS. 2003. Wilderness and biodiversity conservation. *Proceedings of the National Academy of Sciences* 100: 10309-10313.

MOORE, N.P., S.S. ROY & A. HELYAR. 2003. Mink (*Mustela vison*) eradication to protect ground-nesting birds in the Western Isles, Scotland, United Kingdom. *New Zealand Journal of Zoology* 30: 443-452.

PEARSON, O. 1995. Annotated key for identifying small mammals living in or near Nahuel Huapi National Park or Lanín National Park, Southern Patagonia. *Mastozoología Neotropical* 2: 99-148.

PREVITALI, A., M.H. CASSINI & D.W. MACDONALD. 1998. Habitat use and diet of mink in Argentine Patagonia. *Journal of Zoology* 246: 482-486.

REISE, D. 1973. Clave para la identificación de los cráneos de marsupiales y roedores chilenos. *Gayana Zoología* 27: 1-20.

ROZZI, R., F. MASSARDO, A. BERGHOFER, C.B. ANDERSON, A. MANSILLA, M. MANSILLA *et al.* 2006. *Cape Horn Biosphere Reserve: Nomination Document for the Incorporation of the Cape Horn Archipelago Territory into the World Biosphere Reserve Network.* Programa MAB-Unesco. Ediciones de la Universidad de Magallanes, Punta Arenas, Chile.

ROZZI, R. & M. SHERRIFFS. 2003. The mink (*Mustela vison* Schreber, Carnivora: Mustelidae), new alien mammal for Navarino Island. *Anales del Instituto de la Patagonia* 31: 97-104.

VITOUSEK, P.M., C.M. D'ANTONIO, L.L. LOOPE & R. WESTBROOKS. 1996. Biological invasions as global environmental change. *American Scientist* 84: 468-478.

BIOGRAPHICAL SKETCHES

JOSÉ TOMÁS IBARRA studies bird assemblages and the impact of American mink on island ecosystems in the Cape Horn Biosphere Reserve.

LAURA FASOLA studies other aspects of American mink and southern river otter ecology in Argentinean Patagonia.

DAVID MACDONALD is head of the Wildlife Conservation Research Unit at the University of Oxford, which focuses on aspects of fundamental biology relevant to solving problems of wildlife conservation and environmental management. RICARDO ROZZI is a philosopher and ecologist and his research combines both disciplines through the study of the relationships between the ways of knowing and inhabiting the natural world.

CRISTIÁN BONACIC is head of Fauna Australis, a Chilean wildlife conservation research organization, teaching and outreach group for biodiversity conservation and wildlife management.

Received 5 February 2008. Revision requested 28 May 2008. Accepted 4 July 2008.

23 VULNERABILITY OF GROUND-NESTING WATERBIRDS TO PREDATION BY INVASIVE AMERICAN MINK IN THE CAPE HORN BIOSPHERE RESERVE, CHILE

ELKE SCHÜTTLER[1,2,3,*], REINHARD KLENKE[1,4], STEVEN MCGEHEE[2,5], RICARDO ROZZI[2,6], AND KURT JAX[1,2,3]

[1]Department of Conservation Biology, UFZ-Helmholtz Centre for Environmental Research, Permoserstr. 15, 04318 Leipzig, Germany
[2]Parque Etnobotánico Omora (IEB-Instituto de Ecología y Biodiversidad, Universidad de Magallanes y Fundación Omora), Puerto Williams, Provincia Antártica, Chile
[3]Lehrstuhl für Landschaftsökologie, Technische Universität München-Weihenstephan, Am Hochanger 6, 85350 Freising, Germany
[4]Society for Nature Conservation and Landscape Ecology, Dorfstr. 31, 17237 Kratzeburg, Germany
[5]Department of Biology, University of Victoria, P.O. Box 3020,Victoria, B.C. V8W 3N5, Canada
[6]Department of Philosophy, University of North Texas, Denton, TX 76201, USA
*Correspondent author: Department of Conservation Biology, UFZ-Helmholtz Centre for Environmental Research, Permoserstr. 15, 04318 Leipzig, Germany. Tel.: +49 341 2351652; fax: +49 341 2351470 E-mail: elke.schuettler@ufz.de, eschuttler@gmx.de (E. Schüttler), reinhard.klenke@ufz.de (R. Klenke), smcgehee@uvic.ca (S. McGehee), rozzi@unt.edu (R. Rozzi), kurt.jax@ufz.de (K. Jax)

Biological Conservation 142: 1450-1460 (2009)

ABSTRACT. Vulnerability of ground-nesting waterbirds to predation by invasive American mink in the Cape Horn Biosphere Reserve, Chile. Biological invasions constitute one of the most important threats to biodiversity. This is especially true for "naïve" birds that have evolved in the absence of terrestrial predators in island ecosystems. The American mink (*Mustela vison*) has recently established a feral population on Navarino Island (55°S), southern Chile, where it represents a new guild of terrestrial mammal predators. We investigated the impact of mink on ground-nesting coastal waterbirds with the aim of deriving a vulnerability profile for birds as a function of different breeding strategies, habitat, and nest characteristics. We compared rates of nest survival and mink predation on 102 nests of solitary nesting species (*Chloephaga picta*, *Tachyeres pteneres*), on 361 nests of colonial birds (*Larus dominicanus*, *Larus scoresbii*, *Sterna hirundinacea*), and on 558 artificial nests. We calculated relative mink and bird densities at all nest sites. Nests of colonial species showed the highest nest survival probabilities (67-84%) and no predation by mink. Nest survival rates for solitary nesting species were lower (5–20%) and mink predation rates higher (10-44%). Discriminant analyses revealed that mink preyed upon artificial nests mainly at shores with rocky outcroppings where mink were abundant. High nest concealment increased the probability for predation by mink. Conservation planning should consider that invasive mink might severely affect the reproduction success of bird species with the following characteristics: solitary nesting, nesting habitat at rocky outcrop shores, and concealed nests. We recommend that work starts immediately to control the mink population with a priority in the nesting habitats of vulnerable endemic waterbirds.

Key words: artificial nests, breeding birds, management, *Mustela vison*, nest characteristics, nest survival.

RESUMEN. Vulnerabilidad de aves acuáticas de nidificación terrestre a la depredación por el visón norteamericano invasor en la Reserva de Biosfera Cabo de Hornos, Chile. Las invasiones biológicas constituyen una de las amenazas más importantes para la biodiversidad. Esto es especialmente cierto para aves "ingenuas" que han evolucionado en ausencia de depredadores terrestres en ecosistemas tipo isla. El visón norteamericano (*Mustela vison*) ha establecido recientemente una población silvestre en la isla Navarino (55°S), sur de Chile, donde representa un nuevo gremio de mamíferos depredadores terrestres. Investigamos el impacto del visón sobre las aves acuáticas costeras que nidifican sobre el suelo con el objetivo de derivar un perfil de vulnerabilidad de aves en función de diferentes estrategias reproductivas, hábitat y características del nido. Comparamos las tasas de sobrevivencia de nidos y la depredación por visón en 102 nidos de especies que nidifican solitarias (*Chloephaga picta*, *Tachyeres pteneres*), en 361 nidos de aves coloniales (*Larus dominicanus*, *Larus scoresbii*, *Sterna hirundinacea*) y en 558 nidos artificiales. Calculamos las densidades relativas del visón y aves en todos los sitios de nidificación. Los nidos de especies coloniales mostraron las mayores probabilidades de sobrevivencia (67-84%) y ninguna depredación por visón. Las tasas de sobrevivencia de los nidos de especies de nidificación solitaria fueron menores (5-20%) y las tasas de depredación por visón más altas (10-44%). Los análisis discriminantes revelaron que el visón depredó sobre nidos artificiales principalmente en costas con afloramiento rocoso donde el visón era abundante. La cobertura de los nidos aumentó la probabilidad de depredación por el visón. La planificación para la conservación debería considerar que el visón invasor podría afectar seriamente el éxito reproductivo de las especies de aves con las siguientes características: nidificación solitaria, nidificar en hábitats con afloramientos rocosos y nidos ocultos. Recomendamos que el trabajo para controlar la población de visones costeros se comience inmediatamente con prioridad en los hábitats de nidificación de las aves acuáticas endémicas vulnerables.

Palabras clave: aves reproductivas, características del nido, manejo, *Mustela vison*, nidos artificiales, sobrevivencia del nido.

INTRODUCTION

The earth's biota is greatly altered by invasive plant and animal species producing concern and discussion about their ecological consequences (Elton 1958, Vitousek *et al.* 1997, Gobster 2005, Vellend *et al.* 2007). Biodiversity on islands is particularly vulnerable to biotic exchange (Courchamp *et al.* 2003, Sax & Gaines 2008). The survival of introduced species on islands and the significance of their ecological impacts are less a matter of low insular biodiversity (Levine & D'Antonio 1999); rather it depends on the nature of those species that are present or those groups of species that are missing from the islands (Goodman 1975, Simberloff 1995). This is especially true for alien carnivore invasions on islands where terrestrial mammalian predators were absent before. Their impact on insular bird populations can cause extensive population reductions and even local extinctions (King 1985, Atkinson 1996, Macdonald & Thom 2001).

Bird populations are regulated by natural limiting factors like predation, food supply, nest sites, parasites, pathogens, competition, and human-induced factors like hunting, pesticides or pollutants (Newton 1998). The effects of predation depend on the extend to which it is additive to compensation by other losses. In some ground-nesting waterbirds, however, predation can not only reduce egg and chick stages (Opermanis *et al.* 2001, Kauhala 2004, Nordström & Korpimäki 2004), but actually also their breeding numbers (Côté & Sutherland 1997, Newton 1998). Hence, bird species are assumed to develop their own strategies to minimize predation (Martin 1993). It is widely accepted that prey naïvety plays a significant role in the confrontation with the threat of an introduced predator, because native fauna often lack those strategies to minimize predation as behavioral or evolutionary adaptations (Berger *et al.* 2001 Short *et al.* 2002, Nordström *et al.* 2004). Critical factors among those adaptations are (i) social factors like coloniality (Inman & Krebs 1987, Siegel-Causey & Kharitonov 1990), (ii) area-specific factors like habitat (Willson *et al.* 2001, Whittingham & Evans 2004) or nest density (Ackerman *et al.* 2004), and (iii) site-specific factors like nest height (Martin 1995) or nest concealment (Butler & Rotella 1998, Rangen *et al.* 2000). Those factors have been investigated separately or in combination, with artificial and/or natural nests, often with contradictory results (Major & Kendal 1996). Finally, nest predation processes cannot fully be understood without knowledge of the predator community, *i.e.*, abundance and searching behavior of predators (Angelstam 1986, Miller & Knight 1993).

The American mink (*Mustela vison*) is a carnivorous species from North America that has recently established its southernmost reproducing population in the world on Navarino Island, Cape Horn Biosphere Reserve (southern Chile, 54-56ºS). It was first registered on the island in 2001 (Jaksic *et al.* 2002, Rozzi & Sherriffs 2003), but arrived earlier in Tierra del Fuego on the other side of the Beagle Channel in the 1940s and 1950s (Lizarralde & Escobar 2000). Therefore, it is most probable that some individuals crossed the Beagle Channel after escaping from fur farms in Tierra del Fuego (Rozzi & Sherriffs 2003). On Navarino Island, mink represent a new guild (Root 1967) because the island lacks native terrestrial mammalian predators. In this pristine ecoregion the most diverse and abundant group of vertebrates are birds (Rozzi *et al.* 2006). Many of them are ground-nesting, including two songbird species (*Turdus falcklandii, Zonotrichia capensis*) that use ground nests in the Cape Horn region (S. McGehee, unpublished data), while in other parts of Chile the same species nest in trees. Therefore, scientists and public agencies have expressed strong concerns about the impact of mink on the island's avifauna, especially ground-nesting birds (Rozzi & Sherriffs 2003, Anderson *et al.* 2006, Soto & Cabello 2007).

American mink are semi-aquatic mustelids inhabiting marine coasts, flowing waters, and banks with a generalist diet including prey from both aquatic and terrestrial sources (Dunstone 1993). Birds are most exposed to the risks of opportunistic predation by mink during their reproductive period due to the birds' limited mobility (Arnold & Fritzell 1987, Bartoszewicz & Zalewski 2003) in combination with the higher energy requirement of the breeding mink (Dunstone 1993). In Europe, introduced mink have successfully established feral populations (reviews in Macdonald & Harrington 2003; Bonesi & Palazon 2007), which prey significantly upon ground-nesting wetland birds (Ferreras & Macdonald 1999) and seabirds (Collis 2004, Nordström & Korpimäki 2004). Also some cases of surplus-killing of chicks and adults within a colony have been reported (*e.g.*, Craik 1997). In South America, wild mink populations in the southern part of Chile and Argentina also include birds in their diets (Medina 1997, Previtali *et al.* 1998, Fasola *et al.* 2008, Schüttler *et al.* 2008, Ibarra *et al.* 2009). However, studies on the impact of mink on waterbirds in the southern hemisphere are scarce.

The main purpose of our study was to understand the impact of the American mink as a recently introduced terrestrial predator on the nest survival of naïve ground-nesting waterbirds on Navarino Island. We aim to draw an overall vulnerability profile of bird species to predation by mink as a function of their breeding strategy (colonial vs. solitary nesting), as well as area-specific (habitat), and site-specific (nest concealment) factors. Based on this profile, we discuss high priority species of ground-nesting waterbirds for conservation and implications for the management of mink populations in the southernmost tip of the Americas.

METHODS

Study area

The study was carried out on Navarino Island (2528 km^2), Chile, located at the extreme southern tip of South America (Fig. 1). The island forms part of the Cape Horn Biosphere Reserve (Rozzi *et al.* 2006) and belongs to the Magellanic Sub-Antarctic forest ecoregion, recently identified as one of the 24 most pristine wilderness areas of the world Forest Biome (Mittermeier *et al.* 2003). The main habitats include (i) evergreen rainforests dominated by *Nothofagus betuloides* and *Drimys winteri*, (ii) Magellanic deciduous forests of *Nothofagus pumilio*, (iii) peatlands, moorlands, and bogs (*Sphagnum* spp.), (iv) high-Andean communities dominated by cushion plants and lichens, and (v) thickets or scrublands in naturally or anthropogenically disturbed areas (Pisano 1977, Rozzi *et al.* 2006). The climate type is oceanic, with a low annual thermic fluctuation (<5°C), a mean annual temperature of 6°C, and an annual precipitation of 467.3 mm (Pisano 1977). During

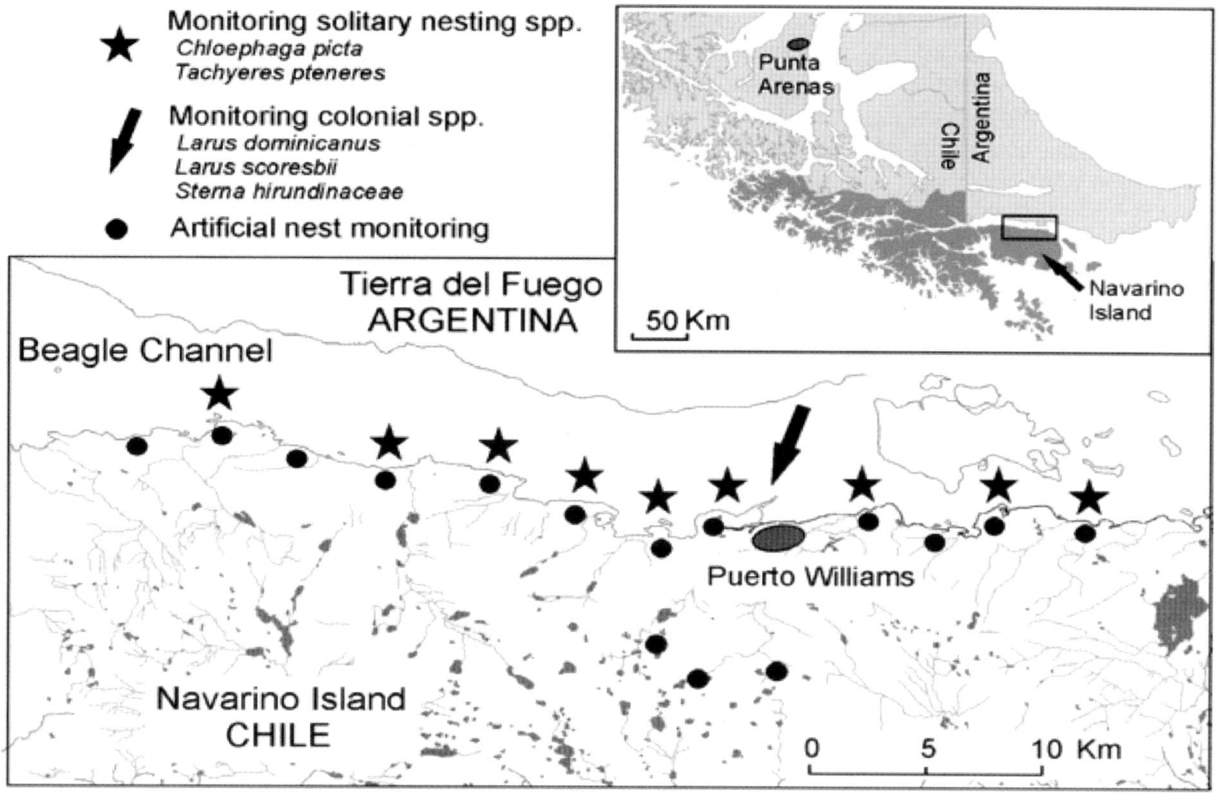

Fig. 1. Map of the nest monitoring study sites. Navarino Island is located within the Cape Horn Biosphere Reserve (54°-56°S, shaded in dark grey, top right) in southern South America.

winter, streams and lakes are ice-bound. Human population is concentrated in the town of Puerto Williams (ca. 2200 inhabitants), the capital city of the Chilean Antarctic Province, on the northern coast of Navarino. Due to the extremely limited infrastructure on Navarino Island – only one dirt road connects the northern coast – our research was mainly restricted to this accessible coast of the island. The interior of the island must be reached by the three existing trekking trails, and western, southern, and eastern coasts rely on water transport.

Our study sites comprised twelve 4-km long transects of marine shoreline and three lakes at a distance of 5.3-7.7 km from the coast and an altitude of 387-510 m. All study sites were separated by more than 3 km in order to cover distinct territories of mink, which occupy on average linear territories of 3 km (Dunstone 1993). We conducted natural and artificial nest monitoring, bird counts, and mink surveys in the same coastal study sites and during the same breeding seasons. At lakes, we only focused on artificial nest monitoring and mink surveys.

Species studied

We concentrated our study on solitary nesting and colonial species that are resident, common, or endemic in the region. The solitary nesting species studied were the Upland Goose (*Chloephaga picta*) and the Fightless Steamer-Duck (*Tachyeres pteneres*). The Upland Goose occurs as a resident on coasts and in wet grasslands of Patagonia and the Falkland Islands (Couve & Vidal 2003). It usually breeds close to water (up to 200 m), along the coast, river valleys, and around ponds (Summers & McAdam 1993). On Navarino Island, Upland Geese were found breeding close to water mainly in scrublands dominated by *Berberis buxifolia*, *Pernettya mucronata* and *Chiliotrichum diffusum* (Moore 1983), and less frequently in forested habitats or meadow communities. The Fightless Steamer Duck is a strictly coastal species endemic to western Patagonia and Tierra del Fuego (Couve & Vidal 2003). The flightless species nests on rocky outcrops, but access to uplands and to the sea must be easy (Weller 1976). This pattern was observed on Navarino Island, too. Among colonial seabirds, we focused on Kelp Gulls (*Larus dominicanus*), Dolphin Gulls (*Larus scoresbii*), and South American Terns (*Sterna hirundinacea*). Only the Dolphin Gull is endemic to Patagonia and the Falkland Islands (Couve & Vidal 2003). Kelp gulls nest in a wide variety of environments along the sea coast and at continental wetlands (Yorio *et al.* 1999; Yorio & Borboroglu 2002), whereas Dolphin gulls and South American terns are more restricted to breeding habitats on bare rocks close to the water's edge or on small offshore islands (Scolaro *et al.* 1996, Yorio *et al.* 1996). On Navarino Island, these three species nested in close vicinity (<20 m) to each other on an exposed marine peninsula composed of bare gravels with marine deposits and meadow patches, the latter were used for nesting by the terns (Fig. 1). Potential autochtonous predators of eggs include the Southern Crested Caracara (*Caracara plancus*), Chimango Caracara (*Milvago chimango*), Chilean Skua (*Catharacta chilensis*), and Kelp Gull (Johnson 1965). Among the mammal species introduced to the Cape Horn Biosphere Reserve, the American mink and feral domestic dog (*Canis familiaris*) prey upon bird eggs. Predation by Norway rats (*Rattus norvegicus*) and feral pigs (*Sus scrofa*) was assumed to be uncommon as rats are only associated with the one town on the island, Puerto Williams (Anderson *et al.* 2006), and tracks of pigs were absent along our transect walks. Humans occasionally take eggs from the nests of Upland Geese and gulls.

Natural nests

Breeding activity of solitary nesting and colonial bird species was monitored along the northern coast of Navarino Island during 74 days of the nesting season of 2005/2006 and during 88 days of the nesting season of 2006/2007 starting on 1st November on each year. We monitored 463 nests: Upland Goose (*N* = 79), Fightless Steamer Duck (*N* = 23), Kelp Gull (*N* = 204), Dolphin Gull (*N* = 83), and South American Tern (*N* = 74). Study sites comprised seven (2005/2006) and nine transects (2006/2007) of 4 km shoreline. In order to detect nests of solitary nesting species, we walked the transects and recorded territorial behavior or presence of a guarding male. Dolphin Gulls and South American Terns were only found breeding in 2006/2007 and in close vicinity to the Kelp Gull colony. Typically, eight days (8.16 d, SD = 1.91 d) elapsed between successive visits to the same nest. At each visit we recorded the number of eggs until hatching, abandonment, or predation occurred. In order to minimize positive or negative observer effects on nest survival as a result of human tracks, nest attendance, or behavior (Götmark 1992), we chose a moderate frequency of nest visits, flagged shrubs decently with short orange and yellow tapes at a distance of 5 m from the nest and reduced nest visit time to the minimum (<1 min). We use the term nest survival to refer to the probability that a nest will hatch at least one young over the entire nesting period (Dinsmore *et al.* 2002, Jehle *et al.* 2004). This definition does not take into account predation events affecting only some eggs of a surviving nest; therefore, it overestimates offspring's survival. However, in our nest data, predation or unknown loss of some eggs in successful nests only happened in 9% of all successful nests (*N* = 318). These nests concerned 7 nests of Upland Geese, 16 nests of Kelp Gulls, 2 nests of Dolphin Gulls, and 3 nests of South American Terns. We guess that possible predators for Upland Geese were autochthonous bird predators or humans rather than the mink, since it destroys many or all eggs at once (Ferreras & Macdonald 1999). In the colonies the gulls themselves were probably responsible for scavenging or removing eggs.

Artificial nests

Although artificial nests are widely used in avian field studies (Moore & Robinson 2004), they have been criticized as not reliably reflecting predators and predation rates of natural nests (*e.g.*, Faaborg 2003). This is because artificial nests differ from real nests in a number of important ways, such as nest type, egg type, concealment, nest spacing, odor, missing adults, etc. (Major & Kendal 1996). In order to maximize the external validity of experimental design, authors recommend using artificial nests primarily in conjunction with active nests and identifying the predators at both types of nest (Mezquida & Marone 2003, Moore & Robinson 2004). So far, a direct comparable context (same data taken at the same time and location, and with the same methods) has been achieved in only a few studies (review in Moore & Robinson 2004). Here, we combined natural nest monitoring with artificial nests in a comparable context as a way to effectively investigate the influence of habitat and nest characteristics on the predator type. Yet, we are aware that the interpretation of our data derived from artificial nests has to be treated with caution.

Artificial nests were constructed to imitate geese nests and were baited with one domestic chicken egg and one clay egg. Artificial ground nests were of approximately 20 cm in diameter, lined with dry plant material and Upland Goose down. In order to reduce olfactory cues that might influence predators, we washed the chicken eggs and used gloves when handling eggs and nests. The nests were marked with flagging tape in the same manner as for the natural nests.

Five hundred and seventy-five artificial nests were installed at twelve study sites in marine coastal habitats (*N* = 500) and at three sites along lakes (*N* = 75). These sites included two different types of shore: (1) "rocky outcrop" (*N* = 225 nests), and (2) "beaches" (*N* = 350 nests). These two shore types were assigned as follows: within each study site, each 200 m we measured the percentage of cliff, rock, pebbles, sand, mud, and vegetation (leaf litter, grasses, mosses) within 10 x 1 m of the shoreline, as well as the incline of the shore within 10 m from the water shed (flat: 0-1 m, medium: 1-2 m, steep: <2 m). The first shore type, rocky outcrop, was assigned when over the half of these measurements were predominated by cliffs, rocks, and steep shores, whereas beaches were characterized by over 50% pebbles, sand, mud, vegetation, and flat shores. The adjacent vegetation in the twelve study sites were majorly shrubs of *B. buxifolia* and *C. diffusum*, but also forested habitats (*Nothofagus* spp.) and meadows (Pisano 1977, Moore 1983, Rozzi *et al.* 2006). In coastal habitats, 200 nests were placed in a 71 day nesting season starting on 29th December 2005, and 300 nests were placed in a 69 day nesting season starting on 4th December 2006. At the lakes, 75 nests were installed starting on 6th January 2007 in order to survey for mink predation in the interior of the island where Upland Geese use wetlands for breeding (Summers & McAdam 1993, S. McGehee, unpublished data).

Each of the 12 artificial nest sites comprised a stretch of 1.25 km, where 25 nests were distributed with a distance of 50 m in between each nest. Nests were placed up to 30 m from the water's edge in different vegetation types, cover and different degrees of nest concealment defined by the vegetation found at each 50 m mark. We monitored nests at 5 day intervals (5.12 d, SD = 0.41 d) for 29-30 days, which is the incubation period of upland geese after completing the clutch (Summers & McAdam 1993). We considered a nest to have been preyed upon when at least one egg was found preyed on or marked with bills or teeth. Humans destroyed 17 nests in the marine coastal habitat. These nests were excluded from our analysis (total *N* = 558).

Predator identification

We categorized predators into five groups: American mink, domestic dogs, humans, birds, and "unknown" for uncertain cases caused by multi-predator visits or the lack of clues. The identification of predators was based on a detailed examination of eggshells and their location, nest material dislocation, and other signs, such as the presence of hairs or scats. Although some authors preclude the identification of nest predators from nest remains (*e.g.*, Larivière 1999), we believe that we classified predators in an unbiased manner. The predator community on Navarino Island is remarkably small and overlaps in predator patterns are rare. However, for the three species nesting in colonies, it was difficult to identify predators, mostly due to the disappearance of egg shells, which were exposed to strong winds and bird activity in the colony. To diagnose mink predation, we followed Craik (1995) and Opermanis *et al.* (2001). Mink predation signs were: canine marks, typically 1-2 mm wide and, if paired, ca 10 mm apart, egg shells often hidden under vegetation, eggs might be untouched, and nests little damaged. Bird predation signs were: small egg fragments, eggshells in nest or vicinity, eggs missing, and nest material lifted or spread to the nearby surroundings. Dog predation signs were: egg punctures 4-5 mm wide and, if paired, >3 cm apart, egg fragments >½ egg, widespread, and nest remains spread out. If all eggs disappeared without signs of fragments, and downy feathers still covered the nest, this pattern was attributed to humans (but only for solitary nesting species). Employing comparison with beak imprints in the clay eggs taken from specimens, 42.3% of artificial clay eggs preyed upon by birds could further be classified into species. A small number of imprints of rodent incisors on artificial clay eggs (2.5%, N = 14 nests) were attributed to mice (either *Abrothrix xanthorhinus* or *Oligoryzomys longicaudatus*).

Area and nest site characteristics

We examined the effect of area and site characteristics of nests on the type of predator (mink vs. birds) they attracted. We chose variables that were important for different search tactics (*e.g.*, Butler & Rotella 1998) (Appendix A). We took measurements at all nest sites of solitary nesting species (N = 102) and at all artificial nests set in the nesting period 2006/2007 (N = 375). Area-specific variables were taken for all artificial nests (N = 558). Measurements were conducted on the day a nest was found or constructed.

Bird and mink abundance

We censused bird populations 2-4 times during each breeding season at seven (2005/2006) and nine (2006/2007) coastal study sites. We counted the target species and avian predators (together eight species) using binoculars (8 x 25) while walking 4-km transects along the shorelines during the morning. We recorded all observations of adult and juvenile animals on both sides of the transect, up to 50 m away (1 transect = 40 ha). Line transects are favored over point counts if targeted species are relatively easy to identify, but mobile, and occurring at low densities (Bibby *et al.* 2000). Gulls and terns occupying large colonies were counted from a larger distance to prevent flushing and were cross-checked by a second observer. However, we are aware that our estimations of colonial birds are approximate. For our analysis we pooled data over study sites, although the abundance of solitary nesting species and predatory birds (colonies excluded) differed significantly between sites (Kruskal-Wallis-Test: χ^2 = 224.7, df = 6, P < 0.001). However, further investigation of the causes of these differences is beyond the scope of this paper.

We systematically surveyed for mink signs (scats, tracks, sightings) twice each breeding season (spring and summer) at seven (2005/2006) to twelve (2006/2007) marine coastal study sites and at three lakes (2006/2007). At three marine sites and at lakes, we relied on data from one survey only (summer 2007). The 4-km transects and lake perimeters (1 km) were divided into 500 m contiguous sections and the proportion of positive sections (with signs) for each transect was calculated (*e.g.*, Bonesi & Macdonald 2004).

Statistical analysis

For estimates of nest survival of natural and artificial nests, we used the Mayfield estimator (Mayfield 1961) with the standard error developed by Johnson (1979). The Mayfield method estimates the daily survival rate (*DSR*) as *DSR* = 1 - *DPR* (daily predation rate). *DPR* is calculated as the number of failed nests divided by the number of exposure days. For the calculation of the number of exposure days of failed nests, we assumed that failure occurred at the midpoint between the final nest checks. The nest survival rate over the nesting period t (days) is calculated as $(DSR)^t$, which can be expressed as a percentage. The durations of egg-laying and incubation periods were taken from the literature (see caption Fig. 2). As we lacked literature for Fightless Steamer Ducks, we used the periods described for Upland Geese. Data for both breeding periods were pooled. As data from the egg-laying period was sparse, we could not stratify by stages (as recommended by Jehle *et al.* 2004) and thus, had to assume constant nest survival.

We used principal component analysis (PCA) and linear discriminant analysis (DA) in order to check for the compatibility of artificial nests with natural nests, and to check for differences in site-specific nest variables (six variables, Appendix A) between nests of the solitary nesting species (classes = species). This first data set (DA 1) combined artificial and natural nests (N = 475). In a second data set (DA 2), we investigated the combination of area-specific and site-specific nest variables (eleven variables, Appendix A), which best separates the type of predator (classes = predators) using artificial nests (N = 375, 2006/2007). We performed a DA based on the results of the PCA using the ade4 package rewritten for the R environment (R Development Core Team, 2008) of the ADE-4 software (Thioulouse *et al.* 1997). The implementation of the ade4 package follows the "French way" (Holmes 2008) and is based on the use of a unifying mathematical tool: the duality diagram (for details see Dray & Dufour 2007). For DA 2, we applied a reduced set of variables based on the PCA results in order to prevent redundancy of information. The significance of the DA was tested by a Monte-Carlo permutation test. Continuous variables that were not normally distributed were transformed. We log transformed distance and height, and arcsine square-root transformed mink density. The PCA routine of the ade4 library applies variables standardized to zero mean and unit variance.

We used non-parametric statistics for comparing proportions (2-sample tests for equality of proportions), sample medians (Wilcoxon's rank sum test), and for testing for independence in contingency tables (Fisher's exact and chi-square tests) with Yates' continuity and Bonferroni corrections, all two-sided. The statistical analyses conducted in R version 2.7.1. (R Development Core Team, 2008) were considered significant when *p*-values were <0.05. The discriminant analysis DA 2 was documented in R source code and submitted as Supplementary material.

RESULTS

Nest survival probabilities

Mayfield constant nest survival rates were comparatively high for species nesting in colonies: 84.2% for Dolphin Gulls, 76.3% for South American Terns, and 67.2% for Kelp Gulls. In contrast, nest survival rates were low for the solitary nesting Upland Goose (20.0%), and very low for the solitary nesting Flightless Steamer Duck (5.2%) (Fig. 2). Artificial nests had the lowest survival rates with only 0.4% surviving. As the 95% confidence intervals of Mayfield nest survival probabilities of species nesting in colonies did not overlap with those of solitary nesting species, differences in the nest survival of these two breeding strategies were significant.

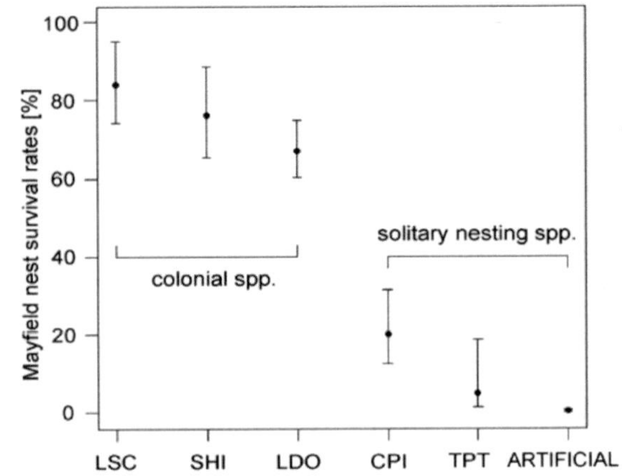

Fig. 2. Nest survival estimates for colonial and solitary nesting species. Survival probabilities followed Mayfield (1961) with 95% confidence intervals (Johnson 1979). Breeding periods were pooled. LSC = *Larus scoresbii*[a], SHI = *Sterna hirundinacea*[b], LDO = *Larus dominicanus*[c], CPI = *Chloephaga picta*[d], TPT = *Tachyeres pteneres*[e], ARTIFICIAL = Artificial nests[f]. Egg-laying and incubation periods (*t*) applied in DSR[t]: [a]26 days (Yorio *et al.* 1996), [b]26 (Scolaro *et. al.* 1996), [c]28.5 (Yorio & Borboroglu 2002), [d]36 (Summers & McAdam 1993), [e]36 (as *C. picta*), [f]30.

Identified predators

We found different patterns of predation for solitary nesting species, colonial species, and the artificial nests (Table 1). Minks were the most important predators of nests of Fightless Steamer Ducks accounting for 52.6% of preyed nests (successful and abandoned nests excluded). On the contrary, the mink was responsible for only 18.2% of the preyed nests of Upland Geese, the difference being significant between the two solitary nesting species (2-sample test for equality of proportions: χ^2 = 6.1, P = 0.01). The predation rate of mink on artificial nests (pooled over breeding periods as there were no significant differences) was 17.3% and thus comparable with values of predation by mink on Upland Geese nests. Predation on artificial nests in rocky outcroppings, however, yielded a significantly higher predation rate (36.4% of total predators) than along beaches (6.2%, 2-sample test for equality of proportions: χ^2 = 77.9, P <0.001). The same trend was detected for Upland Geese, whose nests were preyed upon by mink to a higher proportion at rocky outcrop shorelines (27.3% or 3 out of 11 preyed nests)

TABLE 1. Nest fate of natural and artificial nests. Numbers indicate the percentage of total nests found and, in parenthesis, the percentage of total preyed nests (successful and abandoned nests excluded).

Nest fate	*Chloephaga picta*	*Tachyeres pteneres*	*Larus dominicanus*	*Larus scoresbii*	*Sterna hirundinacea*	Artificial nests
Successful*	36.7	8.7	74.0	90.4	82.4	0.0
Abandoned	7.6	8.7	0.0	0.0	0.0	-
Mink	**10.1 (18.2)**	**43.5 (52.6)**	**0.0 (0.0)**	**0.0 (0.0)**	**0.0 (0.0)**	**17.3 (17.3)**
Bird	17.7 (31.8)	21.7 (26.3)	2.0 (7.5)	0.0 (0.0)	0.0 (0.0)	68.6 (68.6)
Human	13.9 (25.0)	4.4 (5.3)	0.0 (0.0)	0.0 (0.0)	0.0 (0.0)	0.0 (0.0)
Dog	1.3 (2.3)	8.7 (10.5)	0.0 (0.0)	0.0 (0.0)	2.7 (15.4)	0.0 (0.0)
Mouse	-	-	-	-	-	2.6 (2.6)
Unknown	12.7 (22.7)	4.3 (5.3)	24.0 (92.5)	9.6 (100.0)	14.9 (84.6)	11.5 (11.5)
Total nests	79	23	204	83	74	558

* No. of successful nests/no. of total nests found yields into the naïve nest survival estimator which is positively biased (Jehle *et al.*, 2004).

than at beaches (15.2% or 5 out of 33 preyed nests). This was not true for Flightless steamer ducks, though (46.7% or 7 out of 15 at rocky outcrops versus 75.0% or 3 out of 4 at beaches). We did not find any predation by mink in the three colonial species, but they had a quite high rate of unknown nest failure of 84.6-100% (9.6-24.0% of total nests found). Thus predation by introduced mink coincided with those species characterized by rather low Mayfield nest survival probabilities.

Birds as autochthonous predators were greatly responsible for lowering the nest survival of artificial nests (68.6%). We identified bird predators of 236 clay eggs: the Southern Crested Caracara was the most common avian predator with 62.7% predation on 236 clay eggs, followed by the Chimango Caracara (31.3%). *Larus* spp. (5.1%) and the Chilean Skua (0.9%) did not play a major role in the predation of artificial nests. For the natural nests, birds accounted for 31.8% of preyed nests in Upland Geese, 26.3% in Flightless Steamer Ducks, and only 7.5% in colonial species. Finally, humans were identified as an important predator for Upland Geese, causing 25.0% of failed nests.

Factors influencing mink predation

Using discriminant analysis on PCA results, we tested the representativeness of artificial nests ("artificial"), and the explanatory nest site variables which best separated the species classes of Upland Goose ("CPI") and Flightless Steamer Duck ("TPT") (N = 475, DA 1). The first principal component explained 37% of the variance, the second component 19%. Variables with high loadings (>|0.7|) on the first component was height of shrubs around the nest (0.82). Two discriminant functions (the axes) were generated (in general n - 1, n = number of classes), and the first axis accounted for 85% of total inertia. The Monte-Carlo permutation test showed that the discrimination of these two axes was significant (P < 0.001, based on 1000 permutations). The centroid of artificial nests strongly overlapped with the centroids of solitary nesting species (Fig. 3). However, the chosen set of variables discriminated very well between the nests of both "real" species. The first (horizontal) axis was mainly determined by distance to the shore (cosines = -0.77) and top nest cover (0.58, all other cosines <|0.46|) (Fig. 3 circle). Along the second axis (15% of total inertia), side nest cover (0.54, all other cosines <|0.47|) contributed to the separation of classes. Thus Upland Geese built their nests at a greater distance from the shore, whereas nests of Fightless Steamer Ducks were characterized by a high overhead and lateral concealment (see Table 2 for empirical values).

The same multivariate analysis was performed on 375 artificial nests with four predator classes "mink", "bird", "mouse", "unknown", and a fifth class containing "successful" nests (DA 2). The first principal component explained 27% of the variance, the second 23%. High loading variables (>|0.7|) for the first component were lakes (1.3), top nest cover (-0.75), and height of shrubs around the nest (-0.71). This time, we out sorted redundant variables such as relative abundance of mink signs as it highly correlated with the type of shore (Spearman's rho = 0.86). Other redundant variables were temperature, cover, and side. We then performed the discriminant analysis with seven main variables.

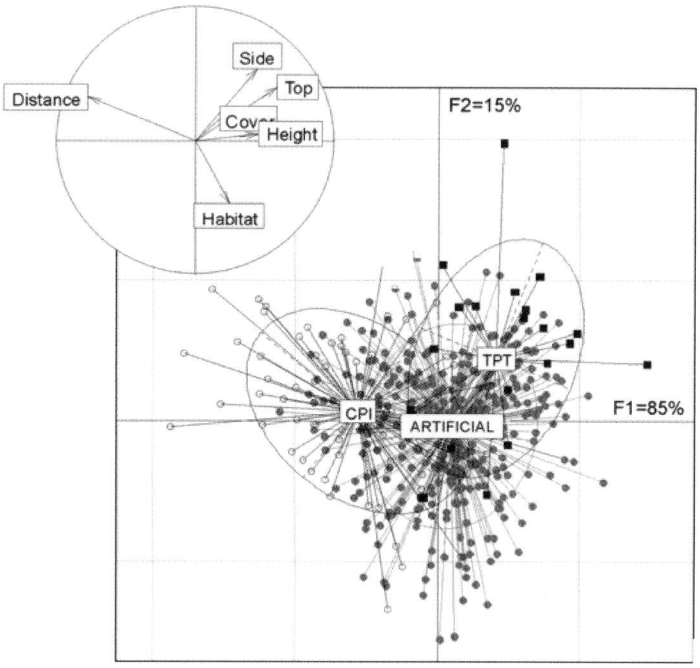

Fig. 3. Results of the discriminant analysis 1. DA 1 was based on PCA results of six site-specific nest variables classified by artificial nests and natural nests of solitary nesting species (N = 475). ARTIFICIAL = Artificial nests, CPI = *Chloephaga picta*, TPT = *Tachyeres pteneres*. Eigen values F1: 0.26 (85% total inertia), F2: 0.05 (15% total inertia). The circle represents the cosines between the variables and the canonical scores. Direction and length of the arrows are a metric of the discriminatory power of the variables.

TABLE 2. Empirical values for variables with discriminatory power in discriminant analysis 1.

Discriminant variables DA 1	*Chloephaga picta*	*Tachyeres pteneres*	Artificial nests
Distance [m] (mean, SD, median, range)	35.8 ± 47.6 6 (1-236)	8.5 ± 6.5 17 (1-24)	9.1 ± 6.1 7 (2-32)
Top (median, range)	1 (1-4)	4 (1-4)	2 (1-4)
Side (median, range)	2 (1-4)	4 (1-4)	3 (1-4)
Total nests	77	23	375

The variables are described in the Appendix A.

Accordingly, the predator classes were significantly separated ($P < 0.001$, Monte-Carlo test based on 1000 permutations) along four discriminant functions. The first function accounted for 50% of total inertia, the second for 38%, and the following axes for 10%, and 2%, respectively. The centroid of successful nests was very well separated from the centroids of preyed nests (Fig. 4). Of the four predator classes, mink formed the most distinct class, although overlapping with the three remaining classes, which were located close together. Nests with unknown reasons for failure were more probably preyed upon by birds and mice rather than by mink, although multi-predator visits cannot be excluded. Along the first axis, the type of shore (cosines = 0.94 for rocky outcrop, and -0.94 for beaches, all other cosines <|0.32|) was the variable contributing most to separating the predator classes (Fig. 4 circle). Along the second axis, nest age (-0.70), height of vegetation at the nest (-0.54), and top nest cover (-0.50, all other cosines <|0.36|) were the three variables contributing most discriminatory power. Thus mink mainly preyed on nests at coastal habitats with rocky outcroppings where relative mink abundance was also higher than along beaches (Wilcoxon rank sum test: $W = 114$, $P < 0.001$, $N = 22$ sites). At rocky outcrops 68.8% 500 m sections contained signs (median, range 25-100%, $N = 9$), whereas at beaches only 12.5% of the sections were positive (0-43.8%, $N = 13$). Nests built towards the end of the breeding season were apparently more successful (see Table 3 for empirical values).

Navarino Island's bird community

As the abundance patterns are different for colonial species, which occur clumped, individual numbers per colony were used, whereas for the other species studied individuals/km was employed. Kelp Gulls as a predatory species on search flights were also counted along our transects. The Kelp Gull colony had a size of 315 individuals (range 168-433, both breeding seasons pooled), Dolphin

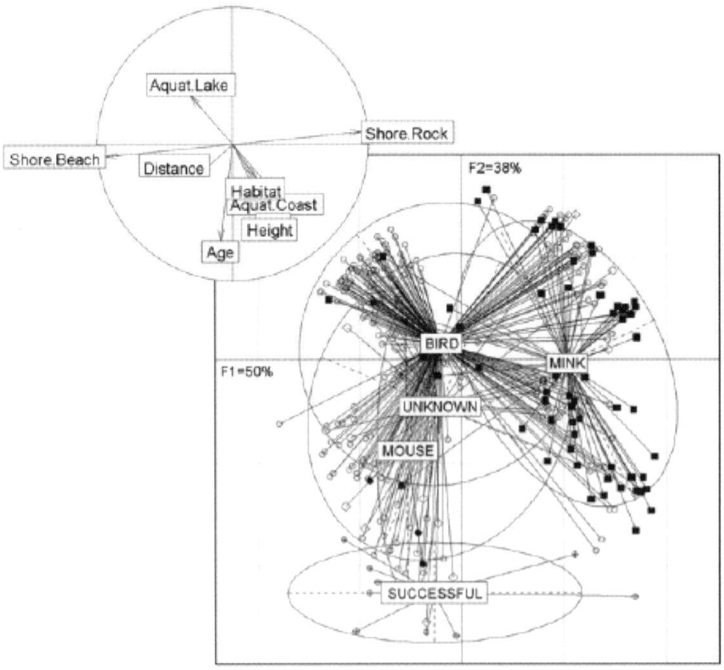

Fig. 4. Results of the discriminant analysis 2. DA 2 was based on PCA results of seven area-specific and site-specific nest variables of artificial nests ($N = 375$), classified by nest fate ("successful", "mink", "bird", "mouse", "unknown"). Mice can either refer to *Abrothrix xanthorhinus* or *Oligoryzomys longicaudatus*. The variable Top is covered by Height. Eigenvalues F1: 0.22 (50% total inertia), F2: 0.17 (38% total inertia). The circle represents the cosines between the variables and the canonical scores. Direction and length of the arrows are a metric of the discriminatory power of the variables.

TABLE 3. Empirical values for variables with discriminatory power in discriminant analysis 2.

Discriminant variables DA 2	Mink	Bird	Mouse	Unknown	Successful
Shore, rocky outcrop (# nests)	55	85	0	8	2
Shore, beaches (# nests)	8	183	5	22	7
Age [days] (mean, SD, range)	28.1 ± 20.8 (5-57)	28.5 ± 18.8 (5-58)	31.4± 23.2 (6-49)	31.7 ± 20.4 (5-59)	68.4 ± 0.73 (67-69)
Height [m] (mean, SD, range)	0.47 ± 0.44 (0-2)	0.38 ± 0.31 (0-2)	0.38 ± 0.08 (0.3-0.5)	0.39 ± 0.39 (0.1-1.7)	0.28 ± 0.08 (0.2-0.4)
Top (median, range)	2 (1-4)	1 (1-4)	4 (2-4)	3 (1-4)	3 (1-4)
Total nests	63	268	5	30	9

The variables are described in the Appendix A. The main focus of DA 2 was to assess differences between predation patterns of invasive mink and autochthonous birds. Sample sizes for artificial nests preyed on by mice and for successful nests were small, but included into the analysis for reasons of integrity.

Gulls of 150 individuals (132-200) and South American Terns of 90 individuals (64-320). Among the predatory birds, Kelp Gulls had a relative abundance of 4.5 individuals/km (median, range 0.5-19.3) per transect, a significantly higher abundance than estimated for the three raptor species (all Wilcoxon-tests: $W = 820$, $P < 0.001$) (Fig. 5). Abundances for raptors ranged from 0 to 0.75 individuals/km for Crested Caracara, 0 to 3.75 for the Chilean Skua, and 0 to 3.25 for Chimango Caracara. Among solitary nesting species, we estimated 4.75 individuals/km (0.5-31.75) for the Upland Goose and 1.75 individuals/km (0.5-13.75) for Flightless Steamer Ducks. Although we made our counts in the same sites where nests were monitored, these counts refer to resting and feeding birds as well as guarding males (cryptic incubating females could not be included) (Fig. 5).

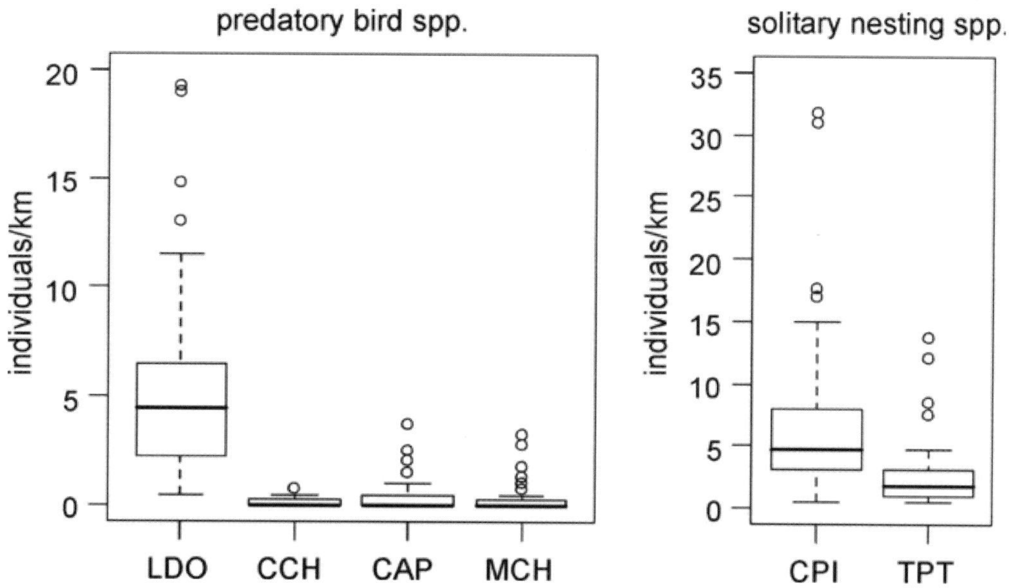

Fig. 5. Bird counts of predatory birds and solitary nesting species. Predatory birds refer to Kelp Gulls on search flights and raptors. Counts were made along the northern coast of Navarino Island and cover all study sites and breeding periods (2005–2007). LDO = *Larus dominicanus*, CCH = *Catharacta chilensis*, CAP = *Caracara plancus*, MCH = *Milvago chimango*, CPI = *Chloephaga picta*, TPT = *Tachyeres pteneres*.

DISCUSSION

Vulnerability profile

With a combined approach of natural and artificial nests a vulnerability profile was drawn for ground-nesting waterbirds under mink invasion on Navarino Island. We found patterns of nest predation by mink among social nesting strategies, habitat, and nest characteristics. Thus, the ground-nesting waterbirds that are especially vulnerable to predation by mink are those that are (i) solitary nesting, (ii) breeding in coastal habitats with rocky outcrop shores, and (iii) concealing their nests. This profile is best illustrated by a high predation rate of mink (44%) on Flightless Steamer Ducks, a species with very low densities (1.75 individuals/km), which perfectly match all the characteristics of our vulnerability profile. Other ground-nesting species to which most of these characteristics apply are: Flying Steamer Duck (*Tachyeres patachonicus*), Crested Duck (*Lophonetta specularioides*), and Kelp Goose (*Chloephaga hybrida*). In consequence, these represent bird species vulnerable to mink predation, and might require special conservation efforts.

How can these patterns characteristic for predation by mink be explained? We start with the first pattern, solitary nesting (i). Depending on group size, nest densities, predator type, and predator size, colonial breeding can lead to a decrease in predation risk due to earlier predator detection and/or higher nest defense efficiency (reviews in Wittenberger & Hunt 1985, Siegel-Causey & Kharitonov 1990). For example, gull colonies successfully show aggressive behavior towards predators (Kruuk 1964), a reason why some bird species are found associated with gull colonies during nesting, *e.g.*, Tufted Ducks (Opermanis *et al.* 2001). The second pattern, the association with rocky outcrop marine shore habitats (ii), apparently is a function of habitat requirements of mink. Dunstone (1993) stated that in coastal habitats sheltered rocky shores are ideal for mink. Our results agree with this. Mink abundance was significantly higher along steep and rocky coastal shores than along beaches. Our results also revealed the importance of concealed nests (iii) as predictors for predation by mink. Many researchers agree about the differences between avian and mammalian predators with respect to the importance of nest concealment (*e.g.*, Butler & Rotella 1998, Opermanis *et al.* 2001). Thus, avian predators appear to visualize nests, whereas mammalian predators primarily depend on olfactory cues and therefore prey upon nests irrespective of concealment (Guyn & Clark 1997). This explains why mink were preying upon both types of nests; open nests (predominantly Upland Geese) and concealed nests (Flightless Steamer Ducks). However, our results on natural and artificial nests indicate that concealed nests in surroundings with dense vegetation were more vulnerable to predation by mink. We suggest two reasons. First, concealing nests requires a minimum amount of dense vegetation, which is a habitat preference of mink (*e.g.*, Yamaguchi *et al.* 2003), and second, concealment is limiting the view of the surroundings of the nest and thus, might prevent appropriate response if predators are not detected in time (trade-off hypothesis by Götmark *et al.* 1995).

Vulnerability and "naïvety"

Behavioral decisions under the risk of predation include escaping from predators, inspecting predators, and mobbing predators depending on the encounter situation and type of prey, *i.e.*, adults or offspring (Lima & Dill 1990). Defense strategies against potential avian predators on their clutches have been described for Upland Geese (Quillfeldt *et al.* 2005) and Flightless Steamer Ducks (Livezey & Humphrey 1985). However, assuming that a mammalian predator will induce different encounter situations than avian predators, this requires different behavioral decisions than those for avian predators. On Navarino Island, bird species were not confronted with a terrestrial mammalian predator until the arrival of mink during the mid-1990s (Rozzi & Sherriffs 2003). There is no evidence for evolutionary isolation from native terrestrial predators as Navarino Island does not harbor endemic bird species (Couve & Vidal 2003). Bird species endemic for Patagonia have been evolved together with native predators like the Fuegian red fox (*Pseudalopex culpaeus lycoides*) in Tierra del Fuego. Although bird species on Navarino Island should have developed evolutionary adaptations to terrestrial predators, we believe that they might lack behavioral adaptations to the recently arrived mink. Animals have the ability to behaviorally influence their risk of being preyed upon in ecological time, *i.e.*, during their lifetime (reviewed in Lima & Dill 1990). Missing confrontation with a terrestrial predator therefore should result in behavioral naïvety for resident bird species as shown for Arctic Terns in mink removal areas (Nordström *et al.* 2004).

Validity of predicting vulnerability

Concerns about the validity of inferences made about nest survival of natural nests from artificial nests basically originate from the differences in parental presence, odor, egg characteristics, and location of these two nest types (Butler & Rotella 1998). In our study, artificial nests had significantly lower survival rates than natural nests. We assume that the main draw-back of artificial nests, "no cryptic female sitting on the eggs" (Angelstam 1986, p. 370), could have attracted more avian predators to the more visible nests. A possible solution for this draw-back could be to cover the eggs of artificial nests with down, just like geese and ducks in our study area do when leaving their nests. Despite these difficulties and possibilities to improve our study design, we think that our data is sufficiently reliable, for three reasons. First, as we were primarily interested in predation by mink, we think that our artificial nests provided sufficient olfactory cues (downy feathers) to attract mammalian predators irrespective of concealment (see the comparable predation rates of mink between artificial nests and Upland Geese). Second, the internal validity of our artificial nest design was maintained carefully in order to accurately measure predator behavior (see Moore & Robinson 2004); and third, an extrapolation of our results to real nests might be justified to a certain degree as we maximized comparability, for example nest characteristics of artificial nests matched well with those of natural nests (results of DA 1) indicating that they were quite "truly" built.

We have been investigating predation of mink in its early colonization stage at the beginning of the 2000s. Since then, mink sightings have been increasing, suggesting a rapid increase in the mink population (Rozzi & Sherriffs 2003). However, trapping and sign surveys in different semi-aquatic habitats of Navarino Island (Anderson *et al.* 2006, E. Schüttler, unpublished data) have shown that the density of mink is still below densities measured in other invaded regions (*e.g.*, Previtali *et al.* 1998, Bartoszewicz & Zalewski 2003). Although we lack knowledge of the carrying capacity for mink on Navarino Island, we assume that the population of mink will grow further. The consequences this might have for bird populations have to be investigated by long-term research and cannot be extrapolated from our short-term results. One possible divergence from our suggested vulnerability profile could be that species nesting in colonies will also be affected by a future increasing population of mink. The higher defense behavior of colonies does not always protect them from predation by mink as various studies have demonstrated (*e.g.*, Kilpi 1995, Clode & Macdonald 2002, Collis 2004, Craik 1997, Nordström *et al.* 2004). Mink predation on bird populations seems to be further influenced by a variety of factors as several studies in Europe (most are mentioned in Bonesi & Palazon 2007) have shown. Among those factors (majorly assessed through mink removal) figure: the body size of waterbirds (smaller sized waterbirds increased in breeding densities after mink removal, *e.g.*, Nordström *et al.* 2002); the timing of breeding (later breeders seem to be stronger affected than earlier ones, because of the increasing food requirements of growing mink kits, Banks *et al.* 2008); the predator community (compensatory predation may occur, *e.g.*, Opermanis *et al.* 2005); the activity of the predator (mink are described as nocturnal predators, and seabird colonies can hardly defend against nocturnal predators, Hunter & Morris 1976 in Banks *et al.* 2008).

Management implications

The observed vulnerability patterns are valuable for decision making and priority setting in the management of invasive mink on Navarino Island. We identified ground-nesting waterbirds under risk from a conservationist point of view. Control programs should focus on preventing mink from establishing territories near breeding areas of vulnerable ground-nesting species and bird colonies. The assessment of coastal breeding habitats can be facilitated by using geographical information systems (GIS) and existing digital data archives (Rönkä *et al.*,2008). Predator removal programs have been shown to have a positive effect on hatching success and post-breeding population size of target bird species (reviews in Newton 1994, Côté & Sutherland 1997). Targeted removal of mink from particular areas, particularly rocky outcrop coasts, during the breeding season (*e.g.*, Clode & Macdonald 2002, Banks *et al.* 2008, Ratcliffe *et al.* 2008) could therefore represent a first task of conservationists. The design of a long-term management plan should include clear objectives, participation of local stakeholders, careful consideration of costs vs. benefits, possible negative effects of target and non-target species, and prevention efforts (*e.g.*, Moore *et al.* 2003, Nordström *et al.* 2003, Baxter *et al.* 2008). However, mink control should not overshadow vigilance against other human-induced factors contributing to mortality in ground-nesting waterbirds, such as dog predation or egg-stealing by humans.

ACKNOWLEDGMENTS

We wish to thank José Tomás Ibarra, Brett Maley, and Christopher Anderson for their contributions to designing a study with artificial nests. Claire Brown, Melisa Gañan, Annette Guse, José Llaipén, and Francesca Pischedda were indispensable for completing our fieldwork. Carsten Dormann, Bernd Gruber, and Michael Gerisch kindly gave statistical advice. We express our gratitude to the Chilean navy for facilitating meteorological data. This work was co-sponsored by the German Academic Exchange Service DAAD (D/04/38329), by the Chilean Millennium Institute of Ecology and Biodiversity Contract (ICM, PO2-051-F), the Magallanes University, the Omora Foundation, and the German-Chilean Research Project BIOKONCHIL (funded by the German Ministry of Education and Research, FKZ 01LM0208).

REFERENCES

ACKERMAN, J.T., A.L. BLACKMER & J.M. EADIE. 2004. Is predation on waterfowl nests density dependent? tests at three spatial scales. *Oikos* 107: 128-140.

ANDERSON, C.B., R. ROZZI, J.C. TORRES-MURA, S.M. MCGEHEE, M.F. SHERRIFFS, E. SCHÜTTLER & A.D. ROSEMOND. 2006. Exotic vertebrate fauna in the remote and pristine sub-Antarctic Cape Horn Archipelago, Chile. *Biodiversity and Conservation* 15: 3295-3313.

ANGELSTAM, P. 1986. Predation on ground-nesting birds' nests in relation to predator densities and habitat edge. *Oikos* 47: 365-373.

ARNOLD, T.W. & E.K. FRITZELL. 1987. Food habits of prairie mink during the waterfowl breeding season. *Canadian Journal of Zoology* 68: 2322-2324.

ATKINSON, I.A.E. 1996. Introductions of wildlife as a cause of species extinctions. *Wildlife Biology* 2: 135-141.

BANKS, P.B., M. NORDSTRÖM, M. AHOLA, P. SALO, K. FEY & E. KORPIMÄKI. 2008. Impacts of alien mink predation on island vertebrate communities of the Baltic Sea Archipelago: review of a long-term experimental study. *Boreal Environment Research* 13: 3-16.

BARTOSZEWICZ, M. & A. ZALEWSKI. 2003. American mink, *Mustela vison* diet and predation on waterfowl in the Słonsk Reserve, western Poland. *Folia Zoologica* 52: 225-238.

BAXTER, P.W.J., J.L. SABO, C. WILCOX, M.A. MCCARTHY & H.P. POSSINGHAM. 2008. Cost-effective suppression and eradication of invasive predators. *Conservation Biology* 22: 89-98.

BERGER, J., J.E. SWENSON & I.L. PERSSON. 2001. Recolonizing carnivores and naïve prey: conservation lessons from Pleistocene extinctions. *Science* 291: 1036-1039.

BIBBY, C., N.D. BURGESS, D.A. HILL & S.H. MUSTOE. 2000. *Bird Census Techniques*, second ed. Academic Press, London.

BONESI, L. & D.W. MACDONALD. 2004. Evaluation of sign surveys as a way to estimate the relative abundance of American mink (*Mustela vison*). *Journal of Zoology* 262: 65-72.

BONESI, L. & S. PALAZON. 2007. The American mink in Europe: status, impacts, and control. *Biological Conservation* 134: 470-483.

BUTLER, M.A. & J.J. ROTELLA. 1998. Validity of using artificial nests to assess duck-nest success. *Journal of Wildlife Management* 62: 163-170.

CLODE, D. & D.W. MACDONALD. 2002. Invasive predators and the conservation of island birds: the case of American mink *Mustela vison* and terns *Sterna* spp. in the Western Isles Scotland. *Bird Study* 49: 118-123.

COLLIS, K. 2004. Breeding ecology of Caspian terns at colonies of the Columbia Plateau. *Northwest Science* 78: 303-312.

CÔTÉ, I.M. & W.J. SUTHERLAND. 1997. The effectiveness of removing predators to protect bird populations. *Conservation Biology* 11: 395-405.

COURCHAMP, F., J.L. CHAPUIS & M. PASCAL. 2003. Mammal invaders on islands: impact, control and control impact. *Biological Reviews* 78: 347-383.

COUVE, E. & C. VIDAL. 2003. *Birds of the Beagle Channel and Cape Horn*. Ediciones Fantástico Sur, Punta Arenas.

CRAIK, J.C.A. 1995. Effects of North American mink on the breeding success of terns and smaller gulls in west Scotland. *Seabird* 17: 3-11.

CRAIK, J.C.A. 1997. Long-term effects of North American mink *Mustela vison* on seabirds in western Scotland. *Bird Study* 44: 303-309.

DINSMORE, S.J., G.C. WHITE & F.L. KNOPF. 2002. Advanced techniques for modelling avian nest survival. *Ecology* 83: 3476-3488.

DRAY, S. & A.B. DUFOUR. 2007. The ade4 package: implementing the duality diagram for ecologists. *Journal of Statistical Software* 22: 1-20.

DUNSTONE, N. 1993. *The Mink*. T & AD Poyser Ltd., London.

ELTON, C.S. 1958. *The Ecology of Invasions by Animals and Plants*. Methuen and Company, London.

FAABORG, J. 2003. Truly artificial nest studies. *Conservation Biology* 18: 369-370.

FASOLA, L., C. CHEHÉBAR, D.W. MACDONALD, G. PORRO & M.H. CASSINI. 2008. Do alien North American mink compete for resources with native South American River otter in Argentinean Patagonia? *Journal of Zoology* 1-9. doi:10.1111/j.14697998.2008.00507.x.

FERRERAS, P. & D.W. MACDONALD. 1999. The impact of American mink *Mustela vison* on water birds in the upper Thames. *Journal of Applied Ecology* 36: 701-708.

GOBSTER, P.H. 2005. Invasive species as ecological threat: is restoration an alternative to fear-based resource management? *Ecological Restoration* 23: 261-270.

GOODMAN, D. 1975. The theory of diversity-stability relationships in ecology. *Quarterly Review of Biology* 50: 237-266.

GÖTMARK, F. 1992. The effects of investigator disturbance on nesting birds. *Current Ornithology* 9: 63-104.

GÖTMARK, F., D. BLOMQVIST, O.C. JOHANSSON & J. BERGKVIST. 1995. Nest site selection: a trade-off between concealment and view of the surroundings? *Journal of Avian Biology* 26: 305-312.

GUYN, K.L. & R.G. CLARK. 1997. Cover characteristics and success of natural and artificial duck nests. *Journal of Field Ornithology* 68: 33-41.

HOLMES, S. 2008. Multivariate data analysis: the French way. Pp. 219-233 *in Probability and Statistics: Essays in Honor of David A. Freedman*, D. Nolan & T. Speed, (eds.). Institute of Mathematical Statistics, Beachwood, Ohio.

HUNTER, R.A. & R.D. MORRIS. 1976. Nocturnal predation by a Black-crowned night heron at a common tern colony. *Auk* 93: 629-633.

IBARRA, J.T., L. FASOLA, D.W. MACDONALD, R. ROZZI & C. BONACIC. 2009. Invasive American mink *Mustela vison* in wetlands of the Cape Horn Biosphere Reserve, southern Chile: what are they eating? *Oryx* 43: 1-4.

INMAN, A.J. & J. KREBS. 1987. Predation and group living. *Trends in Ecology and Evolution* 2: 31-32.

JAKSIC, F.M., J.A. IRIARTRE, J.E. JIMÉNEZ & D.R. MARTÍNEZ. 2002. Invaders without frontiers: cross-border invasions of exotic mammals. *Biological Invasions* 4: 157-173.

JEHLE, G., A.A.Y. ADAMS, J.A. SAVIDGE & S.K. SKAGEN. 2004. Nest survival estimation: a review of alternatives to the Mayfield estimator. *Condor* 106: 472-484.

JOHNSON, A.W. 1965. *The Birds of Chile and Adjacent Regions of Argentina, Bolivia and Peru*, vol. 2. Platt, Buenos Aires.

JOHNSON, D.H. 1979. Estimating nest success: the Mayfield method and an alternative. *Auk* 96: 651-661.

KAUHALA, K. 2004. Removal of medium-sized predators and the breeding success of ducks in Finland. *Folia Zoologica* 53: 367-378.

KILPI, M. 1995. Breeding success, predation and local dynamic of colonial Common gulls *Larus canus*. *Annales Zoologici Fennici* 32: 175-182.

KING, W.B. 1985. *Island Birds: Will Future Repeat the Past?* Technical Publication 3. International Council for bird preservation, Cambridge.

KRUUK, H. 1964. Predators and anti-predator behaviour of the Black-headed gull (*Larus ridibundus* L.). *Behaviour Supplement* 11: 1-130.

LARIVIÈRE, S. 1999. Reasons why predators cannot be inferred from nest remains. *Condor* 101: 718-721.

LEVINE, J.M. & C.M. D'ANTONIO. 1999. Elton revisited: a review of evidence linking diversity and invasibility. *Oikos* 87: 15-26.

LIMA, S.L. & L.M. DILL. 1990. Behavioral decisions made under the risk of predation: a review and prospectus. *Canadian Journal of Zoology* 68: 619-640.

LIVEZEY, B.C. & P.S. HUMPHREY. 1985. Territoriality and interspecific aggression in Steamer-Ducks. *Condor* 87: 154-157.

LIZARRALDE, M.S. & J.M. ESCOBAR. 2000. Mamíferos exóticos en la Tierra del Fuego. *Ciencia Hoy* 10: 52-63.

MACDONALD, D.W. & L.A. HARRINGTON. 2003. The American mink: the triumph and tragedy of adaptation out of context. *New Zealand Journal of Zoology* 30, 421-441.

MACDONALD, D.W. & M.D. THOM. 2001. Alien carnivores: unwelcome experiments in ecological theory. Pp. 93-122 *in Carnivore Conservation*, J.L. Gittleman, S.M. Funk, D.W. Macdonald & R.K. Wayne, (eds.). Cambridge University Press, Cambridge.

MAJOR, R.E. & C.E. KENDAL. 1996. The contribution of artificial nest experiments to understanding avian reproductive success: a review of methods and conclusions. *Ibis* 138: 298-307.

MARTIN, T.E. 1993. Nest predation among vegetation layers and habitat types: revising the dogmas. *American Naturalist* 141: 897-913.

MARTIN, T.E. 1995. Avian life history evolution in relation to nest sites, nest predation, and food. *Ecological Monographs* 65: 101-127.

MAYFIELD, H. 1961. Nesting success calculated from exposure. *Wilson Bulletin* 73: 255-261.

MEDINA, G. 1997. A comparison of the diet and distribution of Southern River otter (*Lutra provocax*) and Mink (*Mustela vison*) in Southern Chile. *Journal of Zoology* 242: 291-297.

MEZQUIDA, E.T. & L. MARONE. 2003. Are results of artificial nest experiments a valid indicator of success of natural nests? *Wilson Bulletin* 115: 270-276.

MILLER, C.K. & R.I. KNIGHT. 1993. Does predator assemblage affect reproductive success in songbirds? *Condor* 95: 712-715.

MITTERMEIER, R.A., C.G. MITTERMEIER, T.M. BROOKS, J.D. PILGRIM, W.R. KONSTANT, G.A.B. DA FONSECA & C. KORMOS. 2003. Wilderness and biodiversity conservation. *Proceedings of the National Academy of Sciences* USA. 18: 10309-10313.

MOORE, D. 1983. *Flora of Tierra del Fuego*. Anthony Nelson, Missouri Botanical Garden, England.

MOORE, R.P. & W.D. ROBINSON. 2004. Artificial bird nests, external validity, and bias in ecological field studies. *Ecology* 85: 1562-1567.

MOORE, N.P., S.S. ROY & A. HELYAR. 2003. Mink (*Mustela vison*) eradication to protect ground-nesting birds in the Western Isles, Scotland, United Kingdom. *New Zealand Journal of Zoology* 30: 443-452.

NEWTON, I. 1994. Experiments on the limitation of bird breeding densities a review. *Ibis* 136: 397-411.

NEWTON, I. 1998. *Population Limitation in Birds*. Academic Press, London.

NORDSTRÖM, M. & E. KORPIMÄKI. 2004. Effects of island isolation and feral mink removal on bird communities on small islands in the Baltic Sea. *Journal of Animal Ecology* 73: 424-433.

NORDSTRÖM, M., J. HÖGMANDER, J. NUMMELIN, J. LAINE, LAANETU, N. & E. KORPIMÄKI. 2002. Variable responses of waterfowl breeding populations to long-term removal of introduced American mink. *Ecography* 25: 385-394.

NORDSTRÖM, M., J. HÖGMANDER, J. LAINE, J. NUMMELIN, N. LAANETU & E. KORPIMÄKI. 2003. Effects of feral mink removal on seabirds, waders and passerines on small islands in the Baltic Sea. *Biological Conservation* 109: 359-368.

NORDSTRÖM, M., J. LAINE, M. AHOLA & E. KORPIMÄKI. 2004. Reduced nest defence intensity and improved breeding success in terns as responses to removal of non-native American mink. *Behavioral Ecology and Sociobiology* 55: 454-460.

OPERMANIS, O., A. MEDNIS & I. BAUGA. 2001. Duck nests and predators: interaction, specialisation and possible management. *Wildlife Biology* 7: 87-96.

OPERMANIS, O., A. MEDNIS & I. BAUGA. 2005. Assessment of compensatory predation and re-colonisation using long-term duck nest predator removal data. *Acta Universitatis Latviensis* 691: 17-29.

PISANO, E. 1977. Fitogeografía de Fuego-Patagonia chilena. I. Comunidades vegetales entre las latitudes 52° y 56°S. *Anales del Instituto de la Patagonia* 8: 121-250.

PREVITALI, A., M.H. CASSINI & D.W. MACDONALD. 1998. Habitat use and diet of the American mink (*Mustela vison*) in Argentinian Patagonia. *Journal of Zoology* 246: 482-486.

QUILLFELDT, P., I.J. STRANGE & J.F. MASELLO. 2005. Escape decisions of incubating females and sex ratio of juveniles in the Upland Goose *Chloephaga picta*. *Ardea* 93: 171-178.

R DEVELOPMENT CORE TEAM. 2008. R: A language and environment for statistical computing. *In R Foundation for Statistical Computing*, Vienna, Austria. ISBN 3900051-07-0, URL <http://www.R-project.org>.

RANGEN, S.A., R.G. CLARK & K.A. HOBSON. 2000. Visual and olfactory attributes of artificial nests. *Auk* 117: 136-146.

RATCLIFFE, N., C. CRAIK, A. HELYAR, S. ROY & M. SCOTT. 2008. Modelling the benefits of American mink *Mustela vison* management options for terns in west Scotland. *Ibis* 150: 114-121.

RÖNKÄ, M., H. TOLVANEN, E. LEHIKOINEN, M. VON NUMERS & M. RAUTKARI. 2008. Breeding habitat preferences of 15 bird species on south-western Finnish archipelago coast: applicability of digital spatial data archives to habitat assessment. *Biological Conservation* 141: 402-416.

ROOT, R.B. 1967. The niche exploitation pattern of the Blue-gray gnatcatcher. *Ecological Monographs* 37: 317-350.

ROZZI, R. & M. SHERRIFFS. 2003. El visón (*Mustela vison* Schreber, Carnivora: Mustelidae), un nuevo mamífero exótico para la Isla Navarino. *Anales del Instituto de la Patagonia* 31: 97-104.

ROZZI, R., F. MASSARDO, A. BERGHÖFER, C.B. ANDERSON, A. MANSILLA, M. MANSILLA, J. PLANA, U. BERGHÖFER, P. ARAYA & E. BARROS. 2006. *Cape Horn Biosphere Reserve: Nomination Document for the Incorporation of the Cape Horn Archipelago Territory into the Word Biosphere Reserve Network*. MaB Program-UNESCO. Ediciones de la Universidad de Magallanes, Punta Arenas.

SAX, D.F. & D. GAINES. 2008. Species invasions and extinction: the future of native biodiversity on islands. *Proceedings of the National Academy of Sciences* USA. 105: 11490-11497.

SCHÜTTLER, E., J. CÁRCAMO & R. ROZZI. 2008. Diet of the American mink *Mustela vison* and its potential impact on the native fauna of Navarino Island, Cape Horn Biosphere Reserve, Chile. *Revista Chilena de Historia Natural* 81: 599-613.

SCOLARO, J., S. LAURENTI & H. GALLELLI. 1996. The nesting and breeding biology of the South American Tern in northern Patagonia. *Journal of Field Ornithology* 67: 17-24.

SHORT, J., J.E. KINNEAR & A. ROBLEY. 2002. Surplus killing by introduced predators in Australia -evidence for ineffective anti-predator adaptations in native prey species? *Biological Conservation* 103: 283-301.

SIEGEL-CAUSEY, D. & S.P. KHARITONOV. 1990. The evolution of coloniality. Pp. 285-330 *in Current Ornithology*, vol. 7, D.M. Power, (ed.). Plenum, New York.

SIMBERLOFF, D. 1995. Why do introduced species appear to devastate islands more than mainland areas? *Pacific Science* 49: 87-97.

INTRODUCTION

The Upland Goose (*Chloephaga picta* Gmelin, 1789) is one of the five species of wild geese of the genus *Chloephaga* present in Chile, and is endemic to the Neotropics (Summers & McAdam 1993). The subspecies *C. p. picta* ranges from the mountains of Santiago to the Cape Horn Archipelago (Martínez & González 2004), constituting the most abundant species in wetlands in Chilean Patagonia (Garay *et al.* 1991). It is also one of the most numerous birds on the island of Tierra del Fuego (Humphrey *et al.* 1970, Earnshaw 1973, Schlatter *et al.* 2002), and in the islands south of the Beagle Channel (Barros 1971, Couve & Vidal 2003). Part of the nesting population has been described from central Chile and Argentina to the south (Couve & Vidal 2003), although its reproduction has been studied only descriptively in Tierra del Fuego (Jory *et al.* 1974). Quantitative studies have been developed in southern areas as the Falkland Islands, on the subspecies *C. picta leucoptera* (*e.g.*, Harradine 1977, Summers 1983, Summers & McAdam 1993, Quillfeldt *et al.* 2005). These last studies have been conducted mainly because of a perceived conflict between this bird and livestock due to the bird's herbivorous diet (Summers & McAdam 1993).

In the case of the southern tip of Chile, specifically Navarino Island, the studies on this species report some of their migratory patterns (Ibarra 2007), estimates of their densities along the north coast and vulnerability to predation by the American mink *Neovison vison* (Schüttler *et al.* 2009). However, its reproductive biology has not been studied in Chilean Patagonia. In this context, there is a need to broaden the knowledge base on this species in the Cape Horn Biosphere Reserve (CHBR), where *C. picta* is the most abundant species in the wetlands of the area (Ibarra *et al.* 2009a).

Predation is the dominant factor affecting reproductive success of birds and may have a critical role in the population dynamics of birds (Wilcove 1985, Klett *et al.* 1988, Newton 1998). Consequently, selection should favor birds that use strategies or selected sites that minimize the risk of predation. A number of factors can affect nest vulnerability to predation, but one of the most important is the degree of concealment of the nest. The selection of habitats that present a greater complexity, micro-relief, and vegetation coverage should reduce the visibility of the nest and, therefore, the risk of predation in species that nest at ground level (*e.g.*, Hill 1984, Jackson *et al.* 1988, Schieck & Hannon 1993). However, Götmark (1992) points out that greater coverage of the nest could obstruct the view of its surroundings of the bird incubating and could limit an effective response to a predator. This is particularly important in oceanic islands where species have evolved free from predators, with the result that most of them nest on the ground and lack behavioral defense mechanisms (Amarasekare 1993). This may be the case with several birds in Navarino Island, where *N. vison* represents a new species of functional land predator (Rozzi & Sherriffs 2003, Schüttler *et al.* 2009), along with other invasive species such as feral dogs (Rozzi *et al.* 2006).

To provide knowledge of the reproductive biology of *C. picta picta*, we developed a study that aimed to: (a) determine clutch size and nesting success, (b) study habitat attributes that this species selects for nesting, (c) assess whether habitats with greater complexity, micro-relief, and vegetation cover, reduce the predation risk for nests of this species on Navarino Island, and (d) discuss the historical background studied, including their reproductive success, with the potential effect *N. vison* could be having on the reproductive success of *C. picta* in the CHBR.

MATERIAL AND METHODS

Study area

The study was conducted on Navarino Island (54°55'S, 67°39'W), which is part of the CHBR, at the southern tip of Chile (Fig. 1). The ecoregion corresponds to the sub-Antarctic Magellanic forest where terrestrial ecosystems are a mosaic of different habitat types, among which are the following: 1. Evergreen forests, dominated by Magellanic Coigüe (*Nothofagus betuloides*) and Winter´s Bark (*Drimys winteri*); 2. Deciduous forests dominated by Lenga (*N. pumilio*) and/or Ñirre (*N. antarctica*); 3. High elevation habitats with cushion plants and lichen pastures; 4. Tundra formations, such as wetlands with rushes and *Sphagnum* spp.; and 5. Freshwater ecosystems (Pisano 1977, Rozzi *et al.* 2006). The average annual temperature is 6°C and the average rainfall reaches 467.3 mm, falling as snow in the colder months (di Castri & Hajek 1976). Navarino Island has an area of approximately 2,528 km^2, where Puerto Williams is the most important human settlement in the area with 2,200 inhabitants (Rozzi *et al.* 2006).

Monitoring nests and clutch size

C. picta usually nests near water in the valleys, bordering wetlands and rivers, and along the sea shore (Summers 1983). In the present study, the reproductive activity of *C. picta* was monitored at nine sites along the northern coast of Navarino Island for 54 days of the reproductive period 2005/06, starting on November 11th and 59 days of the reproductive period 2006/07, starting on November 6th. Thirty five nests were monitored in 2005/06 and 44 nests in 2006/07 (N = 79). Nest searching was conducted along seven (2005/06) and nine (2006/07) transects of 4 km in length along the sea shore, in each of the sites studied. To detect the birds in reproductive activity, transects up to 300 m from the edge of the coast were repeatedly monitored. The reproductive behavior of the birds or the presence of a territorial male was noted. Often, eight days (8.1 d, S.D. = 1.78 d) elapsed between visits to the same nest. To minimize the positive or negative effects of the observer on the survival of eggs by anthropogenic factors associated with footprints, nest visitation, or behavior (Götmark 1992), we selected a moderate frequency of visits to nests and kept the time of nest visits to a minimum. Nests were positioned on a map to ± 25 m, using a Global Positioning System receiver (Fig. 1). Ribbons were installed at a distance of 5 m from each nest. The clutch size for *C. picta* was determined only for those nests that were visited at least twice during the incubation period (Godfray *et al.* 1991). Nests were visited during: (a) the laying period, (b) the incubation period, (c) at hatching, and (d) after hatching (Fig. 2). At each visit we recorded the number of eggs until hatching, abandonment or complete predation occurred.

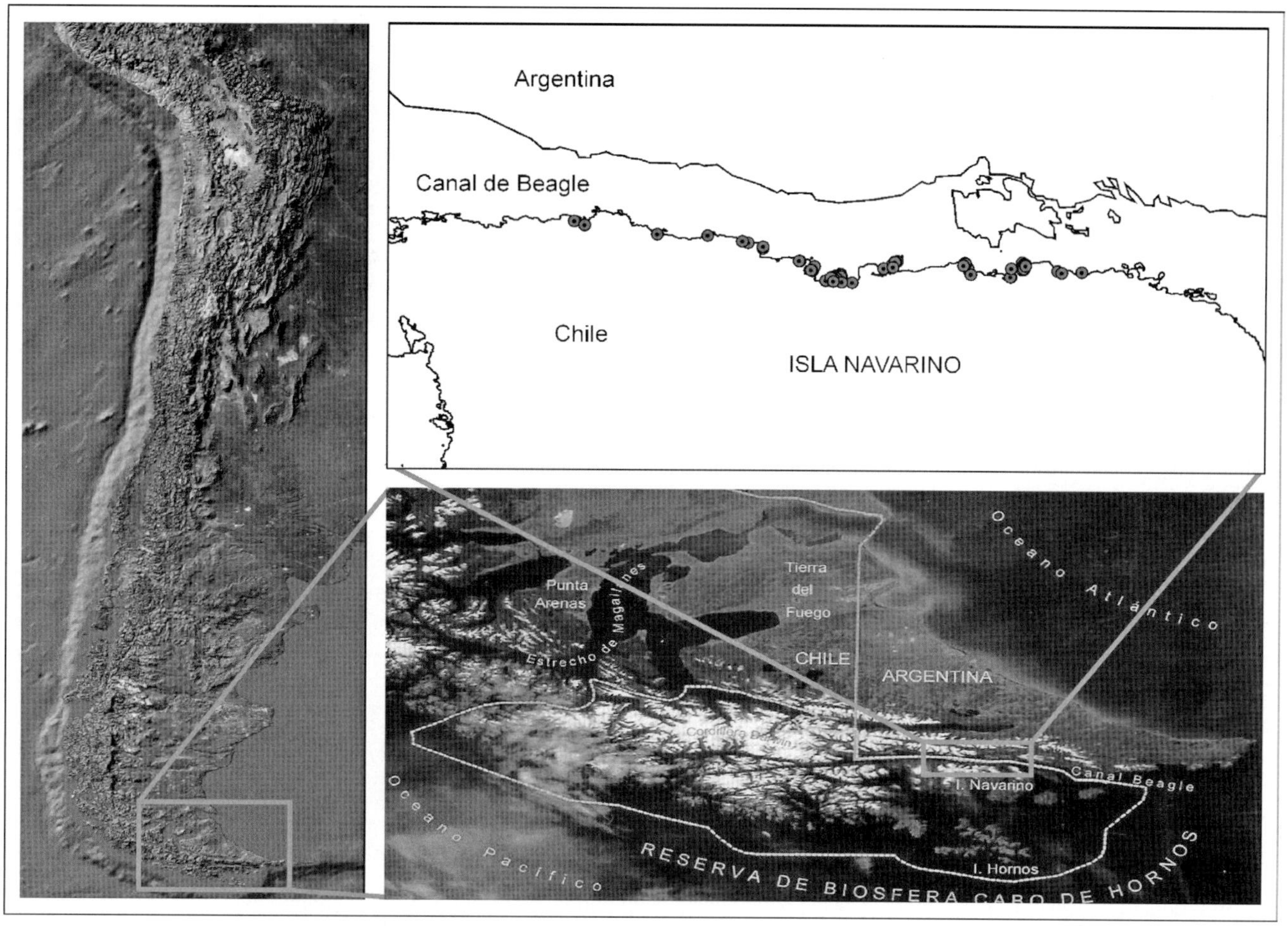

Fig. 1. Map of the southern tip of South America. The Cape Horn Biosphere Reserve is shown, with the distribution of the monitored nests (gray dots) along the northern coast of Navarino Island (54-55°S, 67°W), Chile.

Nesting sites

The following attributes were characterized in nesting habitat: 1) habitat type, classified as bare (no vegetation), uniform (grassland or peat), mixed (shrubs and grassland, but without trees), complex (deciduous, mixed, or evergreen), 2) dominant plant species in an area of 100 m^2 around the nest, 3) vegetation coverage (0.1-1 m high) within 5 m from the nest (in categories <20%, 21-40%, >40%), 4) height (m) of shrubs adjacent to the nest, when present; 5) distance (m) to the water, 6) shoreline slope (0-10°, 11-20°, 21-30°, >31°); 7) proportion (%) of elements in an area of 20 m^2 on the coastal edge (ravine, rocks >10 cm in diameter, stones <10 cm, sand, mud, vegetation); 8) cover above the nest, estimated from 50 cm above it (0-25, 25-50, 50-75, 75-100%); and 9) cover adjacent to the nest from all four sides (0-25, 25-50, 50-75, 75-100%) from 5 m away.

Reproductive success and data analysis

Reproductive success refers to the proportion of those nests in which at least one egg hatched (Jehle *et al.* 2004). To estimate reproductive success we used the traditional method (number of successful nests/total number of nests) and the method proposed by Mayfield (Mayfield 1961, 1975), with the variance estimate developed by Johnson (1979). We calculated two Mayfield estimators. First, considering a period of egg laying and incubation of 36 days (Summers 1983), assuming a constant nest survivorship (Jehle *et al.* 2004). Then we used stratified estimates, as recommended by Jehle *et al.* (2004), considering six days for egg laying and 30 days of incubation (Summers 1983). We used logistic regressions with a *Mixed Stepwise* procedure simultaneously evaluate the effect of the variables and their interactions on nesting success. Reproductive success was analyzed as a binary variable (1 = success, 0 = failure) (Tremblay *et al.* 1997). Statistical analyses were performed using SPSS version 10.0 (SPSS 1999).

RESULTS

Nests of *C. picta* were built on the ground and consisted of a base of twigs of shrubs (*Chilliotrichum diffussum*, *Berberis buxifolia*, etc.), along with grasses of the family Poaceae. The dimensions of the nests (N = 24 described) were 15.2 ± 3.4 cm thick and 50 cm in maximum diameter. In the cup, each nest had a thick layer of down and some small feathers that housed the eggs (Fig. 2). In all cases, the female was the one in charge of incubation (Fig. 2), whereas the

Fig. 2. Reproductive cycle of *C. picta* from the laying period, incubation period, hatching, and nest abandonment.

male usually remained in the territory, defending it from other males of the same species.

We visited a total of 39 nests of *C. picta* on more than one occasion during the incubation period. We determined the clutch size, which was 6.1 ± 1.2 eggs. Clutch sizes observed in this and other previous studies conducted in different areas of South America are presented in Table 1.

The dominant habitat type in which *C. picta* nested was the scrub or mixed (70.5%), followed by grassland-peat land or uniform (15.4%) (Fig. 3). The plant species that dominated the nesting site was the Black Bush (*Chilliotrichium diffussum*) (30.8%), followed by Box-leafed Barberry (*Berberis buxifolia*) (24.4%), unidentified species of the Poaceae family (21.8 %) and Prickly Heath (*Pernettya mucronata*) (17.9%). The plant cover of the habitat within a 5 m radius around the nest was greater than 40% in 39.7% of the nests, followed by a cover between 20-40% in 34.6% of them. Moreover, the average height of the bushes neighboring nests was 0.24 ± 0.13 m (range 0 to 0.5 m). In regard to nesting sites, it was found that the average distance to water was 35.6 ± 47.3 m (range 1 to 236 m). Nests were installed mainly on low-slope sites between 0-10° (62%) of altitude, followed by slopes between 21-30° (21%). Sites with slopes greater than 31° were rarely used (4%). Regarding the proportion (%) of elements dominating nesting sites, stones were the most common (50.1%), followed by rocks (31.2%), and ravines (14.9%). The vegetation cover above the nest was considerably lower, reaching 82.1% of the cases between 0-25%, followed by a 25-50% coverage (12.8%). The coverage adjacent to nests was variable, was 29.5% of the cases between 25-50%, followed by 26.9% of cover between 0-25 and 25-50%.

Fifty-seven percent of nests were found depredated on the last visit. In 36.7% of the nests, eggs hatched (considered as successful according to the traditional method), and 6.3% of them were abandoned (Fig. 4). The reproductive success or survival of the nests, calculated using the Mayfield method (Mayfield 1961, 1975) was 20.0% (95% CI = 12.62 to 31.41%) from the laying of the first egg. Considering only the egg-laying period (six days), the survival of nests was 52.63% (95% CI = 36.96 to 77.36%) and, considering the incubation period, the estimated reproductive success was 33.46% (95% CI = 23.17 to 48.09%). None of the studied habitat attributes was significantly related to the success or failure of the clutches of *C. picta* ($P > 0.05$). In one of the nests already visited a week earlier, there was an accidental encounter with an individual of *N. vison* just as he was about to attack the nest.

DISCUSSION

This paper addresses fundamental aspects of the reproductive biology of *C. picta* in the Cape Horn Biosphere Reserve. The clutch sizes *C. picta* on Navarino Island were similar to those reported in other studies conducted in similar systems as in the Falkland Islands (*e.g.*, Harradine 1977, Summers 1983, Quillfeldt *et al.* 2005) and Tierra del Fuego (Jory *et al.* 1974) (Table 1). However, the data suggest that the nesting success of *C. picta* on Navarino Island is relatively low and the main cause of reproductive failure would be nest predation. The reproductive success of other waterfowl (ducks and geese) normally exceeds 60% for the maintenance of viable populations (*e.g.*, Mickelson 1975, Cooper 1978, Ely & Raveling 1984, Lepage 1998, Guyn & Clark 2000). For *C. picta* in the Falkland Islands, Quillfeldt *et al.* (2005) found 75% reproductive success (according to the traditional method), a value much higher than the 36.7% found in this study on Navarino Island. Summers (1983) found reproductive success values above 61% in the Falkland Islands.

Although this study did not determine the species putting pressure on the nests of *C. picta*, Schüttler *et al.* (2009) have identified the responsible predators in the same set of data (*i.e.*, birds 17.7%, people 13.9%, American mink 10.1%, and dogs 1.3%, of the total nests). Native birds responsible for depredation included *Caracara plancus*, *Milvago chimango*, *Stercorarius chilensis*, and *Larus dominicanus*. However, the other causes of nests failures are anthropogenic (including invasive species and harvesting by people). In this regard, although *N. vison* ranked third among predators, this invader could increase its role in nest predation of *C. picta*. *N. vison* was first recorded in Navarino Island earlier this decade and has shown rapid population growth (Rozzi & Sherriffs 2003). In the northern hemisphere, *N. vison* and *Alopex lagopus* are the main predators of nests of several species of wild geese (Summers & Underhill 1986, Samelius & Lee 1998), which have evolved in conjunction with these waterfowl.

In Navarino Island no native species of terrestrial carnivores exist (Anderson *et al.* 2006). However, *C. picta* would present defense strategies against aerial predators of their clutches (Quillfeldt *et al.* 2005). In the case of *N. vison*, recent studies on its diet have determined that *C. picta* is the bird most often consumed by this predator in moist environments of the Cape Horn Biosphere Reserve (Ibarra *et al.* 2009b). Ultimately, the impact of *N. vison* in addition to the impact of egg harvesting by people, represent two factors that could affect the population of this waterfowl in sub-Antarctic ecosystems. These threats seem to be a priority to understand and consider in future conservation efforts for this species in the CHBR.

At the ecosystem level, *C. picta* is a species considered critical for the growth and maintenance of *Tayloria dubyi*, a moss species endemic to the temperate forests of South America (Jofre 2009). This moss is characterized by attracting flies that land on their sporophytes, which then deposit spores on feces of this goose species in the *Sphagnum* peat lands in Navarino Island, where they can germinate (Jofre 2009). In addition, *C. picta* is the most abundant bird species in moist environments in the Cape Horn Biosphere Reserve (Ibarra *et al.* 2009a), an attribute that makes it an essential part of the group of birds that draw bird watchers and tourists to the area (Rozzi *et al.* 2006).

It was not possible to identify a habitat attribute that was related to the success or failure of the nests studied. Habitat variables do not always relate to the reproductive success of a species, a fact that has also been reported in other wild geese (Tremblay *et al.* 1997). In the present study, this fact suggests that there may be other habitat variables or climatic factors that have not been studied that may affect the success or failure of nests. However, *C. picta* may have low reproductive success due to extrinsic factors, as was noted above (*e.g.*, invasive species).

It has been suggested that reproductive success in geese may be higher in high elevation areas away from the shore rather than close to the shore (*e.g.*, Ely & Raveling 1984). This study was conducted only in coastal areas of Navarino Island. Some of the predators listed above are more abundant in these environments. For example, *N. vison* would prefer and would be more abundant in coastal environments than in inland water bodies of the island (Schüttler *et al.* 2010). The same goes for birds of prey associated with moist environments, which decrease in richness and abundance with distance from the coast and along the watersheds of Navarino Island (Ibarra *et al.* 2009a). In this regard, it is important to test this hypothesis about the reproductive success of *C. picta* in inland environments of the island, where nests may have a reduced predation pressure, as many breeding pairs of *C. picta* disperse towards the island interior to reproduce (Ibarra *et al.* 2009a).

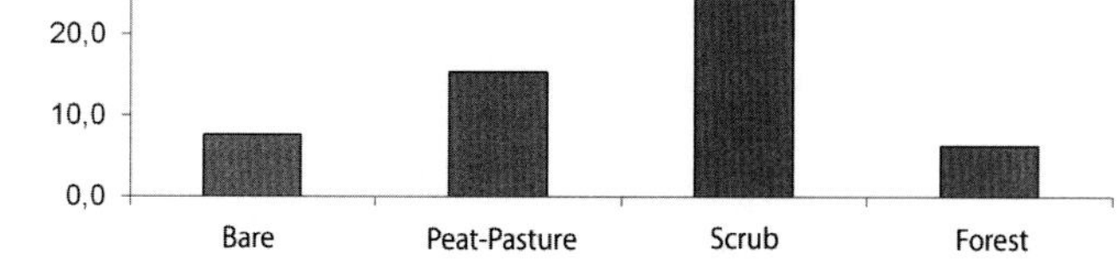

Fig. 3. Dominant habitat types (%) in which *C. picta* nested along the northern coast of Navarino Island.

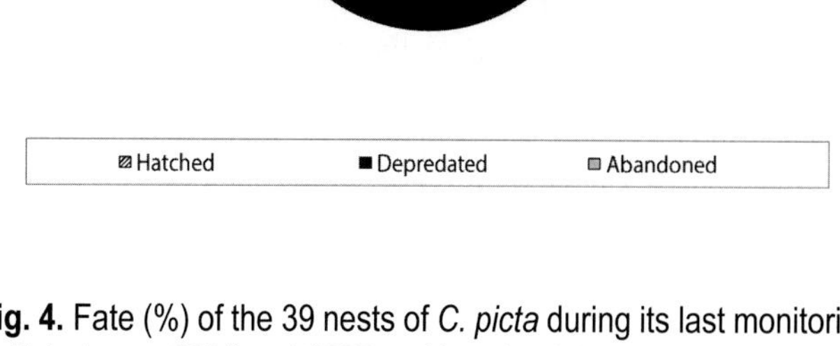

Fig. 4. Fate (%) of the 39 nests of *C. picta* during its last monitoring visit, between 2005 and 2007 on Navarino Island.

TABLE 1. Reproductive biology studies of *Chloephaga picta* at the southern tip of South America and their respective records of clutch size, range of values, number of nests monitored, and study area.

Clutch size	Min.	Max.	Rank	No. monitored nests	Study area	Reference
6.1 ± 1.2	4.9	7.3	3 to 8	39	Navarino Island	Ibarra *et al.* (this work)
5.9 ± 1.5	4.4	7.4	Unknown	126	Malvinas Islands	Harradine 1977
6.2 ± 1.1	5.1	7.3	Unknown	20	Malvinas Islands	Harradine 1977
4.9 ± 0.4	4.5	5.3	3 to 8	16	Malvinas Islands	Quillfeldt *et al.* 2005
6.1 ± 1.1	5	7.2	3 to 8	24	Malvinas Islands	Summers 1983
5		7	1 to 7	41	Tierra del Fuego	Jory *et al.* 1974

From this study, we suggest investigating the effect of exotic predators on the reproductive success of *C. picta*. Also, we consider relevant to study the survival of fledglings, anti-predatory behaviors, fight-or-flight response of the females during incubation, and the effect of climate change on their reproductive biology. This last factor would significantly affect the reproductive success of several species of geese at high latitudes (Dickey *et al.* 2008). Environmental education programs can contribute to promote sustainable attitudes and egg harvesting practices, and appreciation of the birds in the region. All of these elements could be key in identifying the factors that set limits on the population of *C. picta* in southern South America. Furthermore, these would provide the foundation for more comprehensive management plans and conservation of the species in the CHBR.

ACKNOWLEDGMENTS

To the Millennium Institute of Ecology and Biodiversity (IEB) (ICM P05-002) and the Puerto Williams branch of University of Magallanes, for all its logistical and financial support. CONAMA (Magallanes), Wildlife Trust Alliance, and the DAAD co-funded this project. The research was made possible by those who accompanied us in the field: Tomas Altamirano, Antonia Barreau, Claire Brown, Melisa Gañán, Germán González, and José Llaipén. The authors, along with Nicolás Galleguillos, took photographs of the reproductive process. Special thanks to Robert Petitpas for his cartographic support. We appreciate the comments of Professor Claudio Venegas, Dr. Jaime Jiménez, and an anonymous reviewer, who helped to improve earlier versions of the manuscript. José Tomás Ibarra has a CONICYT scholarship. This research is part of the conservation programs and long-term research of Omora Park in the Cape Horn Biosphere Reserve, a member of the Long Term Socio-Ecological Research Sites Network in Chile (www.ieb-chile.cl/LTSER). This work is dedicated to the memory of Paula Ayerdi, a perseverant and enthusiastic naturalist of the coasts of Chile.

REFERENCES

AMARASEKARE, P. 1993. Potential impact of mammalian nest predators on endemic forest birds of western Mauna Kea, Hawaii. *Conservation Biology* 7: 316-324.

ANDERSON, C.B., R. ROZZI, J.C. TORRES-MURA, S.M. MCGEHEE, M.F. SHERRIFFS, E. SCHÜTTLER & A.D. ROSEMOND. 2006. Exotic vertebrate fauna in the remote and pristine sub-Antarctic Cape Horn Archipelago region of Chile. *Biodiversity and Conservation* 10: 3295-3313.

BARROS, A. 1971. Aves observadas en las islas Picton, Nueva, Lennox y Navarino oriental. *Anales del Instituto de la Patagonia* (Chile) 2: 166-180.

COOPER, T.A. 1978. The history and breeding biology of the Canada geese of Marshy Point, Manitoba. *Wildlife Monographs* 61: 3-87.

COUVE, E. & C. VIDAL. 2003. *Aves de Patagonia, Tierra del Fuego y Península Antártica.* Editorial Fantástico Sur Birding Ltda., Punta Arenas, Chile. 656 pp.

DI CASTRI, F. & E. HAJEK. 1976. *Bioclimatología de Chile.* Vicerrectoría de Comunicaciones, Pontificia Universidad Católica de Chile. Santiago, Chile. 129 pp.

DICKEY, M., G. GAUTHIER & M. CADIEUX. 2008. Climatic effects on the breeding phenology and reproductive success of an arctic nesting goose species. *Climate Change Biology* 14: 1973-1985.

EARNSHAW, E.M. 1973. Entre las aves de Tierra del Fuego. *Hornero* 11: 203-208.

ELY, C.R. & D.G. RAVELING. 1984. Breeding biology of Pacific White-fronted geese. *Journal of Wildlife Management* 48: 823-837.

GARAY, G., W.E. JOHNSON & W.L. FRANKLIN. 1991. Relative abundance of aquatic birds and their use of wetlands in the Patagonia of southern Chile. *Revista Chilena de Historia Natural* 64: 127-137.

GODFRAY, H.C.J., L. PARTRIDGE & P.H. HARVEY. 1991. Clutch size. *Annual Review of Ecology and Systematics* 22: 409-429.

GÖTMARK, F. 1992. The effects of investigator disturbance on nesting birds. *Current Ornithology* 9: 63-104.

GUYN, K.L. & R.G. CLARK. 2000. Nesting effort of northern pintails in Alberta. *Condor* 102: 619-628.

HARRADINE, J. 1977. *General Ecology*. Unpublished report to Ministry of Overseas Development, UK. 130 pp.

HILL, D.A. 1984. Factors affecting nest success in the Mallard and Tufted duck. *Ornis Scandinavica* 15: 115-122.

HUMPHREY, P., D. BRIDGE, P.W. REYNOLDS & R.T. PETERSON. 1970. *Birds of Isla Grande (Tierra del Fuego)*. Smithsonian Institute, Washington D.C, U.S.A.

IBARRA, J.T. 2007. *El Visón en Upushwaia: Variaciones Estacionales de la Avifauna en Ambientes de Humedal y de su Depredación por la Especie Exótica* Mustela vison *(Schreber) en Isla Navarino, Reserva de Biosfera Cabo de Hornos, Chile.* Tesis Magister (MSc), Recursos Naturales, Facultad de Agronomía e Ingeniería Forestal, Pontificia Universidad Católica de Chile. 117 pp.

IBARRA, J.T., R. ROZZI, H. GILABERT, C.B. ANDERSON, S.M. MCGEHEE & C. BONACIC. 2009a. Dinámica estacional y patrones de distribución de la avifauna asociada a humedales subantárticos en la Reserva de Biosfera Cabo de Hornos (54-55°S). *Ornitología Neotropical* 20: 321-337.

IBARRA, J.T., L. FASOLA, D.W. MACDONALD, R. ROZZI & C. BONACIC. 2009b. Invasive American mink in wetlands of the Cape Horn Biosphere Reserve, southern Chile: what are they eating? *Oryx* 43: 87-90.

JACKSON S.L., S.H. DAVID & R.F. ROCKWELL. 1988. The influence of nesting habitat on reproductive success of the Lesser Snow geese. *Canadian Journal of Zoology* 66: 1699-1703.

JEHLE, G., A.A.Y. ADAMS, J.A. SAVIDGE & S.K. SKAGEN. 2004. Nest survival estimation: a review of alternatives to the Mayfield estimator. *Condor* 106: 472-484.

JOFRÉ, J. 2009. *Fenología del Musgo* Tayloria dubyi *en las Turberas de la Reserva de Biosfera Cabo de Hornos: Un Caso de Entomofilia?* Tesis de Magister (MSc), Manejo y Conservación de Recursos Naturales en Ambientes Subantárticos, Facultad de Ciencias, Universidad de Magallanes. 111 pp.

JOHNSON, D.H. 1979. Estimating nest success: the Mayfield method and an alternative. *Auk* 96: 651-661.

JORY, J.E., C. VENEGAS & W.A. TEXERA. 1974. La avifauna del Parque Nacional "Laguna de los Cisnes" Tierra del Fuego, Chile. *Anales del Instituto de la Patagonia* (Chile) 5: 131-157.

KLETT, A.T., T.L. SHAFER & D.H. JOHNSON. 1988. Duck nest success in the prairie-pothole region. *Journal of Wildlife Management* 52: 431-440.

LEPAGE, D., G. GAUTHIER & A. DESROCHERS. 1998. Larger clutch size increases fledging success and offspring quality in a precocial species. *Journal of Animal Ecology* 67: 210-216.

MARTÍNEZ, D. & G. GONZÁLEZ. 2004. *Las Aves de Chile: Nueva Guía de Campo.* Ediciones del Naturalista, Santiago, Chile. 620 pp.

MAYFIELD, H.F. 1961. Nesting success calculated from exposure. *Wilson Bulletin* 73: 255-261.

MAYFIELD, H.F. 1975. Suggestions for calculating nest success. *Wilson Bulletin* 87: 456-466.

MICKELSON, P.G. 1975. Breeding biology of Cackling geese and associated species on the Yukon-Kuskokwim delta, Alaska. *Wildlife Monographs* 45: 3-35.

NEWTON, I. 1998. *Population Limitation in Birds.* Academic Press, London, UK. 596 pp.

QUILLFELDT, P., I.J. STRANGE & J.E. MASELLO. 2005. Escape decisions of incubating females and sex ratio of juveniles in the Upland Goose *Chloephaga picta. Ardea* 93: 171-178.

PISANO, E. 1977. Fitogeografía de Fuego-Patagonia Chilena. Comunidades vegetales entre las latitudes 52° y 56°S. *Anales del Instituto de la Patagonia* (Chile) 8: 121-250.

ROZZI, R. & M. SHERRIFFS. 2003. El visón (*Mustela vison* Schreber, Carnivora: Mustelidae), un nuevo mamífero exótico para la isla Navarino. *Anales del Instituto de la Patagonia* (Chile) 31: 97-104.

ROZZI, R., F. MASSARDO, A. BERGHÖFER, C.B. ANDERSON, A. MANSILLA, M. MANSILLA, J. PLANA, U. BERGHÖFER, P. ARAYA & E. BARROS. 2006. *Reserva de Biosfera Cabo de Hornos: Documento de Base para la Incorporación del Territorio Insular del Cabo de Hornos a la Red Mundial de Reservas de Biosfera.* Programa MAB-Unesco. Ediciones de la Universidad de Magallanes, Punta Arenas, Chile. 274 pp.

SAMELIUS, G. & M. LEE. 1998. Arctic fox, *Alopex lagopus*, predation on Lesser Snow geese, *Chen caerulescens*, and their eggs. *Canadian Field-Naturalist* 112: 700-701.

SCHLATTER, R.P., P. VERGARA & M. BRIONES. 2002. El Canquén (*Chloephaga poliocephala*: Anatidae) en bosques de Tierra del Fuego: distribución y depredadores. *Anales del Instituto de la Patagonia* (Chile) 30: 61-66.

SCHÜTTLER, E., R. KLENKE, S. MCGEHEE, R. ROZZI & K. JAX. 2009. Vulnerability of ground-nesting waterbirds to predation by invasive American mink in the Cape Horn Biosphere Reserve, Chile. *Biological Conservation* 142: 1450-1460.

SCHÜTTLER, E., J.T. IBARRA, B. GRUBER, R. ROZZI & K. JAX 2010. Abundance and habitat preferences of the southernmost population of mink: Implications for managing a recent island invasion. *Biodiversity and Conservation* 19: 725-743.

SHIECK, J.O. & S.J. HANNON. 1993. Clutch predation, cover and the overdispersion of nests of the Willow ptarmigan. *Ecology* 74: 743-750.

SPSS INC. 1999. *SPSS Base 10.0 for Windows User's Guide.* SPSS Inc., Chicago, IL.

SUMMERS, R.W. 1983. The life cycle of the Upland goose *Chloephaga picta* in the Falkland Islands. *Ibis* 125: 524-544.

SUMMERS, R.W. & L.G. UNDERHILL. 1986. Factors related to breeding production of Brent geese *Branta bernicla bernicla* and waders (Charadrii) on the Taimyr Peninsula. *Bird Study* 34: 161-171.

SUMMERS, R.W. & J. MCADAM. 1993. *The Upland Goose.* Bluntisham Books, UK. 162 pp.

TREMBLAY, J.P., G. GAUTHIER, D. LEPAGE & A. DESROCHERS. 1997. Factors affecting nesting success in greater snow geese: effects of habitat and association with Snowy owls. *Wilson Bulletin* 109: 449-461.

WILCOVE, D.S. 1985. Nest predation in forest tracts and the decline of migratory songbirds. *Ecology* 66: 1211-1214.

Received: 24th July, 2009. Accepted: September 30th, 2009

PART V

BIRDS THAT PROMOTE CONSERVATION, ECOTOURISM, EDUCATION, AND ENVIRONMENTAL ETHICS AT THE SOUTHERN END OF THE AMERICAS

25 INTRODUCTION TO THE BIRDS OF THE SOUTHERNMOST TIP OF SOUTH AMERICA THAT PROMOTE CONSERVATION, ECOTOURISM, EDUCATION, AND ENVIRONMENTAL ETHICS

J. Cristóbal Pizarro, José Tomás Ibarra, and Ricardo Rozzi

In the Robalo River watershed, on the northern coast of Navarino Island, rivers, streams, and wetlands emerge that are a magnet for birds, humans and other living beings inhabiting Cape Horn Biosphere Reserve (CHBR) that find here a vital sustenance (Ibarra *et al.* 2007). The closeness among different types of habitats (forests, peat lands, high Andean steppes, lakes and rivers, Beagle Channel, etc.) that make up the mosaic of ecosystems of the watershed allow residents and visitors to experience an endless set of feelings following each of encounters with their co-inhabitants, humans and non-humans (*sensu* Rozzi *et al.* 2005). For example, in a day hike through this basin, one can "meet" with aged Lengas (*Nothofagus pumilio*) or deciduous beeches, and stoic Magellanic coigües (*N. betuloides*) or evergreen beeches, which dominate the sub-Antarctic forests. Along the coast, the colorful lichens on rocks indicate us the high tide line, under which live mussels, barnacles, limpets, and other beings of the delicate area that forms this productive habitat in the marine-terrestrial interface. In all these habitats, it is also common to find families enjoying a sunny afternoon and, increasingly, groups of birdwatchers. In the recurrence of these experiences, we interweave our coexistence and cultivate an ethic within a community that extends beyond the human scope, and hosts birds and other living beings with whom we co-inhabit in these unique habitats.[1]

The sound of the wing feathers of an Andean Condor (*Vultur gryphus*) accompanying our hikes on the summits of the mountains, the surprising finding of an Upland Goose (*Chloephaga picta*) nest that we notice when we unintentionally disturb the mother brooding mimetically among bushes, the dive of a Ringed Kingfisher (*Megaceryle torquata*) that with confidence, in front of us, swoops from the branch of a tree to catch a mackerel fish swimming in the estuary, are some of the examples of encounters with birds that remain forever in our memory. Thanks to the good conservation status of the habitats at the southern end of the Americas, these encounters with a variety of birds of different colors, sizes, and voices are common and recurrent in places of the CHBR such as Puerto Williams and its surroundings. This is due to a unique characteristics of this region: birds are the most abundant, diverse, and conspicuous group of vertebrates (Venegas & Sielfield 1998, Rozzi *et al.* 2006).

For humans and birds to continue to co-inhabit these environments, however, it is essential to: (i) conserve the habitats that are home to birds and to all living beings, including humans; (ii) observe our habits and different ways of experiencing encounters with birds and biodiversity. To achieve this goal, the education program of Omora Park proposes an integration of theory and practice and a path to move from science to conservation, emphasizing that it is not enough to just describe and convey these interspecific encounters. Rather, it is necessary to also promote and empower these encounters through interdisciplinary research, environmental education, and biocultural conservation with a long-term horizon and commitment. Thus, the long-term ornithological studies of Omora Park address multiple ways of seeing and engaging with the biocultural diversity, ranging from listening to the birds, the wind , the rain, the rustling of forest streams and slow melting of glaciers that give origin to rivers and streams, to the stories that people have narrated for generations.

Face of development pressures and other threats now reaching remote areas and unique as the CHBR, birdwatching or birding ethics, supported by more than ten years of encounters with birds and conservation practices and education emerges as an activity option economically and environmentally sustainable for the region of Magallanes. The challenge is to cultivate this activity as an integral experience. To achieve this aim, the incorporation of philosophy and environmental ethics is key to addressing complex processes that seek to combine the welfare of humans, the birds, and all living beings in a relationship of co-inhabitation. In this regard, in her afterword to the *Multi-Ethnic Bird Guide of the Sub-Antarctic Forests of South America* (Rozzi *et al.* 2011), the philosopher Irene Klaver states that:

> Birding (bird-watching) has become an increasingly popular activity today in Western culture. For example, in the United States has established an industry with more than 46 million people who practice birding (bird -watchers). What do they see in this activity?

[1] Based on an understanding of the linkages between habitats, inhabitants and habits, Rozzi et al. (2008) has proposed a biocultural ethic that recovers the archaic meaning of the Greek word *ethos*: the den of an animal. This biocultural ethic proposes a co-inhabitation among different co-inhabitants (humans and non-humans), in their habitats, oriented by ecologically conscious habits.

Fig. 6. The biologist Margaret Sherriffs and the Magellanic architect Fernando Haro observe the diverse shapes of lichens of the genus *Cladonia* that grow over the rocks during one of the field environmental ethics workshops at Omora Park. (Photographs: Ricardo Rozzi [left] and Silvina Ippi [right]).

of the site, they draw images of the lichens observed. An interesting fact usually happens in this activity: the colors of pencils do not coincide with the colors of the lichens. This fact leads students to experientially learn about the technical difficulties of "capturing" the colors and other dimensions of the living beings with whom we co-inhabit. This experience helps us to reflect on the limitations of our ability to comprehend biodiversity within our conceptual categories and with our technical tools. At the same time, this experience stimulates our awareness of the unlimited singularities of each living being.

CONCLUDING REMARKS

In summary, Omora Park's methodological approach emphasizes: a) field experiences, *in situ*, with biological and cultural diversity in its integral context (in contrast with today's emphasis on mediated knowledge); b) educational processes (rather than an emphasis on content and information); c) reflection on learned understanding of biocultural diversity and of our ways of relating with other living beings (humans and non-humans); d) the adoption of biocultural conservation actions oriented by continuous learning experiences, experiences of co-inhabitation, and inter-cultural dialogue and reflection. This educational process invites us to evaluate our ways of life and eventually modify them, in a continuous cultivation of an environmental ethic.

ACKNOWLEDGMENTS

Omora Ethnobotanical Park's education and field environmental ethics program has been a collective effort that has involved the participation of many people of the Puerto Williams community, investigators, students, and volunteers. In its development, collaboration with CONAMA and CONAF Region XII and the government of the Chilean Antarctic Province and the program EXPLORA-CONICYT has been essential. We thank the support of the Centro Milenio para Estudios Avanzados en Ecologia y Biodiversidad (N°-P02-051-FICM), BIOKONCHIL (BMBF, N°-FKZ0ILM0208) and the Darwin Initiative UK (N°13024).

REFERENCES

DARWIN, C. 1839. *The Zoology of the Voyage of H.M.S. Beagle, Under the Command of Captain Fitzroy, R.N., During the Years 1832 to 1836: Published with the Approval of the Lords Commissioners of Her Majesty's Treasury.* Smith, Elder & Co., London.

HARGROVE, E. 1997. Ética ambiental y educación ambiental. *Ambiente y Desarrollo* 13: 47-51.

IPPI, S. & R. ROZZI. 2004. Actividad diurna y nocturna del concón (*Strix rufipes*) en los bosques del Cabo de Hornos. *Boletín Chileno de Ornitología* 10: 9-12.

LÓPEZ-CALLEJA, M.V. & F. BOZINOVIC. 1995. Maximun metabolic rate, thermal insulation and aerobic scope in a small-sized Chilean hummingbird (*Sephanoides sephaniodes*). *Auk* 112: 1034-1036.

MARGUTTI, L., R. OVIEDO, M. HERBEL & P. FEINSINGER. 2001. Indagación ecológica en el patio de la escuela. Pp. 526-528 *in Fundamentos de Conservación Biológica: Perspectivas Latinoamericanas*, R. Primack, R. Rozzi, P. Feinsinger, R. Dirzo & F. Massardo. Fondo de Cultura Económica, México.

ROZZI, R. 1997. Enseñar a los niños a ser ciudadanos, no meramente consumidores: el aporte de Hargrove a la educación ambiental en Chile. *Ambiente y Desarrollo* 13: 52.

ROZZI, R. 1999. The reciprocal links between evolutionary-ecological sciences and environmental ethics. *BioScience* 49: 911-921.

ROZZI, R., P. FEINSINGER & R. RIVEROS. 1997. *La Enseñanza de la Ecología en el Entorno Cotidiano.* Ministerio de Educación de Chile, Santiago de Chile.

ROZZI, R., F. MASSARDO, C. ANDERSON, K. HEIDINGER & J. SILANDER, JR. 2005. Ten principles for biocultural conservation at the southern tip of the Americas: the approach of the Omora Ethnobotanical Park. *Ecology & Society*: http://www.ecologyandsociety.org/vol11/iss1/art43/

WEBER, M. 1958. *Essays in Sociology.* H.H. Gerth & C.W. Mills, (eds.). Oxford University Press, New York.

27 DISCOVERY AND IMPLEMENTATION OF THE MAGELLANIC WOODPECKER (*CAMPEPHILUS MAGELLANICUS*) AS A CHARISMATIC SPECIES: A BIOCULTURAL APPROACH FOR CONSERVATION IN THE CAPE HORN BIOSPHERE RESERVE

XIMENA ARANGO[1], RICARDO ROZZI[1,2], FRANCISCA MASSARDO[1], CHRISTOPHER B. ANDERSON[1], AND TOMÁS IBARRA[1,3]

[1]Parque Etnobotánico Omora, Universidad de Magallanes-Sede Puerto Williams e Instituto de Ecología y Biodiversidad (IEB-Chile)
Correo electrónico: xae_arango@yahoo.es
[2]Department of Philosophy and Religion Studies, University of North Texas, Denton, USA
[3]Laboratorio Fauna Australis, Programa de Recursos Naturales, Facultad de Agronomía e Ingeniería Forestal, Pontificia Universidad Católica de Chile

Magallania (Chile) 35: 71-88 (2007) (Originally published in Spanish)

ABSTRACT. Discovery and implementation of Magellanic Woodpecker (*Campephilus magellanicus*) as a charismatic species: a biocultural approach for conservation in the Cape Horn Biosphere Reserve. Cape Horn Biosphere Reserve, located in the southernmost part of South America, protects unfragmented Magellanic sub-Antartic evergreen *Nothofagus* forests. This zone has been considered one of the world´s most pristine regions and is currently subject to increasing development pressures from new connectivity, urban expansion, and tourism development. Using a biocultural approach we found that the Magellanic Woodpecker is the favorite bird for the community of Puerto Williams, the world´s southernmost town and the largest human settlement in the Cape Horn Biosphere Reserve. Although significant differences ($P < 0.05$) exist between the preferences of five interviewed sociocultural groups toward birds, all agreed in their preferences toward the Magellanic Woodpecker. This species of bird was most preferred by people who have lived longest in Cape Horn: the members of the Yahgan Indigenous Community and long time residents. The analysis of these results, as well as the discovery of ten attributes of this species of picid, which is also the largest of South America, have led us to consider it as a charismatic or flagship species with a high potential for the conservation of southern forests in the Cape Horn Biosphere Reserve. For these reasons the Magellanic Woodpecker could be transformed into a symbol, or a charismatic species, for the conservation of the pristine austral forest. Between the year 2005 and 2007 we developed a Program for the Implementation of the Magellanic Woodpecker with the objective to increase its intrinsic value, and valuation of its habitat, and the knowledge within the local community. The discovery of its potential as a flagship species, as well as its subsequent implementation, may contribute to the conservation of the fragile, unique and beautiful austral forests and its unmatched biological and cultural diversity.

Key words: *Campephilus magellanicus*, Cape Horn Biosphere Reserve,onservation, flagship species, cold-growth forests.

RESUMEN. Descubrimiento e implementación del pájaro carpintero gigante (*Campephilus magellanicus*) como especie carismática: una aproximación biocultural para la conservación en la Reserva de Biosfera Cabo de Hornos. Al sur de Sudamérica se encuentra la Reserva de Biosfera Cabo de Hornos donde persisten los bosques siempreverdes subantárticos. Esta zona, considerada una de las más prístinas del mundo, está sometida a presiones crecientes de conectividad, crecimiento urbano y desarrollo turístico. Sin embargo, todavía alberga bosques no fragmentados de *Nothofagus.* Con una aproximación biocultural a la conservación determinamos que el pájaro carpintero gigante es la especie de ave preferida por la comunidad de Puerto Williams, capital de la Provincia Antártica Chilena, ciudad más austral del mundo y mayor asentamiento humano de la reserva. Aunque existen diferencias significativas (P<0,05) en las preferencias hacia las aves por parte de los distintos grupos socioculturales entrevistados, todos coincidieron en sus preferencias hacia el carpintero gigante. Esta especie de ave fue la predilecta para las personas que llevan más tiempo en el Cabo de Hornos: los miembros de la Comunidad Indígena Yagán y los residentes más antiguos. El análisis de estos resultados, así como el descubrimiento de diez atributos que presenta esta especie de pícido, que además es el más grande de Sudamérica, nos han llevado a considerarlo como *una especie carismática* o *emblemática* con un alto potencial para la conservación de los bosques australes de la Reserva de Biosfera Cabo de Hornos. Por las anteriores razones el carpintero gigante puede transformarse en un símbolo o especie carismática para la conservación de los prístinos bosques australes. Entre el año 2005 y 2007 desarrollamos el *Programa para la Implementación del Carpintero Gigante* con el objetivo de aumentar su valoración, de su hábitat y el conocimiento por parte de la comunidad local. El descubrimiento del potencial de esta especie como carismática así como su posterior implementación podrían contribuir a la conservación de los frágiles, singulares y bellos bosques australes y de su diversidad biológica y cultural.

Palabras clave: bosques antiguos, *Campephilus magellanicus*, conservación, especie emblemática, Reserva de Biosfera Cabo de Hornos.

INTRODUCTION

At the southern end of South America, the evergreen and deciduous *Nothofagus* trees dominate the southernmost forests in the world (Rozzi *et al.* 2006a,b). The Magellanic sub-Antarctic forests ecoregion at the austral extreme is one of the few areas of the world where unfragmented, temperate forests are still conserved, and are free from direct modern human impact (Silander 2000, Mittermeir *et al.* 2002, Rozzi *et al.* 2004). This was one of the reasons for the creation of the Cape Horn Biosphere Reserve, created by Chile and UNESCO in June 2005. With a total of 4,884,274 ha (2,967,036 ha of marine areas and 1,917,238 ha of terrestrial areas), Cape Horn is the largest biosphere reserve in the Southern Cone of South America (Uruguay, Argentina, and Chile) (Rozzi *et al.* 2006b). Within this island territory, a mosaic of tundra and forests dominated by trees of the genus *Nothofagus* characterize the landscape. There are distinct types of forest ecosystems: sub-polar Magellanic evergreen rainforests dominated by Magellanic Coigüe (*N. betuloides*) in lowlands on insular areas near the Pacific, deciduous sub-Antarctic forests of Ñirre or Low-deciduous Beech (*N. antarctica*) near the tree-line zone on high slopes and in hyper-humid lowland areas, deciduous forests of Lenga or High-deciduous Beech (*N. pumilio*) on the slopes with good drainage, and mixed forests of Lenga and Magellanic Coigüe in the lowlands of islands bordering the Beagle Channel (Pisano 1977, Rozzi *et al.* 2006a).

Upon the creation of the Cape Horn Biosphere Reserve, the Chilean Government proposed to implement it through the development of sustainable economic activities that simultaneously protect and conserve these unique natural ecosystems in the world. The fact that these ecosystems are conserved in an almost pristine state offers a great economic potential for ecotourism development (Martinic 2006, Rozzi *et al.* 2004). However, the appeal for economic potential in this area could just as easily lead to short-term economic development through unregulated tourism, and to highly negative environmental and social impacts in this southern archipelagic region (Martinic 2006, Rozzi *et al.* 2006a). To conserve these forests, as well as their biological and cultural diversity, it is necessary to develop conservation actions with the participation of the local and regional communities. One way to facilitate participation of the community is to define a charismatic or flagship species serving as a symbol for the value of biodiversity in the region. Thus, a charismatic species can become a stimulus for the integration of ecological and socio-cultural motivations (Dietz *et al.* 1994, Walpole & Leader-Williams 2002).

In this article, we present the research that identified the Magellanic Woodpecker (*Campephilus magellanicus*) as a charismatic species for the Cape Horn Biosphere Reserve. Our approach considers the methodology used for its identification, and the methodology used for its subsequent implementation at the local and regional levels.

CHARISMATIC SPECIES

The term *charismatic species* was first used in the mid-eighties (Mittermeier 1986, Walpole & Leader-Williams 2002), and refers to a popular species that serves as a symbol, stimulates public awareness of the importance of biodiversity conservation (Heywood 1995), and can serve as leading figure in a conservation campaign (Simberloff 1998). On a global scale, the selection of charismatic species has been based on aesthetic criteria and/or its conservation status (Kleiman *et al.* 1987, Dietz *et al.* 1994). Successful examples of the implementation of the *charismatic species* concept are the program for the recovery of the threatened Golden Lion-tamarin (*Leonthopitecus rosalia*) in Brazil (Kleiman *et al.* 1987, Dietz *et al.* 1994, Leader-Williams & Dublin 2001) and the program to conserve the Giant Panda (*Ailuropoda melanoleuca*). Because it captured the attention of millions of people and because it has contributed to secure financial support for its own conservation and that of its habitat (Dietz *et al.* 1994), the Panda was adopted as the symbol and the charismatic species of the World Wide Fund for Nature (WWF).

Generally, *charismatic species* have included birds (such as the Condor) and attention-drawing mammals (large cats or primates). Only very occasionally some plants (such as orchids or cactuses) or invertebrates (butterflies and small insects) have fulfilled this function (Heywood 1995, Meffe & Carroll 1997). In regard to ecotourism and potential economic income, it has been reported that birds are efficient species supporting sustainable tourism in a large number of protected areas (Krüger 2005). Tourists pay large sums of money to visit zones that contain attractive birds (Naidoo & Adamowicz 2005), especially if these zones involve species recognized as charismatic (Walpole & Leader-Williams 2002).

In the Cape Horn Biosphere Reserve, birds are the most diverse group of vertebrates, with more than 90 species of marine and terrestrial birds (Couve & Vidal 2003). Other groups of terrestrial vertebrates, such as amphibians and reptiles, are not present in the area due to the low temperatures in the extreme south; terrestrial-coastal mammals present only 9 species in this area (Anderson *et al.* 2006, Rozzi *et al.* 2006b). Additionally, in the Cape Horn region, birds are distributed in all types of habitats, a relevant aspect from a conservation standpoint: if we protect a *charismatic species*, then its habitat will also be preserved, along with a significant number of other species and natural *in situ* processes (Primack & Massardo 2001).

For our study at the Cape Horn Biosphere Reserve, we decided to focus on searching for a *charismatic species* within the class of birds, and to subsequently implement it in the region to help conserve the old-growth *Nothofagus* forests. A fundamental and innovative aspect in this work is that we used methodologies that allowed us to identify a *charismatic species* taking into account the perceptions of the local community, and not only through aesthetic or ecological criteria defined *a priori*. Although we decided to focus our search for a *charismatic species* on the taxonomic group of birds, the identification of this charismatic bird was determined *a posteriori* based on the results of the survey.

The specific objectives of this research were to: 1) identify the bird species preferred by the community of Puerto Williams; 2) determine which ones have the greatest potential to act as a *charismatic species*; 3) identify the attributes or criteria to nominate a species as charismatic for the conservation of the ecosystems at the Cape Horn Biosphere Reserve; and 4) after defining the charismatic species of the Cape Horn Biosphere Reserve, generate social proactivity, at both local and regional levels, through the design and implementation of education and outreach actions focused on the conservation of the *charismatic species* and its habitat.

MATERIALS AND METHODS

Study area

The study was conducted with the population of the city of Puerto Williams (54°56'S, 67°36'W), capital of the Chilean Antarctic Province. This is the largest human population of the Cape Horn Biosphere Reserve with 2,262 inhabitants (INE 2002 in Rozzi *et al.* 2006b). The city is located on the northern coastline of Navarino Island. The terrestrial habitats in this island include evergreen forests of Magellanic Coigüe, mixed forests of Coigüe and Lenga, diverse tundra formations, high Andean zones, and some scrubland areas and anthropic grasslands. The marine systems include the boulder coasts, rocky coasts, bay formations, fjord and channel systems, and other areas more open to the sea (Rozzi *et al.* 2006b).

Identification of knowledge and preference for bird species

The study of the parameters that allowed us to identify the potential species of charismatic birds was done conducting surveys with the population of Puerto Williams. The population consists of diverse socio-cultural groups. Among them, the following can be identified: members of the Yahgan Indigenous Community of Mejillones Bay including descendants of the Yamana people, also known as Yahgans, native inhabitants of the area included today in the Cape Horn Biosphere Reserve (Alvarez *et al.* 2004, Martinic 2006); also, a group of long-time residents of Navarino Island was distinguished, who are third, second, or first generation settlers and have resided for more than 7 years in Puerto Williams or in other parts of the biosphere reserve; people who have been residents for less than 5 years were also identified, including personnel of the Chilean Navy on the Beagle Naval Base, officials of local public services (National Forest Service, Provincial Governor of Puerto Williams, National Fisheries Service, Municipality of Cape Horn), teachers of the Donald McIntyre Griffiths School – the only educational establishment of the city covering pre-school to full secondary education –, and the students of this school, who are for the most part associated in various ways to the Chilean Navy.

According to the socio-cultural groups identified in the proposal for the creation of the Cape Horn Biosphere Reserve (Rozzi *et al.* 2006b), the following six groups of respondents were defined: 1) long-time residents (*residents*), ii) members of the Yahgan Indigenous Community of Mejillones Bay (*Yahgans*), iii) teachers of the Donald McIntyre Griffiths School (*teachers*), iv) authorities and public services (*authorities*), v) personnel associated to the Chilean Navy (*Navy personnel*), and vi) students of the Donald McIntyre Griffiths School (*students*).

With the objective of having an equal number of respondents in each group, we conducted 14 interviews per group, given that this was the largest number of people that satisfied the criteria for each one of the groups identified. Between June 2004 and May 2005 we gave each respondent a structured questionnaire in written format (Bernard 1995) that included three questions: (1) *Write the name of the first 5 birds that you can think of*, (2) *Write the name of birds that you know live in Navarino Island* (15 blanks), and (3) *Which are the 3 birds in Navarino Island that you like the most?*

The first question, *Write the name of the first 5 birds that you can think of*, had as objective the characterization of the birds that were primarily present in the memory, thoughts or imagination of the respondents, and to determine if these birds corresponded to species that inhabit the area of the biosphere reserve, or species that inhabit other regions of the country or the world. Question two, *Write the name of 15 birds that you know live on Navarino Island*, evaluated knowledge about a fundamental ecological attribute in the life history of the species: its geographical distribution. If a respondent named a species of bird that was not present on Navarino Island, this response expressed a degree of ignorance with respect to that bird species. Question three, *Which are the 3 birds on Navarino Island that you like the most?* is central to the definition of the charismatic species. The analysis of the answers to this question seeks to identify the preferred species. This preference involves not only knowledge of the name, but also expresses valuation of the bird. Valuation for a bird species by a respondent can imply an understanding and/or emotion for its diverse attributes, such as its aesthetic features (for example, color of plumage), utility (for example, palatability, and valoration of poultry birds), popularity (for example association with people or characters on television), symbolic value (for example national emblem birds), rarity (for example, a bird that is difficult to observe) (Arango 2007).

The questionnaires were applied in different ways according to the group interviewed: for the *residents*, *Yahgans*, and *teachers* questionnaires were conducted through personal interviews at each respondent's house, where each person answered the questionnaire in written format. All repondents were of Chilean nationality. Some did not answer in writing format for various reasons (9 *residents*, 13 *Yahgans*, and 12 *teachers*) and were interviewed verbally. The group of *authorities* was interviewed personally at their respective workplaces. The *Navy personnel* were interviewed by giving all questionnaires and explaining procedures to a supervisor of the Beagle Naval Base. The students responded to the questionnaire in a classroom at the school during class hours.

Interviewed individuals were asked to voluntarily indicate in question 3 the reasons that explained their preferences. This allowed recording the qualities and general observations about the preferred species.

The composition of each of the groups of respondents (Arango 2007) depended on the availability of the people, and was characterized in the following way:

- *Residents* had an average age of 42 years (range between 18 and 70). Of the respondents, 21.4% were women, and 78.6% were men who had lived in the Cape Horn area for more than 7 years. Their occupations were artisanal fishermen for the most part, but some were merchants, and others unemployed.
- The *Yahgan* respondents are of Yahgan heritage and one person pure Yahgan, and were born either in Puerto Williams, in other parts of Navarino Island, and in other Islands of the Magallanic Region or adjacent to Argentina (Isla Grande of Tierra del Fuego). The median age of this group was 48 years (with a range between 28 and 84 years old). Of the respondents, 57.1% were women and 42.9% men who were living in the area for 45.1 years. Their occupations were principally artisanal fishermen and craftsmen.

• *Teachers* had a median age of 45 years old (range of age between 34 and 53). In this group of respondents, the percentage of women and men was equal. They had been in Puerto Williams for an average of 13.6 years.

• The *authorities* interviewed are officers with the Chilean Antarctic Government and the Municipality of Cape Horn, and the remaining worked in Public Services. Their median age was 33 years (age range between 25 and 65). The proportion of men and women respondents was equal, and they had been residents of Puerto Williams for an average of 11.6 years.

• The *Navy personnel* worked for the Chilean Navy of the Beagle Naval District. They were on average 38 years old (their ages ranged between 29 and 48). In this group of respondents the percentage of women was 21.4% and 78.6% for men. Their residence in Puerto Williams was 2.4 years on average.

• The *students* were enrolled in their senior year of high school at the Donald McIntyre School in Puerto Williams. The average age of the group was 17 years (age range between 16 and 17). The percentage of female respondents was 71.4%, while male respondents were 28.6%; their average length of stay in Puerto Williams was 6.8 years.

To evaluate the presence of significant differences among the responses between the sociocultural groups, a matrix for each of the questions was elaborated using the names of individual birds to apply an ANOSIM[1] (Analysis of Similarities) test, using the statistical program PRIMER, version 5.2.2 for Windows. When significant differences existed, an *a posteriori* test was done comparing each of the group's responses using the same statistical program. When significant differences existed, an *a posteriori* test was conducted comparing each of the group's responses using the same statistical program.

Implementation of the charismatic species

Based on the results obtained from the analysis described in the previous section, it was determined that the preferred species with charismatic potential was the Magellanic or Black Woodpecker (*Campephilus magellanicus*). For this reason, an extensive literature review was conducted on the biological and cultural attributes of this species, including its potential for ecotourism, conservation status, distribution area, and ethnographic records, complemented by ethnographic research in the field, to determine valuable attributes for its role as a charismatic species.

After determining the Magellanic Woodpecker's charismatic character, a publicity campaign was initiated in order to educate and recreate the existing empathy of the local community towards the species, as well as to increase knowledge and appraisal by all participants. For this purpose, the methodology developed by Dietz *et al.* (1994) for the protection of primates in Brazil was adapted. This consisted of designing a strategic plan that emphasizes the inclusion of demographic groups that are less familiar with the avifauna of the Cape Horn Biosphere Reserve, taking into account the available working resources of the Omora Ethnobotanical Park program (Rozzi *et al.* 2003). This research, education, and biocultural conservation program is coordinated by the Omora Foundation, and the University of Magallanes at its Puerto Williams campus along with the Master in Science program, in collaboration with the Institute of Ecology and Biodiversity (IEB), and the University of North Texas (Rozzi *et al.* 2006). The Dietz *et al.* (1994) model provides a systematic process of seven steps considering: 1) identification of main problems (conservation status of the species and potential threats to its habitat); 2) identification and evaluation of the target population, the resources available, and the socio-ecological context; 3) development of positive interactions with participants; 4) selection and analysis of methods; 5) implementation of activities; 6) evaluation; and 7) recycling of activities. These seven steps were adapted to the case of the Magellanic Woodpecker as a charismatic species in the Cape Horn Biosphere Reserve.

The program for the implementation of the Magellanic Woodpecker as a charismatic species involved several institutions. The field activities were developed on the guided circuit trails of the Omora Ethnobotanical Park, and the indoor activities took place at the local school in Puerto Williams. These activities were part of the weekly Natural History Workshop that the Omora Park's scientific group has maintained continuously in the school since the year 2000. This workshop is offered as an elective course to students from 4th to 8th grades, as well as to the school's Scouts group. Guided by researchers and teachers, students develop ecological research that is presented annually at the Regional School Science Fair EXPLORA-CONICYT Program. In 2006 the topic was the conservation of the Magellanic Woodpecker and its forest habitats on Navarino Island. The government institutions that collaborated with the Omora Park science team were: the Municipality of Cape Horn, the Environmental Protection Fund (FPA) of the regional CONAMA, the EXPLORA-Program of the Chilean National Science Foundation (CONICYT), and the Regional Chile's Forest Service (CONAF). The Government of the Chilean Antarctic Province and the Chilean Navy supported the implementation of the activities in all their phases.

Actions taken to implement the charismatic species

The actions taken to implement the charismatic species considered three areas:

1) *Educational strategies:* wildlife and ecotourism workshops were offered along with multiple guided field tours. For children, in addition to the weekly Natural History Workshop at the local school, puppet shows, theater plays, and painting workshops were created and presented to the public. These activities also involved preschool children from the public Ethnic Kindergarten "Ukika." Educational activities considered the ecology, ethology, and traditional knowledge about the Magellanic Woodpecker, especially the stories documented by early 20th century ethnographers (Gusinde 1986) or Omora Park researchers (Rozzi *et al.* 2003b, Zárraga 2005).

[1] The statistical analysis ANOSIM is a non-parametric method of permutations that is applied to a matrix of ranges with similarities. This test provides a coefficient of global similarity R ($-1 \leq R \leq 1$) and another coefficient for the pairs being compared, thus distinguishing values of significance (Clark & Gorley 2001).

2) *Development of educational materials:* a series of posters, brochures, videos, CDs, gifts, presentations, drawings, embroidery, postcards, and tote bags for grocery shopping with images of the Magellanic Woodpecker were designed and published with students.

3) *Dissemination of the work:* permanent contact was maintained with the community and community leaders, local authorities (Government of the Chilean Antarctic Province, Municipality of Cape Horn, Chilean Navy), directors and officers of local government agencies (CONAF) and regional authorities (CONAF, CONAMA), media (radio, press, television), and private companies (EDELMAG).

RESULTS AND DISCUSSION

Identification of the Magellanic Woodpecker as a charismatic species for the Cape Horn Biosphere Reserve

The ANOSIM analysis found that the major socio-cultural groups of Puerto Williams showed significant difference in their answer to question 1: *Write the names of the first 5 birds that you remember* (R = 0.103, P <0.01). We found significant differences between the old *residents* and *Navy personnel* (P <0.01), old *residents* and *authorities* (P <0.015), and old *residents* and *students* at the local school (P <0.01). There were also significant differences between *Yahgan* people and *Navy personnel* (P <0.01), between *Yahgan* people and *students* (P <0.02), between *Navy personnel* and *students* (P <0.014), and between *authorities* and *Navy personnel* (P <0.033).

Regarding the five birds mentioned in the answers to "Question 1" by members of the sociocultural groups of Puerto Williams, the percentage of answers that included native species present on Navarino Island is shown in Fig. 1. All (100%) birds named by *Yahgans* are present in the area of the Cape Horn Biosphere Reserve. *Residents* also named almost exclusively species present in the area (98.6%). The *authorities* mentioned species present in the area 91.4% of the times, and *Navy personnel* in 87.1%. Both, *teachers* and *students* named 82.9% of native birds; thus, this was the highest percentage (about 17.1%) of species absent from the island, being the species in their answers native to southern Chile, exotic and absent from Navarino Island, or introduced. In short, knowledge of *teachers*, *students*, and *Navy personnel* shows more attention to birds in other regions, whereas the knowledge by *Yahgans* and older *residents* is focused on the regional avifauna. Knowledge by authorities is intermediate between the other two groups.

Regarding the degree of knowledge of avifauna, the analysis of "Question 2" – *Write the name of 15 birds you know from Navarino Island* – showed significant differences among socio-cultural groups (R = 0.172, P <0.01). The *Yahgan* group had the highest degree of knowledge of birds, giving in most cases 15 names of birds that actually live in the Cape Horn Biosphere Reserve area (Fig. 2). The average number of birds named was 14.3, which is 95.7% of the 15 bird species names requested by the question (Fig. 2). The group of *residents* averaged 13.2 names of bird species present in the reserve (87.6%). This number is less than the average names in the response by *Yahgan* descents, but these groups did not differ significantly (P >0.05). *Authorities*, *teachers*, *students*, and *Navy personnel* named about 10 species on average (a number of bird names significantly lower than the previous groups); thus, they answered satisfactorily between 69.5 and 63.8% of the birds requested by this question (Fig. 2). It is important to point out that to count the number of birds named by each respondent, we considered only those living in the Cape Horn Biosphere Reserve, including among these domestic poultry (chickens) or urban (sparrow and pigeon) species that, although not native to the region, they currently inhabit Puerto Williams. In conclusion, the degree of knowledge about regional avifauna by these groups is significantly less than the knowledge possessed by *Yahgan* descents and older residents of the Navarino Island.

Regarding "Question 3" - *What are the 3 birds on Navarino Island that you like the most?* – we found significant differences among the six socio-cultural groups showed (R = 0.038, P <0.04). However, a posteriori tests showed that significant differences were found only between *teachers* and *old residents*, between *teachers* and *Navy personnel*, and between *Navy personnel* and *authorities*. These more uniform results are due in part to the fact that all cultural groups named among their preferred bird species, the Magellanic Woodpecker.

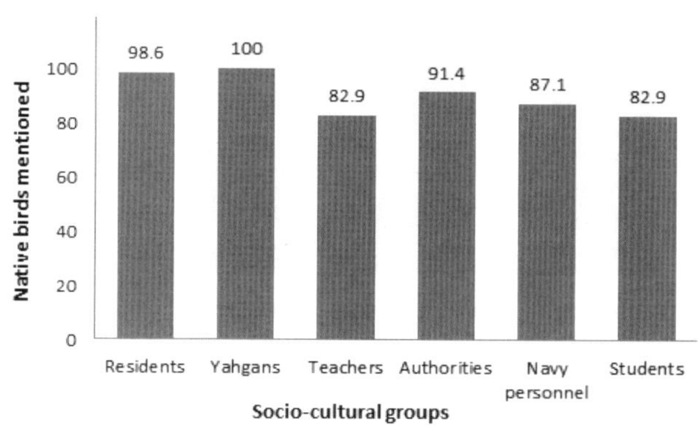

Fig. 1. Percentage of native birds in the Cape Horn Biosphere Reserve named by respondents in each of the sociocultural group in Puerto Williams when asked: *Write the name of the first 5 birds that you remember.*

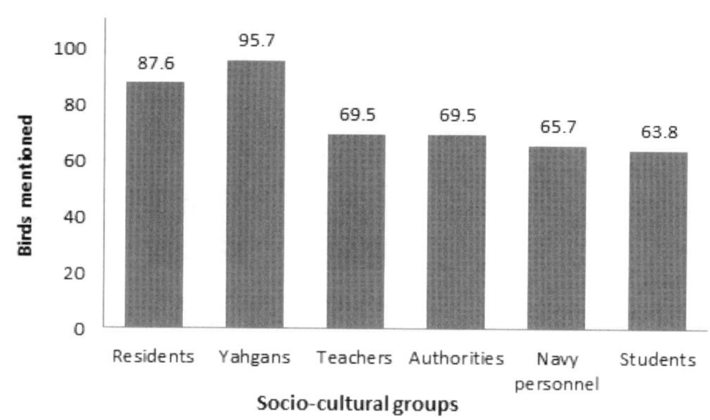

Fig. 2. Percentage of birds (native and introduced) present on Navarino Island named by respondents in each of the sociocultural group in Puerto Williams when asked: *Write the name of 15 birds on Navarino Island that you know.*

of this beautiful bird can be tracked back to ancient times, when birds were still human.

"In ancestral times, when birds were still humans, a boy fell in love with his sister and tried everything to meet and sleep with her. His sister noticed his intentions and dodged all of her brother's aims, thus avoiding the forbidden relationships. However, in her heart, she wanted to be with him, and at the same time did not. Her brother kept thinking of excuses to lure her out of the *akar,* or hut. One day he discovered large red *Chaura* fruits in a clearing of the forest and went swiftly to tell his sister: "I have found huge *Chauras* in the forest; you should go and collect them." His sister took her basket and ventured into the woods, while he followed her without anyone noticing, hiding in a place where his sister would pass. Once she arrived, he lunged and hugged her, and fell together on the ground unleashing his love. When they stood up, they became birds and flew like *lana*. Since then, they live together in the woods. Her brother carries on his head a red tuft reminiscent of those great fruits of *Chaura.*"

This story expresses the closeness between the life of this bird and that of the Yahgan people. The beginning of this story when birds were still humans, as well as the metamorphosis of these siblings into Magellanic Woodpeckers, express a notion of genealogical relatedness between birds and humans. Although expressed differently by scientific evolutionary theories, this common genealogy between birds and humans implies a sense of kinship, which could encourage greater ethical consideration for the intrinsic value of this bird (Rozzi 2004).

From the point of view of popular valoration, the Magellanic Woodpecker is associated with the Woody Woodpecker cartoon, which was a representation of the Pileated Woodpecker *Dryocopus pileatus*.

With regard to its value for biodiversity conservation, *Campephilus magellanicus* could be evolutionarily the closest species to the largest woodpecker in the world: the Imperial Woodpecker (*Campephilus imperialis*), extinct in the Northern Hemisphere due to habitat disturbance and hunting pressures. Therefore, *C. magellanicus* represents a high-value lineage and an inspiration for conservation. Nationwide in Chile, this species presents conservation problems and is categorized as Vulnerable (Glade 1993).

The relationship between the Magellanic Woodpecker and the Imperial Woodpecker, and the fact that this is the largest woodpecker in South America, generates attraction for birdwatchers and thus, importance for ecotourism. Among ecotourists visiting the Magellanic Region, the Magellanic Woodpecker is one of the birds most frequently requested (Ricardo Matus, personal communication). Both, from the standpoint of conservation and tourism, it is interesting that the Magellanic Woodpecker is endemic to the temperate forests of southern Chile and Argentina (Couve & Vidal 2003). For these reasons, its implementation as a charismatic species could increase income from tourism in the Cape Horn Biosphere Reserve.

Finally, from the ecological point of view, three essential attributes make the Magellanic Woodpecker and ideal charismatic species: a) it is a *keystone species* because it builds nesting sites for other bird species, and also controls the populations of insect larvae in trunks of trees of the genus *Nothofagus*; b) it is an *indicator species* of old-growth forests, as it builds its cavities and feeds on larvae requiring ancient trees and old-growth forests and thus, has special value for biodiversity conservation; c) it is an *umbrella species* because it establishes territories with larger home ranges than other bird species in the sub-Antarctic forests. Therefore, the protection of the Magellanic Woodpecker habitat protects also the habitat of many other species with smaller home range requirements.

Table 1 presents a summary of the 10 attributes that describe *C. magellanicus* as an ideal charismatic species in the Cape Horn Biosphere Reserve. Once identified, this set of attributes was communicated to the community of Puerto Williams and of the Magallanic Region. This was done in order to make them aware of the importance of implementing this species for the conservation of biological and cultural diversity, as well as for conserving the regional identity.

Implementation of the Magellanic Woodpecker as a charismatic species in the Cape Horn Biosphere Reserve

The efforts put into the Program for the Implementation of the Magellanic Woodpecker were focused on increasing knowledge of this species and its habitat, along with an enhancement of awareness about ecological interactions. These actions were aimed at various sectors of the Puerto Williams population. The population felt attracted to and interested in environmental issues, and also showed enthusiasm due to the fact that, aside from being a symbol of local pride for being a beautiful and rare bird, the Magellanic Woodpecker could be associated with economic benefits through the promotion of tourism in the community (Fig. 4a,b).

Undoubtedly, one of the most important interactions was the partnership with the Donald McIntyre School and the JUNJI kindergarden. Through this work, in a large portion of the adult population, awareness was raised significantly in each family through the children. Moreover, since there is only one high school in Puerto Williams, children attending this school belong to families included in all sociocultural groups. Therefore, working with children was a key action for the diffusion of the implementation program of this bird species (Fig. 5).

For regional outreach, it was very important to rely on media to reinforce the Magellanic Woodpecker as a symbol of the Cape Horn Biosphere Reserve. Additionally, we designed and published posters, brochures, videos, CDs, gifts, presentations, drawings, embroidery, postcards, and fabric shopping bags with images of the Magellanic Woodpecker. Shopping bags were especially relevant, because it is an item that was part of the program *Let's Clean Puerto Williams*, organized by the Municipality of Cape Horn, seeking to reduce the use of disposable plastic bags. This cleaning program also wanted to raise awareness about the need for protecting and maintaining the planet's southernmost forests (Fig. 6). The Program for the Implementation of the Magellanic Woodpecker interested and encouraged the Chilean Navy authorities to use the Magellanic Woodpecker's image as a symbol in its own environmental campaign to reduce the use of plastic bags in the biosphere reserve. It is important to point out that, according to the interviews, the Navy personnel group showed the lowest percentage of preference for the Magellanic Woodpecker (Fig. 3).

Fig. 4. a) In January 2005, Magellanic Woodpecker photographs were delivered to the local artisan Alejandra Guanel with information about its natural history and ecology. Alejandra used these photographs to develop novel embroidery motifs and sell them to local people and tourists. b) Some people from the local community requested woodpecker photographs to display in their business. The resident and restaurant owner in Puerto Williams, Victor Hugo Henríquez, received photographs to display at *The Albatross*, local shop. These are just some examples of the many activities involving the local community (Photos Ximena Arango and Christopher Anderson).

The main stages for the implementation process of the Magellanic Woodpecker are summarized in Appendix I.

FINAL REMARKS AND CONCLUSIONS

This work shows that there are significant knowledge differences with regard to how different cultural groups view bird diversity in the Puerto Williams Community. However, all six groups converged upon choosing the Magellanic Woodpecker as the favorite among bird species. Therefore, the Magellanic Woodpecker has the potential to appeal to different sectors and bring them together in a common action for the development of sustainable tourism, thus generating income for the community and promoting biodiversity conservation.

Species attracting people's attention, like the Magellanic Woodpecker, have high potential to motivate conservation actions of both the species and its habitat. Conservation of these key species helps also to protect the biodiversity resting under its shade. In this way, protecting the Magellanic Woodpecker also protects the species of plants and animals that live in the ancient forests, its rivers and fertile soils in Navarino Island and in the whole Cape Horn Biosphere Reserve. The program on the Omora Ethnobotanical Park charismatic species had as its first objective of defining which of these species appealed to the local community. The implementation program took place only after the community *discovered* the Magellanic Woodpecker as a charismatic species, thus gaining public support for the *C. magellanicus* through education and outreach activities. This enhanced the local community's awareness and appreciation for this species and the forests that it inhabits. The actions that were undertaken involved schoolchildren and permanent residents of the biosphere reserve for the most part. Through these groups adopting the charismatic species, the effort put in the implementation program will last over time.

In sum, this study shows that:

1. The Puerto Williams sociocultural groups differ significantly in their preferences for birds in the area, but they agree in their preference toward the Magellanic Woodpecker.

2. The Magellanic Woodpecker was especially preferred by the members of the Yahgan Indigenous Community and old residents, who are people living longer in the Cape Horn region and possesing greater knowledge of the local avifauna.

3. The Magellanic Woodpecker is a charismatic species for the population of Puerto Williams.

4. This research identified ten criteria that identify the Magellanic Woodpecker as an ideal charismatic species to foster specific conservation actions towards the old-growth forests of the Cape Horn Biosphere Reserve.

5. The activities conducted as part of the Implementation Program of the Magellanic Woodpecker, together with the specific outreach materials, help spreading ecological knowledge and appreciation for the Magellanic Woodpecker at different scales (local, regional, national, and international), but primarily at the local level.

6. The environmental education activities and the outreach materials used during the Implementation Program of the Magellanic Woodpecker show that it is possible to apply knowledge generated by ecological research. This also emphasizes that the transdisciplinarity between biological and cultural conservation requires collaborative processes in which knowledge is generated and used simultaneously. Here, the process of research and its application promises to build public support for the Magellanic Woodpecker, resulting in the protection of the Magellanic sub-Antarctic forests. This process also generated more options for a sustainable economy in the community of the Cape Horn Biosphere Reserve.

Fig. 6. In April 2007, in coordination with the Municipality of Cape Horn, the Program for the Implementation of the Magellanic Woodpecker marked the beginning of the Earth Month at the Donald McIntyre School of Puerto Williams. The ceremony was attended by all students, their teachers, the directives of the school, and the Mayor of the Community. On that event, the Implementation Program gave out 300 cloth bags printed with the Magellanic Woodpecker drawings that had won the competition *The Magellanic Woodpecker and its habitat.* The purpose of giving this gift to the community was to start a campaign to reduce the use of plastic bags. Each of the families with children attending the School received a bag. In this photograph, a group of teachers posing with bags that were funded by the Outreach Science Program of the Institute of Ecology and Biodiversity and Omora Park (Photos Mauricio Bahamonde and Ximena Arango).

Fig. 5. Steps of the Drawing Contest *The Magellanic Woodpecker and its habitat* organized by the Implementation Program of the Magellanic Woodpecker for the Puerto Williams community. Children repeatedly visited the Omora Park where they became familiarized with the Magellanic Woodpecker habitat and occasionally saw individual woodpeckers. In a classroom afterwards, they drew what they had seen on the field. Later on, they presented their work to the public. Community leaders Jose Soto Passek, Mayor of the Cape Horn County, Julia Gonzalez, a member of the Yahgan Indigenous Community, Francisco Gascogne, Area Manager of Cape Horn's Chile Forest Service (CONAF), and Claudio Yánez, Second Beagle Naval District Commander actively participated as juries of the contest. Winning children received prizes and certificates from the Mayor. The active participation of the different sectors of the local community is critical for a successful Implementation Program (Photos Ximena Arango and Rina Charlín).

ACKNOWLEDGMENTS

We appreciate the contribution of all those who participated in the Program for the Implementation of the Magellanic Woodpecker. The work was supported by the project BIOKONCHIL (Subprojects 1 and 3, BMBF, No.-FKZ01LM0208, German Ministry of Education and Research), and the Institute of Ecology and Biodiversity IEB (No-P02-051-FICM). We appreciate the cooperation of Omora Park, Omora Foundation, Outreach Science Program of the Institute of Ecology and Biodiversity, University of Magallanes, Donald McIntyre School of Puerto Williams, Ethnic Kindergarten of Ukika Village, Municipality of Cape Horn, Chilean Navy, Governor of the Chilean Antarctic Province, CONAMA, CONAF, SERNATUR, and EXPLORA-CONICYT. Ximena Arango Estévez thanks the Institute of Ecology and Biodiversity for the Masters Scholarship ICM Project (P05-002).

REFERENCES

ÁLVAREZ, R., F. MASSARDO, R. ROZZI, U. BERGHÖFER, A. BERGHÖEFER & J. FREDES. 2004. Patrimonio Cultural de la Reserva de Biosfera Cabo de Hornos Propuesta. Pp. 153-177 *in The Cape Horn Biosphere Reserve: A Proposal for Conservation and Tourism to Achieve Sustainable Development at the Southern End of the Americas*, R. Rozzi, F. Massardo & C. Anderson, (eds.). Bilingual English-Spanish Edition. Ediciones Universidad de Magallanes, Punta Arenas, Chile.

ANDERSON, C, R. ROZZI, J.C. TORRES-MURA, S. MCGEHEE, M. SHERRIFFS, E. SCHUETTLER & A. ROSEMOND. 2006. Exotic vertebrate fauna of the remote and pristine sub-Antarctic Cape Horn Archipelago, Chile. *Biodiversity and Conservation* 15: 3295-3313.

ARANGO, X. 2007. *El Pájaro Carpintero Gigante (*Campephilus magellanicus, *King 1828): Una Especie Carismática al Fin del Mundo y su Implementación para la Conservación de los Bosques Antiguos de la Reserva de Biosfera Cabo de Hornos.* Tesis de Magíster en Ciencias, Universidad de Magallanes, Punta Arenas.

ARANGO, X. & R. ROZZI. 2004. El carpintero gigante (*Campephilus magellanicus*): una especie carismática al fin del mundo. P. 239 *in Libro de Resúmenes de la II Reunión Binacional Chile-Argentina de Ecología*, Mendoza.

ARANGO, X. & R. ROZZI. 2005. El carpintero gigante (*Campephilus magellanicus*): una especie carismática al fin del mundo. Pp. 47-54 *in Vivir en la Isla Navarino: Un Mundo Diverso en el Fin del Mundo.* Proyecto BIOKONCHIL (Ministerio de Educación e Investigación, BMBF, N01LM0208).

BERNARD, H.R. 1995. *Research Methods in Anthropology. Qualitative and Quantitative Approaches.* Altamira Press, Inc., Walnut Creek.

CLARKE, K.R. & R.N. GORLEY. 2001. *PRIMER v5: User manual/ tutorial.* PRIMER-E, Plymouth, UK.

COUVE, E. & C. VIDAL. 2003. *Aves de la Patagonia, Tierra del Fuego y Península Antártica.* Editorial Fantástico Sur Birding Ltda., Punta Arenas.

DIETZ, J.M., L.A. DIETZ & E.Y. NAGAGATA. 1994. The effective use of flagship species for conservation of biodiversity: the example of Lion Tamarins in Brazil. Pp. 32-49 *in Creative Conservation: Interactive Management of Wild and Captive Animals.* P.J.S. Olney, G.M. Mace & A.T.C. Feistner, (eds.), Chapman & Hall, London.

GLADE, A. 1993. *Libro Rojo de los Vertebrados Terrestres de Chile.* Segunda Edición Corporación Nacional Forestal (CONAF). Ministerio de Agricultura, Santiago.

GUSINDE, M. 1986. *Los Indios de Tierra del Fuego.* Tomo Segundo: *Los Yámana.* Centro Argentino de Etnología Americana. 1 y 2, Buenos Aires.

HEYWOOD, V.H. (ED.). 1995. *Global Biodiversity Assesment.* Cambridge University Press, Cambridge.

KLEIMAN, D.G., B.B. BECK, J.M. DIETZ, L.A. DIETZ, J.D. BALLOU & A.F. COIMBRA-FILHO. 1987. Conservation program for the golden lion tamarin: captive research and management, ecological studies, education strategies, and reintroduction. Pp. 959-979, *in Primates: The Road to Self-Sustaining Populations*, K. Benirschke, (ed.). Springer-Verlag, New York.

KRÜGER, O. 2005. The role of ecotourism in conservation: panacea or Pandora´s box? *Biodiversity and Conservation* 14: 579-600.

LEADER-WILLIAMS, N. & H.T. DUBLIN. 2001. Charismatic Megafauna as Flagship Species. *In Priorities for the Conservation of Mammalian Diversity. Has the Panda had its Day?* A. Entwistle & N. Dunstone, (eds.). Cambridge University Press, London.

MARTINIC, M. 2006. *Crónica de las Tierras del Sur del Canal Beagle.* Ediciones Hotel Lakutaia, Punta Arenas.

MCBRIDE, P. 2000. *Magellanic Woodpecker (*Campephilus magellanicus*) Habitat Selection in Deciduous* Nothofagus *Forests of Tierra del Fuego.* Unpublished master's thesis, Western Washington University.

MEFFE, G.K. & C.R. CARROLL (EDS.). 1997. *Principles of Conservation Biology.* Sinauer Associates Inc. Publishers, Sunderland, Massachusetts.

MITTERMEIER, R.A. 1986. Primate conservation priorities in the Neotropical region. Pp. 221-240 *in Primates:The Road to Self-Sustaining Populations*, E. Benirschke, (ed.). Springer-Verlag, New York.

MITTERMEIER, R., C. MITTERMEIER, J. PILGRIM, G. FONSECA, W. KONSTANT & T. BROOKS. 2002. *Wilderness. In Earth's Last Wild Places*, P. Robles, (ed.). CEMEX-Conservation International, Washington, DC.

NAIDOO, R. & W.L. ADAMOWICZ. 2005. Economic benefits of biodiversity exceed costs of conservation at an African rainforest reserve. *Proceedings of the National Academy of Sciences* 102: 16712-16716.

OJEDA, V. 2004. Breeding biology and social behaviour of Magellanic Woodpeckers (*Campephilus magellanicus*) in Argentine Patagonia. *European Journal of Wildlife Research* 50: 18-24.

PISANO, E. 1977. Fitogeografía de Fuego Patagonia chilena. I. Comunidades vegetales entre las latitudes 52° y 56°S. *Anales del Instituto de la Patagonia* 8: 121-250.

Step 5: Implementation of activities (*continued*).	*Materials* produced by the Program for the Implementation of the Magellanic Woodpecker as a charismatic species for conservation included: • News, press releases, and interviews on local radio and television; • Chapters in two programs of the National Television of Chile; • Full program on the conservation of*C. magellanicus* for cable television; • DVD with a piece of puppets; • Posters of the research presented at the Regional Science Fair Workshop for Omora Workshop students; • Posters presented at different scales (local, regional, national and international); • Chapters in books; • Four different series of postcards; • CD with lectures to be presented at workshops on ecotourism and education; • Cloth bags with drawings of the Magellanic Woodpecker; • Badges and cards with messages about conservation; • Mounted pictures with images of the Magellanic Woodpecker and texts alluding to preserve the program as a gift to community leaders and local business owners who exhibit in their premises; • Inclusion of the Magellanic Woodpecker logo in the Master of Science Program at the University of Magallanes; • Preliminary design logo of the Cape Horn Biosphere Reserve including*C. magellanicus* and their habitat; • Inclusion in of a Magellanic Woodpecker giant poster of CONAMA produced under the National Action Plan for the Conservation of Biodiversity.
Step 6: Evaluation.	The Magellanic Woodpecker's image is now in the local public discourse as are many cases where government representatives have made public references to the importance of this species for the locality. Two ongoing environmental campaigns of the I. Municipality of Cape Horn and the Chilean Navy, have Magellanic Woodpecker's image as a symbol of a clean and natural environment. Empathy for the species emerged from meetings between CHBR Scientific Committee and the Municipality, who have repeatedly talked about the importance of *C. magellanicus* and ideas that have been raised together that have been able to materialize. At the regional level, a significant achievement of the program has been the inclusion of *C. magellanicus* in the "National Action Plan for the Conservation of Biodiversity, Magallanes and Chilean Antarctic - CONAMA." Locally, public interest in the species has increased as evidenced by the following indicators: • Pictures of the species published by regional SERNATUR to promote tourist visits to Cape Horn in an international journal, National Geographic Magazine, where the giant woodpecker represents the terrestrial fauna of the region, and is in a bilingual brochure; • Paintings of*C. magellanicus* by artists from Puerto Williams, displayed on the premises of the Municipality of Cape Horn and hostels, with family groups of woodpeckers (male, female and chicks) that stand in respect to the biological characteristics of the species, such as color and sexual dimorphism; • Dissemination materials (brochures) on local and regional tourism published by people from the local community including woodpecker photos; • Interest from local merchants to present photographs of*C. magellanicus* and request for sale of promotional material (such as postcards) in their stores; • Interest shown by regional businesses, such as the company that printed calendars EDELMAG 2007 with photographs of the bird; • Adoption of Magellanic Woodpecker's image as a symbol for environmental campaigns of the Municipality of Cape Horn and the Beagle Naval Base of the Chilean Navy.
Step 7: Recycling.	Although this Step is ongoing, institutions such as the Chilean Navy have decided to replicate some initiatives implemented by our program. Educational materials produced in this program may be held by different generations of students, kindergarten teachers, tour operators, etc. Thus, the conservation of this species could be reflected to the north of the country as a national example. An expression of remarkable continuity in the picture postcards of the Biosphere Reserve Cape Horn, and therefore the image of *C. magellanicus* as elements of its habitat, are in the preliminary design of the official logo of the Cape Horn Biosphere Reserve.

PART VI

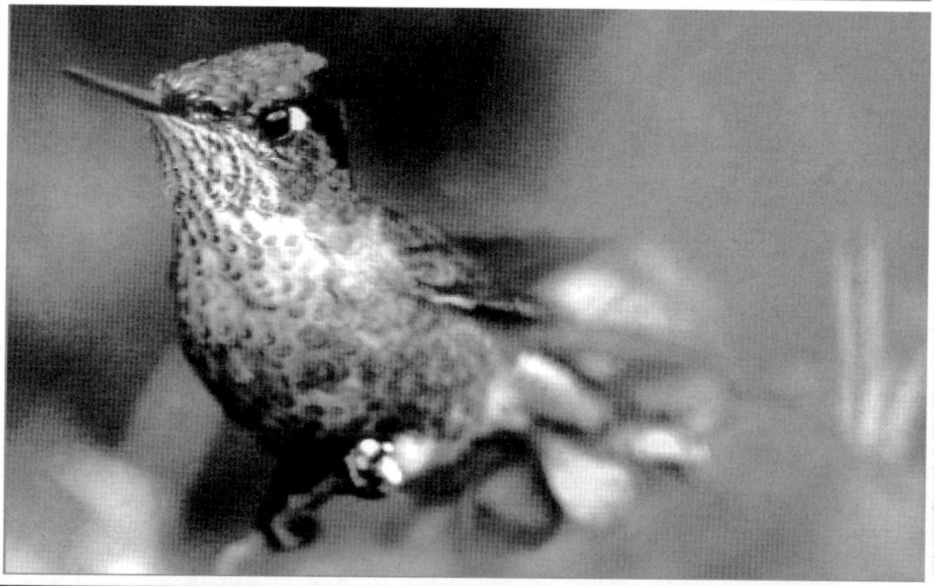

ETHNO-ORNITHOLOGY IN THE SUB-ANTARCTIC FORESTS OF SOUTH AMERICA

28 INTRODUCTION TO ETHNO-ORNITHOLOGY IN THE SUB-ANTARCTIC FORESTS OF SOUTH AMERICA

VICTORIA CASTRO, FRANCISCA MASSARDO, AND JOSÉ TOMÁS IBARRA

And his main delight, when he heard the birds to speak with their own diction,
he felt the same joy that assists the rural people when they converse with them,
as with the water, the stone, and the tree. (Plath 1976:11)

One of the best lessons of how human beings search for connection and meaning with the world that they inhabit is offered by the knowledge hosted by native people and rural, traditional mestizo communities. The humanists dedicated to anthropology and the naturalists dedicated to ecology and biology have become involved in a point of encounter to collaborate among themselves and with the bearers of ancestral knowledge in the Americas. These knowledges build an identity that is deeply rooted in the depth of time of the pre-Columbian people, and are orally transmitted from generation to generation until the present, in a dynamic world of continuities and transformations.

Within this indigenous science, we find refined forms of knowledge embedded in life. Today, the readers and society are convened by ethno-ornithology of the Mapuche, as narrated by Lorenzo Aillapan and Ricardo Rozzi, and by the ethno-ornithology of the Yahgan (or Yamana), as narrated by the grandmothers Úrsula and Cristina Calderon, accompanied by Francisca Massardo and Ricardo Rozzi. As you read this text, you will appreciate these persons for the love they have given to their day-to-day work to make this ethno-ornithology available to us, thereby providing bridges to learn about these substantial narratives. We will be amazed by the richness of the wisdom embodied by these languages and their ability to name, describe, and integrate the birds in all spheres of life. If something is indissoluble in the worldviews of the American people, it is the integration of all spheres of reality, saturated with values and sensibilities that are lived in the walks, and inhabiting all facets of daily life, as well as in the rituals and myths that reaffirm the everlasting life.

From this perspective, these knowledges are complex processes in which both the senses and the rational understanding take part, creating links of social communication that include ideas, feelings, norms, and values. In many of these philosophical and scientific concepts —which rely on worldviews that are different from the Western world—, even the words "nature" and "culture" are not necessary as these concepts are contained and intermixed in the Amerindian conception of the world.

In this understanding of the indissoluble relationship between nature and culture, the articles presented here will show us, tacitly, the historical route of the ethno-zoology, in general, and of the ethno-ornithology, in particular (Santos Fita *et al.* 2009). These articles will teach us about the sensible systems of classification of the animal world found in the forests of southern South America; the inexhaustible presence of the birds in the stories, narratives, poems, and the role that they play in the explanation of reality (cosmogonies and worldviews); and, finally, they will provide evidence as to how the dimensions of knowledge, practices, and beliefs about the birds, are embedded in the austral traditional ecological knowledge (TEK).

So, it should not surprise us that we find horizontal, equal-to-equal relationships established with the birds and the human beings, because they could be interchangeable characters: you could be a bird or you could convert yourself into one, in a sort of enchanted regulator, of balance among things. The respect crosses all these representations, engendering a care for all the living beings and for all that exists in the world that we inhabit: forests, rivers, hills, ponds, soil, and stones. Each element with its value and its meanings is integrated into the daily lives. A sage from another latitude, from the Andes in the north of Chile, said that the land is alive, that the rocks are the bones, the waters the blood, and the soil itself the flesh...

In this way, the landscape that is inhabited by humans is being built and filled with representations, giving names and meanings to each bend, place, flora, and fauna that surrounds and accompanies them. Readers will be able to discover that the wealth of understanding from the Mapuche society that, despite five centuries of struggle to defend their culture, still holds its TEK with vitality. In another example, we will learn from the last bearers of the Yahgan language about the deep knowledge of the birds and sub-Antarctic landscapes that have shaped the culture in these latitudes for about 8,000 years.

Lorenzo and Ricardo open our minds to learn to hear other voices; we are taught a political position that contributes to caring for this chaotic world today. We are given the gift of the possibility to conceive the landscape from an inclusive, holistic, and comprehensive humanity, in which nothing is left to chance. In this context, one might relate to a concept of historical ecology (*sensu* Crumley 2003), as old as it is complex. The implications for the disciplines of education, conservation, and environmental ethics are tangible.

As it always should be, in these dialogues, a conversation is shared in which the affections, the delicacy, and the ethical considerations are embedded in past and present times that are alive. We should appreciate this present and past in all of its unfolding dimensions in order to care for the future of the earth. This is still true on the extreme south of the Americas where the grandmothers, Úrsula and Cristina Calderon, immersed in the habit of telling stories, offer in the early of the 21th century the stories they heard from their father Juan Calderon and from other members of the Yahgan community on Navarino Island. The richness of their stories about the birds is grounded as much in the legacy inherited from an oral tradition as from the careful observation of the life histories of birds such as the Austral Thrush or *Hakasir*, where the human and the ornithological converge in the behavior displayed by parents who feed and educate their young (Rozzi *et al.* 2010: 148). The names of the birds in its languages with its meanings and histories were recorded by grandmother Úrsula with the simplicity of the vivid memories of her childhood in Mejillones Bay, surrounded by thick forests, plethoric with birds and their songs, and nourished by the stories told by her father and later by her grandfather.

These histories express a refined capacity of observation of the natural cycles, of the relationships between the plants and the birds, and among the birds and the humans in the archipelagos of Cape Horn. This emphasis on the observation and the retelling of the stories passed down from generation to generation, contrast with the inventive and playful emphasis of the poet and creator, Lorenzo Aillapan, *Uñumche* or Mapuche Bird-Man. For the grandmothers, the forgotten customs, beliefs, and stories "of the elders" were often yearned for during lengthy conversations in Puerto Williams.

The knowledge passed on by the grandmothers Úrsula and Cristina, together with the poetry of Lorenzo Aillapan, allow us today to relive the permanent encounter with the birds, and give us the option to recuperate an experience that enriches our imagination and inner life in finding this winged world, palpitating, suffering, and many times threatened, is entitled to continue to exist. That option for the conservation of these ecosystems and its birds is therefore ours. This is the option that the Omora Ethnobotanical Park aims to promote, as described in detail by the article about its biocultural research, education, and conservation program, and symbolized by its name and mission, which is inspired by the Yahgan narrative of the hummingbird *Omora*. The ethno-ornithology helps us to understand this option through the vivid exercise of opening our understanding to diverse understandings and visions of the natural world. These worldviews are imprinted in the geographic names and symbolic language that remind us of the instrumental value of the birds for the existence of humans as much as the intrinsic value of the life of each one of the beings with which we co-inhabit in the sub-Antarctic forests of the Americas.

REFERENCES

CRUMLEY, C. 2003. *Historical Ecology: Integrated Thinking at Multiple Temporal and Spatial Scales. World System History and Global Environmental Change*. Lund University, Sweden.

PLATH, O. 1976. *Lenguaje de los Pájaros Chilenos*. Editorial Nascimiento, Santiago, Chile.

ROZZI, R., F. MASSARDO, C. ANDERSON, S. MCGEHEE, G. CLARK, G. EGLI, E. RAMILO, U. CALDERÓN, C. CALDERÓN, L. AILLAPAN & C. ZÁRRAGA. 2011. *Multi-ethnic Bird Guide of the Subantarctic Forests of South America*. 2nd Edition. University of North Texas Press & Ediciones Universidad de Magallanes, Denton, TX and Punta Arenas, Chile.

SANTOS FITA, D., E.M. COSTA NETO & E.J. CANO-CONTRERAS. 2009. El quehacer de la etnozoología. Pp. 23-44 *in Manual de Etnozoología: Una Guía Teórica-práctica para Investigar la Interconexión del Ser Humano con los Animales*, E.M. Costa Neto, D. Santos Fita & M. Vargas Clavijo (eds.). Tundra Ediciones, Valencia, España.

29 A CONTEMPORARY MAPUCHE ETHNO-ORNITHOLOGY: TWENTY WINGED POEMS FROM THE NATIVE FORESTS OF CHILE

LORENZO AILLAPAN[1,2] AND RICARDO ROZZI[2,3,4]

[1]Academia Mapuche *PÜLLÜMAPUKIMUNWEFTUY*, Puerto Saavedra, IX Región, Chile
[2]ONG Omora, Puerto Williams, Región de Magallanes y Antártica, Chile. *E-mail:* fundacion@omora.org
[3]Universidad de Magallanes, Casilla 113-D, Punta Arenas, XII Región Chile
[4]Department of Philosophy and Religion Studies, University of North Texas, P. O. Box 310980, Denton, Texas 76203-0980, USA

Ornitología Neotropical 15 (Suppl.): 419-434 (2004) (Originally published in Spanish)

ABSTRACT. A contemporary Mapuche ethno-ornithology: twenty winged poems from the native forests of Chile. This paper presents a particular view of the current indigenous Mapuche ornithology, which is developed on the basis of the poetry of Lorenzo Aillapan and the analysis of this poetry from the perspective of ecological sciences and environmental ethics. Aillapan is a Mapuche poet who lives in the countryside and forests of the Lafkenche Region, inhabited by Mapuche communities associated to the coast and ocean (= *lafken*) in southern Chile. At the age of 8 to 9 years old, Aillapan was initiated through a dream as the Birdman, initiation which was guided by renowned Mapuche masters. Today, Aillapan is a Mapuche Birdman or *Üñümche* (*üñüm* = bird, *che* = human being), and as such he communicates with the birds. Before doing that, he asks for permission from Mother Nature and the birds to talk about them and for them. In the austral summer 2001, Aillapan worked with the sisters Úrsula and Cristina Calderón of the Indigenous Yahgan Community, and with a team of scientists of the Omora NGO on Navarino Island, Cape Horn County. A result of that work are these twenty winged poems from the native forests of Chile, which we present at this symposium. This article exposes and analyzes five of these poems: the *lloyka* or Longtailed Meadowlark (*Sturnella loyca*), the *traru* or Crested Caracara (*Polyborus plancus*), the *triuki* or Chimango Caracara (*Milvago chimango*), the *raki* or Black-faced Ibis (*Theristicus melanopis*), and the *pinda* or Green-backed Firecrown (*Sephanoides sephaniodes*). This selection of poems allows us to associate the forest birds with the Mapuche cosmogony and their onomatopoeic names in the Mapudungun language. The ethno-ornithological analysis of the poems illustrates how interrelated biological and cultural diversities are in the Mapuche natural medicine, social and environmental ethics, language, traditional ecological knowledge, and worldview. The poems are based on refined observations in the field and the cultural tradition of *Üñümche*, and they allow us to understand some key signs about the relations established by the Mapuche communities with their natural environment and avifauna. Interestingly, the biocultural approach of the poem analysis shows not only differences, but also fundamental similarities between the ornithological knowledge rooted in the indigenous tradition and the biological sciences. *Accepted February 5, 2004.*

Key words: Avifauna, Chile, environmental ethics, ethno-ornithology, Mapuche, poems, poetry, temperate forests.

RESUMEN. Una etno-ornitología Mapuche contemporánea: poemas alados de los bosques nativos de Chile. Este trabajo expone una mirada particular de la ornitología mapuche actual, elaborada a partir de la poesía de Lorenzo Aillapan y del análisis de esta poesía desde las ciencias ecológicas y la ética ambiental. Aillapan es un poeta mapuche que vive en los campos y los bosques de la Región Lafkenche, donde habitan comunidades mapuches asociadas a las costas o mar (= *lafken*) en el sur de Chile. A la edad de 8 a 9 años, Aillapan es iniciado a través de un sueño como el hombre pájaro, consagración que es guiada por célebres maestros mapuches. Hoy don Lorenzo es un Hombre-Pájaro-Mapuche o *Üñümche* (*ünüm* = ave, *che* = ser humano) y como tal puede comunicarse con las aves. Antes de hacerlo, pide permiso a la Madre Naturaleza y a las aves para hablar de ellas y por ellas. En el verano del 2001 don Lorenzo trabajó con las hermanas Úrsula y Cristina Calderón de la Comunidad Indígena Yagán y con científicos de la ONG Omora en la isla Navarino, Comuna Cabo de Hornos. Fruto de ese trabajo son los veinte poemas alados de los bosques nativos del sur de Chile que se presentan en este simposio. Este trabajo expone y analiza cinco de esos poemas: la *lloyka* o Loica (*Sturnella loyca*), el *traru* o Traro (*Polyborus plancus*), el *triuki* o Tiuque (*Milvago chimango*), el *raki* o la Bandurria (*Theristicus melanopis*), y la *pinda* o el Picaflor (*Sephanoides sephaniodes*). Esta selección de poemas nos permite asociar a las aves con la cosmogonía mapuche y sus onomatopéyicos nombres en la lengua mapudungun. El análisis etno-ornitológico de los poemas ilustra cuán interrelacionadas están la diversidad biológica y cultural en la medicina natural, la ética ambiental y social, el lenguaje onomatopéyico, el conocimiento ecológico tradicional y la cosmovisión mapuche. Los poemas que surgen a partir de las atentas observaciones en los campos y la tradición cultural de *Üñümche* nos llevan también a comprender algunos signos acerca de las relaciones establecidas por las comunidades mapuches con su entorno natural y las aves. A su vez, la aproximación biocultural del análisis de los poemas muestra no sólo diferencias, sino que también interesantes similitudes entre los saberes ornitológicos tradicional indígena y científico.

Palabras clave: avifauna, bosques templados, Chile, ética ambiental, etno-ornitología, mapuche, poemas, poesías.

INTRODUCTION

During the last decades, traditional ecological knowledge [TEK] (Inglis 1993) has increasingly captured the attention not only of anthropologists, but also of scientists in many areas of biology (Berkes 1993) and conservation (Cunningham 1993, Balick *et al.* 1994). Regarding the importance of this latter dimension, the Convention of Biological Diversity and Agenda 21, both of which resulted from the UN Earth Summit in Rio de Janeiro (1992), underline the importance of traditional indigenous knowledge and conservation of both cultural and biological diversity (Rozzi & Feinsinger 2001). In this article, we study a particular aspect of the intimate tie that the Mapuche people and their culture maintain with both their land, which today includes the south of Chile and Argentina, and the beings that inhabit that land: We look at the traditional Mapuche knowledge about birds from the point of view of a current poetic expression (Aillapan 1994).

Both the name of the Mapuche people (*mapu* = land, *che* = people) and of their language mapudungun (*mapu* = land, *dungun* = language) express the ancestral identification that these people maintain with their land (Coña & Moesbach 1930). This communication of the Mapuche culture with the diversity of living beings in the region of South American temperate forests is expressed by and captured through very diverse cultural practices (Ovalle 1646, Gusinde 1917a,b, Coña & Moeshbach 1930, Domeyko 1971, Bulnes 1980, Massardo & Rozzi 1996). Among these practices, poetry constitutes an expression that allows us to get particularly close to the birds, thanks to the rich onomatopoeic nature of Mapudungun and to the close attentive bird watching of Mapuche poets (Rozzi 2001). Mapuche poetry today acquires a bilingual form, Mapudungun and Spanish, that allows a poet like Lorenzo Aillapan to express himself. In turn, through this expression of the Mapuche poet, also the birds and the community of co-inhabitants that coexist with birds and humans in the Mapuche land can "speak" to us.

Aillapan lives in the countryside and forests of the Lafkenche region in the coastal zone of Puerto Saavedra in southern Chile. He is a "Mapuche-Bird-Man" and as such can communicate with the birds and speak for them, especially in his native language: Mapudungun (Aillapan 2003). Before doing this, on each occasion Aillapan asks permission of the birds and Mother Nature to speak about them and for them. The capacity of the Mapuche people to listen to the birds and "the land that speaks to us" (*la tierra que nos habla*) is expressed in his poems. Through his texts, Aillapan keeps alive the profound interconnections between his culture and the birds, as well as between these birds and the ecosystems of southern Chile (Aillapan & Rozzi 2001).

In this work, we present a selection of poems about the birds of southern Chilean forests. The poems were recorded, edited and analyzed together with Lorenzo Aillapan. In particular, we analyze the similarities and differences between the Mapuche ornithological worldview and the scientific one. We also discuss some implications of the poems and their interpretations for biocultural conservation, intercultural education, and environmental ethics.

METHODS

During February 2001, a team of biologists worked with Lorenzo Aillapan on Navarino Island (55°S), in an ethno-ornithology research project described in Rozzi (2002). The work was carried out in two locations, Bahía Mejillones and the Omora Ethnobotanical Park, alongside members of another indigenous culture that has inhabited the South American temperate forests for millennia, the Yahgans. During the development of our investigation, the poet Aillapan showed us his manuscripts of a series of poems on birds that he was working on in the Mapudungun and Spanish languages. We spoke extensively about the poems, taking advantage of a good climate of communication generated from being together in the field, around the fire camp, sharing sunrises and sunsets. As a result of these conversations, we decided that the best way to present the poems was to record and edit them as a team, being as faithful as possible to the original oral text and concepts. Of all the birds Aillapan wrote about, we decided to concentrate on those which were most peculiar to the native forests of Chile, since we have centered our ornithological and ethno-ornithological research on this group of birds for the last ten years (Rozzi *et al.* 1996a,b, Anderson & Rozzi 2000, Anderson *et al.* 2002, Rozzi *et al.* 2003a). As a result of our work, we co-designed educational material that we called "Twenty Winged Poems from the Forests of Southern Chile" (Aillapan & Rozzi 2001), an analogy to the well-known "Twenty Poems of Love" ("Veinte Poemas de Amor") of Pablo Neruda, because both cases are a matter of poems about love. In the "winged poems" love is between humans and birds. Finally, this article about the winged poems offers an analysis for consideration, which aims is to expose the unfolding of a conversation between the perspectives of both Aillapan the Lafkenche poet, and Rozzi the biologist.

The poems, as any oral or written text, contain multiple levels of expression and message. Hence, they admit multiple interpretations and analyses dependent upon the interests of the person who reads them. It seemed to us, therefore opportune to select and integrally include in this article five representative poems of the Twenty Winged Poems: the *lloyka* or Long-tailed Meadowlark (*Sturnella loyca*), the *traru* or Crested Caracara (*Polyborus plancus*), the *triuki* or Chimango Caracara (*Milvago chimango*), the *raki* or Black-faced Ibis (*Theristicus melanopis*), and the *pinda* or Green-backed Firecrown (*Sephanoides sephaniodes*). In this way, we offer the poems in order to enable each individual reader his or her own analysis of them. At the same time, based on these poems, we develop the major part of our ethno-ornithological analysis. We focus on the connotation of the winged poems relative to Mapuche natural medicine, environmental and social ethics, the onomatopoeic language, traditional ecological knowledge, and Mapuche cosmogony. The birds contained in the other fifteen poems, which texts we do not reproduce here, include the *pideñ* or Plumbeous Rail (*Rallus sanguinolentus*), the *tregül* or Southern Lapwing (*Vanellus chilensis*), the *küreu* or Austral Blackbird (*Curaeus curaeus*), the *chinkol* or Rufous-collared Sparrow (*Zonotrichia capensis*), the *maykoño* or Eared Dove (*Zenaida auriculata*), the *okory* or American Kestrel (*Falco sparverius*), the *wilki* or Austral Trush (*Turdus falcklandii*), the *tiftifken* or Andean Tapaculo (*Scytalopus magellanicus*), the *kaniñ* or Turkey Vulture (*Cathartes aura*), the *ñamku* or Red-backed Hawk (*Buteo polyosoma*), the *kekereke challwafe üñüm* or Kingfisher (*Megaceryle torquata*), the *pitriu* or Chilean Flicker (*Colaptes pitius*), the *killkill* or Austral Pygmy-Owl (*Glaucidium nanum*), the *chiwüd* or Barn Owl (*Tyto alba*), and the *kongkong* or Rufous-legged Owl (*Strix rufipes*). Nevertheless, this article considers essential elements of this group of birds and of the forest ecosystems where they live, analyzing the ecological and cultural relations of these avifauna and with the Mapuche people. The nomenclature for Mapudungun bird names and other Mapuche terms have been defined by Lorenzo Aillapan, corresponding almost completely with linguist María Catrileo's (1998) systematization.

Each one of the twenty poems is comprised of three stanzas. Each stanza is composed of six verses followed by an onomatopoeia or Mapudungun onomatopoeic text alluding to the song of the bird to which the poem refers. Each verse is presented in paired form in the English and Mapudungun languages. Our analysis of the poems is centered on their implications for biocultural conservation, ethno-ecology, multicultural education, and environmental ethics.

RESULTS AND DISCUSSION

Natural medicine and Mapuche culture

Poem 1 integrates elements of the past and present that are alive in Mapuche culture. In the Long-tailed Meadowlark poem it says, "I have always healed my people with pure herbs. For this, my name is full-time healer," alluding to a central element in the culture: natural medicine based in medicinal plants (San Martín 1983, Aldunate 1996, Foerster & Gundermann 1996, Massardo & Rozzi 1996, Massardo 1997, Villagrán 1998). Already in the 18th century, the priest Juan Ignacio Molina admired the Mapuche herbalist, writing that "vegetables, especially herbs, form the capital of the Mapuche pharmacy… and their doctors, called *machi* and *ampife*, are expert herbalists that possess, through tradition, the secret of a large number of simple [plants], adaptable to every kind of illness, with which they make marvelous cures every time; and although…they try to hide what they know in this matter, nevertheless, moved by friendship, they have demonstrated up through today the medical virtues of many trees and more than 200 healing herbs that they use with much success…" (Molina 1978).

Just as the Mapuche shaman (*machi*), the brother bird (the *lloyka*) is a full-time healer who brings his medicine made of the forest plants. Moreover, regarding the Mapudungun name *lloyka*, Aillapan indicates that this name stems from the terms *llako* (= healing) and *lawen* (= remedy). Linking rural knowledge about this bird in other regions of Chile, Aillapan suggests that the bird wishes to cure its chest carrying the blood of the Mapuche people and of the *Copihue* flower or *kolkopiw* (*Lapageria rosea*). In the countryside of central Chile, it is believed that the song of the Long-tailed Meadowlark says, "with the knife it was, with the knife it was," alluding to the moment when the bird would have been stabbed with a knife in its chest that shows still bloodied, covered with its gaudy red color .

POEM 1. The *lloyka* or Long-tailed Meadowlark (*Sturnella loyca*).

Dramatist bird, red-breasted challenger of people. It used to perch in a native tree, and now only in the plantations. In its challenging language, it seems mad. It collects news. With certainty it presents itself to the police. Singing, it says: "Visitors will come, and you should clean your house well. You should wash your pot and also wash your serving spoon." *¡Wichin wichin küpalu piam pu lonko traruuu* *wichin wichin küpalu piam lautuuu kangey ñi trekan lautuuu!*	*Tachi kayñe lukatufe che kelü pütra weda üñüm* *Anülepuy kiñe aliwen mew feula anümka pingey* *Wakeñ wakeñ mülepuy illkuley reke fill dungu witraniefi* *Akualu piam pu lonko traru pipingey wakeñün* *Feypiley ülkantulu, akualu piam pu witran liftuaymi tami rukaaa* *Küchaymi tami challa ka fey küchay tami wütrüuuggg.* *¡Wichin wichin küpalu piam pu lonko traruuu* *wichin wichin küpalu piam lautuuu kangey ñi trekan lautuuu!*
"It is certain that a widow will come who has a peculiar walk, looking for a new suitor because the dead one will never return. That woman is a great witch. Take care!" the bird says. "That man barks like a Culpeo Fox. Great mouth, it wags and diverts with its tail: You all are advised by the four winds of the earth!" *¡Wichin wichin kalku domo wüna weda fütra kalkutiiii* *wichin wichin wentru wang piley weda ngürü rekeee!*	*Rüf küpalu feyti lantu domo fey ta kangey ñi trekan* *Kintupelu wentru layemfei fotrü fütayem doy wüñolay* *Feyti domo fütra kalkutiiii, tüfikay, piley* *Ka feyti wentru fütra wang NGÜRÜ KALLFÜLEY* *Fütra wüntuley mütreng ka tümül wüneu fütra külen* *Deunga amuldungun meli kürüf wallpa mapu püle.* *¡Wichin wichin kalku domo wüna weda fütra kalkutiiii* *wichin wichin wentru wang piley weda ngürü rekeee!*
Have no fear Great Father and Great Mother, I will have the remedy in time, abundant and good. For this, I call myself another red-breasted being, and I have always healed my people with pure herbs. For this, my name is "full-time healer." The same as another bird brother, I sing from birth until death… *¡Wichin wichin küchay tami challaaa lif tuay tami rukaaa* *wichin wichin küchay tami wütrüggg wütrüggg!*	*Llükalayaymi turpu fütra chaw ka eymi kude ñuke* *Inche ta fey lawentuayu feula fachante frentren ka tutelu* *fey mew ta llapüm chefe kelü pütra pingeken* *Ka fey rumel llakotukey pu che re mapu lawen mütem* *Inche ta fey mew mongelchefe rumel lloyka pingen* *Fey ñi chauwüngen wakeñ ülkantuken puwülu fey feutekünün.* *¡Wichin wichin küchay tami challaaa lif tuay tami rukaaa* *wichin wichin küchay tami wütrüggg wütrüggg!*

The *lloyka* is also a bird that possesses special powers to discern a person's nature and foresees future events. It is a bird capable of connecting the human with the divine. Mapuche cosmogony distinguishes four cardinal points that involve four central divinities, the "Four Wind-Spirits of the Earth." In agreement with Aillapan, they come from the west with the guardian of the sea called *Dumpall*, from the southeast the guardian of the volcanoes or *Pillan*, from the northeast the princess of the sun or *Anchümallen*, and from the north the great visitor who had the riches or *Witranalwe* (see Faron 1968). In these poems —like in the Mapuche culture— the natural, human, and divine worlds live together.

As in many of his poems, Aillapan passes on through the *lloyka*, "dramatist bird, red-breasted challenger of people", a clear message of conservation, indicating in the first stanza that "It used to perch in a native tree, and now only in the plantations." This message takes on full force in the verse "Have no fear Great Father and Great Mother, I will have the remedy in time, abundant and good." Thus, the poet and the *lloyka* affirm that one should not be afraid for one's health when medicinal plants exist in the region. The plants are the source of life and health for the birds and the Mapuche people.

Environmental ethics

Poem 2 deals with a friendly predatory bird, simple and not very selective. The *triuke* or Chimango Caracara perches in a diversity of places and eats everything, even lice. The poem coincides with studies on the diet of the Chimango Caracara, which demonstrate that this species is an omnivorous bird, mainly scavenging and carnivorous. It consumes toads, lizards, rodents, small fishes, insects, worms, earthworms, caterpillars, and slugs (Yáñez *et al.* 1982, Rozzi *et al.* 1996a, Jaksic 1997). For this reason, it is a beneficial bird for agriculture, contributing to control diseases (eliminating the carrion) and plagues (by acting as predatory and insectivorous), and to the fertility of the soil by scratching the earth while accompanying the farmers in the sowing.

The poem signifies, among other things, that this beneficial bird of simple habits and a non-selective diet is also a distinguished forecaster of the weather and has the capacity to bring the rain —rain that favors human agriculture as well as the birds themselves. The *triuke* is seen by the Lafkenche poet in a way similar to the Huilliche (see Cárdenas 1994), it is a powerful bird. At the same time, it possesses a lunatic appearance and enhances the life of the Lafkenche by accompanying them in their labors of the field. When the plow is passed, groups of about one-hundred Chimango Caracaras can be seen eating insects in the soil and emitting their sequence of strong descending yells "*triiiiuuuu, triu, triu, triu, triu ...*" that is the origin of their onomatopoeic Mapudungun name *triuke*. The onomatopoeias constitute the most frequent mode for naming birds in the Mapudungun language (see more below).

POEM 2. The *triuke* or Chimango Caracara (*Milvago chimango*).

Of raptor legs, talons and jaw,
carnivore and insectivore its basic diet.
In the fable it uses a cloak made of a brown sack.
It also eats all types of lice from animals.
In the furrow of the plow it pursues grubs and worms.
With insistence it pecks all classes of insects.
¡Triuuuuki triuuuuki triuuuu triuuuu!

The best perch is the dry limb of the mountain.
Also it goes to perch on the house, stumps and trees.
The onomatopoeic call of the Chimango Caracara is a strident concert.
In the course of the year, in the old moon and the new it appears a lunatic.
It is a foreteller that invokes the rain during the day.
As a reward it has abundant grubs and worms for food.
¡Triuuuuki triuuuuki triuuuu triuuuu!

When there are plantings in the furrows of the sifted soil,
in the open countryside it brings happiness to the festivities of the planting.
The pig-like plow seems a blessing, and
the abundance of grubs and worms is the divine call.
In spring and summer time its major instinct is to bring happiness.
Of nickname "licey" because of looking for animal lice – birdy fable.
¡Triuuuuki triuuuuki triuuuu triuuuu!

Wünkufe wün niey ka tañi wili namun
llotufe ka fill idike mongengey
Daku makuñ kolü femngey ñi tukun
Pütrar fill Kulliñ rume kümentukey
ketramapungen weflu piru engu düllwi
Müna ayiukey fey pitrontuku wera idike.
¡Triuuuuki triuuuuki triuuuu triuuuu!

Rume wedañma anülekey kochay mauwidantumeu
Malal ünkoü fill anümka wente ruka rume
Triuuuutriuuumekey ñi wakennün wera triuki
Mülen meu ñam küyen ka we küyen amulen tripantu
Mütrümniekey mauwün kiñe komantü
Fey ñi wefael fetren piru ka fentren düllwi.
¡Triuuuuki triuuuuki triuuuu triuuuu!

Ka feyti pu che koni ñi ketran tuwe mapu
Lelfünketrawemeu amulelu ngakangen tuwepüllü
Feyti timuñ ngakapiyum fey wefkülekey
Rume fentren düllwi ka feyti piru llemay
Mületulu antüngen tachi triuki kintukey
Pütrar fill kulliñ feymeu apo pütrar pingey wedañma.
¡Triuuuuki triuuuuki triuuuu triuuuu!

The poem of the Chimango Caracara concludes with an important moral, derived from this "bird fable": it would be an error to dismiss this bird by a superficial judgment based solely on its simple appearance and non-selective dietary habits. The fable of the Chimango Caracara is a praise to the austere, wise, and happy life. As is the case among humans, it is frequent that the artists and the wise will be handled as lunatics and that the modest people will be generous and friendly. The winged poems project images as well as values that are, at the same time, environmental and social. The folly and the wisdom of the *triuke* are those of "birds as well as humans." The winged poems draw us not only toward the ecology of southern Chile forests, but also towards the value-rich Mapuche world that involves an ethic that integrates ecological and cultural realities.

Onomatopoeia and natural history

Poem 3 illustrates how even in such a small and bright bird as the Green-backed Firecrown, the onomatopoeic character of the Mapudungun language to name the birds is expressed. The names *pinda*, *pinguera*, *pigda* or *piñuda* are related to the verb *pigudcun* (to scrub one thing with another thing) and refer to the sound that is emitted by the wings scrubbing during rapid flight. It is interesting to note that the peculiar sound of this bird's flight also gives origin to the English name "Hummingbird." Just as in the case of the term *pinda*, the English name alludes to the humming sound that this bird emits upon beating its wings. In the case of the poem, the onomatopoeia is used further in the refrain "*pin pin pin pin daaaa pin pin pin pin daaaa! Müpümüpütuy ka ülkantuy.*" The onomatopoeia imitates the sound of this bird on the basis of its name, and it suggests that the song (*ülkantun*) of the Hummingbird is emitted with the wings (*müpü*).

The Hummingbird poem also expresses the detailed observation and knowledge of natural history contained in Mapuche traditional ecological knowledge. The poem signifies that the Hummingbird "sleeps or nods off in April, and awakes or revives in October, when there are many flowers." Thus, it alludes to a central aspect of the hummingbirds' life history in the forests of southern Chile: winter migration (Rozzi *et al.* 1996a). The Hummingbirds obtain their energy from plant nectar and when those disappear, so do the birds (Smith-Ramírez 1993). At the same time, the Hummingbirds fall into torpor during the night and they reactivate themselves every day in the morning (Skutch 1974, Lopez-Calleja & Bozinovic 2003). Such a type of sleepiness or torpor is a frequent explanation in the south of Chile for the disappearance of the Hummingbird during the winter (Rozzi *et al.* 2003a). Both farmers and fishermen of southern Chile claim to have encountered Hummingbirds sleeping in concavities of trees during the winter. This explanation is intriguing

POEM 3. The *pinda* or Green-backed Firecrown (*Sephanoides sephaniodes*).

It has a long and thin beak, like a pin, and it is a little bird, as tiny as the head of a thumb. It nourishes itself with the morning dew, the honey and the liquor of the flowers, without sitting on the rose. Tiny, little bird with so much speed, that while flying its wings cannot be seen. *¡Pin pin pin pin d a a a a pin pin pin pin d a a a a!* *Müpümüpütuy ka ülkantuy.*	Fütra pichi rume wüntuley damituwe reke müna pichi ünüm fütra changüll küwüg feuteni ñilonko mongekey mülüngmew ka modkoñomew ka llumed pezdkiñmew anükelay wente rosa pezdkiñmew kürüf trüri ñi miyawün pichi ünüm upünüyüm pefalkelay ñi müpü. *¡Pin pin pin pin d a a a a pin pin pin pin d a a a a!* *Müpümüpütuy ka ülkantuy.*
It sleeps or nods off in April, and awakes or revives in October, when there are many flowers. Its nest is made of cotton flecks, and on a gold balance they weigh barely anything, it and its nest. Subtle little birds, that in the borders of the hours of prayer hang lightning bugs from their nest. *¡Pin pin pin pin d a a a a pin pin pin pin d a a a a!* *Müpümüpütuy ka ülkantuy.*	Umagtukey peuwün küyen püle Nepetu mongetukey tromüngen küyem püle Mülelu fentren pezdkin fillpüle Lüg Chillkeñ takun feley ñi dañe pichike ünüm kam ñi femngeken üyyümkey küzdemallo reke pelomtuwe. *¡Pin pin pin pin d a a a a pin pin pin pin d a a a a!* *Müpümüpütuy ka ülkantuy.*
Its feathers are of very beautiful colors, golden and green and of other colors, and it uses them to fashion gold. At sunset, the fireflies take flight, like lanterns. Their eyes gleam like fire from the nest. By this light the Mapuche women spin, sew, knit and dance, and with one of these nests a girlfriend´s letter can be read. *¡Pin pin pin pin d a a a a pin pin pin pin d a a a a!* *Müpümüpütuy ka ülkantuy.*	Pütrümüna azdi ñi pichuñ Kelü karüg ka karüg ka itrofill kelütuwün ka millatuwe kimfalual ella punlu wepümi ñi pelomtuwe kuzdemallo reke fey ñi epu kuralnge wilüf wilüfngey relmu reke feyti pelomtuwe mew fügwü ka nizdüfkay kawitral ka pürüyche. Feyti pelomtwe mew kiñe ünam ñi wirin chillkatungey. *¡Pin pin pin pin d a a a a pin pin pin pin d a a a a!* *Müpümüpütuy ka ülkantuy.*

for science since each day the Hummingbirds fall into torpor at sunset and during winters it could occur that some individuals have more prolonged torpors for which they would seek refuge in holes of trunks and other protected places. Thus, the poem and the traditional ecological knowledge present an intriguing question for scientists.

Another interesting observation of natural history is the poet's allusion to the nest of mosses made by the Hummingbird. In southern Chile, the Hummingbird is used to utilize *Acrocladium auriculatum*, one of the most abundant mosses in the austral forests, to make its nest (Rozzi & McGehee in prep.). In these nests fireflies can perch, poetically evoking with their lights routine activities of Mapuche women, such as spinning, sewing, weaving, and dancing. Thus, as in the previous poems, Aillapan integrates natural history and poetry to express how interwoven the lives of the Mapuche are with the birds.

Intrinsic value of the birds

Poem 4 asserts that the *traru* or Crested Caracara is a thousand-year-old warrior who belongs to the lineage of eagles and gave origin to the grandest of Mapuche warriors, Lautaro, whose name signifies the fast Crested Caracara (*lef*): "Traru is transformed into Lautaro." In this way, the genealogical histories of the birds are intermingled with those of human beings. Both, Mapuche cosmogony and scientific evolutionary biology converge by affirming kinship and a common origin for all live beings (Rozzi 2001).

The verse "Several Caracara detect visions among the people" suggests a communication between the birds and humans. This figure of communication reinforces the notion of a common genealogy in which there is identification between the birds and humans. Both belong to the same biotic community. The feeling of belonging to the same community, as much as the belief in a common genealogy for birds and Mapuche, leads to the consideration of birds like persons, similar to human beings.

Analogies between birds and humans acquire a humorous touch beneath the view that the Lafkenche community has of the Crested Caracara. In the poem, Aillapan incorporates a funny Lafkenche analogy between the crest of the Crested Caracara and the cap of Chilean police or riflemen. The percussive vocalization struck by the Crested Caracara evokes an announcement of the war of the Mapuche, and the current "warriors" are policemen. Allusions to actual relations between humans and this bird conclude with sad verification that the Crested Caracara is being extinguished in the Lafkenche territory as a result of the disappearance of native forests where they nest. Habitat protection is now indispensable both for the conservation of this predatory bird, and for the preservation of the values bequeathed by Mapuche hero *Leftraru*. In this sense, Aillapan's poem reinforces through its genealogical, warlike, humorous verses that biological and cultural conservation should go hand-in-hand.

POEM 4. The *traru* or Crested Caracara (*Caracara plancus*).

This Caracara is a millenary bird.
Of eagle its variety pertains,
Carnivorous bird of raptor legs and talons.
Today it is on its way to extinction.* Few are yet left
due to absence of native forest, as the only reason.
For its reproduction there is no appropriate tree.
¡Trarrrr trarrrr traruuuu traruuuu!

Its form makes him known as the police bird,
of police hat with an ancestral hat-band,
figure and form equal to an armed police officer.
In addition, its speed of flight with contagious thinking,
its onomatopoeic, rough call resounding in the territory
announce war, when they noisily appear in flocks.
¡Trarrrr trarrrr traruuuu traruuuu!

For people in the territory it announces misfortune.
Several Caracaras detect visions among the people,
then the people of the community comment and shout; "Like
that the outsiders come to cause war in the land."
A young man named Traru, incorporates its fleet spirit.
A novel figure, his first name becomes Lautaro.**
¡Trarrrr trarrrr traruuuu traruuuu!

Kuyfi ünüm may tachi traru
Feyti ñamku pingey ñi küpal kuyfideuma
Weñefe wili namun niey ka itro fill ilotufe
Newe ngewelay ella müfü mütem miyaukey
Mauwidantu kuyfi aliwen newe ngewelay
feymeu aftuy reke ñi rumel chauken.
¡Trarrrr trarrrr traruuuu traruuuu!

Miyaukechi tralkatufe ka che pingekey üytungen
Nienmeu allangechi trarilonko kuyfi chetuley
Pidpid meñkituley tralkatufe lonkotraru üytungey
Kimfaluwi ngenikayauwün rakiduam reke pünerpuy
Fütrarume wakeñ/ülkantun wepümniey lofmeu
Muñkupüle nomnom ñi wefün müleyaludauma aukan.
¡Trarrrr trarrrr traruuuu traruuuu!

Wefünmeu weda inautun dungu tachi Lofmeu
Femngechi wüne kimfaluwi rupayawiyum wera traru
kiñeina may pi pu che mülerkeyay tachi fütra
Aukan akulu deuma ka tripaye che Lofmeu
Fey kiñe weche tuy traru ünüm ñi püllü pepilkawi
Wüne Leftraru che wefrumey, feychi üy tukuy.
¡Trarrrr trarrrr traruuuu traruuuu!

*here extinction means local extinction in the Mapuche lands.

**Lautaro is the greatest of the Mapuche warriors (see text).

Mapuche worldview

Poem 5, the Black-faced Ibis or *Raki*, involves essential elements of the Mapuche worldview, such as the number four that appears in these verses: "The *raki*-bird is the one whose ability is to count best. From one to four they go along like wise counselors," "Flying overhead, in pairs of two and pairs of four." The number four is very important for the Mapuche worldview: there are four cardinal points, four guardians of the Earth, and there are four spaces in the vertical axis of the world. These last, in agreement with Aillapan, are: 1) *Wenumapu*, 2) *Ankawenu*, 3) *Mapu*, and 4) *Minchemapu*.

•*Wenumapu*. This is the upper land (= *wenu*), that is like a reflection of the lower land.

•*Ankawenu*. This is the space between (= *anka*) the upper and lower lands. This is the sidereal space where the Black-faced Ibis and other birds fly and where wind, air, clouds, rain, and thunder all dwell.

•*Mapu*. This is the land "of here" where human beings live. Aillapan emphasizes that it is here where we suffer, we pay, and also where we love. *Mapu*, also called *ñukemapu* (= mother earth), is delicate and cannot be sold or perforated. The land must be loved and farmed with care. In the ñukemapu the land is worked freely and fraternally, like *peñi* (= brothers). In the *ñukemapu*, birds and trees should be treated as equals. To achieve this, the ancestors consumed only what they needed.

•*Minchemapu* . This is the land of below (= *minche*), it encompasses all that is underneath, the rivers and the soil, it is like the subsoil.

With their metallic voice and "song- language," the Black-faced Ibis cross the sidereal space communicating with the divine, the human, and the natural. Their sounds are heard in the worlds of *wenumapu*, *ankawenu*, and *mapu*, and over the latter Black-faced Ibis situates themselves in pasturelands, lakes, rivers, or shallow streams. The poem refers to a place where for a very long time the Black-faced Ibis have perched, giving origin to the name of the Mapuche community *Ruka Raki* (home of the Black-faced Ibis). In that same place there today a school also called *Ruka Raki* exists. The school's name ("House of the Black-faced Ibis") manifests the profound identification of Mapuche culture with the birds. At the same time, the name *Ruka Raki* expresses stripping the human ego by identifying the site of teaching with a bird. Such a gesture emphasizes the feeling that belonging to the Mapuche culture has with the biotic community. All along southern Chile, many places preserve their original Mapuche names, referring to the Black-faced Ibis, the Pygmy Owl, and other birds. This ornitho-toponimy (names of places based on bird names) not only shows us the biocultural richeness of the Mapuche language, but also helps non-Mapuche people to understand how interwoven our human lives are with those of the birds (Table 1).

POEM 5. The *raki or Buff-necked Ibis* (*Theristicus melanopis*).

The Buff-necked Ibises have metalic sounds,
it is known anciently and well by this characteristics.
Now and forever a whole community bears its name,
since it came into being, it has been called the Home of the Ibis.
Now a days, the high school is called Ruka Raki,
where students and professors come together.
¡Kiraki kiraki kiraki kiraki rakitruli rakitruli
truliraki truliraki trulirakí truliraki!

The raki-bird is the one whose ability is to count best.
From one to four they go along like wise counselors.
By persistent sounds speaks the male ibis.
In resounding sound speaks the female ibis.
The metalic sounding concert in language of a song.
It is the most accurate, divine and happy melody, and dancing.
¡Kiraki kiraki kiraki kiraki rakitruli rakitruli
truliraki truliraki trulirakí truliraki!

Flying overhead, in pairs of two and pairs of four.
In grunting voice they sing their language.
In the heavenly space from far away by land their coming is heard.
Before perching, they make a circle in the air with their yellow neck,
of black tail, ashy dorsal plumage, reddish-black legs,
they prefer to live in lakes, rivers, estuaries, where there are
grasslands
¡Kiraki kiraki kiraki kiraki rakitruli rakitruli
truliraki truliraki trulirakí truliraki!

Tachi raki ünüm may pañillwe dungun niey
Fütra kuyfi ñi kimngen tañi mülen
Fey meu ta kiñe Lof üytungey rumel feleyal
Feyta Ruka Raki pingey kuyfimel
Feula CHILLKATUWE RUKA RAKI üykoni
Fentren pichikeche ka feyti pu kimelchefe reyülepuy.
¡Kiraki kiraki kiraki kiraki rakitruli rakitruli
truliraki truliraki trulirakí truliraki!

Feyti Raki ünüm doy küme Rakiniey ñi duam
Kimche reke epuke melike miyaukeyngün
Nütramkan nentunentu mekey fütra RAKI
Ka feyti kude RAKI raurau ñi dungun
Wakeñ wakeñ wellün pañillwe metawe ül
Allangechi aiyumafal pürün ül wünülniekey.
¡Kiraki kiraki kiraki kiraki rakitruli rakitruli
truliraki truliraki trulirakì truliraki!

Doy epuke doy melike üpünkiyauwi *wenu* kürüf meu
Konküll reke wepümi ñi pin dungun ülkantun
Wenu kürüf meu fütra ka mapu kimfaluwi ñi küpalen
Walloyaukey cheuñi anüaal chod peltuley
Kürü külen pichuñ, trufken furi, ka kelü namuntuley
Cheuñi mülen trükon leufü, mangiñko, wera kachu mülekey.
¡Kiraki kiraki kiraki kiraki rakitruli rakitruli
truliraki truliraki trulirakì truliraki!

TABLE 1. Examples of Mapuche ornitho-toponyms (names of places based on bird names) in southern Chile (data from Rozzi *et al.* 2003a)

Bird names				Location	Meaning	Region
Spanish (Chile)	English	Scientific	Mapudungun			
Bandurria	Buff-necked Ibis	*Theristicus melanopis*	*Raki, Raqui, Raquin, Rakiñ*	*Ruka Raki*	House (= *ruka*) of the Buff-necked Ibis (= *raki*)	On the coast of Temuco
Choroy	Slender-billed Parakeet	*Enicognathus leptorhynchus*	*Choroi*	Choroico	Water (= *ko*) of the parrot *choroy* (= *choroi*)	In the mountains of Temuco, in the valley of Aillipén river
				Choroihue	Place (= *we* or *hue*) of *choroyes*	Chiloé Island
Chuncho	Austral Pygmy-Owl	*Glaucidium nanum*	*Chuchu, Conchon, Chucho, Chuco, Kill kill*	Chuchunco	Water (= *ko*) of the *chuncho* (= *chucho*)	Valley of the Cachapoal river
Golondrina chilena	Chilean Swallow	*Tachycineta leucopyga*	*Pillmaykeñ, Wüshükon, Pillmayken*	Río Pilmaiquén	River of Chilean Swallows (= *pillmaykeñ*, river = *lewfü*)	The river originates at Lake Puyehue, which has a beautiful waterfall (Salto del Pilmaiquén)
Torcaza Paloma araucana	Chilean Pigeon	*Columba araucana*	*Kono, Konun, Turcasa, Cono*	Conumo	Means “with (= *meu*), Chilean Pigeons (= *kono* o *konun*)”	Arauco Mountain Range
				Pucón	Denotes Chilean Pigeons (*pu* = prefix of the plural, *kon* = apocope de *kono* or *konun*)	Villarrica Lake shore
Tórtola	Eared Dove	*Zenaida auriculata*	*Maykoño, Cullpo, Culpo, Mayconu, Muikoño, Cono*	Muicolpué	Place (= *we*) of feathers (= *lp* apocope of *lepi*) of Eared Doves (= *muiko* apocope of *muikoño*)	Coastal Range in Osorno
Zorzal	Austral Thrush	*Turdus falcklandii*	*Wilki, Huilqui, Huilque*	Huilquío	Place of Austral Trushes (of *wilki* = Austral Trush, or = apocope of *we*, place)	Cachapoal Valley
				Huilquilemu	Forest (= *lemu*) of the Austral Trush (= *wilki*)	Talca Mountain Range
				Río Huilqueco	Water (= *ko*) of the Austral Trush (= *wilki*)	Chiloé Island
Huala	Andean Condor	*Podiceps major*	*Wala*	Hualakura	Names where a stone (= *kura*) with the shape of a huala (= *wala*)	Araucanía Region (IX Administrative Region), North of Temuco

TABLE 1. Continued.

Bird names				Location	Meaning	Region
Spanish (Chile)	English	Scientific	Mapudungun			
Picaflor	Green-Backed Firecrown	*Sephanoides sephaniodes*	*Pinda, Pinguera, Pinuda, Piñuda, Pigda, Pichi pinda*	Pindapulli Pindaco	Hill (= *pulli*) with hummingbirds (= *pinda*) Water (= *ko*) of the hummingbirds (= *pinda*)	Chiloe Island Pichué Mountain Range Chiloe Island, forested place with vines through where the Chonchi river flows
Picaflor gigante	Giant Hummingbird	*Patagonas gigas*	*Fütra pinda*	Pindal	Place with many hummingbirds (= *pinda*)	Chiloe Island, North of Chonchi
Diuca	Common Diuca-Finch	*Diuca diuca*	*Diuka, Shiwka, Fiuca, Diwka, Viuca, Diuca*	Diucalemu	Forest (= *lemu*) of the Common Diuca Finch (= *diuka*)	Araucanía Region (IX Administrative Region)
Peuco	Harris's Hawk	*Parabuteo unicinctus*	*Peucu, Kokoriñ, Peuco, Kokori*	Melipeuco	Shows the place of the encounter of the four (= *meli*) Harris's Hawks (= *peucu*)	Malleco River Mountain Range
Cóndor, Cóndor andino	Andean Condor	*Vultur gryphus*	*Mañke, Manque*	Cerro Manquehue	Place (= *hue*) of Andean Condor (= *mañke*)	Pre-cordillera of Santiago

FINAL CONSIDERATIONS

The twenty winged poems lead us not only toward biological diversity, but also toward the Mapuche cultural world and their intimate living together with nature.

Birds form part of the community of beings with whom we spend our existence. Birds and human beings can sing together, enjoying, taking care of, and loving one another. This identification and communication with birds encourages our living together with them and with nature. We share the forests and the Earth with birds, who also form part of everyday human life. Our sister birds inhabit in their own languages and songs, as much as they inhabit in the multiplicity of languages and dreams of our human species. The poetic dialogue that arises through perception of our kinship with the birds recovers a sensibility toward the profound relations and interdependencies that exist between human life and nature.

The verses of the Crested Caracara reveal to us how the poems speak of both the birds and their intertwined history with Mapuche culture. This integration helps not only conservation, but also current intercultural educational programs that aim to recover an integrated view of the dimensions of knowledge and values, based on interdisciplinary and transversal curricula. Such integration between sciences and the humanities is urgently necessary to reconstruct a sense of history and the social and environmental responsibility in the austral landscapes and Latin American countries (Primack *et al.* 2001). This actual moment in history confronts us with processes of globalization and such fast growth, which surpass our capacity to carefully weigh their consequences (Rozzi *et al.* 2003b). Rapaciousness and intensity of the market and globalization also surpass our comprehension of the meaning that these processes have for a fulfilled life in the continent inhabited by the world's highest bird diversity. It is the task of students, teachers, and lovers of biocultural diversity to decipher the metaphorical language of the Mapuche-Bird-Man, and to project it in this search for meaning. Rooted in the deep link between the Mapuche and nature, these twenty poems generate a dialogue between birds of the southern Chilean forests and the Mapuche poet that offers a bridge for communication between persons from different cultures, and between these persons and the birds.

ACKNOWLEDGMENTS

The authors thank Francisca Massardo for her comments on the manuscript. The Ethno-ornithology symposium in the VII Congress Neotropical Ornithology and the preparation of this article have been supported by the Centro Milenio para Estudios Avanzados de Ecología y Biodiversidad (CMEB, project Nº P02-051-F ICM) and by the BIOKONCHIL Project (FKZ 01 LM 0208, German Ministry of Education and Research, BMBF). This is a contribution of the research, education, and biocultural conservation program of the Omora Ethnobotanical Park (Universidad de Magallanes and Omora NGO), Puerto Williams, Provincia Antártica, Chile.

REFERENCES

AILLAPAN, L. 1994. *El Hombre Pájaro Mapuche*. Ediciones Populares, La Habana, Cuba.

AILLAPAN, L. 2003. Breve biografía del Hombre Pájaro Mapuche o Üñümche. P. 33 *in Guía Multiétnica de Aves de los Bosques Templados de Sudamérica Austral*, R. Rozzi, F. Massardo, C. Anderson, S. McGehee, G.Clark, G. Egli, E. Ramilo, U. Calderón, C. Calderón, L. Aillapan & C. Zárraga (eds.). Fantástico Sur - Univ. de Magallanes, Punta Arenas, Chile.

AILLAPAN, L. & R. ROZZI. 2001. *Veinte Poemas Alados de los Bosques Nativos del Sur de Chile*. Editorial Plaza y Valdés, México, México.

ALDUNATE, C. 1996. Mapuche: gente de la tierra. Pp. 111-134 *in Etnografía: Sociedades Indígenas Contemporáneas y su Ideología*, J. Hidalgo, V. Schiappacasse, H. Niemeyer, C. Aldunate & P. Mege (eds.). Editorial Andrés Bello, Santiago, Chile.

ANDERSON, C. & R. ROZZI. 2000. Bird assemblages in the southernmost forests in the world: methodological variations for determining species composition. *Anales del Instituto de la Patagonia, Serie Ciencias Naturales* 28: 89-100.

ANDERSON, C., R. ROZZI, C. ELPHICK & S. MCGEHEE. 2002. El programa Omora de anillamiento de aves en los bosques subantárticos: estandarización del tamaño de anillos apropiados para las aves de la Región de Magallanes. *Boletín Chileno de Ornitología* 9: 2-11.

BALICK, M., R. ARVIGO & L. ROMERO. 1994. The development of an ethnobotanical forest reserve in Belize: Its role in the preservation of biological and cultural diversity. *Conservation Biology* 8: 316-317.

BERKES, F. 1993. Traditional ecological knowledge in perspective. Pp. 1-9 *in Traditional Ecological Knowledge: Concepts and Cases*, J. Inglis, (ed.). International Program on Traditional Ecological Knowledge, International Development Research Centre, Ottawa, Ontario.

BULNES, G. 1980. *Los Mapuches y la Tierra*. Pequeñas ediciones INC, Rotterdam, Holanda.

CÁRDENAS, R. 1994. *Diccionario de la Lengua y de la Cultura de Chiloé*. Olimpho, Santiago, Chile.

CATRILEO, M. 1998. *Diccionario Linguístico-etnográfico de la Lengua Mapuche. Mapudungun-Español-English*. Editorial Andrés Bello, Santiago de Chile.

COÑA, P. & E. WILHELM DE MOESBACH. 1930. *Vida y Costumbres de los Indígenas Araucanos en la Segunda Mitad del Siglo XIX*. Imprenta Cervantes, Santiago, Chile.

CUNNIGHAM, A. 1993. *Ethics, Ethnobiological Research, and Biodiversity*. World Wildlife Fund International Publication, Gland, Suiza.

DOMEYKO, I. 1971. *Araucanía y sus Habitantes*. Editorial Francisco de Aguirre, Buenos Aires, Argentina.

FARON, L.C. 1968. *The Mapuche Indians of Chile: An Ethnography*. Waveland Press, Inc., New York, New York.

FOERSTER, R. & H. GUNDERMANN. 1996. Religiosidad mapuche contemporánea: elementos introductorios. Pp. 189-239 *in Etnografía: Sociedades Indígenas Contemporáneas y su Ideología*, J. Hidalgo, V. Schiappacasse, H. Niemeyer, C. Aldunate & P. Mege (eds.). Editorial Andrés Bello, Santiago, Chile.

GUSINDE, M. 1917. Medicina e higiene de los antiguos araucanos I. *Revista Chilena de Historia y Geografía* 22: 376-379.

GUSINDE, M. 1917. Medicina e higiene de los antiguos araucanos II. *Revista Chilena de Historia y Geografía* 23: 137-145.

INGLIS, J. 1993. *Traditional Ecological Knowledge: Concepts and Cases*. International Program on Traditional Ecological Knowledge, International Development Research Centre, Ottawa, Ontario.

JAKSIC, F. 1997. *Ecología de los Vertebrados de Chile*. Ediciones Univ. Católica de Chile, Santiago, Chile.

LÓPEZ-CALLEJA, V. & F. BOZINOVIC. 2003. Dynamic energy and budgets in hummingbirds: a study in *Sephanoides sephaniodes*. *Comparative Biochemistry and Physiology* 134A: 283-295.

MASSARDO, F. 1997. Plantas medicinales y medio ambiente. Pp. 189-211 *in 2° Congreso de Plantas Medicinales*. Ministerio de Salud, Santiago, Chile.

MASSARDO, F. & R. ROZZI. 1996. Plantas medicinales chilenas: un recurso subvalorado. *Ambiente & Desarrollo* 12: 76-81.

MOLINA, J.I. 1978. *Historia Natural y Civil de Chile*. Editorial Universitaria, Santiago, Chile.

OVALLE, A. 1646. *Histórica Relación del Reyno de Chile*. Editorial Universitaria, Santiago, Chile.

PRIMACK, R., R. ROZZI, P. FEINSINGER, R. DIRZO & F. MASSARDO. 2003. *Fundamentos de Conservación Biológica: Perspectivas Latinoamericanas*. Fondo de Cultura Económica, México. México.

ROZZI, R. 2001. Éticas ambientales latinoamericanas: raíces y ramas. Pp. 311-362 *in Fundamentos de Conservación Biológica: Perspectivas Latinoamericanas*, R. Primack, R. Rozzi, P. Feinsinger, R. Dirzo & F. Massardo (eds.). Fondo de Cultura Económica, México, México.

ROZZI, R. 2002. *Biological and Cultural Conservation in the Archipelago Forest Ecosystems of Southern Chile*. Ph.D. diss., Univ. of Connecticut, Storrs, Connecticut.

ROZZI, R. & P. FEINSINGER. 2001. Desafíos para la conservación en Latinoamérica. Pp. 619-688 *in Fundamentos de Conservación Biológica: Perspectivas Latinoamericanas*, R. Primack, R. Rozzi, P. Feinsinger, R. Dirzo & F. Massardo (eds.). Fondo de Cultura Económica, México, México.

ROZZI, R. & D. MARTÍNEZ, M.F. WILLSON & C. SABAG. 1996a. Avifauna de los bosques templados de Sudamérica. Pp. 135-152 *in Ecología de los Bosques Nativos de Chile*, J.J. Armesto, C. Villagrán & M.T. Kalin (eds.). Editorial Universitaria, Santiago, Chile.

ROZZI, R., J.J. ARMESTO, A. CORREA, J.C. TORRES MURA & M. SALLABERRY. 1996b. Avifauna de bosques primarios templados en islas deshabitadas del archipiélago de Chiloé. *Revista Chilena de Historia Natural* 69: 125-139.

ROZZI, R., F. MASSARDO, C. ANDERSON, S. MCGEHEE, G. CLARK, G. EGLI, E. RAMILO, U. CALDERÓN, C. CALDERÓN, L. AILLAPAN & C. ZÁRRAGA. 2003a. *Guía Multiétnica de Aves de los Bosques Templados de Sudamérica Austral*. Fantástico Sur-Universidad de Magallanes, Punta Arenas, Chile.

ROZZI, R., F. MASSARDO, J. SILANDER, JR., C. ANDERSON & A. MARIN. 2003b. Conservación biocultural y ética ambiental en el extremo austral de América: oportunidades y dificultades para el bienestar ecosocial. Pp. 51-85 *in Biodiversidad y Globalización*, E. Figueroa & J. Simonetti, (eds.). Editorial Universitaria, Santiago, Chile.

SAN MARTÍN, J. 1983. Medicinal plants in central Chile. *Economic Botany* 37: 217-127.

SKUTCH, A.F. 1974. *The Life of the Hummingbirds*. Octopus Books Ltd., London, UK.

SMITH RAMÍREZ, C. 1993. Los picaflores y su recurso floral en el bosque templado de la isla de Chiloé, Chile. *Revista Chilena de Historia Natural* 66: 65-73.

VILLAGRÁN, C. 1998. Botánica indígena de los bosques de Chile: Sistema de clasificación de un recurso de uso múltiple. *Revista Chilena de Historia Natural* 71: 245-268.

YÁÑEZ, J., H. NÚÑEZ & F. JAKSIC. 1982. Food habits and weight of Chimango Caracaras in central Chile. *Auk* 99: 170-171.

30 YAHGAN AND LAFKENCHE ETHNO-ORNITHOLOGY IN THE TEMPERATE FORESTS OF AUSTRAL SOUTH AMERICA

FRANCISCA MASSARDO[1,2] AND RICARDO ROZZI[1,2,3]

[1]Universidad de Magallanes, Casilla 113-D, Punta Arenas, XII Región, Chile. E-mail: francisca.massardo@umag.cl
[2]ONG Omora, Puerto Williams, XII Región, Chile
[3]Department of Philosophy and Religion Studies, University of North Texas, P.O. Box 310980, Denton, Texas 76203-0980, USA

Ornitología Neotropical 15 (Suppl.): 395-407 (2004) (Originally published in Spanish)

ABSTRACT. Yahgan and Lafkenche ethno-ornithology in the temperate forests of austral South America. The temperate forests of southern South America are inhabited by a unique ornitho-fauna and mosaic of indigenous cultures. Among the latter, the Yahgan indigenous people live in the insular territory of Cape Horn, representing the southernmost ethnic group of the world. Toward the northern forested region, communities of Lafkenche (people of the sea or *Lafken)* live in the coastal forests of south-central Chile, forming part of the Mapuche indigenous group *(mapu* = earth; *che* = people). Between 2000 and 2002, we recorded Yahgan and Mapuche bird names and traditional ornithological knowledge by working with the grandmothers Úrsula and Cristina Calderón, and the Mapuche poet Lorenzo Aillapan. For this work, we considered a set of 58 bird species of the native forests of Chile. Among them, 42 species live in the Yahgan territory of the insular subantarctic forests of Cape Horn Archipelago (54.5-56°S), and 57 species live on the Lafkenche territory on the coast of Temuco (39°S). This study recorded 10 new Yahgan bird names which, in combination with names recorded by other researchers yield a total 38 names, *i.e.*, 93% of the 42 species bird species found in the Yahgan territory. Mapuche names were found for 55 species, *i.e.*, 97% of the 57 species living in the Lafkenche territory. The work with Aillapan yield more than 10 new names for birds, indicating the high variety of Mapudungun names for each bird species. Both, Yahgan and Mapuche bird names express onomatopoeic, morphological, behavioral, and habitat characteristics of the birds. The record of new Yahgan and Mapuche names indicates the need for ethno-ornithological research and conservation efforts of the cultural diversity inhabiting the temperate forests of South America.

Key words: Avifauna, biocultural conservation, Chile, ethno-ornithology, Lafkenche, Mapuche, temperate forests, traditional ecological knowledge, Yahgan.

RESUMEN. Etno-ornitología Yagán y Lafkenche en los bosques templados de Sudamérica austral. Los bosques templados de Sudamérica austral están habitados tanto por una ornito-fauna como por culturas indígenas únicas. Entre estas últimas se encuentran: i) el pueblo Yagán o Yamana, que constituye la etnia más austral del planeta habitando los archipiélagos y canales del Cabo de Hornos; y ii) las comunidades Lafkenches (gente del mar o *Lafken),* que habitan los bosques costeros del centro-sur de Chile, y forman parte del pueblo *Mapuche (mapu* = tierra; *che* = gente). Entre el 2000 y 2002, se trabajó con las abuelas yaganes Úrsula y Cristina Calderón, y el poeta Lorenzo Aillapan, en el registro de nombres de las aves y del conocimiento ornitológico tradicional yagán y mapuche. Para el estudio, se consideró un conjunto de 58 especies de aves propias de los bosques nativos de Chile. Entre ellas, 42 especies habitan en el territorio yagán de los bosques subantárticos en la región insular del Cabo de Hornos (54,5-56°S), y 57 especies habitan en el territorio lafkenche a la costa de Temuco (39°S). Este trabajo registró 10 nuevos nombres yaganes para aves, que junto a los registros de otros investigadores, determinan que, del total de 42 especies, 38 especies (93%) reciben nombre yagán. Entre las 57 especies consideradas para el análisis de los nombres mapuches, se encontraron nombres para 55 de especies de aves (97%). El trabajo con Aillapan aportó más de 10 nombres nuevos para aves, indicando la alta variedad de nombres mapuche para cada especie. Los nombres yaganes y mapuches aluden a aspectos onomatopéyicos, morfológicos, de comportamiento y hábitat de las aves, y el registro de nuevos nombres indica la necesidad de realizar investigación etno-ornitológica y esfuerzos de conservación de la diversidad cultural que habita en los bosques templados de Sudamérica.

Palabras clave: Avifauna, bosques templados, conocimiento ecológico tradicional, conservación biocultural, Chile, etno-ornithología, Lafkenche, Mapuche, Yagán.

INTRODUCTION

The relations between the biological and cultural diversity constitute a field of increasing theoretical and practical interest for ecologists and conservation biologists (Maffi 2001, Primack *et al.* 2001). The current loss of biodiversity has been recognized as a phenomenon as dramatic as the event of mass extinctions documented in the fossil record, such as the disappearance of the dinosaurs (Raup 1979). It is estimated that during the XXI century up to 20% of biological species could disappear from the face of the planet (Primack *et al.* 2001). Less known, but even more dramatic, is the current loss of cultural diversity. It is estimated that during the XXI century 90% of the languages spoken in the world today will disappear (Maffi 2001).

At a global scale, there are about 6809 languages spoken today (Grimes 2000). However, most of these languages belong to small indigenous communities or other minorities. A recent study demonstrated that two thirds of the languages are found in the 238 eco-regions identified as priorities for biodiversity conservation (Oviedo & Maffi 2000). The language and traditional ecological knowledge hosted by the diverse cultures are embedded in the interwoven webs of relationships of biodiversity (Rozzi *et al.* 2003), and both diversities —biological and cultural— are threatened today (Oviedo & Maffi 2000).

This essay deals with a particular aspect of biological and cultural diversity of South American temperate forests: the diversity of Yahgan and Mapuche names for birds and its relationship to traditional ornithological knowledge. Birds are the most diverse group of vertebrates in the austral forests (Rozzi *et al.* 1996). The Mapuche and Yahgan (or Yamana) indigenous people inhabit the extreme north and south of this biome. To address the analysis of relationships among birds, cultures, and forest habitat types, we first present a brief summary about the mosaic of forest types and cultures, and their conservation status in southwestern South America.

South American temperate forests are the planet's southernmost forests, extending for about 2000 km from Central Chile (~ 35°S) to the southern tip (56°S), and for a short distance along the eastern slope of the Andean Cordillera in Argentina (Armesto *et al.*1996). The topography and patterns of temperature and precipitation vary markedly across the area (Arroyo *et al.*1996). In addition, the biogeographic origins of the biota have contrasting sources: the Neotropical (warm) and Gondwanic (sub-Antarctic cold temperate) (Hinojosa & Villagrán 1997); the Andean and the bipolar, sub-Arctic/sub-Antarctic distributions (Rozzi R. & J. Silander in prep.). The combined effect of climatic, topographic, and biogeographic variations determine the presence of a mosaic of diverse forest ecosystems along the latitudinal gradient (Veblen & Schlegel 1982).

South American temperate forests constitute a biome isolated by large orographic and climatic barriers, separated by at least 2000 km for the nearest tropical forests (Arroyo *et al.* 1996). The Atacama Desert (north), the Pacific Ocean (west), the Andes and the Patagonian steppe (east), determine a true biogeographical island in which a highly endemic biota has evolved. Almost 90% of the woody species are endemic (Arroyo *et al.* 1996). Among vertebrates, 147 species are endemic, including: 50% of fish (26 spp.), 80% of amphibians (30 spp.), 36% of reptiles (16 spp.), 30% of land-based birds (44 spp.) and 33% of mammals (33 spp.) (Armesto *et al.* 1996). Regarding the birds, the percentage of endemism rises to 66% if all the habitats of the southern end of South America are included. This elevated degree of endemism in birds is much higher than in other continental forest regions of the world and is comparable to that found in oceanic islands (Rozzi *et al.* 1996).

It is not just the biota inhabiting the mosaic of ecosystems of temperate forests of South America that is unique, but also its indigenous cultures (Hidalgo *et al.* 1997). Among these Amerindian people, the southernmost pre-Columbian inhabitants of the world have been found: the Yahgan, who occupied the southern end of the archipelagic territory of America (55-56°S), from the southern coast of Tierra del Fuego to Cape Horn (Orquera & Piana 1990). The Yahgan had a great tradition of canoeing and a refined traditional ecological knowledge (Gusinde 1961).

The northern region of the biome, between 35 and 42°S, is inhabited by indigenous Mapuche (people = *che*; land = *mapu*). These people speak Mapudungun (language or "talk" = *dungu*; land = *mapu*), expressing deep roots with their environment and a great ability to communicate with its beings, especially birds (Rozzi *et al.* 2003). At present, there are three Mapuche groups living in areas dominated by three types of characteristic forests: in the forests of the Monkey-puzzle tree (*Araucaria araucana*), the Pehuenche (people of the Monkey-Puzzle trees = *pehuen*); in coastal forests, the Lafkenche (sea people= *lafken*); in the Valdivian evergreen forest between the Toltén River (38°S) and Archipelago of Chiloé (42°S), the Huilliche (people of the south = *huilli*) (Hidalgo *et al.* 1997).

Regarding the state of conservation of this intricate biocultural mosaic, it must be considered that the forest landscapes and culture of the Mapuche communities have suffered a noticeable transformation since the arrival of the Spaniards in Chile in 1537. The processes of contact and conquest of the Spaniards during the XVII and XVIII centuries entailed the arrival of virulent diseases, wars, introduction of exotic species such as horse and cow, burning and clearing of forests, which modified both the native ecosystems and original cultures (Donoso & Lara 1996). Later, during the second half of the XIX century, the southern tip (53-55°S) also became a magnet for Chilean and European settlers, who began the large-scale sheep farming that drastically changed the sub-Antarctic landscape (Martinic 1992). Then, gold diggers, sealers and whalers, and large sawmills promoted alcoholization and acculturation, which determined a drastic reduction of indigenous populations of the Ona, Alacalufe, and Yahgan people during the XIX century (Mayorga 1972).

During the second half of the twentieth century, the development of industrial tree plantations had generated a new wave of transformation of the natural landscape and the social order in south-central Chile (Armesto *et al.* 2001). The conversion of native forests to homogeneous plantations of exotic species of *Eucalyptus* and Monterrey Pine had forced many Mapuche people to migrate to the urban fringe (Aylwin 1994). From a cultural standpoint, during parts of the 1970s and 1980s, the Mapuche language, Mapudungun, was eliminated from formal education and everything was taught only in Spanish (Berchidewsky *et al.*1977). This accelerated the loss of the language and an important part of the cultural heritage for young Mapuche.

The type of relationship established by the conquerors, the settlers and more recently, the developers with indigenous people and the biodiversity of southwestern South America has led to a serious loss of indigenous languages and biocultural diversity. Among the twelve pre-Columbian languages that existed in Chile, six (50%) have disappeared, two (17%) (Yahgan and Kaweshkar or Alacalufe) are almost extinct, two (17%) (Rapanui and Quechua) have conservation problems, and only two (17%) (Mapudungun and Aymara) are spoken by more than 5000 people (Grimes 2000).

In summary, both the cultural and linguistic diversity faces a threat level that is higher than the biodiversity in Chile and the world, and requires urgent research and conservation efforts. In the context of this urgent biocultural task, this paper seeks to contribute, from the ethno-ornithology point of view, both to the continuity of Yahgan and Mapudungun languages, and the conservation of birds and their ecological interactions in the temperate forests of South America.

METHODS

Between January 2000 and November 2002, work was done recording the bird names and the traditional Yahgan ornithological knowledge on Navarino Island (55ºS). During the time we conducted this work, the sisters, Úrsula and Cristina Calderón, were the last descendants of both Yahgan father and mother. In January 2003, Úrsula Calderón, who was born in 1925 and had vast experience and knowledge on the Yahgan traditions, passed away. Cristina was born in 1928 and is currently dedicated to traditional crafts. The sisters were 76 and 73 years old when this study was conducted. They spent their childhood in different locations of the archipelagoes of Cape Horn, and their knowledge about nature involves a wide range of recollections.

The study was conducted along the northern coast of Navarino Island in Mejillones Bay, on the land belonging to the Yahgan indigenous community and in the Omora Ethnobotanical Park, a biocultural conservation reserve (Rozzi *et al.* 2002). In February 2001, Lorenzo Aillapan joined the working group. Aillapan is a Mapuche poet, who is known as the Mapuche Birdman, *Üñümche*, and a master of language and culture in the Lafkenche region (39ºS) (see Aillapan & Rozzi 2004).

To study and record the names and traditional knowledge about the avifauna, a set of 58 species of birds of Chilean native forests was considered (see Rozzi *et al.* 1996, Table 1). These species breed or feed in the forest interior or forest margins in the South American temperate forests biome. The bird species belong to 11 different orders.

The study of bird names and the traditional knowledge about them was conducted using the methodology of semi-structured interviews (Russell 1994). By means of bird sightings during walks in the forest, or by asking questions about the names of bird species illustrated with photographs, interviews were conducted individually or collectively. In the case of the Calderón sisters, several recordings were done with both present simultaneously, in order to allow greater fluidity of the Yahgan word, and their memories associated with forest birds. Each interview was recorded with a Sony Minidisc recorder Recorder MZ R70 equipped with a Sony ECM-MS907 microphone.

In addition to the study and recording of names and TEK in the field, a literature review of Yahgan and Mapudungun names was conducted, and for the transcription of these names the original script was preserved. For transcription of Mapudungun names recorded by Aillapan, the unified Mapuche alphabet was used, and for the Yahgan names recorded with the Calderón sisters, the pronunciation closest to Spanish was recorded. The original recordings can be heard on CDs included in the bird guide by Rozzi *et al.* (2003).

RESULTS AND DISCUSSION

The common names in Spanish (*i.e.*, Chilean), Yahgan and Mapudungun, along with the scientific and English names of the investigated species are listed in Table 1.

Yahgan ornithological knowledge

Only 42 of the 58 investigated species of birds, inhabit the Yahgan territory (54.5-56ºS). Regarding Yahgan names for forest birds of the Cape Horn region, this study recorded 10 new species with Yahgan names: *Kulalapij* for *Accipiter bicolor* and *Buteo polyosoma*, *Kikinteka* for *Geranoaetus melanoleucus*, *Kujúrj* for *Strix rufipes*, *Pilit* for *Vanellus chilensis*, *Shakóa* for *Gallinago paraguaiae*, *Tularatachij* for *Cinclodes fuscus*, *Tachij* for *Cinclodes patagonicus*, *Tatajurj* for *Pygarrhichas albogularis* and *Samakéar* for *Anairetes parulus*. These new records, along with those of other researchers, determine that out of the total 42 species of birds that inhabit the forests of the island region of Cape Horn, 38 species (93%) have received a Yahgan name (Table 1).

Only three species of predatory birds, *Buteo ventralis*, *Falco peregrinus* and *Phalcoboenus australis*, and one passerine, *Colorhamphus parvirostris*, lack Yahgan names. The first two birds of prey belong to species that visit the area very occasionally, and the third is a rare resident (Venegas & Sielfeld 1998). *C. parvirostris*, is considered a small scarce insectivorous bird of the south of Tierra del Fuego, and its presence on Navarino island has been detected only recently (S. McGehee *et al.* in prep.). Therefore, we can say that the Yahgan language has recognized most species of birds detected by naturalists or scientists in the forests of Cape Horn.

The Yahgan names for birds also reveal aspects of their behavior, habitat, onomatopoeia, and morphology. For example, regarding behavior, *A. bicolor* and *B. polyosoma* are two aggressive birds of prey, and because of that their name is *Kulalapij* (*kulala* = angry and *pij* = bird). The Magellanic Woodpecker *Campephilus magellanicus* is called *Lana*, alluding to his long and strong tongue (= *lan*) that extracts the beetle larvae to feed. The names also relate to the type of habitat used by the bird. For example in the case of *tularatachij* or *C. fuscus*, it usually inhabits the hills along watercourses as its Yahgan name indicates: *tulara* (= hill) and *táchij* (= Bar-winged Cinclodes). Onomatopoeia is present in Yahgan bird names in the form of mimicking the sound produced by the bird. For example, *Kujurj* imitates the nucturnal sounds of the *S. rufipes*, *tatchikáchina* imitating vocalizations of *Aphrastura spinicauda* and *Chakatej* that mimics the sound of the Ringed Kingfisher *Megaceryle torquata*. The morphology of birds is yet another important factor in Yahgan names such as *Samakear*, meaning "very little" which is applied to

TABLE 1. Names in Mapudungun and Yahgan languages of the birds inhabiting the temperate forests of South America.

Scientific	Spanish	English*	Mapudungun*	Yahgan*
Accipiter bicolor	Peuquito	Bicolored Hawk	Pichi kokoriñ (LA), Peuco (3)	Kúlalapij (UC)
Buteo polysoma	Aguilucho	Red-backed Hawk	Ñancu (1), Nangu (1), Ñamku (2,LA)	Kúlalapij (UC)
Buteo ventralis	Aguilucho de cola rojiza	Rufous-tailed Hawk	Unknown	Unknown
Geranoaetus melanoleucus	Águila	Black-chested Buzzard-Eagle	Caltch (1), Calquin (1,3), Kalkiñ (LA)	Kíkentéka (CC), Kíkintéka (UC)
Parabuteo unicinctus	Peuco	Harris´s Hawk	Peuco (1), Peucu (1), Kokoriñ (2,LA) , Kokori (4)	Kilakimös (7), (Cazapollo) (CC)
Cathartes aura	Gallinazo, Jote de cabeza colorada	Turkey Vulture	Queluy (1), Quelvoni (1), Kanin (2,4), Kelüwún Kaniñ (LA)	Iloea (UC, CC), Ilói (8), Iliuaia (7), Aiskalaix (7), Akixlaiix (7), Ilaóea (6)
Coragyps atratus	Jote de cabeza negra	Black Vulture	Queluy (1), Quelvoni (1), Kaniñ kürüwún (LA)	Out of the bird's range
Vultur gryphus	Cóndor andino	Andean Condor	Manque (1), Mañke (2,4,LA)	Weziyau (CC), Wariáo (6)
Falco sparverius	Cernícalo	American Kestrel	Ocori (1), Lleullequen (1), Lilpillañ (3), Lleylleykeñ (4), Okori (LA)	Kikinteka (7), Akimakaia (7)
Falco peregrinus	Halcón peregrino	Peregrine Falcon	Peuco (3)	Unknown
Milvago chimango	Tiuque, Chimango	Chimango Caracara	Tiuque (1), Chiuque (1), Triuquem (1), Chiwkü (2), Triwkü (4),Triuki (LA)	Yoskaléa (UC,CC), Yookalía (5) Yoskalía (3), Yoakília (7)
Phalcoboenus albogularis	Carancho cordillerano del sur	White-throated Caracara	Out of the bird's range	Unknown
Polyborus plancus	Traro, Carancho	Southern Crested Caracara	Traro (1), Taru (1), Traru (2,4)	
Bubo virginianus	Tucúquere	Great Horned Owl	Nuco (3), Tuco (3), Raiquen (3), Tukuu (LA)	Ketéla (CC,UC,5), Yoakília (7) Yohutela (CC), Yahutéla (UC), Kuhúrux (5)
Glaucidium nanum	Chuncho	Austral Pygmy-Owl	Chuco (1), Conchon (1), Chucho (3) Kill kill (4,LA)	Lafkéa (UC), Lefkéa (CC), Lefkóiya (5), Lí(f)kwia (7)
Tyto alba	Lechuza blanca	Barn Owl	Chiwüd (2, LA), Coa (3), Yarquen (3), Chralchral (3)	Sírra (UC), Síta (5), Lufkié (8) Lafkéja (6)
Strix rufipes	Concón	Rufous-legged Owl	Concon (1), Coa (1), Ñeque (3), Colcon (3), Kong kong (LA)	Kujúrj (UC)
Theristicus melanopis	Bandurria	Buff-necked Ibis	Raqui (3), Raquin (3), Rakiñ (4), Raki (LA)	Lejúwa (UC, CC), Léxuwa (5)
Rallus sanguinolentus	Pidén	Plumbeous Rail	Piden (1), Pideñ (1,4, LA), Cotuta (3)	Wéásh (UC), Heuhs
Vanellus chilensis	Queltehue, Treile, Tero	Southern Lapwing	Queltehue (1), Queltreu (1), Treile (1), Tregül (2,4, LA)	Pílit (UC)
Gallinago paraguaiae	Becacina	Common Snipe	Carcaren (3), Cadcadeñ (3), Porotu (LA), Kaikayem (PC)	Shakóa (UC,CC)
Megaceryle torquata	Martín pescador	Ringed Kingfisher	Queto (3), Quete (3), Kedküchan (3), Challwafe üñüm (LA)	Chákatej (CC,UC), Ségetex (5)
Zenaida auriculata	Tórtola	Eared Dove	Cullpo (1), Culpo (1), Mayconu (1), Maykoño (2,4, LA), Cono (3)	Out of the bird's range
Columba araucana	Torcaza	Chilean Pigeon	Turcasa (1), Cono (1), Kono (2,4,LA)	Out of the bird's range
Enicognathus ferrugineus	Cachaña	Austral Parakeet	Cachaña (1), Rawikma (1), Cachuña (1), Tricau (3), Choroy (3, LA), Rawilma (4), Chilkeñ (PC)	Kínan (UC, CC, 6, 7)
Enicognathus leptorhynchus	Choroy	Slender-billed Parakeet	Choroy (1,2,4, LA)	Out of the bird's range

TABLE 1. Continued.

Scientific	Spanish	English	Mapudungun*	Yahgan*
Patagona gigas	Picaflor gigante	Giant Hummingbird	Pinguera (1), Pinuda (1), Piñuda (1,3) Pinda (3,4), Fütra *pinda* (LA)	Out of the bird's range
Sephanoides sephaniodes	Picaflor	Green-backed Firecrown	Pinguera (1), Pinuda (1), Piñuda (1,3) Pinda (3,4), Pichi *pinda* (LA)	Sámakéär (CC), Samakéa (UC), Omóra (UC,5)
Campephilus magellanicus	Carpintero negro	Magellanic Woodpecker	Concona (1), Concoma (1), Rere (1,4), Pitigüe (3)	Lána (CC,UC,5), Lana (7), Kanára (6)
Colaptes pitius	Pitío	Chilean Flicker	Pitio (1), Pitihue (1), Pitiu (1), Pitigüe (3), Pütiw (4), Pütriu (LA)	Out of the bird's range
Picoides lignarius	Carpinterito	Striped Woodpecker	Pichi rere (LA)	Out of the bird's range
Curaeus curaeus	Tordo	Austral Blackbird	Cureu (1), Chihuanco (1), Quereu (1), Furare (1), Kürew (4), Küreu (LA)	Tátapuj (CC), Kátapuj (UC), Tetapux (7), Totapux (8)
Zonotrichia capensis	Chincol	Rufous-collared Sparrow	Chincol (1), Chingol (1), Mencu Toque (3), Mencu Lonco (3), Meñkutoki (4), Utruftoki (LA), Chinkol (LA)	Chámuj (CC, UC), Camux (5), Júi (8), Kqaiaminix (7)
Sturnella loyca	Loica	Long-tailed Meadowlark	Loica (1), Lloica (1), Lloyka (2, LA), Loyka (4)	Wichóa (UC), Wicoa (7)
Diuca diuca	Diuca	Common Diuca Finch	Diuca (1), Fiuca (1), Viuca (1), Diwka (2,4), Shiwka (4), Diuka (LA)	Out of the bird's range
Sicalis luteola	Chirihue	Grassland Yellowfinch	Chirihue (1), Chirigüe (1), Chirio (1), Chuduf (3), Milliu (3), Chiriu (LA)	Out of the bird's range
Carduelis barbatus	Jilguero	Black-chinned Siskin	Quechan (1), Chirihüe (3), Dihue (3), Chidüf (LA)	Tuín (UC, CC), Tuwín (5), Dwiin (7), Duiin (7)
Phrygilus patagonicus	Cometocino	Patagonian Sierra-Finch	Chuchan (3)	Tashúr (CC), Tachurj (UC), Tösur(7)
Aphrastura spinicauda	Rayadito	Thorn-tailed Rayadito	Pidpid (3), Pishpish (3), Yuquiyuqui (3), Pidpidwiriñ (LA)	Tatchikáchina (UC), Tachikáchina (CC), Tecikásina (5), Tulératécix (5)
Cinclodes fuscus	Churrete acanelado	Bar-winged Cinclodes	Chiuchihuen (3), Chiuchiu (LA)	Tularatáchij (UC)
Cinclodes patagonicus	Churrete	Dark-bellied Cinclodes	Churrete (1), Churreta (1), Churete (1), Chiuchihuen (3), Chiuchiu (LA)	Táchij (UC, CC)
Leptasthenura aegithaloides	Tijeral	Plain-mantled Tit-Spinetail	Yuquiyuqui (3), Pishpish (3)	Out of the bird's range
Pygarrichas albogularis	Comesebo	White-throated Treerunner	Pishonquillu (3), Pelchokiñ (LA)	Tatajúrj (UC,CC)
Sylviorthorhynchus desmursii	Colilarga	Desmur´s Wiretail	Epukudén (3, L), Changkihue (3), Futrakülen (LA)	Out of the bird's range
Pygochelidon cyanoleuca	Golondrina de dorso negro	Blue-and-white Swallow	Pillmayken (2), Pillmaykeñ (4, LA)	Lásij (CC,UC), Láöix (5), Láöixkipa (5), Lasix (7)
Tachycineta leucopyga	Golondrina chilena	Chilean Swallow	Pillmayken (2), Pillmaykeñ (4, LA)	Lásij (CC,UC), Lásix (5), Lásixkipa (5), Lasix (7)

TABLE 1. Continued.

Scientific	Spanish	English	Mapudungun*	Yahgan*
Turdus falcklandii	Zorzal	Austral Thrush	Huilque (1), Huilqui (1), Wilki (2,4, LA)	Jakásirj (UC), Hakásirj (UC), Hákasír (CC), Hakásij (CC), Xakacir (6)
Phytotoma rara	Rara	Rufous-tailed Plantcutter	Rara (1,4), Cadi (3), Kamtrü (3, LA)	Out of the bird's range
Eugralla paradoxa	Churrín de la Mocha	Ochre-flanked Tapaculo	Unknown	Out of the bird's range
Pteroptochos castaneus	Hued-hued castaño	Chesnut-throated Huet-Huet	Wed-wed (LA)	Out of the bird's range
Pteroptochos tarnii	Hued-hued	Black-throated Huet-Huet	Hued-hued (1), Huahueta (1), Huid-huid (1), Wed-wed (LA)	Out of the bird's range
Scelorchilus rubecula	Chucao	Chucao Tapaculo	Chucao (1), Chucau (1), Tricau (1), Chiduco (3), Chukaw (4), Chukau (LA)	Out of the bird's range
Scytalopus magellanicus	Churrín	Magellanic Tapaculo	Churin (1, LA), Chercan (3), Chircan (3), Tiftifken (LA)	Túto (CC,UC), Tútu (5)
Troglodytes aedon	Chercán	House Wren	Chercan (1), Chircan (1), Chedquen (1), Chedüf (4), Chüdüf (4), Chedkeñ (4), Charkan (LA)	Chílij (UC,CC), Huluöénuwa (5), Apuscilix (7), Apustöcix (7), Ulusönuwa (7)
Anairetes parulus	Cachudito	Tufted Tit-Tyrant	Pichpich (3)	Sámakéär (UC)
Colorhamphus parvirostris	Viudita	Patagonian Tyrant	Peutren (3), Lantu (LA)	Unknown
Elaenia albiceps	Fío-fío	White-crested Elaenia	Fio (1), Fio-fio (1), Vio-vio (1), Peutren (3), Wiyu (4, LA)	Feo (UC), Fuiiu (7), Puiu (7)
Pyrope pyrope	Diucón	Fire-eyed Diucon	Diucon (1), Püdko (4, LA)	Hashpúl (UC,CC), Jashpúl (CC), Héspul (5), Uspul (7)

* English names according to Alvaro Jaramillo (2005). Aves de Chile, Ingoprint S.A., Barcelona.

1. Lenz (1905-1910), 2. Catrileo (1998), 3. Gorgoglione (1997), 4. Hernández et al. (1997), 5. Gusinde (1961), 6. Salas & Poblete (1997), 8. Aguilera (2000), 7. Bridges (1933), LA: Lorenzo Aillapan (este estudio), UC: Úrsula Calderón (este estudio), CC: Cristina Calderón (este estudio).

the Green-backed Firecrown *Sephanoides sephaniodes* and to the Tufted-tit Tyrant *A. parulus*.

It is noteworthy that this research is the first to record Yahgan names for nearly 25% of the forest bird assemblage in the region of Cape Horn. The lack of knowledge on Yahgan language is illustrated by the cumulative number of registered names for Yahgan birds by researchers over time. Darwin, who visited Cape Horn in the 1830s, left no record of the Yahgan names of the forest birds. The missionary Thomas Bridges, who cohabited with the Yahgan in the second half of the XIX century, recorded 16 Yahgan names for bird species. Later, in the early XX century, the anthropologist Martin Gusinde increased the knowledge about Yahgan names for forest birds, adding 10 names to Bridges' list. Afterward, at the end of XX century with studies by Poblete & Salas (1997), the number rose to 28 Yahgan names. Finally, this study added 10 new names, suggestive of the amount of knowledge yet to be learned from Yahgan language.

Yahgan language contains over 33,000 words (Bridges 1987), many of which express an understanding of the nature of forests and sub-Antarctic ecosystems. This wealth of TEK has been lost due to a strong process of acculturation. However, the interviewed Yahgan grandmothers show that there is still a wealth of linguistic and TEK that lies dormant. Ms. Úrsula Calderón recalled the Yahgan names for 36 species, which accounts for 86% of the total. Among the seven species whose names she could not remember, one belongs to the passerine *C. parvirostris* and six to birds of prey. In contrast to the diurnal raptors, grandmother Úrsula recalled the names for all forest owls. Meanwhile, Ms. Cristina Calderón recalled Yahgan names for 32 species, *i.e.*, 76% of the total. Like her sister Úrsula, grandmother Cristina had the most difficulty with raptors (eight species without a Yahgan name). Our research on traditional Yahgan knowledge on southern end birds seeks to contribute to the rescue of this knowledge and conservation of the language through gathering material for younger generations (Zárraga *et al.* 2003).

Lafkenche ornithological knowledge

Among the 58 investigated bird species, *P. australis* does not live in the Lafkenche region. Thus, the analysis of Mapudungun names considers 57 species. The Mapudungun names for 55 of these species (97%) were found. Only for two species, *B. ventralis* and *Eugralla paradoxa*, no Mapudungun names were recorded. The first bird is a rare species throughout its range (Glade 1993). The second species is a Tapaculo, and is difficult to observe because it inhabits the bush foliage, prefebly bamboo (*Chusquea* sp.) in forest habitats with well-developed understory (Rozzi *et al.*1996).

For the Falconiformes, other than the *B. ventralis*, Aillapan could not suggest a specific name for *F. peregrinus*, a species generically called *Peuco*. This Mapudungun name was also reported for *A. bicolor* (Gorgoglione 1997) and *Parabuteo unicinctus* (Lenz 1905-1910). For *A. bicolor*, Aillapan provided the new name *Pichi Kokoriñ*, which means small (= *pichi*) peuco (= *Kokoriñ*). Aillapan also contributed new names for the species of vultures (= *kaniñ*): *Kürüwún Kaniñ* for *Coragyps atratus* and *Kelüwún Kaniñ* for *Cathartes aura*, alluding respectively to the black (= *kürü*) and red (= *kelü*) color of their heads and the bare skin on the basis of their beaks (= *wún*). For the Strigiformes, like the Yahgan grandmothers, Aillapan has a vivid knowledge. The names for these nocturnal birds are onomatopoeic with their vocalizations, and Aillapan added new variations (Table 1).

For species of Ciconiiformes, Gruiformes and Charadriiformes, the names given by Aillapan for *Theristicus melanopis*, *Rallus sanguinolentus* and *Vanellus chilensis* have been reported by previous studies (Table 1). For the Coraciiformes, Aillapan introduced a new name for *M. torquata*: *Challwafeüñüm*, which means Fish (= *challwa*) Bird (= *üñüm*) (Catrileo 1998). For the species of Psittaciformes and Columbiformes the names match those reported in other studies. For Apodiformes, Aillapan uses the familiar Mapudungun name of *pinda* (= hummingbird), but makes a distinction in size, consistent with the common names for these species in Castilian: *Fütra pinda* (*fütra* or *füta* = large) for the Giant Hummingbird *Patagona gigas*, and *Pichi pinda* (*pichi* = small) for the Green-backed Firecrown *S. sephaniodes*. Aillapan introduces a similar innovation between Piciformes, by naming the Striped Woodpecker *Picoides lignarius* as *Pichi rere*, which is the smallest of the woodpeckers (= *rere*) inhabiting southern Chile (Table 1).

Regarding the Passeriformes, Aillapan contributed nine names that had not been previously recorded: 1) *Porotu*, which is onomatopoeic with the sound emitted by male *Gallinago paraguaiae* during the time of courtship. 2) *Utruftoki* for *Zonotrichia capensis* which refers to the movement of his head that looks like the action of "throwing" (= *utruf*) an ax (= *toki*). This new name represents a variation from another Mapudungun name for this bird: *meñkutoki* that means "to carry on the head" (= *meñku*) an ax. Both names refer to the tuft that this bird carries on his head. 3) *Chidüf* for *Carduelis barbatus* represents a generalization of this Mapudungun name given to other small songbirds such as *Troglotyes aedon* (Hernández *et al.* 1997). 4) *Pidpidwiriñ* for *A. spinicauda* represents an interesting combination consisting of two Mapudungun terms: *pidpid* which is onomatopoeic with the repeated vocalizations of the bird, and *wirin* which refers to the appearance of the bird; it refers to the bands or stripes of their plumage. 5) *Chiuchiu* for both Cinclodes species is onomatopoeic with the vocalizations of these birds, and is a variation of the previously registered Mapudungun name by Gorgolione (1997): *Chiuchihuen*. 6) *Pelchokiñ* for *P. albogularis* had not been recorded previously, and Aillapan had no explanation for the etymology of this term. 7) For *Sylviorthorhynchus desmursii*, Aillapan contributed the new name *Futrakülen* alluding to the large (= *futra*) tail (= *Külen*) of this species. 8) For *Scytalopus magellanicus*, Aillapan incorporated the Lafkenche name *Tiftifkeñ*, citing Magellanic Tapaculo vocalizations that resemble the regularity of a winding watch (= *tiftif*). 9) For *C. parvirostris*, Aillapan uses two names that are interrelated: *Peutren* that means knowing the ancestors, and *Lantu* that means widow. This second Mapudungun name is similar to the common Chilean name (little widow) for this bird, and refers to the vocalization that resembles the weeping of a suffering woman wandering through the forests of southern Chile.

CONCLUDING REMARKS

As in the case of the Yahgan names recorded for the birds of the Magellanic sub-Antarctic forests, this study provides new Mapudungun names for the birds of the temperate forests of Central and southern Chile and Argentina. These new names express various dimensions of the Mapuche ornithological observations and traditional ecological knowledge. In coincidence with studies by Lenz (1905-1910) and Villagrán et al. (1999), it seems that many of the Mapudungun names for birds are derived from onomatopoeias. The variety of Mapudungun names for each bird species is also noteworthy. In this sense, the recording of new names based on the work with Lafkenche poet Lorenzo Aillapan shows that the study of Mapuche ethno-ornithology is an open field for research. The continuity of ethno-ornithological studies will enrich our knowledge about both the birds and the cultural diversity that co-inhabits the temperate forests of South America.

ACKNOWLEDGEMENTS

The authors acknowledge and thank of the Indigenous Yahgan Community of Mejillones Bay and Mapuche Academy *PÜLLÜMAPUKIMUNWEFTUY* and the commentaries by Raymond McNeil on the manuscript. The Ethno-ornithology Symposium, VII Neotropical Ornithological Congress, has been supported by the Millennium Center for Advanced Studies in Ecology and Biodiversity (CMEB, Project No. P02-051-F ICM) and the project BIOKONCHIL (FKZ 01 LM 0208, German Ministry of Education and Research, BMBF). This is a contribution to the research and conservation of the Omora Ethnobotanical Park (University of Magallanes and Omora NGO), Puerto Williams, Antarctic Province, Chile.

REFERENCES

AGUILERA, O. 2000. *En torno a la estructura fonológica del yagán*. Fonología de la palabra. http://www.uchile.cl/cultura/lenguas/yaganes.

AILLAPAN, L. & R. ROZZI. 2004. Una etno-ornitología mapuche contemporánea: Poemas alados de los bosques nativos de Chile. *Ornitologia Neotropical* 15 (Suppl.): 419-434.

ALWYN, J. 1994. *Pueblos Indígenas de Chile: Antecedentes Históricos y Situación Actual. Nº 1*. Instituto de Estudios Indígenas, Universidad de la Frontera. Temuco, Chile.

ARMESTO, J.J., C. VILLAGRÁN & M.T. KALIN. 1996. *Ecología de los Bosques Nativos de Chile*. Editorial Universitaria, Santiago, Chile.

ARMESTO, J.J., R . ROZZI & J. CASPERSEN. 2001. Past, present, and future scenarios for biological diversity in South American temperate forests. Pp. 223-249 *in Future Scenarios for Biological Diversity*, T. Chapin & O. Sala (eds.). Springer Verlag, New York, New York.

ARROYO, M.T.K., L. CAVIERES, A. PEÑALOZA, M. RIVEROS & A.M. FAGGI. 1996. Relaciones fitogeográficas y patrones regionales de riqueza de especies en la flora del bosque lluvioso templado de Sudamérica. Pp. 71-99 *in Ecología de los Bosques Nativos de Chile*, J.J. Armesto, C. Villagrán & M.T.K. Arroyo (eds.). Editorial Universitaria, Santiago, Chile.

BERDICHEWSKY, B., I. LJUBETIC, A. SAAVEDRA & M. MUENZEL. 1977. *Chiles Mapuche: Armee Gegen Indianer*. POGROM Report 25, Vienna, Austria.

BRIDGES, T. 1987. *Yamana-English Dictionary*. Zaglier & Urruty Publicaciones, Buenos Aires, Argentina.

CATRILEO, M. 1998. *Diccionario Lingüístico-Etnográfico de la Lengua Mapuche*. Editorial Andrés Bello, Santiago de Chile.

DONOSO, C. & A. LARA. 1996. Utilización de los bosques nativos en Chile: Pasado, presente y futuro. Pp. 363-387 *in Ecología de los Bosques Nativos de Chile*, J.J. Armesto, C. Villagrán & M.T.K. Arroyo (eds.). Editorial Universitaria, Santiago, Chile.

GLADE, A. 1993. *Libro Rojo de los Vertebrados Terrestres de Chile*. Corporación Nacional Forestal (CONAF), Santiago, Chile.

GORGOGLIONE, E.C. 1997. *Guía de Campo para la Identificación de las Aves del Neuquén*. Editorial Hemisferio Sur, Montevideo, Uruguay.

GUSINDE, M. 1961. *The Yamana: The Life and Thought of the Water nomads of Cape Horn*. Volumes I-V. New Haven Press, New Haven, Connecticut.

GRIMES, B. 2000. *Ethnologue: Languages of the World*. 14th ed. Summer Institute of Linguistics, Dallas, Texas.

HERNÁNDEZ, A., N. RAMOS & C. CÁRCAMO. 1997. *Diccionario Ilustrado* Mapudungun. *Pehuén*, Santiago, Chile.

HIDALGO, J., V. SCHIAPPACASSE, H. NIEMEYER, C. ALDUNATE & I. SOLIMANO. 1997. *Culturas de Chile. Prehistoria: Desde sus Orígenes hasta los Albores de la Conquista*. Andres Bello, Santiago, Chile.

HINOJOSA, F. & C. VILLAGRÁN. 1997. Historia de los bosques del sur de Sudamérica, I: Antecedentes paleobotánicos, geológicos y climáticos del Terciario del cono sur de América. *Revista Chilena de Historia Natural* 70: 225-239.

LENZ, R. 1905-1910. *Diccionario Etimológico de las Voces Chilenas Derivadas de las Voces Indígenas Americanas*. Imprenta Cervantes, Santiago, Chile.

MAFFI, L. 2001. *On Biocultural Diversity: Linking Language, Knowledge, and the Environment*. Smithsonian Institution Press, Washington, D.C.

MARTINIC, M. 1992. *Historia de la Región Magallánica*. Universidad de Magallanes, Punta Arenas, Chile.

MAYORGA, P. 1972. *Costumbres y Extinción de los Indios del Extremo Austral*. Arancibia Hermanos, Santiago, Chile.

ORQUERA, L. & E. PIANA. 1990. Canoeros del extremo austral. *Ciencia Hoy* 1: 18-27.

OVIEDO, G. & L. MAFFI. 2000. *Indigenous and Traditional Peoples of the World and Ecoregion Conservation*. WWF World Wide Fund for Nature, Gland, Switzerland.

POBLETE, M.T. & A. SALAS. 1997. Fonemas Yámana (yagán). *Estructura Fonológica de la Palabra*. Dirección de Investigación, Universidad de Concepción, Concepción, Chile.

PRIMACK, R., R. ROZZI, P. FEINSINGER, R. DIRZO & F. MASSARDO. 2001. *Elementos de Conservación Biológica: Perspectivas Latinoamericanas*. Fondo de Cultura Económica, Mexico, Mexico.

RAUP, D.M. 1979. Size of the Permo-Triassic bottleneck and its evolutionary implications. *Science* 206: 217-218.

ROZZI, R., D. MARTÍNEZ, M.F. WILLSON & C. SABAG. 1996. Avifauna de los bosques templados de Sudamérica. Pp. 135-152 *in Ecología de los Bosques Nativos de Chile*, J.J. Armesto, C. Villagrán & M.T.K. Arroyo (eds.). Editorial Universitaria, Santiago, Chile.

ROZZI, R., C. ANDERSON, F. MASSARDO & J. SILANDER, JR. 2002. Diversidad biocultural subantártica: El programa del Parque Etnobotánico Omora, Puerto Williams, Chile. *Chloris Chilensis* 4(2). http://www.chlorischile.cl

ROZZI, R., F. MASSARDO, C. ANDERSON, S. MCGEHEE, G. CLARK, G. EGLI, E. RAMILO, U. CALDERÓN, C. CALDERÓN, L. AILLAPAN & C. ZÁRRAGA. 2003. *Guía Multiétnica de Aves de los Bosques Templados de Sudamérica Austral.* Fantástico Sur-Universidad de Magallanes, Punta Arenas, Chile.

RUSSELL, B. 1994. *Research Methods in Anthropology: Qualitative and Quantitative Approaches.* Sage Publications, Thousand Oaks, California.

VEBLEN, T.T. & F.M. SCHLEGEL. 1982. Reseña ecológica de los bosques del sur de Chile. *Bosque* 2: 73-115.

VENEGAS, C. & W. SIELFELD. 1998. *Catálogo de los Vertebrados de la Región de Magallanes y Antártica Chilena.* Ediciones de la Universidad de Magallanes, Punta Arenas, Chile.

VILLAGRÁN, C., R. VILLA, F. HINOJOSA, G. SÁNCHEZ, M. ROMO, A. MALDONADO, L. CAVIARES, C. LATORRE, J. CUEVAS, S. CASTRO, C. PAPIC & A. VALENZUELA. 1999. Etnozoología *mapuche*: un estudio preliminar. *Revista Chilena de Historia Natural* 72: 595-627.

ZÁRRAGA, C., F. MASSARDO, T. HAKIN & R. ROZZI. 2003. *Hen Harmameuti Hen Maruti Intien: Diccionario infantil Yagán.* Fondo de Desarrollo de las Artes y la Cultura (FONDART), Punta Arenas, Chile.

Accepted March 28, 2004.

31 ETHICAL IMPLICATIONS OF YAHGAN AND MAPUCHE NARRATIVES ABOUT BIRDS OF THE TEMPERATE FORESTS OF SOUTHERN SOUTH AMERICA

RICARDO ROZZI[1,2,3]

[1]Universidad de Magallanes, Casilla 113-D, Punta Arenas, XII Región Chile. *E-mail:* ricardo.rozzi@umag.cl
[2]ONG Omora, Puerto Williams, Región de Magallanes y Antártica, Chile
[3]Department of Philosophy and Religion Studies, University of North Texas, P.O. Box 310980, Denton, Texas 76203-0980, USA

Ornitología Neotropical 15 (Suppl.): 435-444 (2004) (Originally published in Spanish)

ABSTRACT. Ethical implications of Yahgan and Mapuche narratives about the birds of the austral temperate forests of South America. This paper analyzes the ethical implications of Yahgan and Mapuche stories about forest birds of southern Chile and Argentina, from the perspective of biological conservation and environmental philosophy. To allow comparisons among notions of traditional ecological knowledge, evolutionary-ecological sciences, and environmental ethics, I focus on two well known metaphors: the *tree of life* and the *web of life*. The analysis of the first metaphor allows to conclude that both modern sciences and the Yahgan and Mapuche indigenous cosmogonies affirm a common origin for birds and humans. This notion supports the *intrinsic value* of the avifauna, because birds are regarded as our evolutionary relatives. This implies that, to a certain degree, the life of birds can be subject to moral considerations based on ontological and ethical judgments commensurable with those involved in assessing the value of human life. The analysis of the metaphor of the *web of life* also reveals essential correspondences between contemporary scientific knowledge and Yahgan and Mapuche traditional ecological knowledge regarding the net of biotic interactions and ecosystem processes. Bird stories such as the Green-backed Firecrown (*Sephanoides sephaniodes*) or *omora* (in Yahgan) and the Magellanic Tapaculo (*Scytalopus magellanicus*) or *tiftifken* (in Mapuche) affirm, as much as sciences, that it is necessary to preserve the community of birds, and biodiversity in general, to ensure water supply and other ecosystem services and goods in the long term. The ethical imperatives, implicit in the second metaphor, are consistent with the notion of *instrumental value*. According to it, the conservation of birds can be regarded as an instrument for human survival. Traditional ornithological knowledge and modern sciences provide support for the instrumental and intrinsic value of biodiversity; today both values appeal for a respectful living together with birds.

Key words: Avifauna, Chile, environmental ethics, ethno-ornithology, instrumental value, intrinsic value, Mapuche, metaphors, temperate forests, Yahgan.

RESUMEN. Implicaciones éticas de narrativas yaganes y mapuches sobre las aves de los bosques templados de Sudamérica austral. Este trabajo analiza las implicancias éticas de las historias yaganes y mapuches sobre las aves de los bosques del sur de Chile y Argentina, desde el punto de vista de la conservación biológica y la filosofía ambiental. Para permitir comparaciones entre nociones del conocimiento ecológico tradicional, las ciencias ecológicas y evolutivas, y la ética ambiental actual, el análisis se centra en dos metáforas ampliamente conocidas: el *árbol de la vida* y la *red de la vida*. El análisis de la primera metáfora permite concluir que tanto las ciencias como las cosmogonías indígenas yagán y mapuche proponen un origen común para las aves y los humanos. Esta noción genealógica o evolutiva realza el *valor intrínseco* de la avifauna, porque las aves son consideradas como nuestros parientes evolutivos lejanos. Esto implica que, en algún grado, la vida de las aves puede estar sujeta a consideraciones morales fundadas en juicios ontológicos y éticos conmensurables con aquellos con que juzgamos el valor de la vida humana. El análisis de la metáfora de *la red de la vida* también revela concordancias esenciales entre el conocimiento actual de las ciencias ecológicas y el conocimiento ecológico tradicional yagán y mapuche respecto a las tramas de interacciones bióticas y abióticas de los ecosistemas. Historias de aves como la del picaflor (*Sephanoides sephaniodes*) u *omora* (en yagán) y la del Churrín (*Scytalopus magellanicus*) o *tiftifken* (en mapuche) afirman, tanto como las ciencias, que es necesario conservar la comunidad de aves y la biodiversidad en general para mantener la continuidad del abastecimiento de agua y otros servicios y bienes ecosistémicos en el largo plazo. Los imperativos éticos implícitos en la segunda metáfora son consonantes con la noción de *valor instrumental*, donde la conservación de las aves puede ser apreciada como un instrumento para la sobrevivencia humana. Tanto el conocimiento ornitológico tradicional como el científico afirman los valores instrumental e intrínseco de la diversidad biológica; ambos valores apelan hoy a una convivencia respetuosa con las aves.

Palabras clave: avifauna, bosques templados, Chile, ética ambiental, etno-ornitología, Mapuche, metáforas, valor instrumental, valor intrínseco, yagán.

INTRODUCTION

Indigenous narratives correspond to oral traditions whose concepts are open to multiple interpretations. The narratives possess many layers of significance (Turner 1967, Geertz 1974), whose decodification and interpretation vary depending to the analyst's questions and perspectives (Hunt 1977). The present analysis of Yahgan and Mapuche stories about the birds of the forests of southern South American aims to essay a communication bridge among notions of traditional ecological knowledge, ecological and evolutionary sciences, and environmental ethics. For this purpose, the analysis centers on two metaphors extensively used in environmental and evolutionary sciences, both of which involve an ethical dimension: the *tree of life* and the *web of life* (see Rozzi 1999).

METHODS

Storytelling formed an integral part of Yahgan and Mapuche life. According to Gusinde (1961), in Yahgan communities there were men who narrated stories around the fires at night in the austral archipelagooes of Cape Horn. However, the stories recorded in this work, as well as the majority of those recorded by ethnographers at the beginning of the 20th century, were told by women. In the early 21st century, sisters Úrsula and Cristina Calderón were considered the major repositories of Yahgan traditional ecological knowledge. The stories recorded with them provide the basis for the analysis of Yahgan traditional ornithological knowledge presented in this paper. Regarding the Mapuche narratives, these originate principally from stories and poems of the Lafkenche poet Lorenzo Aillapan (see Aillapan & Rozzi 2004). Field methodology and area of study are described in Massardo & Rozzi (2004).

RESULTS AND DISCUSSION

Yahgan ornithological narratives associated with the *tree of life* metaphor

From an environmental ethics perspective, the central concept of the *tree of life* metaphor is the notion "of evolutionary kinship" (*sensu* Leopold 1949). Let's consider the Yahgan story of the Green-backed Firecrown (*Sephanoides sephaniodes*) or *Omora* as told by Cristina Calderón (Story 1).

Processes of transformation from human beings into birds -like those evoked in the story about the Green-backed Firecrown or

STORY 1. Yahgan narrative of the Green-backed Firecrown (*Sephanoides sephaniodes*) or *Omora*, narrated by Cristina Calderón (Yahgan and English audio recording in Rozzi *et al.* 2003).

In ancestral times, when birds were still humans, a great drought occurred in the region of Cape Horn and the people were dying of thirst. Only the wily fox, known as *Cilawáia*, found a lake. He told no one and built, instead, a solid fence around it so that no one could enter. Hidden as such, he drank lots of water, concerned only with himself. However, after some time the other people discovered this lake and went as a group to ask the fox for a little bit of water. *Cilawáia* did not even listen to their supplications and expelled them with rude words. The condition of the people got increasingly worse and when they were at the point of death, they decided in their desperation to send a message to the hummingbird or little *Omora*, an occasional visitor who had saved them before in times of crisis. *Omora* was always prepared to help others and arrived quickly. Although diminutive, this little tiny bird-man-spirit is braver and bolder than any giant. When he arrived, the downcast people told him about their great sufferings, and, indignant, *Omora* raised himself and rapidly undertook the flight towards where the egotist *Cilawáia* was, and he confronted him directly saying: "Listen! What is it that the people are telling me? Have you had access to a lake full of water, but you don't want to share it with other people who soon will die of thirst!" The fox replied, "Why should I worry myself about the others? This lake has only a little water, and it barely is enough for me and my closest family. I cannot give anything to other people because soon I'll suffer thirst myself."

Upon hearing this, *Omora* became furious and without replying to the fox, returned to the settlement. He reflected briefly and rapidly rose again, took his sling and flew back to the fox. Along the way he collected several sharp stones, and when he was within sight of the fox and sufficiently near, he shouted to him: "Are you going to share your water for now and for always with the other people? Come on, don't be so selfish. If you don't spare water, they sorely will die of thirst!"

The fox answered indifferently, "They can all die of thirst. Why should I worry? I can't give water to each one of the people who live here, or else my family and I will also die of thirst." *Omora* was so offended that he could not control himself any longer and furiously fired a sharp stone with his sling, killing the fox with his first throw.

The rest of the community had been watching and became happy, running to the place. Rapidly they broke down the fence, went to the lake and drank, satiating their thirst. They drank so much that the lake became completely empty and a few birds that arrived too late barely found a few drops with which they moistened their throats. Then, the wise owl *Sirra* or *Sita*, the grandmother of *Omora*, said to the birds that arrived late, "collect mud from the bottom of the lake and fly to the peaks of the mountains, and fling the mud over them." The little birds flew, and their balls of mud gave birth to springs that became water courses that spouted from the mountains, forming small streams and large rivers that flowed through the ravines. When all the people saw this, they were extremely happy and drank great quantities of fresh, pure water, that was much better than the water of the lake, and now all find themselves safe again. Today all of those water courses still flow down from the mountains and provide exquisite water. Since that time no one has died of thirst.

omora- are encountered in numerous Yahgan stories, and constitute a primary foundation for the notion of a relationship between birds and humans. Grandmother Úrsula Calderón related stories like that of beautiful young women transforming into Black-chinned siskins (*Carduelis barbatus*) or *twin*, or of two Yahgan siblings who fell in love and were transformed into a couple of Magellanic woodpeckers (*Campephilus magellanicus*), or *lana*, while collecting fruits from the *chaura* (*Pernettya mucronata*). Such transformations imply a common ancestral lineage.

Transformations of humans into animals, or animals into humans, were conceived by numerous Amerindian cultures. Hunt (1997) referred to this type of cosmogony as a "pre-Hispanic theory of evolutionary biology." In scientific evolutionary theory, as well as in pre-Hispanic evolutionary perspectives, humans, birds, and other living creatures share a common evolutionary origin and, as a consequence, are linked through the notion of a genealogical kinship.

In the Yahgan story of the hummingbird omora, we also find a notion of "inter-species kinship" relationships among birds. For example, the Barn Owl (*Tyto alba*) or *sirra* is the grandmother of the hummingbird or *omora*. In other Yahgan narratives, this spectrum of kinship and transformations is projected onto the physical world: rocks are transformed into human beings, and humans transformed into stars, the moon or the sun (see Gusinde 1961).

Interspecies kinship relationships, as well as transformations from one species to another, support a notion of a common nature that presents a parallel to the biological theory of the unity of life. Such a notion is further affirmed when the story speaks of "the time of the ancestors, when birds were still humans." This affirmation begins the *omora* narrative and many other Yahgan stories. For example, the grandmother Úrsula Calderón relates that, in ancestral times, the inventor of canoes or *anan* was the widowed *tuto*. He then transformed into the little Magellanic Tapaculo (*Scytalopus magellanicus*), and today this bird collects and weaves vegetable fibers of roots, branches, lichens, and mosses to carefully build his nest among roots or in tree-trunk cavities (Rozzi *et al.* 2003).

In summary, Yahgan ornithological stories present three elements that, in a broad sense, imply a notion of kinship between human beings and birds:

1) Yahgans note that human beings have given origin to various species of birds through diverse processes of transformations;

2) Yahgans recognized kinship relationships among diverse species of birds, such as those between the Barn Owl and the Green-backed Firecrown;

3) Yahgan cosmogony refers to an ancestral time when birds were still humans.

Yahgan ornithological narratives associated with the *web of life* metaphor

The main ecological concept of the *web of life* metaphor refers to the numerous and complex interactions among biological species, and also between those species and the physical environment they inhabit (see Carson 1962). Humans and other biological species form part of this network of interactions. Futhermore, from a biocultural perspective, the *web of life* metaphor allows one to compare the "ecological order" and the "social order." To this end, we consider the story of the Black-faced Ibis (*Theristicus melanopis*) or *lewuja* (Story 2).

This story evokes the association between the migratory behavior and the seasons of the year. For the Yahgans, the arrival of the Black-faced Ibis signifies the beginning of spring, and its departure involves snowstorms and cold weather. This bird has a large size and color, and emits powerful metallic vocalizations. Thus, at the end of autumn upon its departure it announces very clearly the changes of the seasons and marked climatic changes. This kind of relationship between the Yahgans and the birds involves a type of inter-specific coupling of behaviors, where birds act as signs to predict the climate.

STORY 2. Yahgan Narrative of the Buff-necked Ibis (*Theristicus melanopis*) or *Lejuwa,* narrated by Úrsula Calderon (Yahgan and English audio recording in Rozzi *et al.* 2003).

In the times of the ancestors, one day when spring arrived, a *yamana* [= man] leaned out of his *akar* or hut and saw a Buff-Necked Ibis or *lejuwa* flying in the sky. The *yamana* was so happy that he yelled to the other members of the community, "An ibis is flying over our *akar*. Look!" Immediately the others came out of their huts, shouting loudly, "The spring has come, the ibises are already in their return flight." They jumped for joy, and they jumped hard.

Lejuwa is, however, a very delicate and sensitive woman and likes to be treated with special deference. Upon hearing the noise of the shouts, she became furious, and profoundly offended, causing a strong snow storm. It snowed incessantly; it was freezing cold; all the land was covered with ice; and all the waters froze. Many people died because they could not navigate in their canoes to search for food. Neither could they leave from their *akars* to look for firewood because all was covered with snow. More and more people continued to die. Only after a long time did it stop snowing, and the sun begin to shine.

The sun warmed so much that begin to melt the ice and snow that had completely covered the land, and consequently floods began to flow towards the channels and the oceans. The ice that covered the wide and narrow channels broke apart and melted, and the Yahgans could finally go to the shores and navigate in their canoes to collect food. On the big mountainsides and deep valleys, however, the ice was so thick that not even this sun could melt it. Even today, glaciers can be observed coming down to the sea, that remind us of the severe frost and snowstorm that *lejuwa*, who is a very delicate and sensitive woman, provoked in those times. Since then, the Yahgans treat the Buff-Necked Ibis or *lejuwa* with much respect, and when they come near the houses or *akars*, the people keep quiet, particularly the little children who are not allowed to imitate them or to cry.

The Buff-necked Ibis story also expresses a social rule that prohibits to disturb the birds. Such a rule is found in numerous Yahgan bird stories. For example, the grandmother Úrsula observes that the Fire-eyed Diucon (*Xolmis pyrope*) or *hashpúl* is a powerful shaman who should be treated with respect, and never have stones thrown at him. If this rule is broken, *hashpúl* gets irritated and provokes storms and tempests with south winds or *ilan*. This Yahgan prohibition of thrown stones at certain birds has been documented since the first registered ethnographies (Lothrop 1928). Grandmother Cristina says that no one is permitted to imitate the hum of the South American Snipe (*Gallinago paraguaiae*) or *shakóa*. Whoever imitates or bothers *shakóa* will awaken with their toes cut by the fibers of the rushes (*Marsippospermum grandiflorum*) or *ushkulampi*. This plant grows in the southern wetlands inhabited by the Snipe; wetlands where Yahgan women collect rush fibers for weaving baskets (Rozzi *et al.* 2003). The narratives thus express an awareness about ecological interactions with these birds that should not be disturbed. Through the stories, rules and norms are established for interacting with the birds. Whoever disobeys these rules will suffer the consequences.

Stories like those of the Buff-necked Ibis, the Fire-eyed Diucon, and the South American Snipe are narrated by Yahgan grandmothers and grandfathers. In turn, the bird stories also express respect toward elders. For example, grandmothers Úrsula and Cristina tell how when the Austral Thrush (*Turdus falcklandii*), or *hakasir*, has chicks, the mother first educates the oldest, and then together they educate and feed the other sibling chicks in the trees and rivers of the forest habitats. *Hakasir* moves from branch to branch talking, whistling, and educating her child. Úrsula remembers one day when "I was laying on the ground watching, when the mother hakasir said to her oldest chick 'when you have a little brother you should educate him, enliven him to work, to bathe himself.' And then the little bird went singing, flying toward the river where together they soaked and washed themselves, then shook and returned to the same branch. This is how the *hakasir* grows up." Úrsula adds that "if the son Thrush does not listen to its mother, if it does not obey, it is abandoned. The little Thrush remains alone and has nothing to eat. It eats bad dihueñes (mushrooms of the genus Cyttaria), rotten ones, that make him sick and he dies. On the other hand, if he obeys his mother, then he grows healthy, feeding himself on good fruits, such as the Prickly Heath (*Gaultheria mucronata*) or *amai*."

Yahgan social relations are projected onto the narratives about the birds. At the same time, observation of the birds' conduct inspires Yahgan habits and social norms. In order to survive, both the Thrush and the young Yahgans should obey their parents and grandparents, and learn to distinguish good from bad foods. Respect toward elders is also present in the story of *omora*, since only with the wisdom of its grandmother, the Barn Owl, is the Green-backed Firecrown able to achieve success in maintaining: (1) the *Yahgan social order*, through punishing greediness, and (2) the *ecological order*, through the conservation of the community of birds and the reestablishment and creation of freshwater streams. These stories express a close parallel between the social order of the Yahgans and the ecological order observed between the birds.

Mapuche ornithological narratives associated with the *tree of life* metaphor

In order to analyze Mapuche ornithological stories from the point of view of the *tree of life* metaphor, we primarily consider the Andean Condor (*Vultur gryphus*) or *mañke* (Story 3) who is the king of birds in traditional Mapuche ornithology, as well as in other Andean cultures (Primack *et al.* 2001).

With its wingspan of more than 3 meters, the Andean Condor is the largest flying bird in the Americas. For the Mapuche culture, *mañke* symbolizes the Andean Mountain Range through its grand size and by its colors: white like the snow and black like the rocks. It also flies to grand heights over the Andean peaks, embodying the fundamental Mapuche virtues. Today, verifies Aillapan, both the *mañke* and the Mapuche virtues –wisdom (*kim*), justice (*nor*), kindness (*küm*), and discipline (*newen*)– are threatened with extinction. Just like the Andean Condor is hard to see because it soars so high, also the essential Mapuche values, which elevate our spiritual lives, are difficult to appreciate and cultivate in our days.

The Mapuche have a special admiration for birds of prey. Among those, the Crested Caracara (*Polyborus plancus*) or *traru* has immense historical significance because its lineage gave origin to the greatest of Mapuche warriors, *Leftraru* (*lef* = fast; *traru* =Caracara). This brave warrior, known by the Spanish conquerors as Lautaro (1535-1557), took advantage in the battles of his knowledge of the topography of the land, his ability to predict weather events, and the speed to attack with his warriors from different flanks (Barella 1971).

STORY 3. Mapuche Narrative of the Andean Condor or *Manke,* narrated by the Mapuche poet Lorenzo Aillapan (Mapudungun and English audio recording in Rozzi *et al.* 2003).

The Condor or the *mañke* is for the Mapuche culture the King of the Birds. It is a symbol of the mountains, not only due to its great size and impressive wing-length, but also its black and white colors, which signify the snow-capped peaks and the black mountain rocks. For the Mapuche, the condor unites the virtues of being *Kimche*, a wise person, *Norche*, a person who loves justice, *Kümeche*, a good-natured person, and *Newenche*, a powerful, lordly person. This King of Birds is the national bird of Colombia, Ecuador, Bolivia, and Chile. *Mañke* was once abundant on *Manquehue* Mountain, which means place (*hue*) of the condors (*mañke*), in the center of Santiago, the capital of Chile. It is still found today throughout the territory of the austral forests, but paradoxically, this king of birds, this national symbol of four countries, finds itself now threatened with extinction.

The story of Lautaro shows how certain essential attributes, such as quickness and wisdom, are conserved through the common lineage of birds and human beings. These attributes belong to both. Thus, for example, the Black-throated Huet-huet (*Pteroptochos tarnii*), or *wed-wed*, is thought of like a buffoon or a crazy man that, with his vocalizations, narrates stories while accompany travelers through the forests along ravines (Rozzi *et al.* 2003). The Southern Lawping (*Vanellus chilensis*) or *tregul* is a bird that acquires various human or animal forms: "it is a man that comes with his sex dangling, it is a woman that walks another way, it has four legs, it can be a dog or a fox" (Aillapan & Rozzi 2001). The flight and song of the *tregul* courtship are mimicked during the Mapuche ceremony, *ngillatun*. Participants dance in circles as the *tregul* does, with arms open toward the north, the west, the south and the east, in order to invoke the Grand Spirit of Universal Fertility (Aillapan & Rozzi 2001).

The identification between birds and humans is expressed, then, in multiple aspects of the Mapuche culture, generating a living sense of familiarity in the bird-human relationships. Furthermore, representation of birds as humans, and vice-versa, implies that these beings possess common characteristics, an assertion consistent with the scientific paradigm of the "unity of life." Analogous to scientific evidence based on the similarities of cellular structure, embryological development, and anatomical patterns found among vertebrates, the Mapuche worldview establishes similarities between human beings and birds based on observation of certain behavioral patterns.

Mapuche ornithological narratives associated with the *web of life* metaphor

Among the ecological interactions expressed through the Mapuche stories, the trophic relations stand out that generate matter and energy flows between the "beings of air, water, and earth" (Aillapan & Rozzi 2001). For example, for the Mapuche worldview, "the hypnosis that the Ringed Kingfisher (*Megaceryle torquata*) or *challwafeüñüm* exercises over the fishes unchains the flow between beings of the water and those of the air through the food chain." For Aillapan, the woodpeckers are seen as "birds that pass the day attached to a tree, pecking at and removing larvae that, in turn, are fed by the sap and tissues of the trees." Thus, a "cord of life" is established that flows from the minerals in the mountain soil through the sap of the trees, the blood of the insects and the birds.

The Mapuche worldview pays attention to all the members of the food chain. Carrion birds such as vultures or *kañin* are admired as "full time cleaners of the cities, beach and open countryside" (Aillapan & Rozzi 2001). Birds of prey and men are notable hunters. The ecological webs between birds and humans is finely interwoven and have a rhythm that imposes an ethical imperative: the rhythm of the birds must be heard and respected (Story 4).

Based on the call of the Magellanic Tapaculo (*Scytalopus magellanicus*), Aillapan questions the notion of progress. At the end of his poem about this millenary bird, Lorenzo sings: "it is said that the native people are behind the times, so I will set my watch forward" (Aillapan & Rozzi 2001). In this way, this poem based on the call of the Tapaculo or pocket-watch bird, *Tiftifken* –which, like the Mapuche, has inhabited the forests of southern Chile since its beginnings– ridicules the notion of progress that is so sacred to the conquistadors. The hurry and acceleration of the rhythm of life is as easy and as mechanical as it is accelerating the "tic-tock" of a clock, but why should we do such a thing? What sense does it make to abandon the forests' own rhythm, in order to accelerate it in a race that dissociates from the forests thereby driving to suicide all their living beings, including birds and humans? (Aillapan & Rozzi 2001). Thus, the notion of the *web of life* acquires in the Mapuche worldview a normative character: one must listen and respect the rhythm of nature.

Implications for environmental ethics and biocultural conservation

The metaphors of the *tree of life* and the *web of life* synthesize central concepts for evolutionary and ecological sciences, and for environmental ethics (Rozzi 1999). Darwin (1859) utilized the first metaphor (*tree of life*) at the end of his chapter on natural selection in "The Origin of the Species" in order to communicate that: "The affinities of all the beings of the same class have sometimes been represented by a great tree. I believe this simile largely speaks the truth. The green and budding twigs may represent existing species; and those produced during each former year may represent the long succession of extinct species... As buds give rise by growth to fresh buds, and these, if vigorous, branch and overtop on all sides many a feebler branch, so by generation I believe it has been with

STORY 4. Mapuche Narrative of the Magellanic Tapaculo (*Scytalopus magellanicus*) or *Tiftifken,* narrated by Lorenzo Aillapan (Mapudungun and English audio recording in Aillapan & Rozzi 2001).

Millenary bird that begins giving the hour from the daybreak, throughout the morning, to the noon-time and into the afternoon. Its song gives the hour in accordance with nature, the cosmovision, the universe and relaxation, great mountain range, plains, rivers, lakes, volcanoes and seas, great spirit, the principal one, our four guardians that take care of the Mapuche world as an 'Earthly Universe." that is the simple territory of the four winds, that is the north, south, east and west without ends where live *Dumpall*, power of the sea, *Pillan*, Mapuche spiritual force, *Anchumallen*, female heat, *Witranalwe*, the treasure. These four guardians of the Sacred Earth tell the hour with the "pocket-watch bird".....The pocket-watch bird, *Tiftifken,* indicates the human pair, always in action and harmony with nature.

the great Tree of Life, which fills with its dead and broken branches the crust of the earth, and covers the surface with its ever branching and beautiful ramifications" (Darwin 1859, pp.129-130). These metaphorical lines communicate synthetically that:

1) all living creatures share a common origin, and
2) humans are part of a evolutionary course of diversification, where some groups of organisms are born, others are transformed, and others disappear.

The ethical implications of this notion are captured lucidly by Leopold (1949), who expresses that: "it is a century now since Darwin gave us the first glimpse of the origin of species. We know now what was unknown to all the preceding caravan of generations: That men are only fellow-voyagers with other creatures in the odyssey of evolution. This knowledge should have given us, by this time, a sense of kinship with fellow creatures; a wish to live and let live; a sense of wonder over the magnitude and duration of the biotic enterprise" (1949, pp.116-7).

This ethical sense based in a notion of evolutionary kinship discovered by modern sciences seems to us now close to the ethical sense contained in the indigenous Yahgan and Mapuche cosmogonies. From the point of view of contemporary environmental ethics, the three cultural perspectives (Yahgan, Mapuche, and scientific) highlight the intrinsic value of the birds, because the birds are our distant evolutionary relatives. This implies that, to some degree, the existence of birds and that of humans are comparable. Consequently, the lives of the birds can be subjected to moral considerations based on ontological and ethical judgments commensurable with those through which we judge the value of human life. This conclusion has great relevance for conservation biology arguments based on the intrinsic value of the life of human beings, the birds, and the diverse biological species that co-inhabit the planet.

The *web of life* metaphor complements the evolution arguments with ecological notions. In "The Origin of Species," Darwin (1859) utilized the metaphor of the *web of life* to illustrate the "complex network of [ecological] relationships," describing how: "bumble-bees alone visit the red clover (*Trifolium pratense*) as other bees cannot each the nectar. Hence I have very little doubt, that if the whole genus of bumble-bees became extinct or very rare in England, the heartsease and red clover would become very rare, or wholly disappear. The number of bumble-bees in any district depends on the number of field-mice, which destroy their combs and nests... Now the number of mice is largely dependent, as every one knows on the number of cats" (Darwin 1859, pp. 73-4).

In this web of biotic interactions, the pink clover leaf plants depend directly on the bumble-bee for their pollination, and indirectly on predators of mice, like cats, to hunt these rodents that destroy the nests of bumble-bee pollinators. Comprehension of the interdependency between species currently represents a key notion for conservation biology's approach to the levels of community and ecosystem (Thompson 1997). More recently, conservation sciences have emphasized the role of human participation in this type of network of biotic interactions (McDonnell & Pickett 1993). Poetically, Carson (1962) expresses that: "the earth's vegetation is part of a web of life ... [and] although the modern man seldom remembers the fact, he could not exist without plants that harness the sun's energy and manufacture the basic foodstuff he depends upon for life" (pp. 63-4).

This evocative image of the *web of life* is analyzed today in quantitative terms by ecological economics (Daily 1997). More extensive applications of this metaphor establish that biological species as well as physical components of ecosystems (such as the water, the soil, and the atmosphere) must be protected in order to allow the continuation of the flow of goods and services that emanate from ecosystems, and that are essential for the well-being of current and future generations of humans (Norton 1991).

Considering again the Yahgan story of the Green-backed Firecrown or *omora*, we can now appreciate clearly how this relation teaches us that:

1) social well-being and conservation of biological species go hand-in-hand, and
2) solidarity ends up being better than selfishness; it leads us to a better life for everyone.

The story of the hummingbird *omora* reveals substantial coincidences between concepts from the ecological sciences and Yahgan traditional ecological knowledge. From the perspective of contemporary science, conservation of the community of birds contributes to ecological processes such as pollination, seed dispersal, insect population control, predation of rodent populations, or nutrient cycles through the guano of the birds that transport mineral salts between marine and mountainous ecosystems. In turn, maintenance of these ecosystemic processes contributes to the conservation of vascular and non-vascular plants in the hydrographic basins. This vegetation plays a key role in the regulation of hydrological flows, preventing floods, and avoiding droughts. Therefore, in order to maintain the water supply in the long-term, both the Yahgan story of *omora* and the scientific ecosystem analysis recommend the protection of biodiversity in ecosystems of the southern extreme of the Americas.

From the perspective of contemporary environmental ethics, the ethical imperatives implicit in the Yahgan narrative (to conserve the community of birds in order to ensure the supply of water) are consistent with the notion of instrumental value. In accordance with the traditional story and the notion of instrumental value, the conservation of the birds is an instrument for human survival. Traditional ecological knowledge contained in the story of the hummingbird *omora*, as well as in other Yahgan and Mapuche narratives of birds, like the actual knowledge of ecological and evolutionary sciences, nourishes and informs the notions of instrumental and intrinsic values of biological diversity. Today, both environmental values appeal to global society for a respectful and conscious co-inhabitation with birds.

ACKNOWLEDGMENTS

The author deeply thanks the teaching of grandmothers Cristina and Úrsula Calderón, and from the poet Lorenzo Aillapan, as well as the support from the Comunidad Indígena Yagán de Bahía Mejillones and of Academia Academia Mapuche *PÜLLÜMAPUKIMUNWEFTUY* from Puerto Saavedra. Francisca

Massardo did valuable comments to the manuscript. The Ethno-ornithology Symposium at the VII Congress of Neotropical Ornithology and the preparation of this article had the support from the Milennium Center for Ecology and Biodiversity Studies (CMEB, project Nº P02-051-F ICM) and of BIOKONCHIL project (FKZ 01 LM 0208, German Ministry of Education and Research, BMBF). This is a contribution to the research program and conservation activities of the Omora Ethnobotanical Park (Universidad de Magallanes and Omora NGO), Puerto Williams, Chilean Antarctic Province, Chile.

REFERENCES

AILLAPAN, L. & R. ROZZI. 2001. *Veinte Poemas Alados de los Bosques Nativos del Sur de Chile*. Editorial Plaza y Valdés, México, México.

AILLAPAN, L. & R. ROZZI. 2004. Una etno-ornitología mapuche contemporánea: Poemas alados de los bosques nativos de Chile. *Ornitología Neotropical* 15 (Suppl.): 419-434.

BARELLA, C. 1971. *Lautaro Guerrillero*. Nueva Universidad, Santiago, Chile.

CARSON, R. 1962. *Silent Spring*. Houghthon Mifflin, Boston, Massachusetts.

DAILY, G. 1997. *Nature's Services: Societal Dependence on Natural Ecosystems*. Island Press, Washington, D.C.

DARWIN, C. 1859. *The Origin of Species by Means of Natural Selection*. Murray, London, UK.

GEERTZ, C. 1974. *Myth, Symbol and Culture*. Norton, New York.

GUSINDE, M. 1961. *The Yamana: The Life and Thought of the Water Nomads of Cape Horn*. Volumes I-V. New Haven Press, New Haven, Connecticut.

HUNT, E. 1977. *The Transformation of the Hummingbird*. Cornell Univ. Press, Ithaca, New York.

LEOPOLD, A. 1949. *A Sand County Almanac*. Oxford Univ. Press, New York, New York.

LOTHROP, S.K. 1928. *The Indians of Tierra del Fuego*. The Heye Foundation, New York.

MASSARDO, F. & R. ROZZI. 2004. Etno-ornitología yagán y lafkenche en los bosques templados de Sudamérica austral. *Ornitología Neotropical* 15 (Suppl.): 395-407.

MCDONNELL, M.J. & S.T.A. PICKETT. 1993. *Humans as Components of Ecosystems: The Ecology of Subtle Human Effects and Populated Areas*. Springer-Verlag, New York, New York.

NORTON, B. 1991. *Toward Unity Among Environmentalists*. Oxford Univ. Press, New York, New York.

PRIMACK, R., R. ROZZI, P. FEINSINGER, R. DIRZO & F. MASSARDO. 2001. *Fundamentos de Conservación Biológica: Perspectivas Latinoamericanas*. Fondo de Cultura Económica, México, México.

ROZZI, R. 1999. The reciprocal links between evolutionary-ecological sciences and environmental ethics. *BioScience* 49: 911-921.

ROZZI, R., F. MASSARDO, C. ANDERSON, S. MCGEHEE, G. CLARK, G. EGLI, E. RAMILO, U. CALDERÓN, C. CALDERÓN, L. AILLAPAN & C. ZÁRRAGA. 2003. *Guía Multiétnica de Aves de los Bosques Templados de Sudamérica Austral*. Fantástico Sur Univ. de Magallanes, Punta Arenas, Chile.

THOMPSON, J.N. 1997. Conserving interaction biodiversity. Pp. 171-190 *in The Ecological Basis of Conservation*, S. Pickett, R. Ostfeld, M. Shachak & G. Likens, (eds.). Chapman and Hall, New York, New York.

TURNER, V. 1967. *The Forest of Symbols*. Cornell Univ. Press, Ithaca, New York.

Accepted in March 16th, 2004..

EPILOGUE

THE SECOND DECADE OF LONG-TERM ORNITHOLOGICAL STUDIES AT OMORA PARK: NEW ORNITHOLOGICAL PERSPECTIVES FROM THE SOUTH

JAIME E. JIMÉNEZ AND RICARDO ROZZI

The diversity and peculiarities of the birds of the southern cone of South America have drawn the attention and have fascinated many laymen and scholars of birds. Since the beginning of European explorations to the ecoregion of the Magellanic sub-Antarctic forests, observations on birds were recorded by Sarmiento de Gamboa, James Cook (Beaglehole 1955), Parker King (1830), Charles Darwin (1939), and Oustelet (1891), among others. Some made collections and descriptions of the first species, most of them around the Strait of Magellan and Cape Horn. These were followed by observations of Crawshay (1907) in Tierra del Fuego. Then, in the second half of the twentieth century, local researchers such as Olrog (1948, 1950), Reynolds (1934, 1935), and Humphrey *et al.* (1970) continued with collections of natural history and descriptions of the bird species, both, in Tierra del Fuego and on other Fuegian islands. These were supplemented in recent decades by observations, censuses, and species lists by Bernath (1965), Sielfeld (1979), Barros (1971, 1976), Venegas (1981, 1991), Schlatter & Riveros (1987) and others. During this period, some works also focused on more detailed and specific aspects of the biology of species such as the Magellanic Plover (*Pluvianellus socialis*) (Jehl 1975), the Ashy-headed Goose (*Chloephaga poliocephala*) (Schlatter *et al.* 2002), the Ruddy-headed Goose (*Chloephaga rubidiceps*) (Blanco *et al.* 2001, Madsen *et al.* 2003) or the Magellanic Woodpecker (*Campephilus magellanicus*, McBride 2000, Vergara & Schlatter 2004, Schlatter & Vergara 2005). At the community level, the ecology of forest birds on the island of Tierra del Fuego (Deferrari *et al.* 2001, Vergara & Schlatter 2006) and of seabirds in the Beagle Channel (Schiavini & Yorio 1995, Schiavini *et al.* 1999, Raya Rey & Schiavini 2000) have been studied. All these works have been limited, however, to short periods, mainly during the breeding season of the austral summer.

In this book we have presented the first synthesis of long-term ornithological studies developed by the interdisciplinary team of the Omora Ethnobotanical Park in the Cape Horn Biosphere Reserve. We have compiled a selection of papers with systematic research, focused primarily on forest birds, and some wetland and coastal environment birds. These papers contribute to the knowledge of the seasonal variations in diversity and abundance, natural history and ecology of the bird species of the Cape Horn Biosphere Reserve, mainly of those from Navarino Island. The topics covered are diverse and include detailed morphometric studies, the autecology of some species, parasitism, descriptions of population dynamics, impacts of predators, distributions and the study of birds also considering the culture and ethno-ornithological knowledge of the ancestral Yahgan native people.

Despite the advances presented in this book and the milestone in the understanding of the birds of the planet's southernmost forests, our understanding of the processes and patterns of bird's ecology and biogeography is still very fragmentary and far from complete. There are still many aspects that are completely unknown for most of the species and that deserve to be studied. The questions are many and diverse. For example, How much do we know about the migratory routes of the species of both terrestrial and aquatic birds, both freshwater and marine? How far do they move and what are their wintering grounds? Our knowledge of most species' population biology is almost nil, as is our knowledge of how the biological processes are governed by internal and external mechanisms, biotic and abiotic factors, and to what extent these can be modified by environmental changes such as global climate change. What are the magnitudes and how are vital rates such as reproduction, survival, mortality, and others are regulated? What is, and what will be the influence that human activities such as fisheries and salmon farming, alteration of forests and landscapes, the potential mass tourism or special interest tourism, including high mountain trekking have on southern birds? It is essential to evaluate the effect of these activities, which have been operating for some decades, but have increased steadily. May human activities have impacts not only on local bird populations, but also on the biota in general. In the context of rapid global and regional environmental change, birds may serve as sentinels of the responses of biodiversity and ecosystems. The pristine condition of much of the archipelagoes could provide a good opportunity to make comparisons with those areas where human impacts have been heavy and lengthy. It is therefore essential to maintain a baseline monitoring that provides information to quantify environmental changes.

Figure 2. The charismatic Magellanic Woodpecker (*Campephilus magellanicus*). A male (upper left) feeding on wood-boring larvae and a female (upper right) perching on a younger branch of a High-deciduous beech or Lenga (*Nothofagus pumilio*) in the Cape Horn Biosphere Reserve, observed by groups of visitors at Omora Park. Photos Jaime E. Jiménez.

HUMPHREY, P.S., D. BRIDGE, P.W. REYNOLDS & R.T. PETERSON. 1970. *Birds of Isla Grande (Tierra del Fuego)*. Preliminary Smithsonian Manual, University of Kansas Museum of Natural History, Lawrence, Kansas, USA. 411 pp.

JEHL, J.R. 1975. *Pluvianellus socialis*: biology, ecology, and relationships of an enigmatic Patagonian shorebird. *Transactions San Diego Society of Natural History* 18: 29-73.

KING, P.P. 1830. *Proceedings of the Committee of Science and Correspondence of the Zoological Society of London*. Part 1: 14-16.

MADSEN, J., R. MATUS, O. BLANK, L. BENEGAS, G. MATEAZZI & D.E. BLANCO. 2003. Populations status of the Ruddy-headed Goose (*Chloephaga rubidiceps*) in Tierra del Fuego and mainland Patagonia (Chile and Argentina). *Ornitología Neotropical* 14: 15-28.

MCBRIDE, P. 2000. *Magellanic Woodpecker (*Campephilus magellanicus*), Habitat Selection in Deciduous* Nothofagus *Forests of Tierra del Fuego*. Thesis, Western Washington University.

OLROG, C.C. 1948. Observaciones sobre la avifauna de Tierra del Fuego y Chile. *Acta Zoológica Lilloana* 5: 437-531.

OLROG, C.C. 1950. Mamíferos y aves del Archipiélago de Cabo de Hornos. *Acta Zoológica Lilloana* 9: 505-532.

OUSTALET, E. 1891. *Oiseaux*. Mission Scientifique du Cap Horn, 1882-1883. Zoologie 6(B): 1-341.

RAYA REY, A. & A.C.M. SCHIAVINI. 2000. Distribution, abundance and associations of seabirds in the Beagle Channel, Tierra del Fuego, Argentina. *Polar Biology* 23: 338-345.

REYNOLDS, P.W. 1934. Apuntes sobre aves de Tierra del Fuego. *Hornero* 5: 339-353.

REYNOLDS, P.W. 1935. Notes on the birds of Cape Horn. *Ibis* 5, Ser. 13: 65-101.

SCHLATTER, R.P. & G.M. RIVEROS. 1987. *Historia Natural del Archipiélago Diego Ramírez, Chile*. Serie Científica INACH (Chile) 47: 87-112.

SCHLATTER, R.P. & P. VERGARA. 2005. Magellanic Woodpecker (*Campephilus magellanicus*) sap feeding and its role in the Tierra del Fuego forest bird assemblage. *Journal of Ornithology* 146: 188-190.

SCHLATTER, R.P., P. VERGARA & M. BRIONES. 2002. El Canquén (*Chloephaga poliocephala*) en bosques de Tierra del Fuego: distribución y depredadores. *Anales del Instituto de la Patagonia* 30: 61-66.

SCHIAVINI, A.C.M. & P. YORIO. 1995. Distribution and abundance of seabird colonies in the Argentine Sector of the Beagle Channel, Tierra del Fuego. *Marine Ornithology* 23: 39-46.

SCHIAVINI, A., E. FRERE, P. YORIO & A. PARERA. 1999. Las aves marinas de la Isla de los Estados, Tierra del Fuego, Argentina: revision histórica, estado poblacional y problemas de conservación. *Anales del Instituto de la Patagonia, Serie Ciencias Naturales* 27: 25-40.

SIELFELD, W.H. 1977. Reconocimiento macrofaunístico terrestre en el área de Seno Ponsonby (isla Hoste). *Anales del Instituto de la Patagonia* 8: 275-296.

SIELFELD, W. & C. VENEGAS. 1980. Poblamiento e impacto ambiental de *Castor canadensis* en Isla Navarino, Chile. *Anales del Instituto de la Patagonia* 1: 247-257.

VENEGAS, C. 1981. Aves de las islas Wollaston y Bayly, Archipiélago del Cabo de Hornos. *Anales del Instituto de la Patagonia* 12: 213-219.

VENEGAS, C. 1991. Ensambles avifaunísticos estivales del archipiélago Cabo de Hornos. *Anales del Instituto de la Patagonia* 20: 69-82.

VERGARA, P.M & R.P. SCHLATTER. 2004. Magellanic woodpecker (*Campephilus magellanicus*) abundance and foraging in Tierra del Fuego, Chile. *Journal of Ornithology* 145: 343-351.

VERGARA, P.M. & R.P. SCHLATTER. 2006. Aggregate retention in two Tierra del Fuego *Nothofagus* forests: Short-term effects on bird abundance. *Forest Ecology and Management* 225: 213-224.

ACKNOWLEDGMENTS

This book synthesizes the work on ornithological research, education, and conservation conducted at the Omora Ethnobotanical Park between January 2000 and December 2010. During these first eleven years of our Long-Term Ornithological Research (LTOR) Program, many people have participated in the Bird Mist-Netting and Banding Program, bird studies, and educational workshops in the Omora Ethnobotanical Park and other sites of the Cape Horn Biosphere Reserve, as well as in the preparation of this book. When Ricardo Rozzi created the Omora Park LTOR Program in 2000, ornithologists Claudio Venegas (Universidad de Magallanes), Chris Elphick (University of Connecticut), and Juan Carlos Torres-Mura (former ornithology curator of the Museo Nacional de Historia Natural, and current of the board of directors of Omora Park), provided invaluable guidance and advice, which has continued to this day. In the field, the work by Francisca Massardo, Steven McGeehe, Christopher Anderson, Ezio Firmani, and Maurice van de Maele was crucial in helping consolidate this LTOR Program during its first years. Ornithologist Ricardo Matus, from Natura Patagonia, biologist Leonard Yanelli, former teacher at the Naugatuck Valley Community College, Teresita Gómez, former teacher at the school Villa Las Nieves, and Carlos Soto, history teacher at Puerto Williams' school and director of the Cape Horn Boy Scouts group, also participated actively in the first field campaigns developed at Omora Park. During these years, support in the field provided by Lorena Peñaranda (lawyer), Germán González (Yahgan handcrafter), Oliver Vogel (photographer), as well as Cristobal Raín, Franco Gómez, and other elementary school students was essential. Primary and secondary school students from the local school, Liceo Donald McIntyre Griffiths, the pre-school children from the city of Puerto Williams and their parents, and the Puerto Williams school teachers María Anguita and Nolberto González, and the administrative directors of the school, Nelson Cárcamo, Luis González, and Francisco Fernández, have accompanied and stimulated us everyday to continue working with the children on ornithological research, as well as guided tours and in situ workshops. We truly appreciate their support.

During the first decade of Omora Park LTOR Program, field work and mist-netting was conducted and/or coordinated by Ricardo Rozzi and Ezio Firmani (in 2000), by Steven McGehee and Ricardo Rozzi (in 2002, 2004-2005), by Steven McGehee and Silvina Ippi (in 2003), by Ximena Arango (in 2006-2008), by Sebastian Dardanelli and Rodrigo Molina (in 2009-2010). After these first eleven years of Omora Park LTOR Program covered by this book, the general direction of this program has been passed from Ricardo Rozzi to Jaime Jiménez in January 2011. During this second-decade period, the field work and mist-netting at Omora Park was conducted and/or coordinated by Ronnie Reyes and Cristián Celis (in 2011-2012), and since 2013 this role has been undertaken by Omar Barroso. We sincerely thank all the volunteers, graduate and undergraduate students, and researchers assistants for their valuable, continuous work in the rigorous environment, including summers and winters, in the Cape Horn Biosphere Reserve.

We acknowledge the Universidad de Magallanes (UMAG) through its authorities: the President Dr. Víctor Fajardo, the Academic Vice-President Dr. Juan Oyarzo, the Administration and Finances Vice-President Dr. José Maripani, and the Director of the Research Office and Graduate School Dr. Andrés Mansilla, for all the institutional support to the Omora Park initiative in general and the LTOR Program in particular. Dr. Mary Kalin-Arroyo and Dr. Juan Armesto, former Director and current Director of the Institute of Ecology and Biodiversity (IEB), respectively, have provided a continuous intellectual and institutional support to develop transdisciplinary research and to integrate research and education. Thanks to the IEB support, since 2008 the Omora Park is one of the three research sites of the Chilean Long-Term Socio-Ecological Research network (LTSER-Chile). Since 2011, LTSER-Chile, including Omora Park, is part of the International LTER network (ILTER), including Bonanza Creek LTER site associated with the University of Alaska–Fairbanks, with which Omora Park established a collaboration supported by UMAG and the University of North Texas (UNT).

We thank Alfredo Prieto, Director of the Editorial Board of the Ediciones Universidad de Magallanes, and Ronald Chrisman and Karen DeVinney, Director and Managing Editor from the University of North Texas Press, respectively, who have made possible the international series of publications coordinated by the University of North Texas Press & Ediciones Universidad de Magallanes, in association with the Sub-Antarctic Biocultural Conservation Program. For this endeavor, Eugene Hargrove and Alexandria Poole, director and associate director of the Center for Environmental Philosophy, respectively, have provided continuous support and advice.

The institutional support of the Agriculture and Livestock Bureau (Servicio Agrícola y Ganadero, SAG), with the National office in Santiago, especially Horacio Merlet, Agustin Iriarte and Charif Tala during the first years, and with the Regional Office in Punta Arenas, especially with Nicolás Soto (Director of Wildlife Management) from 2000 until today, has been essential to consolidate an institutional platform for the long-term research program. The National Commission for the Environment (CONAMA) also provided valuable support for the design and implementation of educational programs and community participation, especially through the leadership of Juan Manuel Draguicevic, CONAMA-Magallanes. For for all their support to the research programs developed at Omora Park, we sincerely thank the local authorities at the moment of the creation of the Omora Park LTOR Program, former Governor of the Antarctic Province, Mr. Eduardo Barros González, former Major of Cape Horn County, Mr. José Soto-Passek, and former Commandant of the Navy Basis of the Armada de Chile in Puerto Williams, Vice-Admiral Francisco Guzmán.

Many grants have supported the development of the Bird Banding Program throughout the first decade. Major support for ornithological research has been provided by the BIOKONCHIL project (German Ministry of Education and Research) for the years 2002-2005, and by the Institute of Ecology and Biodiversity though the grants ICM-P05-002 (Iniciativa Científica Milenio, Ministry of Economy), PFB-23 (Programa de Financiamento Basal para Centros de Excelencia Científica y Tecnológica, CONICYT), and Sub-Antarctic Biocultural Conservation Program for the years 2006-2013. EXPLORA-CONICYT and the Environmental Protection Fund from the Ministry of the Environment (Fondo de Protección Ambiental-Ministerio del Medio Ambiente, has supported educational and outreach projects. Chile's Ornithologists Union (Unión de Ornitólogos de Chile, UNORCH), Fauna Australis (associated to P. Universidad Católica de Chile), the Center for Biodiversity and Conservation of the University of Connecticut, the Peregrine Fund (Boise Idaho, USA), the Manomet Center for Conservation Sciences (USA), the National Audubon Society (especially through the support by Alejandro Grajal and Ricardo Stanoss), and the Earthwatch program have provided valuable technical support as well some basic photography equipments, binoculars, and mistnets. FONDECYT grant Nº 1131133 is supporting the current ecological and ethological research on the Magellanic Woodpecker, which has allowed us to emphasize in this book, especially in the epilogue its role as a biocultural keystone species. A National Geographic grant is supporting the current research on migration of *Elaenia albiceps*, which has allowed us to highlight in this book, especially in the epilogue, the singularity of the only long-distant migrant bird species of the Magellanic sub-Antarctic forests.

We are very grateful to the journals *Anales del Instituto de la Patagonia*, *Austral Ecology*, *Biological Conservation*, *Boletín Chileno de Ornitología*, *Boletín SOA*, *Ciencia e Investigación Agraria*, *Condor*, *Magallania*, *Ornitología Neotropical*, *Oryx*, *Revista Chilena de Historia Natural*, which graciously gave us permission to reprint the papers with some minor format changes, or in the form of English translation for those articles that were originally published in Spanish.

We sincerely thank Melodi Baybordi, Benjamin Christ, and Charlotte Tidrick for the English translation work (see details in page 368), and the revision of the translations conducted by Kelli Moses, Alexandria Poole, Irán Román, and Rachel Weaver, together with the editors. We appreciate the careful revision of the content and/or proofreading by ornithologist James Bednarz and wildlife biologist John Greetenberger. Proofreading also benefited from revision by ornithologists Julie Hagelin and Víctor Cueto, wildlife biologists Ana Piñeiro and Julia B. Smith, and philosopher Eugene Hargrove (see details in page 368).

We are most grateful to all photographers who generously provided images for this book (see details in page 367). Regarding other figures, we acknowledge the following persons: María Rosa Gallardo (Centro de Estudio de Recursos Energéticos CERE-UMAG) for the maps in chapters 1 and 2, Rajan Rijal for histograms and graphs in chapters 1 and 2, Gerardo Soto for Figure 12, and Francisca Massardo for figures of Chapter 1.

Alejandra Calcutta prepared the computer layout of the book, with the collaboration of Stephanie Noell. Kelli Moses, Managing Editor, Sub-Antarctic Biocultural Conservation Program, provided valuable design suggestions, in addition to her continuous coordination work with UNT Press. Francisca Massardo is responsible for the final design and editing of the book, working closely together with Alejandra Calcutta and the authors, in coordination with Ediciones Universidad de Magallanes.

We hope that this synthesis of ornithological research at Omora Park will: (1) stimulate new research projects that contribute to filling the gaps in the understanding of the autecology of bird species and their ecological interactions; (2) foster education and sustainable tourism programs that promote the conservation of the biological and cultural diversity of one of the planet's most unique sub-polar forests, which are remarkably well preserved at the beginning of the 21st century; and (3) provide access to a new body of knowledge that can be of interest to environmental decision makers, educators, tourism operators, researchers, and bird enthusiasts.

Our most proper way to thank this generous support for the Long-Term Ornithological Research Program at Omora Park, is to make the generated scientific knowledge available to the educational community, authorities, and the general public, and to continue working together with the aim of contributing to the welfare of present and future generations of humans, birds, and of the diversity of living beings co-inhabiting the Magellanic sub-Antarctic ecoregion and beyond.

EDITORS

Ricardo Rozzi
Department of Philosophy and Religion Studies
Sub-Antarctic Biocultural Conservation Program
University of North Texas, Denton, Texas 76201, USA
Parque Etnobotánico Omora
Institute of Ecology and Biodiversity
Universidad de Magallanes, Puerto Williams, Chile
rozzi@unt.edu

Jaime E. Jiménez
Department of Biological Sciences
Department of f Philosophy and Religion Studies
Sub-Antarctic Biocultural Conservation Program
University of North Texas, Denton, Texas 76201, USA
Parque Etnobotánico Omora
Institute of Ecology and Biodiversity
Universidad de Magallanes, Puerto Williams, Chile
Jaime.Jimenez@unt.edu

CONTRIBUTORS TO INTRODUCTIONS

Christopher B. Anderson
Universidad Nacional de Tierra del Fuego
Centro Austral de Investigaciones Científicas
Ushuaia, Argentina
canderson@alumni.unc.edu

Victoria Castro
Parque Etnobotánico Omora
Puerto Williams
Departamento de Antropología
Universidad Alberto Hurtado; Universidad de Chile
Santiago, Chile
vcastror53@gmail.com

Sebastián Dardanelli
Grupo Biodiversidad
INTA, EEA Paraná
Entre Ríos, Argentina
s_dardanelli@yahoo.com

Christopher S. Elphick
Department of Ecology and Evolutionary Biology
University of Connecticut
Storrs, Connecticut 06269, USA
chris.elphick@uconn.edu

José Tomás Ibarra
Parque Etnobotánico Omora,
Puerto Williams, Chile
Fauna Australis &
Centro Interdisciplinario de Estudios
Interculturales e Indígenas,
Pontificia Universisad Católica, Chile
The Peregrine Fund, Boise, Idaho, USA
jtibarra@alumni.ubc.ca

Francisca Massardo
Parque Etnobotánico Omora
Programa de Conservación Biocultural Subantártica
Institute of Ecology and Biodiversity
Centro Universitario Puerto Williams
Universidad de Magallanes, Chile
massardorozzi@yahoo.com

J. Cristóbal Pizarro
Department of Environment and Resource Studies
University of Waterloo, Guelph N2L3G1, Canada
jcpizarrop@gmail.com

Ronnie Reyes
Depto. Científico Académico, Dirección de Extensión
Universidad Austral de Chile
Valdivia, Chile
ronnie.reyes@uach.cl

Rajan Rijal
Departments of Biological Sciences and Philosophy
Sub-Antarctic Biocultural Conservation Program
University of North Texas, Denton, Texas 76201, USA
rajan.rijal@unt.edu

Juan Carlos Torres-Mura
Parque Etnobotánico Omora
Puerto Williams, Chile
torresmura@gmail.com

Rodrigo A. Vásquez
Institute of Ecology and Biodiversity
Departamento de Ciencias Ecológicas
Universidad de Chile, Santiago, Chile
rvasquez@uchile.cl

IMAGE CREDITS

Many photographers kindly contributed to this book. In Chapter 1 and Epilogue authors of the photographs are mentioned in the caption of each figure. For full-page photographs, and other pages where the pictures are displayed without credits, the authors of the photographs are listed below.

Tomás Altamirano	Page:	43 (Thorn-tailed Rayadito).
Omar Barroso	Page:	70 (Bar-winged Cinclodes).
Roberto Cañete	Page:	76 (Blue-and-White Swallow).
Tamara Contador	Page:	50 (House Wren).
Gabriela Correa	Page:	297 (Magellanic Woodpecker drawing, First Prize Winner in the Puerto Williams School, 2009).
Werner Flück	Page:	68 (Green-backed Firecrown).
Emilio García de la Huerta	Page:	52 (Black-chinned Siskin).
José Tomás Ibarra	Page:	74 (Austral Blackbird).
Jaime E. Jiménez	Pages:	1 (Magellanic Woodpecker, Fire-eyed Diucon, Austral Thrush, Tuftet Tit-Tyrant, and Patagonian Sierra-Finch), 44 (Patagonian Sierra-Finch), 48 (White-crested Elaenia), 54 (Rufous-collared Sparrow), 56 (Tufted Tit-Tyrant), 58 (Austral Thrush), 60 (Chilean Swallow), 64 (Fire-eyed Diucon), 66 (Patagonian Tyrant), 72 (Dark-bellied Cinclodes), 82 (Austral Pygmy-Owl), 84 (Magellanic Woodpecker), 86 (Austral Parakeet), 88 (Chimango Caraca), 93 (mis-netting in the high-Andean ecosystems) 94 (Dark-faced Ground-Tyrant), 100 (House Sparrow).
Steven Morello	Page:	78 (Magellanic Tapaculo).
Carlos Olavarría	Page:	62 (White-throated Treerunner).
Jordi Plana	Pages:	x (Beagle Channel), 40 (researcher in the field), 80 (Magellanic Woodpecker), 99 (Puerto Williams street), 116 (researcher Ximena Arango banding a bird), 128 (Holland Glacier), 130 (mist-netting in winter), 184 (deciduous forests in winter on Navarino Island), 234(Upland Goose), 292 (deciduous forest in winter), 317 (birds flying), 318 (nocturnal fly).
Sebastian Rosenfeld	Page:	106 (children at Omora Park).
Ricardo Rozzi	Pages:	viii (Sub-Antarctic forest interior), 124 (Navarino Island landscape).
Eduardo Silva	Page:	90 (Rufous-legged Owl).
Elke Schuettler	Page:	233 (American mink).
Steff van Dongen	Pages:	129 (researcher), 352 (ornithologist Steve McGehee handling a Patagonian Sierra-Finch).
Paola Vezzani	Page:	vi (Magellanic sub-Antarctic forest).

Thomas H. White	Page:	96 (Yellow-bridled Finch).
Adam M. Wilson	Pages:	81 (Austral Parakeets), 92 (Navarino Island high-Andean ecosystems), 98 (Puerto Williams from the Beagle Channel), 183 (Measuring a White-crested Elaenia).
Günther Ziesler	Page:	v and 319 (Green-backed Firecrown).
Alfredo Zúñiga	Page:	46 (Thorn-tailed Rayadito).
	Page:	320 (cover of the Multi-Ethnic Bird Guide, Rozzi *et al.* 2010).

TRANSLATORS

Melodi Baybordi	Chapters: 3, 25, 30	Melodi Baybordi; Freelance translator; melodibaybordi@gmail.com
Benjamin Christ	Chapters: 4, 7, 16, 24	Benjamin Christ; Department of Biological Sciences; University of Missouri, Columbia, MO 65211, USA; jbc2nf@mail.missouri.edu
Jaime Jiménez	Chapters: 19, 27	
Ricardo Rozzi	Chapters: 26, 28	
Charlotte Tidrick	Chapters: 29, 31	Charlotte Tidrick; Department of Political Science; University of Virginia, Charlottesville, VA 22904, USA; cet9sx@virginia.edu

Content and English editions

James Bednarz	Chapters 6-15	Department of Biological Sciences, University of North Texas
John Greetenberger	Chapters 1-2	U.S. Fish and Wildlife Service, Washington (retired)
Julie Hagelin	Parts I-VI	Institute of Arctic Biology, University of Alaska Fairbanks
Kurt Heidinger	Chapter 26	Biocitizen School of Field Environmental Philosophy & Practice
Julia B. Smith	Chapters 20-25	Department of Zoology, Southern Illinois University Carbondale

English editions

Eugene Hargrove	Preface, Chapter 28	Department of Philosophy and Religion Studies, Center for Environmental Philosophy, University of North Texas
Kelli Moses	Preface, Epilogue	Department of Biological Sciences, Sub-Antarctic Biocultural Conservation Program, University of North Texas
Alexandria Poole	Chapter 28	Department of Philosophy and Religion Studies, Center for Environmental Philosophy, University of North Texas
Irán Román	Chapters 6-15	Department of Biological Sciences, Sub-Antarctic Biocultural Conservation Program, University of North Texas
Rachel Weaver	Chapters 26, 28-29	Department of Philosophy and Religion Studies, Sub-Antarctic Biocultural Conservation Program, University of North Texas

INDEX

COMMON AND SCIENTIFIC NAMES OF BIRD SPECIES

Parque Etnobotánico Omora
Provincia Antártica Chilena

IEB CHILE
INSTITUTO DE ECOLOGÍA Y BIODIVERSIDAD

The Center for Environmental Philosophy

Book design and editing: Francisca Massardo.

Computer Layout: Alejandra Calcutta and Stephanie Noell.

Cover Design: Alejandra Calcutta.

Cover images: Jaime E. Jiménez (Fire-eyed Diucon, Austral Pygmy-Owl, Tufted Tit-Tyrant), Jordi Plana (Magellanic Woodpecker), Steff van Dongen (measuring a Patagonian Sierra-Finch), World satellite image prepared by by Maria Rosa Gallardo at the GIS Laboratory, CERE-Universidad de Magallanes, based on "The MODIS Rapid Response System."

Full citation of the book: Rozzi R. & J. E. Jiménez. (Eds). 2013. *Magellanic Sub-Antarctic Ornithology. First Decade of Long-Term Bird Studies at the Omora Ethnobotanical Park, Cape Horn Biosphere Reserve, Chile*. University of North Texas Press - Universidad de Magallanes.

SALVIAT IMPRESORES S.A.
Seminario 739, Santiago, 56(2) 23418410 esalviat@salviatimpresores.cl
Printed in Santiago - Chile